INCLUSION PHENOMENA AND MOLECULAR RECOGNITION

INCLUSION PHENOMENA AND MOLECULAR RECOGNITION

Edited by

Jerry L. Atwood

University of Alabama
Tuscaloosa, Alabama

PLENUM PRESS • NEW YORK AND LONDON

Library of Congress Cataloging-in-Publication Data

International Symposium on Inclusion Phenomena and Molecular
  Recognition (5th : 1988 : Orange Beach, Ala.)
    Inclusion phenomena and molecular recognition / edited by Jerry L.
  Atwood.
        p.   cm.
    "Proceedings of the Fifth International Symposium on Inclusion
  Phenomena and Molecular Recognition, held September 18-23, 1988, in
  Orange Beach, Alabama"--T.p. verso.
    Includes bibliographical references and index.
    ISBN-13:978-1-4612-7887-0   e-ISBN-13:978-1-4613-0603-0
    DOI: 10.1007/978-1-4613-0603-0

    1. Clathrate compounds--Congresses.   I. Atwood, J. L.   II. Title.
  QD474.I59   1988
  541.2'242--dc20   90-36551
                                                              CIP

Proceedings of the Fifth International Symposium on
Inclusion Phenomena and Molecular Recognition,
held September 18-23, 1988, in Orange Beach, Alabama

© 1990 Plenum Press, New York
Softcover reprint of the hardcover 1st edition 1990

A Division of Plenum Publishing Corporation
233 Spring Street, New York, N.Y. 10013

PREFACE

The Fifth International Symposium on Inclusion Phenomena and Molecular Recognition was held September 18-23, 1988 at Orange Beach, Alabama. This followed previous very successful symposia in Warsaw (1980), Parma (1982), Tokyo (1984), and Lancaster (1986). The overall tone of the event at Orange Beach was expressed elegantly by Fraser Stoddart at the close of his lecture:

"At a meeting like this, I think we should be asking ourselves more openly where we have come from and where we are going to. I am certainly willing to put my head on the block. Chemistry, as I see it, is entering a golden age of opportunity and those of us here who respond to the multidisciplinary challenge of the subject will perhaps start the movement to reunite the chemical sciences for the first time in more than a century. Given the recognition granted through Charles Pedersen, Donald Cram, and Jean-Marie Lehn to our field from Stockholm last year, there are many here who are surely poised – if they have not already done so – to capture the academic high ground and intellectual leadership of our subject. And what is more – it will be on the back of our fundamental science that many of the exciting technological advances of the twenty-first century will be forged."

In order to capture the flavor and excitement of the symposium, herein we present reviews by thirty-eight of the invited lecturers.

The program was shaped by the Program Committee: Jerry L. Atwood, Richard A. Bartsch, George W. Gokel, Fredric M. Menger, Yukito Murakami, and Galen D. Stucky. The local arrangements were overseen by Jerry L. Atwood, Duane C. Hrncir, William E. Hunter, Gregory H. Robinson, and Robin D. Rogers. The accompanying persons program was set up by Tracey M. Atwood. Financial support was generously provided by:

Amoco Chemical Company,
the Alabama Department of Economic and Community Affairs,
Allied Signal,
American Maize-Products Company,
Dow Chemical Company,
Eastman Kodak Company,
Enraf Nonius,
Nalco Chemical Company,
Nicolet Instruments Corporation,

the Office of Naval Research,

Serpentix, Inc.,

Technicon Instruments Corporation,

and The University of Alabama.

General guidance for the symposium was in the hands of the International Organizing Committee: G. D. Andreetti, J. L. Atwood, R. Breslow, J. E. D. Davies, Yu. A. Dyadin, T. Iwamoto, J.-M. Lehn, J. Lipkowski, D. D. MacNicol, W. Saenger, J. Szejtli, F. Vögtle, and R. L. Wife.

Special thanks go to my Assistant Editor for this volume, Dr. Simon G. Bott, who oversaw all phases of the production of the camera-ready copy.

November, 1989   Jerry L. Atwood

NOTE: Asterisks appear in the author listings of certain chapters to designate the author to whom correspondence should be addressed.

CONTENTS

NEW SHAPES FOR CATALYSIS AND MOLECULAR RECOGNITION

Julius Rebek, Jr.

Department of Chemistry
University of Pittsburgh
Pittsburgh, PA 15260 (USA)

Any discussion of molecular recognition should begin by acknowledging the debt that it owes to macrocyclic chemistry. These molecules have been the workhorses of bioorganic chemistry and have influenced literally and figuratively the shape of things in molecular recognition. Cyclodextrins, for example, have been useful for showing how small molecules can imitate the essential steps of enzyme catalyzed reactions using acyl transfers as a probe [1]. Crown ethers have been successful at revealing aspects of binding and transport of metal ions; at the molecular level, complexation of ammonium species has been used as a vehicle for several bioorganic processes [2]. Allosteric effects, especially that of cooperativity were first demonstrated with crown ether derivatives [3]. More recently, interest in cyclophane derivatives has grown. Their ease of synthesis and their ability to complex aromatic molecules in aqueous media can be used to sort out the intrinsics of hydrophobic and aromatic stacking interactions. Some of the rules for predicting complexation especially those involving optimal rigidity and collapsibility have emerged from examination of cyclophane-derived systems.

Despite their popularity, macrocyclic structures present some disadvantages. While such a shape is ideal for complexation of spherical substrates such as metal ions, these structures are quite difficult to functionalize on their interiors. That is, functional groups attached to macrocyclic structures tend to diverge and be directed away from the substrates held inside. In contrast, natural receptors such as enzymes and antibodies feature functional groups which converge on the substrates held inside. This convergence of functional groups, providing acids, bases, as well as shapes and sizes is, we believe, the key to molecular recognition in biological systems. It is no mere accident that nature often selects a structure resembling a cleft for this purpose. Such a shape offers easy access of the substrate into and out of it and provides a means by which selectivity can be engineered into it. By altering the lining of the cleft it is possible to tailor it into an exquisite, complementary fit for the substrate (Scheme 1).

Inclusion Phenomena and Molecular Recognition
Edited by J. Atwood
Plenum Press, New York, 1990

1

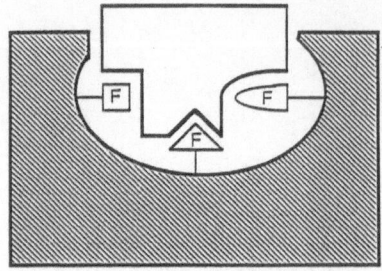

Scheme 1

Nowhere is this convergence of functional groups more important than in the chemistry of the carboxylic acid group. Gandour [5] pointed out that carboxyl functions at the active sites of enzymes present their more basic *syn* lone pairs towards the substrate (Scheme 2). Enzymes of such diverse families as the serine proteases, the aspartic proteases and lysozyme adhere to this rule.

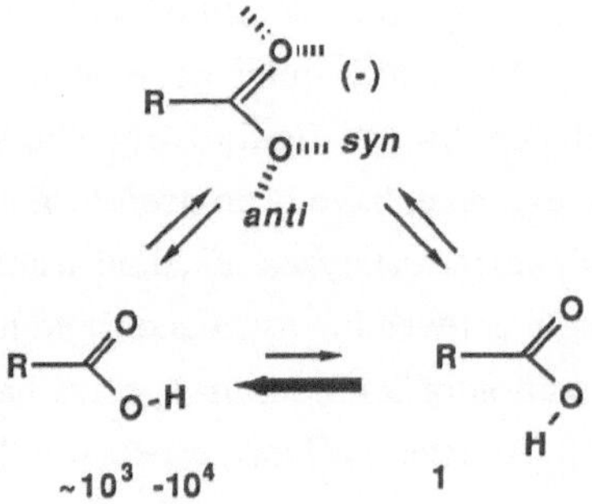

R—⟨⟩ (-) *syn* *anti*

R—⟨⟩O–H   R—⟨⟩O–H

~10^3 -10^4 1

Scheme 2

The convergence of carboxyl groups invariably involves the *syn* lone pairs directed towards the catalytic apparatus. Gandour's observations helped crystallize my thinking on related problems in organic chemistry involving the carboxyl function. The specific task involved developing a selective epoxidising agent for olefins. The more general problem involved controlling the microenvironment near the hydrogen of a carboxylic acid so that it could be used for, say, asymmetric protonation. In order to make progress on some of these problems a new molecular scaffold was required. An ideal shape would involve a molecular cleft in which the OH bond of the carboxylic acid was constrained to be directed toward some group that could be altered at will (Scheme 3). Since the α-carbon of a carboxylic acid is attached in a direction opposite to the one desired, it became clear that a U-turn would have to be engineered somewhere into the structure of the molecule.

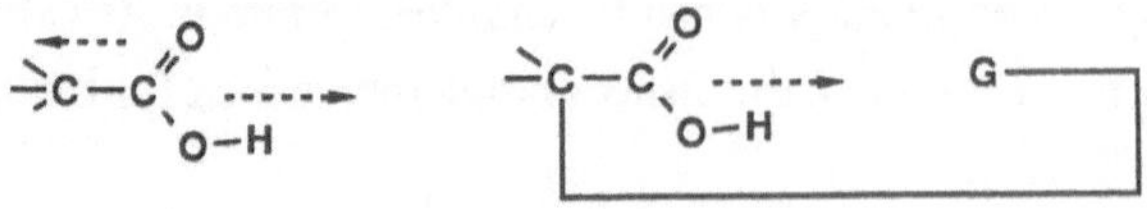

Scheme 3

Fortunately, this was eventually accomplished using a material introduced in 1981, Kemp's triacid [6]. This molecule was made originally from the oxidative degradation of trimethyl adamantol; more conveniently it can be prepared on a large scale by the alkylation of hydrogenated trimesic esters [7]. The Kemp triacid owes its unique shape to the larger A values of methyl groups compared to carboxyl functions. The three equatorial methyl groups force the triaxial conformation of the carboxyl groups and ensure that a U-shaped relationship exists between any two of them (Scheme 4). This relationship is the fundamental one of our studies in molecular recognition. We use the Kemp triacid repeatedly as an architectural cliche or structural motif whenever a U-turn needs to be expressed in a molecular structure. That the U-turn is not due to intramolecular hydrogen bonding was established by X-ray studies of the triacid and its triester derivative. In the solid state all of the carboxylic acids are involved in

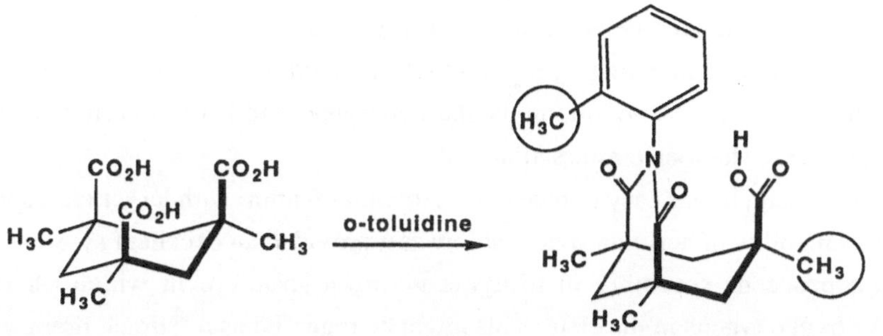

Scheme 4

conventional intermolecular hydrogen-bonded structures [8]. In the latter, the methyl groups of the ester present a slightly divergent arrangement in a triaxial sense.

We found that condensation of the Kemp triacid or its derivatives with anilines led to imides. The temperature dependence of the NMR spectra of these imides suggested hindered rotation about the N_{imide}-C_{aryl} bond. This rotation could be frozen out by the use of *ortho*-substituents. The condensation product with *ortho*-toluidine for example gives a fairly rigid structure in which the seemingly passive methyl groups prevent epimerization and rotations. We dwell on these conformational refinements because we use them continuously to control stereochemistry and stereoelectronics in our systems.

The first molecular cleft was prepared by condensation of a diamine, *meta*-xylidine diamine, with two molecules of the Kemp triacid. The condensation proceeds exceedingly well by merely performing a mixed melting point between the two solids. The resulting dicarbox-

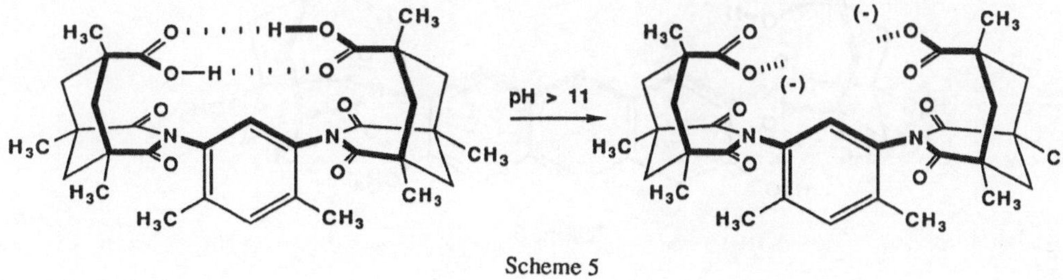

Scheme 5

ylic acid, a structure bristling with methyl groups, gave a unique intramolecular hydrogen-bonded dicarboxylic acid (Scheme 5). That is, it showed all the spectroscopic earmarks of a conventional carboxylic acid dimer but the methyl groups constrain the structure into a shape from which it can escape only by breaking bonds. It is inappropriate here to expound upon the unique chemistry provided by this dicarboxylic acid in its reactions with electrophiles and nucleophiles. These have been chronicled elsewhere [8] and provide some of the first experimental observations concerning stereoelectronic effects at carbonyl oxygen. But one aspect is certainly worth mentioning and that is the extraordinary acidity of these species. The second pK_a of the acid is >11 and the δpK_a (>6) reflects the price paid for forcing two negative charges into a small volume of space [9]. The convergence of the *syn* lone pairs in the dianion leads to an ideal microenvironment for divalent metal ions, and their action as chelating agents will be discussed later. The hydrogen bonding in the mono-anion must also contribute to the large δpK_a. The stereoelectronic effect might be assessed by comparing this to the pK_a's in the Kemp triacid in which 3 pK_a units are involved in the roughly parallel or slightly divergent dianion. However, no intramolecular hydrogen bond is present in this structure, so its effect is not due to stereoelectronics alone.

Quite fortunately, it was easy to repeat this structural feature with larger cleft species. The commercial availability of acridine yellow at low cost provided an excellent system. Again, the condensation proceeds smoothly in triglyme giving a structure in which ≈8 Å separate opposing carboxyl oxygens in the cleft. This might be regarded as a "stretch limousine" model whereas a "mid-size" version was available with naphthalene as the spacer (Scheme 6). The relevant diamine was prepared by Bucherer reaction of the diol, and the condensation again proceeded smoothly to give dicarboxylic acid with ≈5.5 Å opposing the two carboxyl oxygens. The structures and distances involved were obtained through crystallographic studies of the solids or their appropriate derivatives.

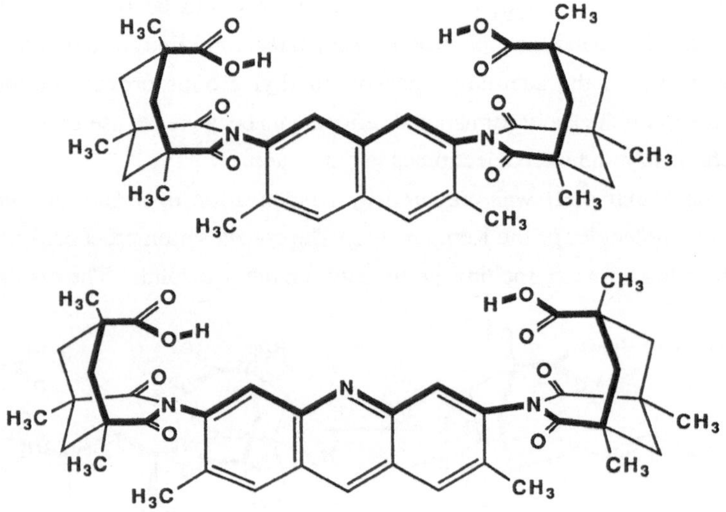

Scheme 6

Because the two carboxyls of the naphthalene version are dying to collapse into an intramolecular hydrogen bond system but are prevented from doing so by the rigid aromatic spacer, the molecule binds tenaciously to small structures that can just bridge this gap with complementary acid and base functions [10]. For example, 2:1 complexes are made with alcohols, presumably involving cooperative hydrogen bonds. Amines likewise formed 2:1 complexes and diamines or diols formed 1:1 complexes. There is not much room inside this molecular cleft, but that which exists can be used to advantage in asymmetric recognition. For example, the phenethyl amide derivative provides the first synthetic structure in which an asymmetric microenvironment confronts a carboxylic acid OH bond. The resulting cavity binds racemic alcohols nicely, and acts as an unusually effective chiral solvating agent for such species (Scheme 7). The effect is due to the placement of the two asymmetric centers near one another. One can appreciate that a conventional carboxylic acid with asymmetry at, say, the α-carbon, would not be nearly as effective; indeed, such materials often need to be used as solvents before sufficient anisotropy is expressed in the NMR spectra.

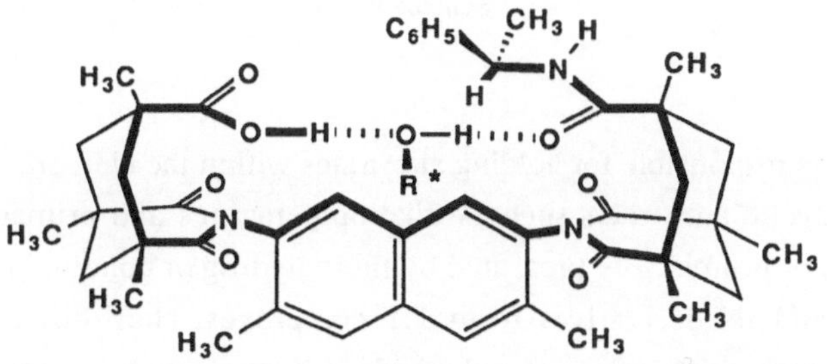

Scheme 7

The superior dimensions of the acridine version drew most of our attention to that species for studies of complexation. The diacids orient their acidity in a convergent sense so they bind molecules that express basicity in a complementary, divergent sense [11]. Accordingly, structures such as DABCO form 1:1 complexes with these materials (Scheme 8). The selectivity is quite high; for example, pyrazine can be removed from its solutions in pyridine. Though the latter base is in higher concentration and has much stronger intrinsic basicity, it does not fit into the cleft appropriately. We have worked through a number of heterocyclic bases in probing the collapsibility and adaptability of the acridine diacid. Small dicarboxylic acids such as oxalic acid also express divergent acidity and their hydrogen-bonding capabilities are complementary to the convergent diacid of the acridine derivative. Even the fluorene derivative (Scheme 8) shows some selectivity for dicarboxylic acids [12]. It shuns smaller C_2 and C_3 diacids but with the larger guests such as glutaric or camphoric acids, 1:1 complexes are formed. Presumably the complexation is driven by the chelate effect.

The acid/base component and its attendant ambiguities in the complexation events can be removed by converting the acridine-derived diacid to its corresponding diamide. Now only

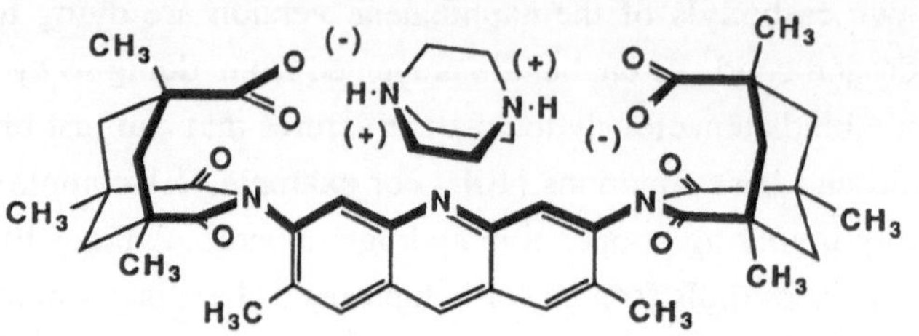

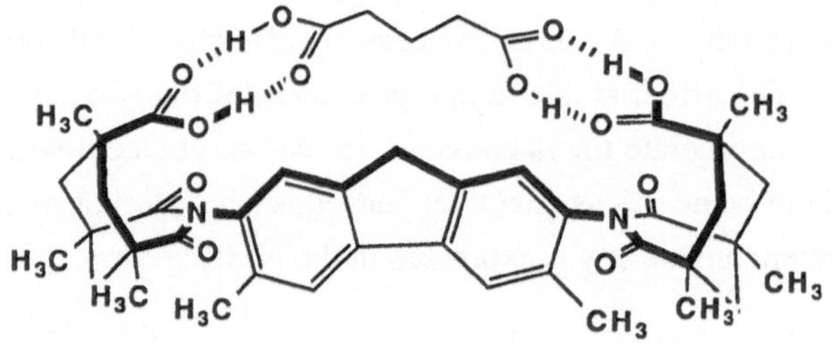

Scheme 8

hydrogen-bonding is responsible for holding substrates within the cleft and we have explored a number of heterocyclic substrates such as diketopiperazines and primadone derivatives to establish the limits of promiscuity presented by these hydrogen bonding patterns [10]. While diketopiperazine and its derivatives form 1:1 complexes, short-circuiting the hydrogen bonding, as is the case with sarcosine anhydride, leads to much-diminished binding or no binding at all.

The central nitrogen of the acridine nucleus presents a singular element in that its lone pair is directed at a point between the two carboxylic acids. In polar media such as methanol, the molecule is zwitterionic and the interior of the cleft is a highly polar microenvironment. This environment is combined with a very lipophilic coating in the methyls and methylenes on the outer surface of the molecule. The large aromatic surface of the spacer is yet another domain. The unusual juxtaposition of these three molecular domains led us to examine some possibilities of binding zwitterionic species within the cleft. Astonishingly, amino acids were extracted from their neutral aqueous solutions into chloroform by the acridine diacid, but only if they bore a suitably disposed aromatic surface [13]. This latter is the key to the recognition and transport of amino acids by these materials; the binding involves aryl stacking interactions between the aromatic side-chain (particularly those in a β-position with respect to the ammonium function) and the electron-deficient acridine nucleus. The stacking interaction involves another point of attachment and we have used this feature for asymmetic recognition of phenethyl amines. For example, the monofunctionalized acridine provides one acidic site for binding a phenylalanine derivative, then aryl stacking interactions pin the substrate in such a manner that the asymmetric centers are placed in intimate contact (Scheme 9). This permits the centers to express their taste (or distaste) for one another in an energetically optimal way.

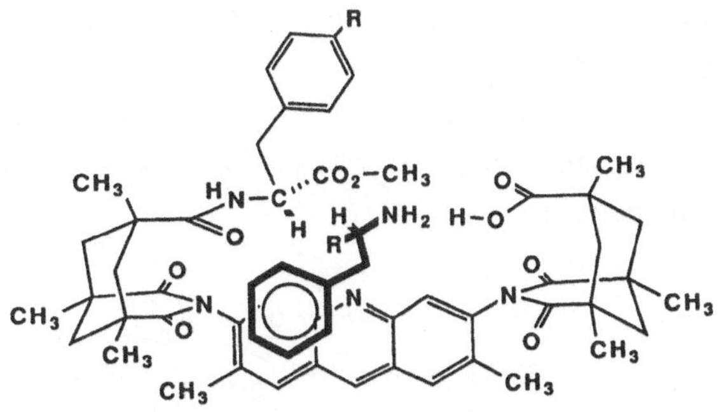

Scheme 9

In recent times, we have explored the use of larger spacers and those bearing even more branched arms for molecular recognition. A porphyrin spacer is now available, thanks to the extraordinarily effective synthesis of tetraaryl porphyrins developed by Jonathan Lindsey [14]. In this structure up to four carboxylic acids can converge on a single point in the center of the molecule (Scheme 10). Since rotation of the aryl functions is permitted, isomerism still exists but the molecule forms 2:1 complexes with 4,4'-bipyridyl [15]. Smaller amine bases such as DABCO or pyridine fail to become chelated between the carboxyl arms of the molecule.

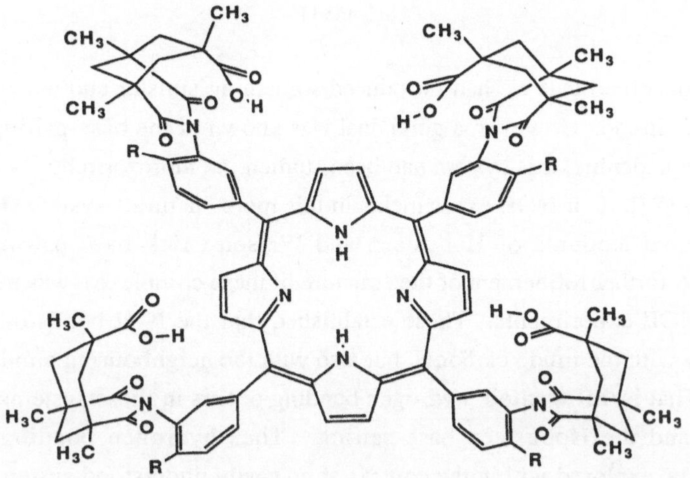

Scheme 10

It occurred to us that the imide derived from Kemp's triacid and ammonia presents a hydrogen bonding edge similar to that of thymine or uracil. Nucleic acid components such as adenine could hydrogen bond and base pair with these imides and could be further bound through aryl stacking interactions. Again, the U-turn within the molecule was applied to this system by attaching aromatic surfaces such as amides or esters in such a way that they extended beyond the hydrogen bonding edge. The hydrogen bonding and aryl stacking interactions converge from perpendicular directions to present a microenvironment complementary to simple adenine derivatives (Scheme 11). By using a number of such

7

systems we were able to sort out the relative contributions of the two intermolecular forces in the recognition event [16]. Moreover, homo- and heteronuclear NOE experiments made it possible to map out the geometric details of complexation. Surprisingly, not only Watson-Crick base pairing but also Hoogsteen base pairing was expressed in the interaction of the two components in chloroform.

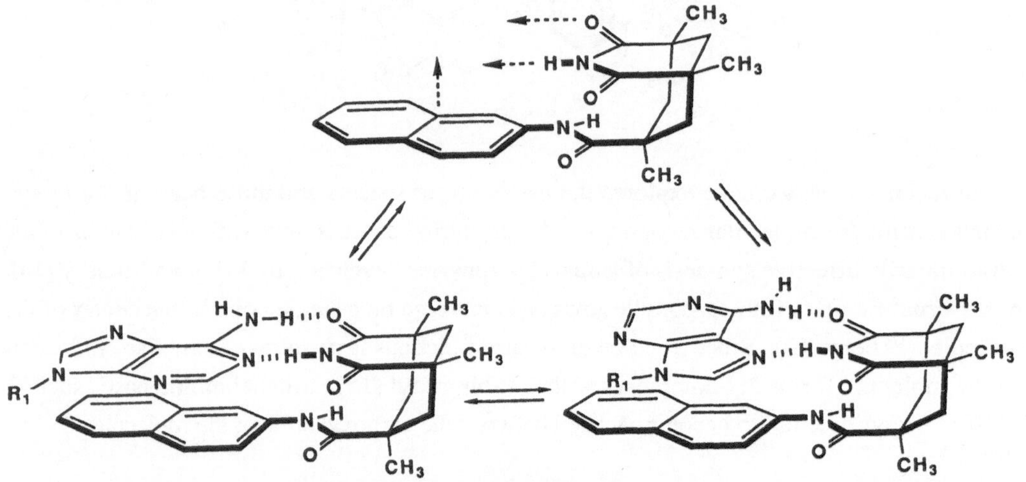

Scheme 11

We were concerned that we had introduced something sinister and unexpected with our entirely synthetic imide. However, a good deal was known of the base-pairing of cyclohexyl uracil with 9-ethyladenine; this system had been studied in chloroform by Rich and others by IR spectroscopy [17]. Our NOE experiments in this more "natural" system showed that here also roughly equal amounts of Hoogsteen and Watson-Crick base pairing occur in the complexation. A further refinement of the structure of these complexes was made possible by heteronuclear NOE experiments. These established that the N-H bonds of adenine are in contact not only with the imide carbonyl, but also with the neighbouring amide carbonyl [18] (Scheme 12). That is, **bifurcated** hydrogen bonding occurs in these systems along with the Watson-Crick and the Hoogsteen base-pairing. Thus hydrogen bonding, stacking and bifurcation can be explored within the context of an easily understood system. These forces are the major contributors to molecular recognition of nucleic acids by drugs and are responsible for the very structure of double-stranded nucleic acids.

It was possible for us to influence Watson-Crick vs. Hoogsteen base pairing by altering the nature of the aromatic stacking surface in the artifical receptor. For example, long, extended aromatics such as the anthryl unit favor Watson-Crick base pairing because better overlap in an aryl-aryl contact is achieved. On the other hand, remote steric influences such as expressed in t-butyl naphthalene derivatives promote Hoogsteen base pairing because aryl stacking interactions can be achieved without unpleasant contacts between the side chains of the two species involved [19].

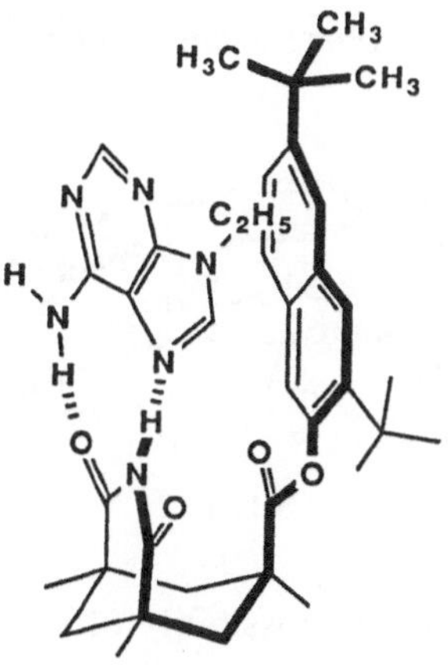

Scheme 12

Selectivity between different bases could also be engineered into these systems through rather modest synthetic investments. For example, reduction of the imide with sodium borohydride changes the acid base pattern of the hydrogen bonded edge. The resulting lactam alcohol now shows a high preference for binding to cytosine derivatives [20]. For example, cyclohexyl cytosine is preferred to adenine by more than a factor of 10. This is a reversal from the original imide system which prefers adenine to cytosine by more than a factor of 20. Thus a single hydrogen bond can effectively alter selectivities (Scheme 13). Its ability to do so is a consequence of the very short distances over which these forces act. It underlines the advantage of weak intermolecular forces for use in selective recognition events.

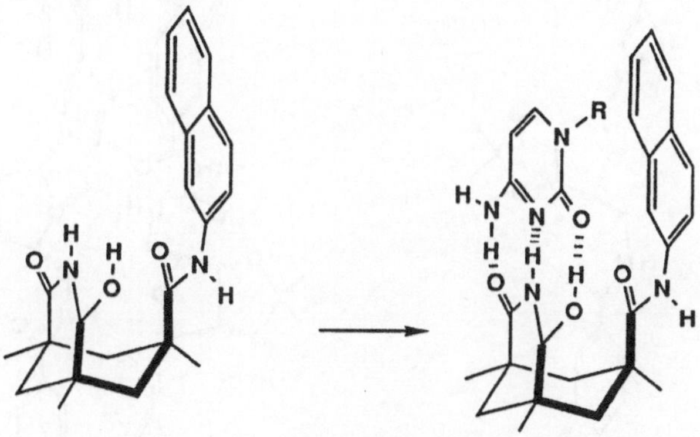

Scheme 13

By merely condsidering two of the imides on an appropriate aromatic surface, such as a naphthalene spacer, it is possible to prepare a receptor in which simultaneous Watson-Crick and Hoogsteen base pairing occur along with stacking interactions. Indeed, this molecule has an extraordinary affinity for adenine; it will extract simple adenine out of aqueous solution into chloroform. The high selectivity extends to its derivatives adenosine and deoxyadenosine (Scheme 14). Thus the molecular recognition of the adenine base is sufficient to pull the carbohydrate moiety into chloroform where it can then be transported to another aqueous phase [21].

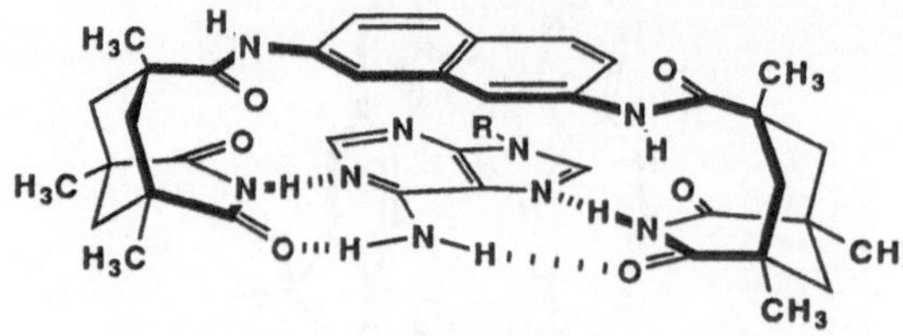

Scheme 14

We have also been examining the possibility of using the base pairing event to drive chemical reactions. Consider the consequences of covalent binding of adenine to a receptor for adenine. Provided that the two binding edges are arranged in a parallel manner, the system is self-complementary. That is, it will tend to dimerize (or polymerize). The free monomeric product, however, has some unusual properties in that it is complementary to both the reaction components from which it is made. Accordingly, it can act as a template for its own formation

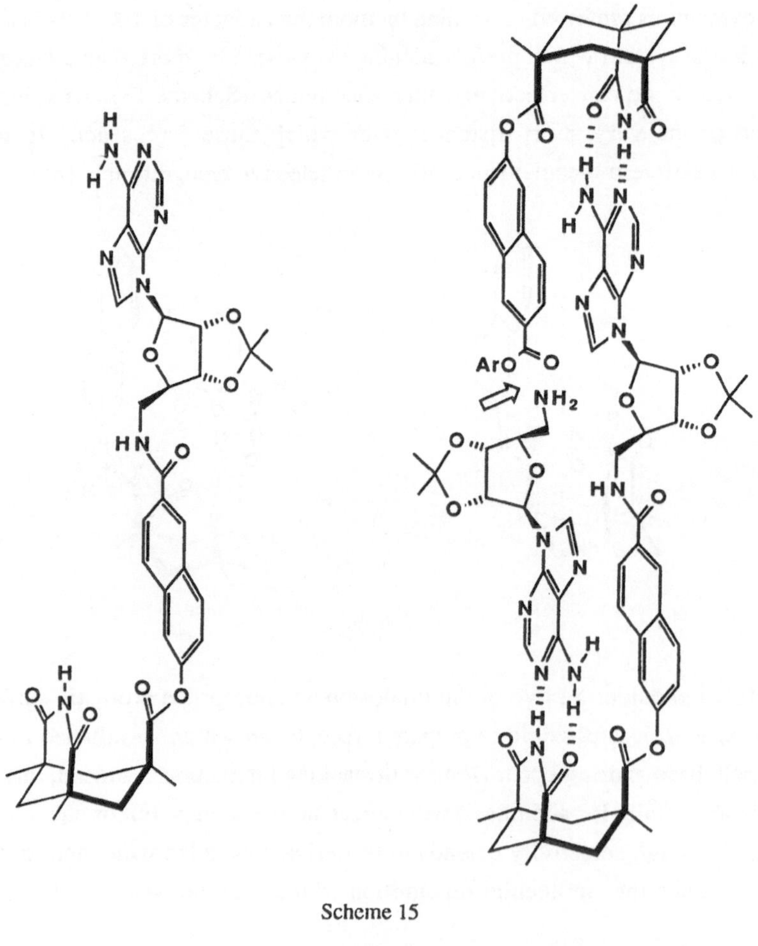

Scheme 15

(Scheme 15), and thereby exhibit autocatalysis. It can complex both the imide and adenine components in such a way that their reactive centers are brought into close contact.

We now have several such systems in hand and they show the earmarks of self-replicating systems [22]. The kinetic analysis of a system capable of autocatalysis and product inhibition is not easy [23] and we are working to sort out the observable effects of such phenomena using suitable control systems. At the very least it is worth emphasizing that model systems are now quite useful in exploring the rules of phenomena such as recognition, transport selectivity, and replication. Whether they do so in chloroform or in aqueous solution is quite beside the point; the rules are not expected to change from one solvent to another. With model systems, it is possible to examine the intrinsics of the interaction in greater spectroscopic detail than can be brought to bear in the natural system. If these models express behavior that is yet unknown in natural systems, so much the better.

Altogether different shapes for receptors within this series can be had from simple spacers such as TREN. These molecules can adopt a number of conformations, one of which is seen to resemble a molecular "tool chuck" in which three imide groups are seen to converge [15]. Not surprisingly, this arrangement provides an ideal microenvironment for melanine (Scheme 16). Indeed, a 1:1 complex is formed at room temperature in chloroform. Exchange between the complex and the free receptor is slow on the nmr time scale. Dilution studies have shown that the association constant must be at least in the micromolar range, possibly even the nanomolar range as might be expected if all nine hydrogen bonds are formed.

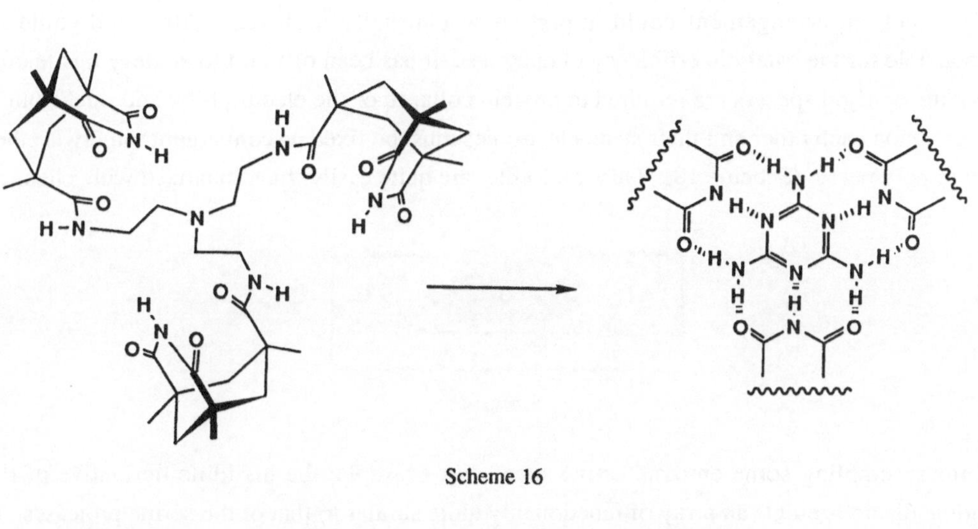

Scheme 16

While the larger spacers have emphasized their unique properties as molecular chelates, we have used the smallest versions as more conventional chelating agents towards divalent metal ions. Their advantages in this respect derive from their highly organized structures and ability to present the more basic lone pairs towards the metal ion. While much of the binding is of electrostatic nature, and stereoelectronic effects are relegated to a minor role, the very instability of the dianion of the benzene or similarly spaced molecule suggests its ability to grasp divalent metals very tightly. Moreover, the binding is necessarily *trans*, and the metals in this environment are expected to show different behavior than those in more conventional environments such as provided by imidodiacetate (Scheme 17). Molecules such as the one represented by the pyridine spacer show astonishing affinity to Ca^{2+} and Mg^{2+} [24]. We have made considerable progress in developing this new family of chelating agents for applications in industry, medicine and catalysis. For example, it is possible to attach the diacids to polystyrene beads and use them as ion exchange resins to remove calcium from concentrated brine solutions.

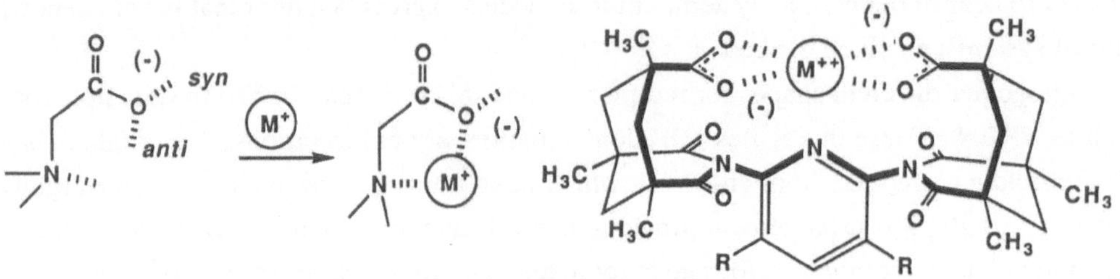

Scheme 17

Another area in which convergent functional groups can be applied to unique advantage is in the study of concerted catalysis. These sytems were envisioned by Swain more than 35 years ago to describe a substrate fixed between, say, nucleophilic and electrophilic centers [25]. Such an arrangement could impart some unusually high reactivities and could be responsible for the catalytic efficiency of enzymes. It has been difficult to achieve this feature in systems: rigid spacers are required to prevent collapse of the electrophilic and nucleophilic centers upon each other and their stereoelectronics must be fixed as convergent otherwise they tend to polymerize (Scheme 18). Our molecules are quite easily functionalized with chemical

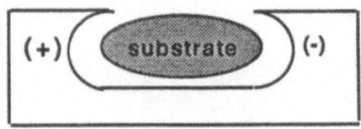

Scheme 18

linings resembling some enyzme active sites. For example, the histidine derivative of the acridine diacid presents an array of functionality quite similar to that of the serine proteases. In this model system the *syn* lone pair of the carboxylate is directed toward the imidizolium moiety (Scheme 19). The intrinsics of this zwitterionic interaction can be examined in even a simpler model. Again the U-turn presents a chance to place an imidazole into contact with the more basic carboxylate lone pair. The acid-base characteristics of the two functional groups are modified accordingly [26].

12

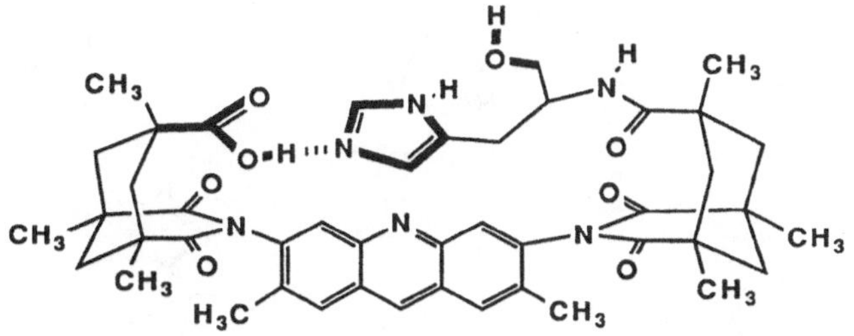

Scheme 19

Condensation of the Kemp triacid with amines obtained from the Dakin-West reaction of amino acids leads to structures in which the carboxylate *syn* lone pair is directed at an enolizable hydrogen atom (Scheme 20). We have studied the efficiency of this process compared to systems involving the less basic lone pair and find about an order of magnitude in rate enhancement [27]. It is quite likely that further optimizing this geometry of proton transfer could lead to even more kinetic advantages with this stereoelectronic arrangement.

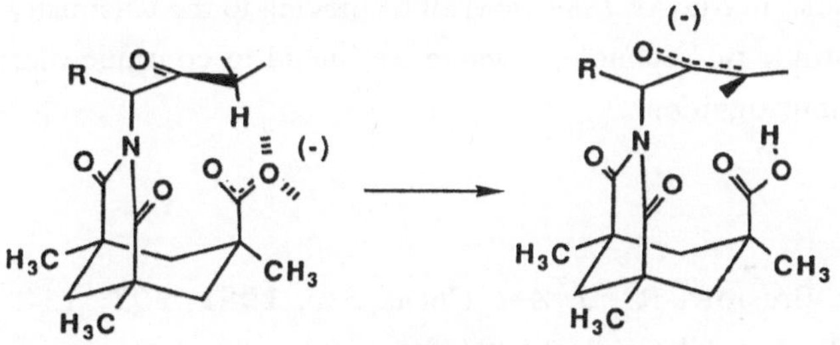

Scheme 20

Finally, the convergent arrangement of acids in the acridine derived material resembles closely the active site of lysozyme. We were most pleased to see the substrates which fit within the cleft become highly susceptible to hemiacetal cleavage (Scheme 21). Specifically, the glycol aldehyde dimer dissociates quite rapidly when in contact with this array of functional groups [28]. Simple acids and bases or their mixtures are quite ineffective in this hemiacetal cleavage. Even the divergent diacid, which incorporates all of the functional groups of the effective catalysts is without effect. This is because it does so in a manner that the groups are unable to act concertedly. This material is not without its intrinsic charm, that of economy. It will be recognized as derived from the minor (unwanted) isomer of the alkylation during the synthesis of the *cis-cis* (Kemp's) triacid. Indeed, we have used this isomer many times as an ideal control to explore the many facets of molecular recognition that are described above.

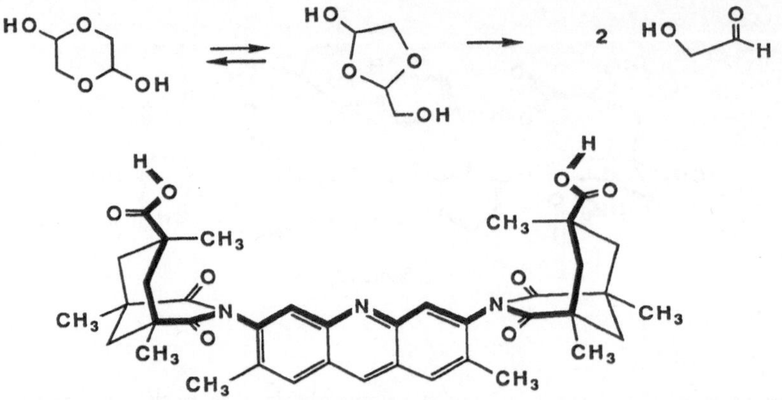

Scheme 21

Acknowledgements. I am grateful to the expert experimental assistance and intellectual support of the many associates that have worked on this project. Their names appear on the original publications. Financial support from the National Institutes of Health and the National Science Foundation is also gratefully acknowledged. Finally, I am grateful to the superb scientific environment provided by the Chemistry Department of the University of Pittsburgh over the last 13 years. In May of 1989 we shall be moving to the Chemistry Department at the Massachusetts Institute of Technology where we intend to continue our program in model studies in molecular recognition.

References

1. Trainor, G.; Breslow, R. *J. Am. Chem. Soc.* **1981**, *103*, 154; D'Souza, V. T.; Bender, M. L. *Acc. Chem. Res.* **1987**, *20*, 146.

2. Lehn, J.-M. *Science* **1985**, *227*, 846; Cram, D. J. *ibid.* **1983**, *219*, 1177; Lehn, J.-M.; Sirlin, C. *J. Chem. Soc., Chem. Commun.* **1978**, 949; Cram, D. J.; Lam, P. Y.-S.; Ho, S. P. *J. Am. Chem. Soc.* **1986**, *108*, 839.

3. Rebek, J. Jr.; Costello, T.; Marshall, L.; Wattley, R.; Gadwood, R. C.; Onan, K. *J. Am. Chem. Soc.* **1985**, *107*, 7481.

4. Stetter, H.; Roos, E. E. *Chem. Ber.* **1955**, *88*, 1390, 1395; Odashima, K.; Itai, A.; Iitaka, Y.; Koga, K. *J. Am. Chem. Soc.* **1980**, *102*, 2504; Miller, S. P.; Whitlock, H. W. Jr. *ibid* **1984**, *106*, 1492.; Winkler, J.; Coutouli-Argyropoulou, E.; Leppkes, R.; Breslow, R. *ibid* **1983**, *105*, 7198; Diederich, F.; Griebel, D. *ibid* **1984**, *106*, 8037; Gutsche, C. D. *Acc. Chem. Res.* **1983**, *16*, 161.

5. Gandour, R. *Biorg. Chem.* **1981**, *10*, 169.

6. Kemp, D. S.; Petrakis, K. S. *J. Org. Chem.* **1981**, *46*, 5140.

7. Rebek, J. Jr.; Askew, B.; Killoran, M.; Nemeth, D.; Lin, F.-T. *J. Am. Chem. Soc.* **1987**, *109*, 2426.

8. Rebek, J. Jr. *Science* **1987**, *235*, 1478; Rebek, J. Jr.; Marshall, L.; Wolak, R.; Parris, K.; Killoran, M.; Askew, B.; Nemeth, D.; Islam, N. *J. Am. Chem. Soc.* **1985**, *107*, 7476. For other clefts in molecular recognition see: Wilcox, C. S.; Greer, L. M.; Lynch, V. *J. Am. Chem. Soc.* **1987**, *109*, 1865; Kelly, T. R.; Maguire, M. P. *ibid*, **1987**, *109*, 6549.

9. Rebek, J. Jr.; Duff, R. J.; Gordon, W. E.; Parris, K. *J. Am. Chem. Soc.* **1986**, *108*, 6068.

10. Rebek, J. Jr.; Askew, B.; Islam, N.; Killoran, M.; Nemeth, D.; Wolak, R. *J. Am. Chem. Soc.* **1985**, *107*, 6736.

11. Rebek, J. Jr.; Nemeth, D. *J. Am. Chem. Soc.* **1986**, *108*, 5637.

12. Rebek, J. Jr.; Nemeth, D.; Ballester, P.; Lin, F.-T. *J. Am. Chem. Soc.* **1987**, *109*, 3474.

13. Rebek, J. Jr.; Askew, B.; Nemeth, D.; Parris, K. *J. Am. Chem. Soc.* **1987**, *109*, 2432.

14. Lindsey, J. S.; Schreiman, I. C.; Hsu, H. C.; Kearney, P. C.; Marguerattaz, A. M. *J. Org. Chem.* **1987**, *52*, 827.

15. Lindsey, J. S.; Kearney, P. C.; Duff, R. J.; Tjivikua, P. T.; Rebek, J. Jr. *J. Am. Chem. Soc.* **1988**, *110*, 6575.

16. Rebek, J. Jr.; Askew, B.; Ballester, P.; Buhr, C.; Jones, S.; Nemeth, D.; Williams, K. *J. Am. Chem. Soc.* **1987**, *109*, 5033. For another system using macrocyclic structures for nucleic acid recognition see: Hamilton, A. D.; Van Engen, D. *J. Am. Chem. Soc.* **1987**, *109*, 5035.

17. Kyogoku, Y.; Lord, R. C.; Rich, A. *Science* **1966**, *154*, 518.

18. Rebek, J. Jr.; Askew, B.; Ballester, P.; Buhr, C.; Costero, A.; Jones, S.; WIlliams, K. *J. Am. Chem. Soc.* **1987**, *109*, 6866.

19. Rebek, J. Jr.; Williams, K.; Parris, K.; Ballester, P.; Jeong, K. -S. *Angew. Chem., Int. Ed. Engl.* **1987**, *26*, 1244.

20. Jeong, K.-S.; Rebek, J. Jr. *J. Am. Chem. Soc.* **1988**, *110*, 3327.

21. Benzing, T.; Tjivikua, T.; Wolfe, J.; Rebek, J. Jr. *Science* **1988**, *242*, 266.

22. Rebek, J. Jr. *Pure Appl. Chem.* **1989**, *61*, 1517.

23. von Kiedrowski, G. *Angew. Chem., Int. Ed. Engl.* **1986**, *25*, 932.

24. Marshall, L.; Parris, K.; Rebek, J. Jr.; Luis, S. V.; Burguete, M. I. *J. Am. Chem. Soc.* **1988**, *110*, 5192.

25. Swain, C. G.; Brown, J. F. Jr. *J. Am. Chem. Soc.* **1952**, *74*, 2538.

26. Huff, J.; Askew, B.; Duff, R. J.; Rebek, J. Jr. *J. Am. Chem. Soc.* **1988**, *110*, 5908.

27. Tadayoni, B. M.; Parris, K.; Rebek, J. Jr. *J. Am. Chem. Soc.* **1988**, *110*, 4503.

28. Wolfe, J.; Nemeth, D.; Costero, A.; Rebek, J. Jr. *J. Am. Chem. Soc.* **1988**, *110*, 983.

"ARTIFICIAL OLIGONUCLEOTIDES": MOLECULAR RECOGNITION BY BASE PAIRING

Jonathan L. Sessler* and Darren Magda

Department of Chemistry
University of Texas
Austin, Texas 78712 (USA)

SUMMARY

The synthesis of a novel cytosine-guanine ribose- and phosphate-free "artificial dinucleotide" **1** and its cytosine-cytosine analogue **2** are described. The syntheses of derivatives of these core compounds bearing porphyrin and polymeric side chains are also presented.

INTRODUCTION

The specific hydrogen bonding interactions which occur between DNA bases provide perhaps the premier example of molecular recognition: they are essential to life and inspirational to chemists (Figure 1). Indeed, we have long been fascinated by the efficiency and apparent simplicity whereby information is transferred and stored in natural nucleic acid based systems, and have sought to reproduce these essential features in synthetic systems. In particular, we have been intrigued by the question of whether it might be possible to construct artificial self-replicating systems from simple ribose and phosphate-free purine and pyrimidine precursors.

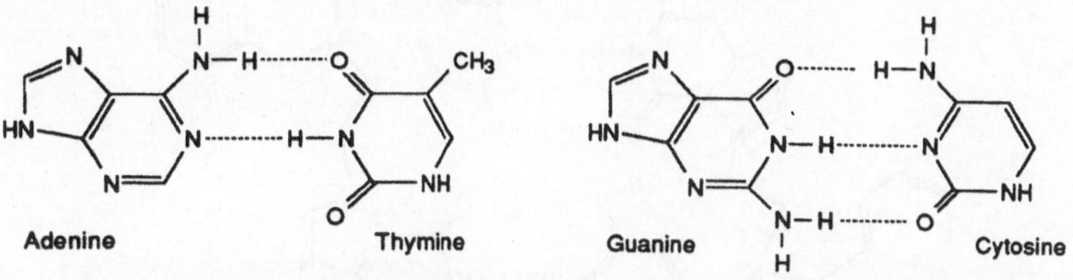

Figure 1. Watson-Crick base-pairing interactions for common nucleic acid heterocycles.

A generalized representation of a possible template-based self-replicating system is shown in Scheme 1. Here **a** and **b** could represent complementary base-pairs and **c** a rigid spacer group. As discussed in detail earlier [1], this bio-inspired chemistry is predicated on a number of assumptions, one of which is the successful development of a base pairing approach to molecular recognition. Recently, a number of research groups have reported approaches to molecular recognition based on hydrogen bonding [2-4]. Few of the extant systems, however, reproduce the essential base pairing aspects of natural oligonucleotides [4]. We have therefore developed a new series of model ribose and phosphate-free "artificial oligonucleotides" in which molecular recognition by base pairing might be possible, and wish to report here the results of initial studies of these systems.

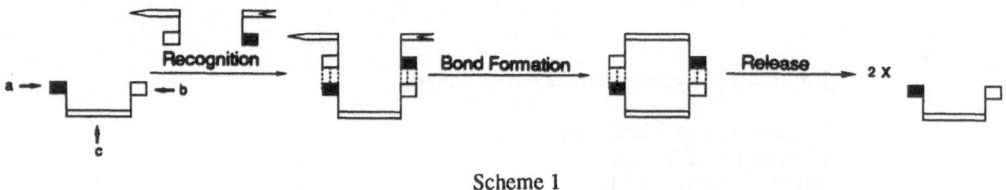

Scheme 1

We recently reported [1] the synthesis and spectral properties of dinucleotide analogues **9** and **10** (Chart 1) in which pendant guanine or cytosine bases are joined by a tertiary aromatic amine linker. The chemical composition of these systems, which bear a distinct resemblance to dimeric DNA fragments, led us to propose that molecular self-recognition by base pairing might be possible in aprotic solvents. The available data is suggestive, but not conclusive. For instance, the chemical shift of N1 of the guanine subunit in the heterodimer **9** is shifted downfield by roughly 40 Hz as compared to that observed in the homodimer **10**, a result that is consistent with base-pairing to form the supramolecular complex shown in Figure 2. Unfortunately, an unequivocal proof of such hydrogen bonding interactions by proton magnetic resonance spectroscopy (^{1}H NMR) techniques can only be made when the dimers are sufficiently soluble in the desired solvent (DMSO-d6, pyridine-d5, chloroform-d1, etc.) to study over a range of concentrations, and the existing dimers, **9** and **10**, are too insoluble to

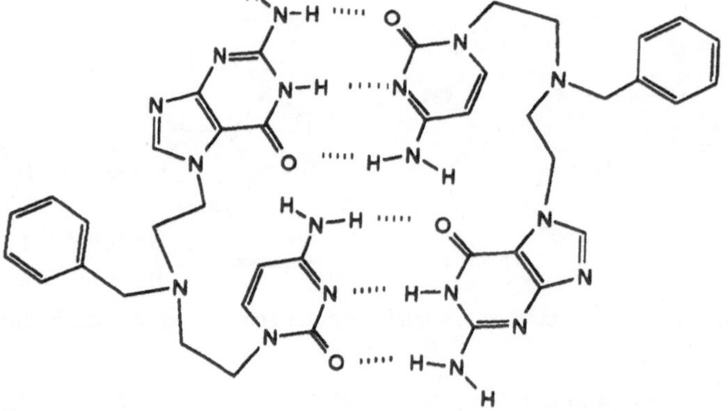

Figure 2.   Schematic representation of the possible supramolecular complex which might be formed from the "artificial dinucleotide" 9.

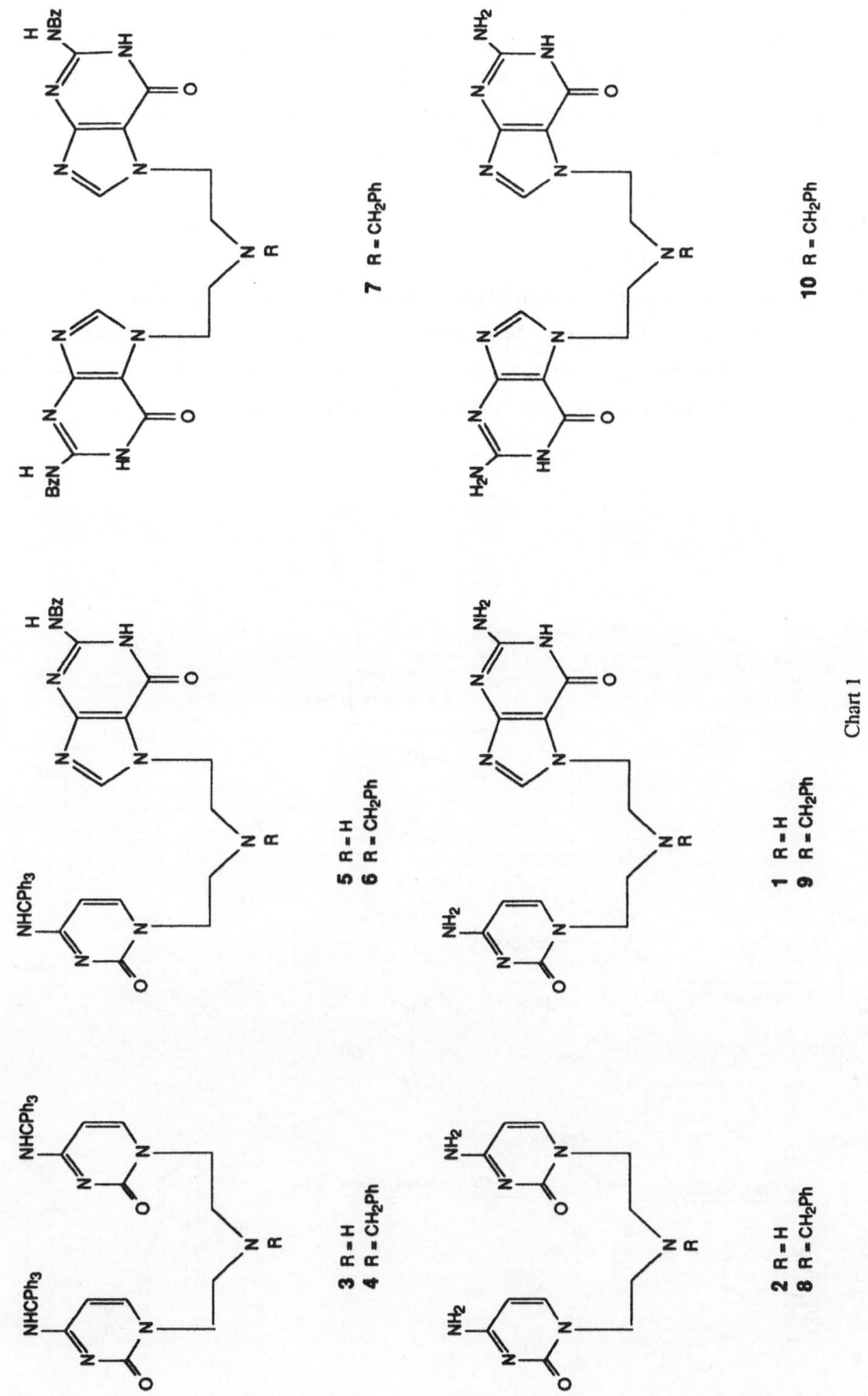

7 R = CH₂Ph

10 R = CH₂Ph

5 R = H
6 R = CH₂Ph

1 R = H
9 R = CH₂Ph

3 R = H
4 R = CH₂Ph

2 R = H
8 R = CH₂Ph

Chart 1

19

allow such detailed studies. Due to these difficulties we therefore decided to modify our synthetic approach to provide secondary amine-linked dinucleotide analogues (e.g. **1** or **2**). These molecules, as nucleophiles, may be alkylated or acylated and therefore offer access into a wide variety of synthetic systems. These may be tailored to meet solubility considerations, as well as to produce interesting intermolecular proximity relationships following the formation of a base-paired supramolecular complex.

SYNTHESIS

The original dinucleotide analogue **9** (and **10**) was produced via a "tertiary amine approach" in which benzylamine was alkylated sequentially with the alkyl iodide derivatives of the guanine and cytosine bases (e.g. **11** and **13**) (Scheme 2). In the modified, "secondary amine approach", the synthetic features remain the same except that a cytosine derived primary amine, **19**, is used as the central point of attachment (Scheme 3).

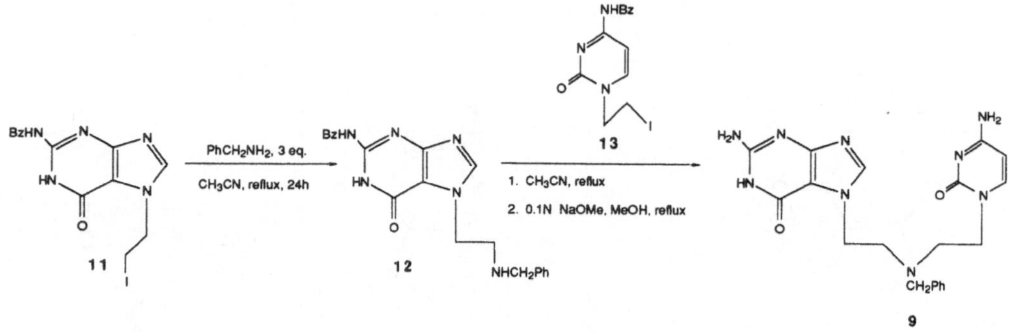

Scheme 2

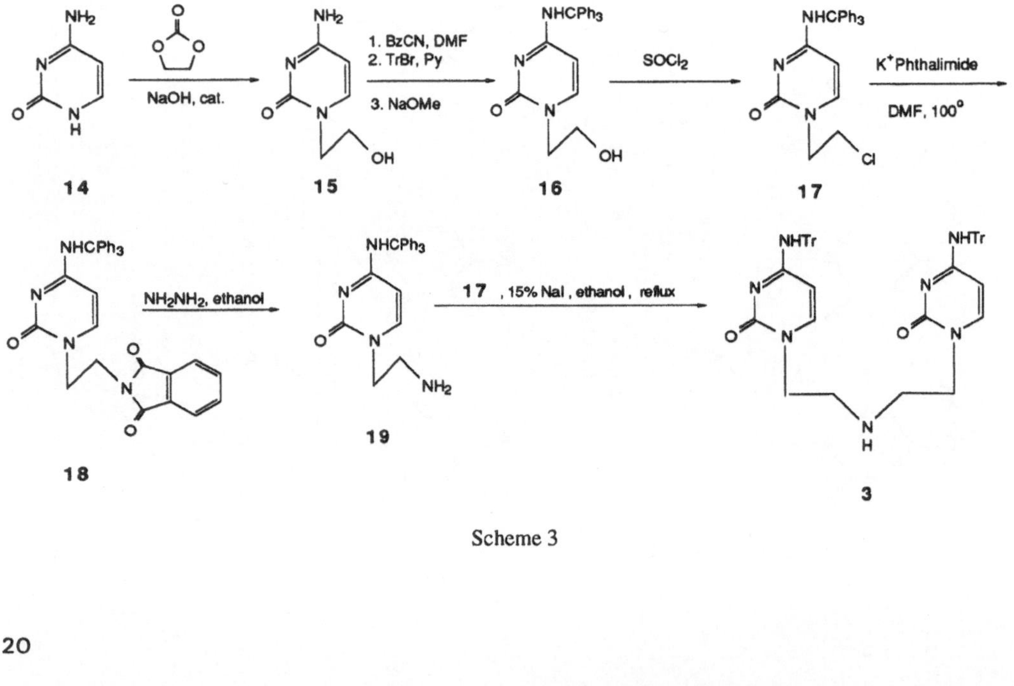

Scheme 3

In order to obtain the requisite cytosine derivative **19**, however, it proved necessary to deviate substantially from the synthetic approach used to obtain **13**. In particular, the benzamide protective group for N4 was found to be too labile for the conditions required for amine synthesis. After testing a variety of potential candidates (dimethylformamidine, anisoyl amide, p-toluenesulphonamide, tetrahydropyranyl) it was determined that the trityl amine best fulfilled the requirements of an ideal protective group: stability, removability, and high solubility. Unfortunately, as will be discussed later, it will be seen that even this protective group is not free of problems.

As before, the syntheses of the various required cytosine derivatives begins with the conversion of cytosine to its sodium salt and alkylation with ethylene carbonate to produce the 1-hydroxyethyl heterocycle **15** [5]. (We note that this reaction is now being carried out on a 100g scale!) Protection of the exocyclic amino group could then be accomplished as follows: First, selective benzoylation of the 2'-hydroxyl group was effected using benzoyl cyanide in N,N-dimethylformamide (DMF). Heating the resulting benzoic ester with trityl bromide in pyridine at 100° for 24 h caused tritylation of the amino group. Finally stirring this intermediate in 0.1 N NaOMe in methanol released the trityl-protected hydroxyethylcytosine **16** in about 50% overall yield from **15**. In spite of the acidic conditions involved, the trityl protective group is stable enough to allow conversion of the alcohol **16** into the chloride **17** in about 90% yield when heated at reflux in thionyl chloride. Chloride **17** could then readily be converted to the amine **19** in 96% overall yield by treatment with potassium phthalimide in DMF, followed by hydrazinolysis of the resulting phthalimide **18**.

When amine **19** was reacted with the chloride **17** in refluxing acetonitrile, a chloroform soluble compound having a lower R_f on silica gel than the starting chloride (**17**) was isolated; however, none of the amine had apparently reacted or been incorporated into this product. A similar product was obtained upon heating **17** at reflux in 15% NaI/acetone in the absence of **19**, conditions which convert it to the iodide (Scheme 4). Based on spectral data and literature precedent [6], this product was assigned structure **20**, a result of an internal cyclization. It is apparent that the trityl protective group does not prevent the loss of the amino proton in this circumstance. Fortunately, conditions were found (15% NaI/ethanol) under which the intermolecular reaction would proceed to give **3** in reasonable yield (62%). This trityl-protected aminoethylcytosine **19** could also be alkylated with the guanine electrophile **11** to afford the diprotected, "mixed dimer" **5** (Scheme 5). Although no cyclized by-products have been observed in the course of this reaction, the yields have yet to exceed 33%. Efforts to improve this modest yield are currently underway.

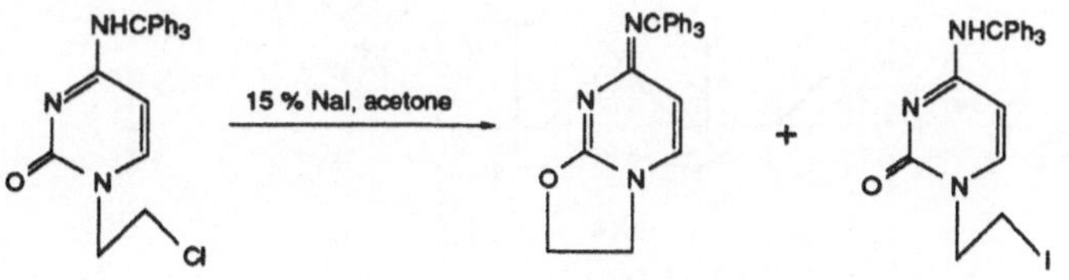

Scheme 4

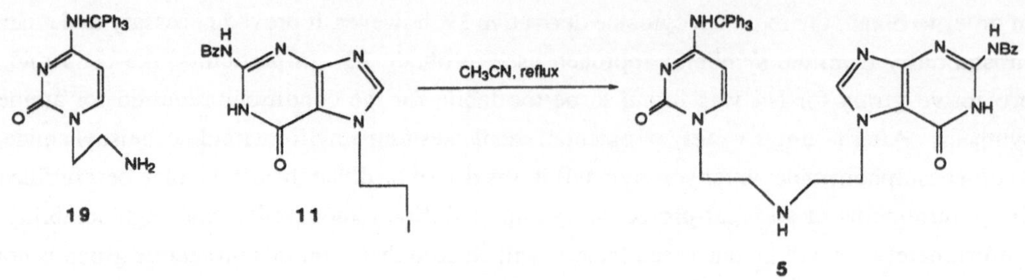

Scheme 5

The homodimer **3** and heterodimer **5** have proven to be excellent nucleophiles in typical alkylation reactions. The protective groups lend favorable solubility characteristics to the amines, and may be removed selectively either before, or after, further functionalization. For instance, the trityl group in compound **5** may be hydrolyzed in the presence of the guanine benzamide by stirring overnight in 30% HBr/acetic acid, while stirring at reflux in 0.1 N NaOMe in methanol for 3 h is sufficient to cleave the benzamide on guanine. The resultant deprotected mixed dimer **1** is water soluble at pH 7, where it presumably exists as the cation.

In order to facilitate ^{1}H NMR studies, we have recently prepared the poly(ethyleneglycol) functionalized iodide **23** (Scheme 6). We selected as a starting material the commercially available alcohol **22**, having an average molecular weight of 750, which was converted easily into the iodide **23** using standard methods. The mixed dimer **5** was alkylated in reasonable ($\approx$ 50%) yield with **23** in refluxing acetonitrile. Simple deprotection (currently in progress) should provide a tertiary amine **24** having the solubility characteristics required to test quantitatively the formation of a putative base-paired complex (such as that presented in Figure 2).

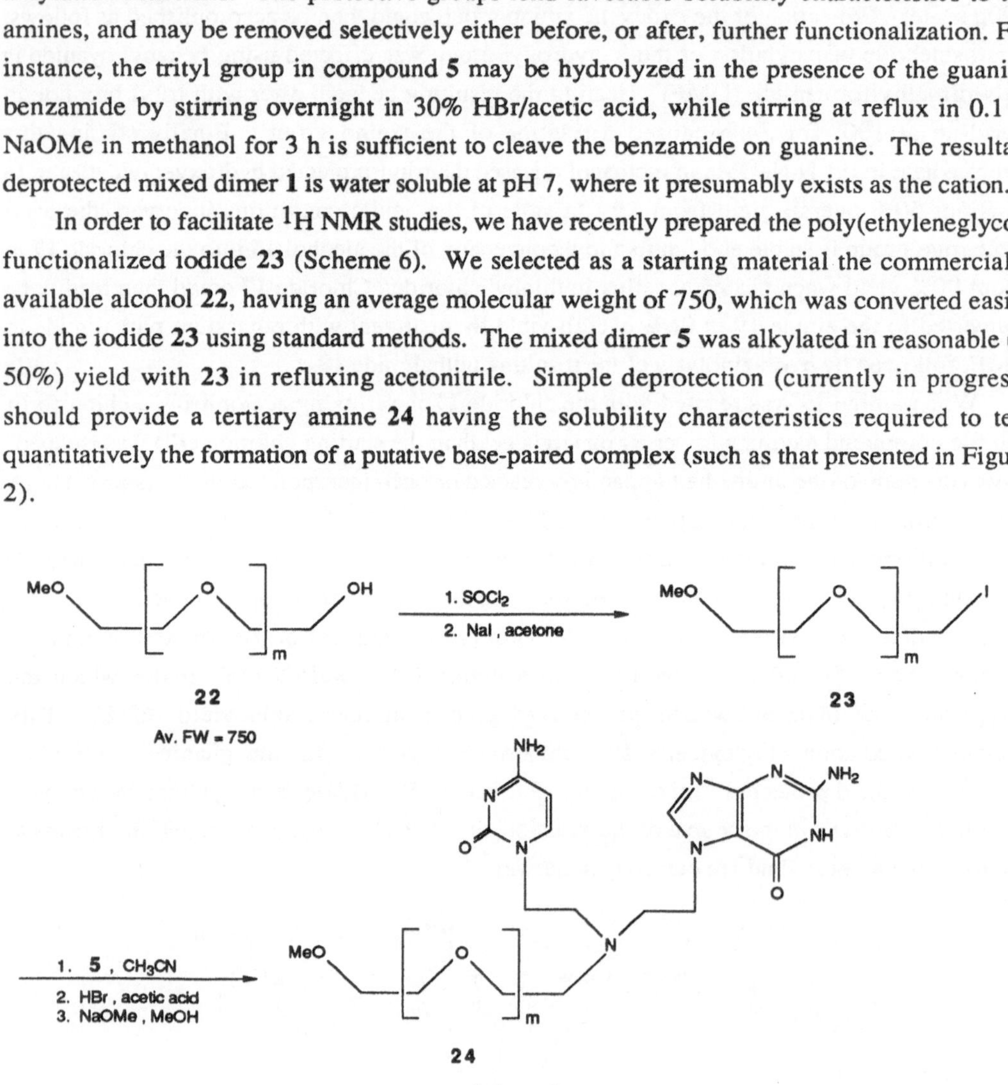

Scheme 6

An alternative approach to solubilizing the artificial dinucleotide **1** is to couple it with the highly soluble electrophilic porphyrin **25** recently developed in our laboratory (Scheme 7). Again, alkylation of the mixed dimer **5** proceeds readily (70%) to give the protected porphyrin **26**. Deprotection of the cytosine moiety results in a monoprotected porphyrin in nearly quantitative yield which may be purified by chromatography. Deprotection of the guanine amino group likewise proceeds routinely to produce the porphyrin-substituted dinucleotide analogue **27**, which was readily metalated by treating with zinc acetate in methanol. As expected, characteristic spectral changes are observed in the Q-band region of the optical spectrum (Figure 3) upon metalation. Interestingly, the spectra also show features in the 200-300 nm region ascribable to the heterocyclic bases. Unfortunately, only a small amount of this material is presently in hand and it is impossible to determine as yet whether the porphyrin will impart sufficient solubility to these dimers to enable ^{1}H NMR studies. Nonetheless, we anticipate that these deprotected products will allow for the formation of self-associating porphyrin dimers (e.g. **28** in Scheme 7). We expect that such organized "supramolecular" structures will be of tremendous interest. For instance, long range inter-chromophore energy and electron transfer reactions between covalently linked porphyrin dimers have been studied in a number of groups [7], including our own [8] in an effort to understand natural electron transfer processes. To date, however, it has not proved possible to study such processes in artificial systems where the donor and acceptor are oriented through means of *noncovalent interactions*. The supramolecular complex **28** would offer such a possibility.

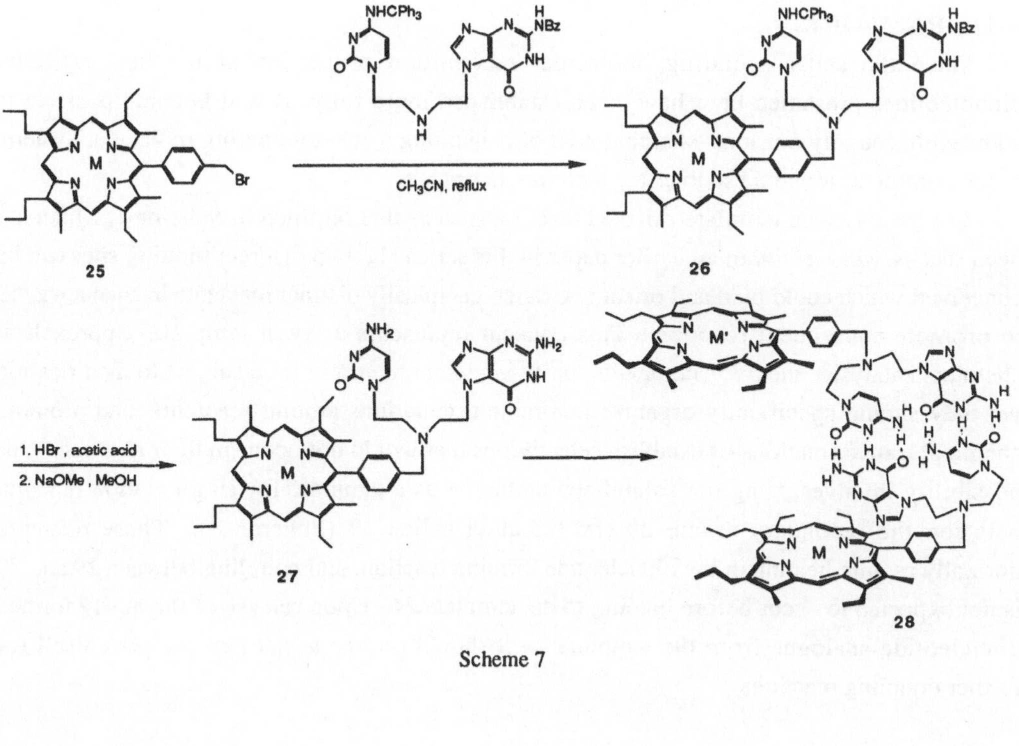

Scheme 7

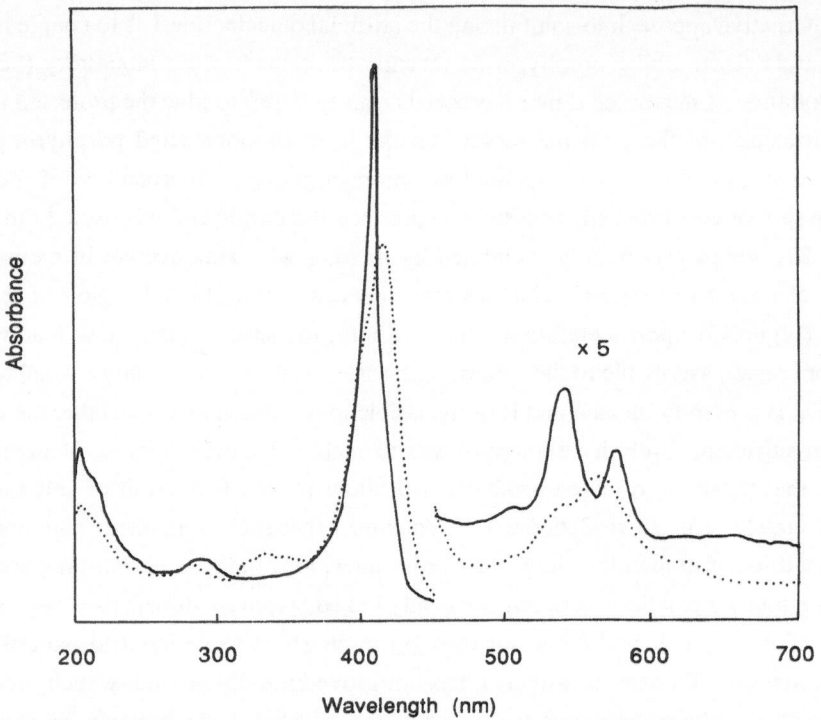

Figure 3.   Absorption spectra in methanol of the protonated form of compound **27** (M = 2H) (.........), and its zinc(II) complex (_____).

FUTURE WORK

Once the self-associating, molecular recognition properties of the new artificial dinucleotides presented here have been established more fully, it will become possible to address thoroughly our long-standing goal of generating a self-replicating molecular system. Some comments along these lines are therefore in order.

In a prototypical template-directed reaction such as that outlined in Scheme 1, which has been discussed at length in an earlier paper in this series [1], two discreet binding sites can be conceived which could bind and orient two other, chemically distinct reactants in such a way as to promote subsequent reaction. This concept represents a "twin template" approach to chemical catalysis: the two-site binding units would not themselves be subject to modification yet they should specifically organize favorable orientations among reactants, and promote chemical transformations or establish selectivities that would not occur in their absence. One possibility involves using the solubilized amine **24** as a template for an alkylation reaction between the secondary amine **30** and the alkyl halide **29** (Scheme 8). These reagents normally require heating in the dinucleotide forming reaction, and coupling between **29** and **30** is not expected to occur before binding to the template **24**. Upon release of the newly formed dinucleotide analogue from the template **24** it should be free to act as a template itself for further coupling reactions.

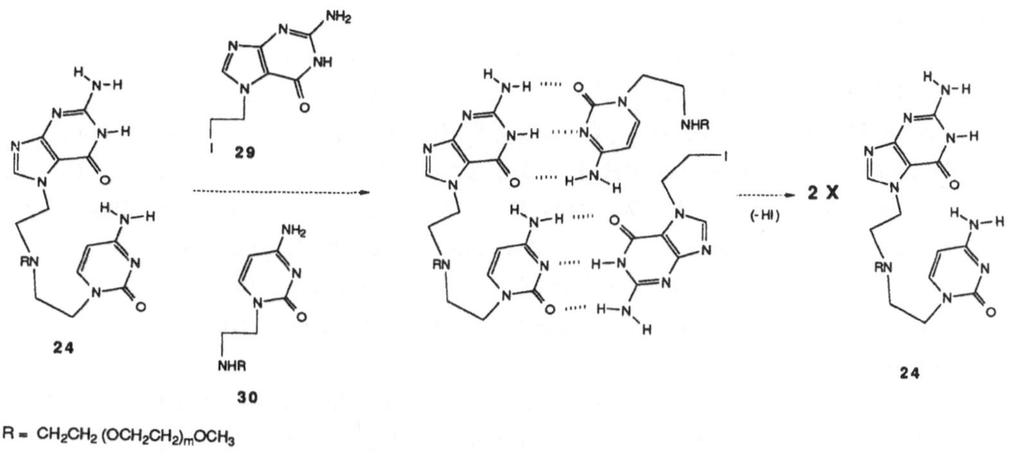

R = CH2CH2 (OCH2CH2)mOCH3

Scheme 8

A similar approach currently being tested involves base pairing between cytosine residues (Scheme 9). Since cytosine residues contain both proton donating and accepting functionalities, it is conceivable that this residue can form a Watson-Crick type complex with itself in dimethylsulfoxide solution. Indeed, the splitting of the exocyclic amino protons in ^{1}H NMR spectra of cytosine derivatives (36-50 Hz) strongly supports this hypothesis. Unfortunately, experiments designed to test for autocatalysis, in which the reactants of Scheme 9 were mixed in the presence and absence of "template" 2, have so far proved inconclusive: the alkyl chloride employed (32) failed to react, even at elevated temperatures. The synthesis of the analogous alkyl iodide, which will allow us to test further this approach, is in progress.

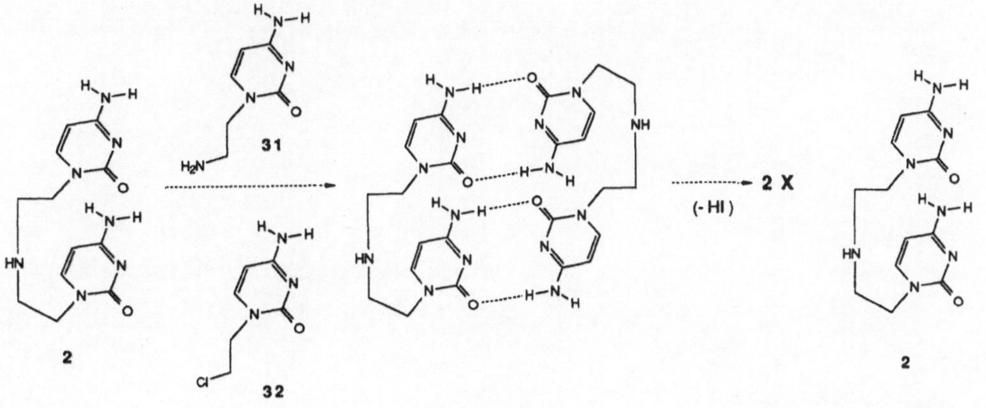

Scheme 9

References

1. Sessler, J. L.; Magda, D. J.; Hugdahl, J. *J. Incl. Phenom.* **1989**, *7*, 19.

2. (a) Rebek, J. Jr.; Nemeth, D. *J. Am. Chem. Soc.* **1986**, *108*, 5637; (b) Rebek, J. Jr.; Askew, B.; Ballester, P.; Buhr, C.; Jones, S.; Nemeth, D.; Williams, K. *J. Am. Chem. Soc.* **1987**, *109*, 5033; (c) Rebek, J. Jr.; Askew, B.; Ballester, P.; Buhr, C.; Costero, A.; Jones, S.; Nemeth, D.; Williams, K. *J. Am. Chem. Soc.* **1987**, *109*, 6866.

3. Hamilton, A. D.; Van Engen, D. *J. Am. Chem. Soc.* **1987**, *109*, 5035.

4. Kim, M. S.; Gokel, G. W. *J. Chem. Soc., Chem. Commun.* **1987**, 1686.

5. Ueda, N.; Kondo, K.; Kono, M.; Takemoto, K.; Imoto, M. *Die Makromolekulare Chemie* **1968**, *120*, 13.

6. Takemoto, K. *Polym. Prepr., Am. Chem. Soc., Div. Polym. Chem.* **1979**, *20*, 215.

7. (a) Schwarz, F. P.; Gouterman, M.; Muljiani, Z.; Dolphin, D. *Bioinorg. Chem.* **1972**, *2*, 1; (b) Anton, J. A.; Loach, P. A.; Govindjee *Photochem. Photobiol.* **1978**, *28*, 235; (c) Mialocq, J. C.; Giannotti, C.; Maillard, P.; Momenteau, M. *Chem. Phys. Lett.* **1984**, *112*, 87; (d) Brookfield, R. L.; Ellul, H.; Harriman, A.; Porter, G. *J. Chem. Soc., Faraday Trans. 2* **1986**, *82*, 219.

8. (a) Sessler, J. L.; Johnson, M. R.; Tzuhn-Yuan, L.; Creager, S. E. *J. Am. Chem. Soc.* **1988**, *110*, 3659; (b) Gubelmann, M.; Harriman, A.; Lehn, J.-M.; Sessler, J. L. *J. Chem. Soc., Chem. Commun.* **1988**, 77.

OPTIMIZING THE NEUTRAL MOLECULAR RECEPTOR PROPERTIES OF WATER SOLUBLE CYCLOPHANES [1]

Craig S. Wilcox* [2], Marlon D. Cowart, Irving Sucholeiki,
Rudolf R. Bukownik, and Vincent Lynch

Department of Chemistry
University of Pittsburgh
Pittsburgh, PA 15260 (USA)

INTRODUCTION

It is now well established that some water soluble cyclophanes can bind to neutral organic molecules in aqueous solutions. Benzenoid, naphthalenoid, and larger aromatic molecules will form inclusion complexes with properly constructed cyclophanes [3]. The phenomenon is important because such host-guest systems might provide a foundation for constructing selective water soluble catalysts, selective reagents, or new separation methods [4].

Water soluble cyclophanes (for example hosts **1**, **2**, and **3**) have invariably been designed with at least one C_2 axis of symmetry passing *across* the macrocyclic ring. It is interesting to realize that such symmetry has never been observed in the solid state for an inclusion complex derived from such a host. Why? The reason might lie in the requirements for crystal lattice formation, but an important possibility is that the symmetrical conformers of the host (or the

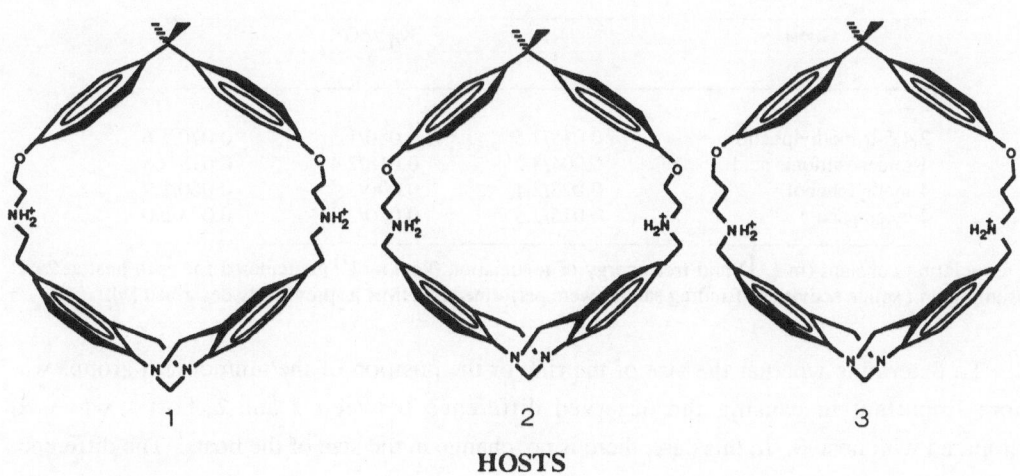

1   **2**   **3**

HOSTS

Inclusion Phenomena and Molecular Recognition
Edited by J. Atwood
Plenum Press, New York, 1990

complex) are too unstable to contribute to the binding phenomenon. If true, this has important consequences for those who wish to optimize the properties of a cyclophane derived host for aromatic guests.

Our interest in understanding and optimizing the properties of water soluble cyclophanes based on Tröger's base derivatives led us to undertake this investigation of the conformations available to **1**, a symmetrical cyclophane host which binds to benzenoid substrates [5]. This paper describes some of the properties observed for hosts **1**, **2**, and **3** and presents a general method for systematic representation and evaluation of potential host conformers. The results for such evaluation of host **1** are examined and some tentative conclusions about the "active forms" of the host are presented. We conclude that there are no conformers with a C_2 axis transverse to the cavity that are of low enough conformational energy to be considered as likely contributors to the observed properties of the host.

HOST PROPERTIES FOR WATER SOLUBLE CYCLOPHANES CONTAINING ONE DIBENZODIAZOCENE UNIT

The dissociation constants for several complexes derived from the three hosts **1**, **2**, and **3** are presented in Table 1. There are some interesting comparisons that can be made among these three molecules. Host **1** binds most strongly to any given guest. Host **2**, which is similar to host **1** but smaller (one atom has been removed from each connecting chain) binds more weakly to each guest. The difference may be due to a size difference (host **2** is smaller than host **1**) or may be due to the change in the position of the ammonium group relative to the aromatic rings. Host **1** is a phenethylamine derivative but host **2** is a benzylamine derivative. The benzylamine derived macrocycle requires at least one ammonium group to be oriented with the N-H bonds directed into the cavity. Obviously, strong hydrogen bonds from the host to water molecules within the cavity would mediate against binding of lipophilic guests. With host **1**, the ammonium groups are more mobile and are not required to take up positions with the N-H bonded directed into the cavity.

Table 1. Dissociation constants and free energy of association of some complexes of **1**, **2** and **3**.

Guest	$K_d / \Delta G$[a]		
	1	2	3
2,4,6-trimethylphenol	0.015/2.5	0.080/1.5	0.070/1.6
4-toluenesulfonic acid	0.004/3.2	0.018/2.4	0.015/2.5
4-methylphenol	0.023/2.1	0.090/1.4	0.060/1.7
4-cyanophenol	0.015/2.5	0.020/2.3	0.033/2.0

[a]Dissociation constant (m L^{-1}) and free energy of association (kcal mol^{-1}) calculated for each host at 25°C assuming unit solute activities. Binding studies were performed by NMR as previously described [5].

To determine whether the size of the ring or the position of the ammonium groups was more important in causing the observed difference between **1** and **2**, host **1** was next compared with host **3**. In this case, there is no change in the size of the hosts. The difference between **1** and **3** is that the positions of a carbon and a nitrogen atom have been reversed. In **1**, the connecting chain sequence is C-C-N-C-C-O, in **3** the sequence is C-N-C-C-C-O. It is interesting that **3** is as poor a host as was **2** and that both **2** and **3** bind more weakly than **1** to benzenoid substrates. It may be concluded that for this type of host phenethylamine derived hosts are superior to benzylamine derived hosts. The observed drop in binding of guests to hosts **2** and **3** may be due to stronger binding of solvent to these hosts or weaker binding of guests to these hosts or may be due to a combination of these factors.

Prior to any attempt to prepare second or third generation hosts based on these simple beginnings, it was important to consider the allowed shapes of these macrocyclic molecules.

WHAT'S GOING ON? A GRAPHICAL REPRESENTATION OF HOST SHAPE

Most host cyclophanes can be described as four benzenoid subunits mutually linked by "spacer groups" to provide the macrocyclic structure. The spacer groups are often simple linear chains of from one to seven atoms composed of carbon, nitrogen, and/or oxygen. More rigid spacer groups include cyclohexane rings, diynes, and the bicyclic spacer groups represented by the atoms linking the aromatic portions of anthracene Diels-Alder adducts and Tröger's base derivatives.

Dale created a useful system for analyzing medium and large ring molecules [6]. An important paper from the Dougherty group suggested that Dale's rules for macrocycle conformation, and the "bird's-eye" graphical representations favored by Dale, are very helpful when applied to cyclophane conformational analysis [7] In agreement with this, we find that the known conformations of many host cyclophanes can be readily depicted by such Dale representations.

A rigid structural unit can be represented in the Dale system by a single extended bond [6]. A 1,4-disubstituted benzene ring, and the atoms attached directly to C-1 and C-4, constitute such a rigid unit. This is a very common structural unit among the host cyclophanes and a convenient representation of this structure is required. We suggest here that the 1,4-disubstituted benzene ring and the two atoms (one at each end) attached to C-1 and C-4 be represented as a single long, filled wedge. The ends of the wedge represent the atoms connected to the benzene ring. The wedge is therefore not equivalent to a benzene ring, but is longer than a benzene ring by two bonds. This simple unit is illustrated in Figure 1(a).

Dale projections of the shapes of several cyclophane substructures are presented in Figure 1. It is clear from this figure and from the examples to be presented below (Figure 6) that the abstract representations of Dale are simplified diagrams of the real structure. The wedges represent an idealized geometry that may be a strained form of the real molecule in question.

These ideas can be applied to generate useful graphical representations for conformations of many of the known cyclophane host molecules. Figure 2 illustrates the relationship between a ball-and-stick representation of a new macrocyclic host (**4**) and the Dale projection representing that conformation. Several important features of the conformation are easily

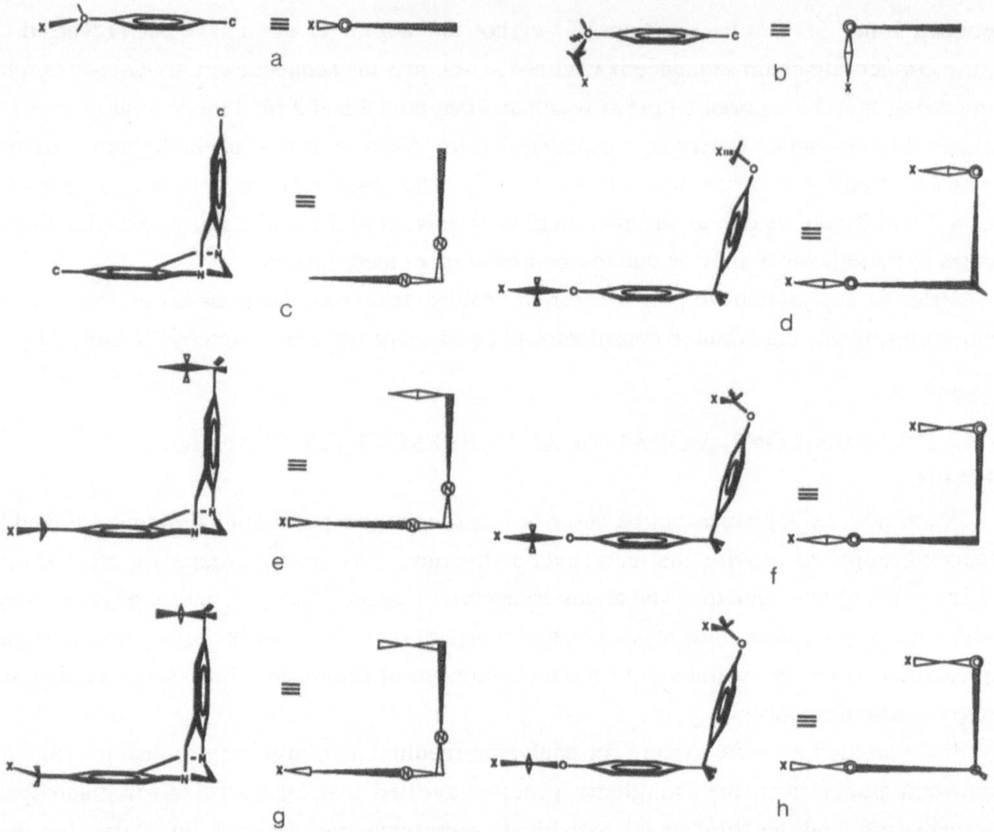

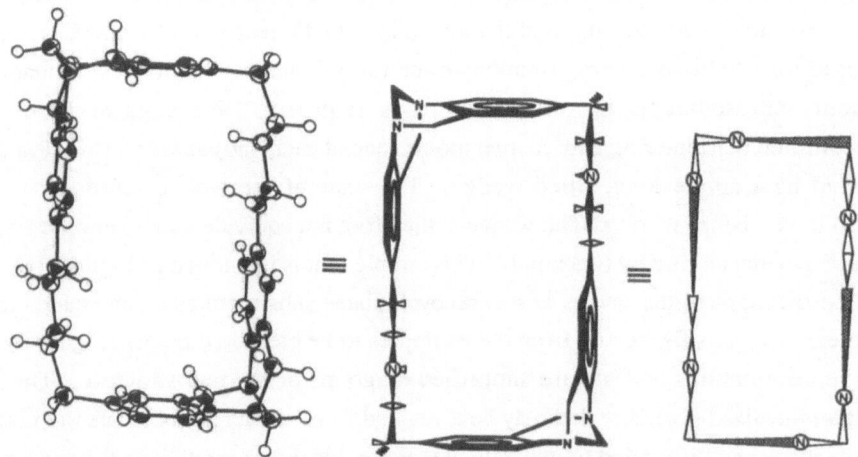

Figure 1. Key to conformational representations used in this paper. The filled long wedges represent a 1,4-disubstituted phenyl group and the atoms attached to that group. Tröger's base (c) is represented by two long and two short wedges.

Figure 2. Three representations of the same conformation of host cyclcophane 4. The final simplified drawing clearly reveals the centrosymmetry of the conformer.

perceived in these representations. The exposure of the side chain nitrogen atoms (i.e., the accessibility of hydrogen atoms attached to the side chain nitrogen atoms) is clearly and correctly represented. In Figure 2, the nitrogen atom on the top side and the two nitrogens on the right-hand side are exposed to the back of the ring and the other three nitrogen atoms are exposed to the front side of the ring. The fact that this conformer has a symmetry element (center of inversion) is also clearly revealed. Parallel bonds, gauche bonds, and antitorsional angles are clearly shown, and the overall shape of the macrocycle is simply represented.

The wedge representation of the Tröger's base subunit clearly illustrates both the chirality of this unit and the absolute configuration of the particular base illustrated. The C_2 axis of symmetry (local to the Tröger's base unit) is represented in the abstraction. The information that the macrocycle is centrosymmetric and the two Tröger's base units (dibenzodiazocenes) are mirror image modifications are both clearly represented in the abstraction.

THE GRAPHICAL REPRESENTATION SUPPORTS A SYSTEMATIC APPROACH TO GENERATING CONFORMERS

There is no exact algorithm for finding the global minimum of a complex function which has many local minima. The general approach to this problem involves an exhaustive search of the potential solution space (the multidimensional space wherein the answer is presumed to lie). The search may be simplified if heuristic principles are available which help to reduce the size of the solution space and therefore reduce the number of potential answers which must be examined.

Dale recognized that a corner in a polymethylene chain or macrocycle can be formed by two consecutive gauche bonds [6]. Preliminary conformational analyses of host 1 revealed that there were several conformational minima which contained only two corners (one in each of the connecting chains). Many local minima were located which had more than this minimum number of gauche bonds, but these other conformers were consistently higher in energy than conformers containing only two corners. This agrees entirely with Dale's rules, with Dougherty's assertions about simpler cyclophanes, and with common sense. With rare exceptions, the lowest energy conformers of any macrocycle are those which minimize the number of gauche interactions.

There are 84 unique conformations of 1 which contain only two Dale "corners". This can be determined by a systematic listing. In each connecting chain, there are six possible positions for a gauche bond. There are therefore 36 possible ways to place the two corners. These 36 possibilities can be represented in matrix form (Figure 3) using a nomenclature similar to Dale's. The four digit numbers simply represent the number of wedges (big or small) to be counted along the four sides of the macrocycle representation [8]. There is, however, one other choice to be made for each of these 36 possibilities. For each corner, the torsional angles may be either positive or negative. There are therefore four possible ways to generate each of the 36 positional isomers. These four choices will be designated as "A", "B", "C", or "D", according to the rule outlined in Table 2 [9].

Table 2. Nitrogen atom relative positions.

Notation	Pair on long side.	Other pair
A	+	+
B	-	+
C	+	-
D	-	-

A "+" indicates that the two nitrogens are both directed towards the same face of the molecule; A "-" indicates that the nitrogens are directed towards opposite faces of the molecule [9].

2617	3517	4417	5317	6217	7117
2626	3526	4426	5326	6226	7126
2635	3535	4435	5335	6235	7135
2644	3544	4444	5344	6244	7144
2653	3553	4453	5353	6253	7153
2662	3562	4462	5362	6262	7162

Figure 3. A matrix representing the 36 possible ways that two corners can be positioned (one in each connecting chain) in cyclophane **1**. The underlined conformers were found to be the most stable.

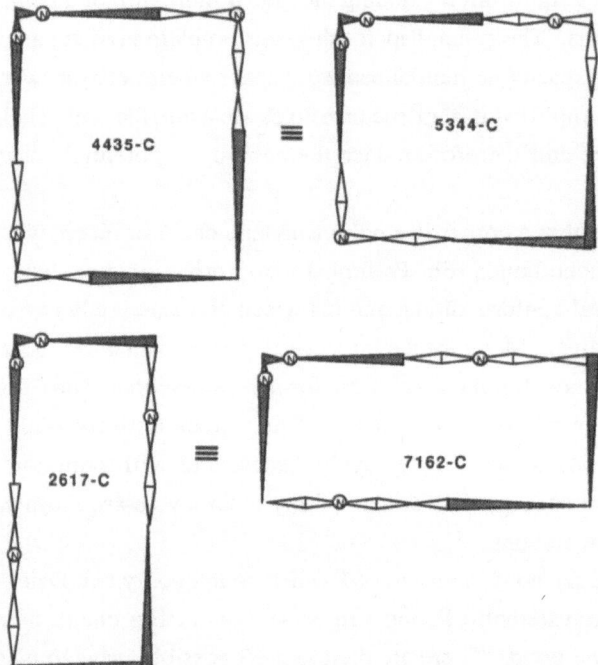

Figure 4. The matrix illustrated in Figure 3 has redundant entries. Two examples of such equivalent entries are illustrated here.

A maximum of 144 (or 4 x 36) conformers each containing only one corner in each connecting chain is therefore possible. The number of unique conformers is less than 144 because, as illustrated for two examples in Figure 4, the matrix is symmetrical. In Figure 3, the conformers in boldface make up the unique, non-redundant set. There are 84 *unique* conformations which contain two corners.

It is notable that of these 84 conformers only 12 have a C_2 axis of symmetry which crosses the binding site. These are the "A" and "D"rotamers of the six shapes represented by boldface italic numbers in Figure 3.

A SYSTEMATIC METHOD IS A VALUABLE COMPLEMENT TO AUTOMATED CONFORMATIONAL SEARCH METHODS

The methodical process just described affords a closed set containing all possible idealized conformers of **1** which have a single gauche bond in each connecting chain. These conformers are rationally chosen starting conformations each of which can be submitted to a procedure for automated conformational energy minimization. The results of such a study are described in the next section.

This set of conformers is important because it systematically identifies and names many shapes which lie in separate "valleys" in the conformational energy surface. A search for a local minimum starting from any one of these conformers results in a relaxed conformer very similar to the initial shape. Put another way, the 84 shapes described in Figure 3 represent 84 "valleys" on the conformational energy surface. These shapes are therefore more important than 84 randomly chosen shapes, because in a randomly chosen set many conformers will minimize to the same shape.

There are other reasons why such a systematic listing of conformers is important. The list can be used to test for completeness following a random automated search. Following such a random search one must ask whether all conformers were properly considered. Was the algorithm used to generate potential solutions complete? This systematic list provides an important set of heuristically chosen examples for comparison with the results of a more automated and perhaps less guided search.

Critical thought suggests that any systematic approach might overlook important conformations. Should systematic approaches be abandoned when satisfactorily fast computers and proven algorithms for randomly searching out local minima become available? The answer is no, because structural systematization allows further order to be imposed on conformational energy information. The best random search would provide a one dimensional array, a list of local minima in order of increasing energy. Systematic structural descriptions are always of value because they provide a basis for pattern recognition and for structure-property generalizations.

THE BEST SHAPES ARE RECTANGULAR AND COMPLEMENTARY TO THE BENZENOID SHAPE

Results of molecular mechanics calculations based on the MM2 or the AMBER force field lead to the same conclusions [10]. The procedure for each calculation was to enter the starting conformer represented by the matrix notation (Figure 3) and to use canonical iterative procedures to find the nearest conformational minimum. In all cases a local conformer was located which closely matched the starting structure. In other words, and not unexpectedly, a torsional angle which was gauche in a starting conformer remained gauche throughout the minimization and anti torsional angles remained anti. This reemphasizes the importance of creating a complete set of starting conformers when such a study is undertaken.

For each conformer the calculated energy (nearest local minimum) of the free base was first determined. The lowest energy conformers were found to be the 20 conformers represented by the underlined code numbers in Figure 3. The larger group within the outlined area of Figure 3 includes these 20 key candidates. All within that group were found to be more stable than those outside that group.

The conformers outside the outlined area of Figure 3 are considered to be unlikely contributors to any binding interactions with a benzenoid guest because they are of high energy and they have shapes which are not complementary to benzenoid guests. The conformers outside the outlined area (e.g. **2644**, **4417**, etc.) represent structures in which the sides of the box formed by the molecule are of oddly alternating lengths. Conformations which have these shapes are indeed minima, but all that were examined were at least 35 kJ/mol less stable than the best of the favored 20 shapes described in Figure 5.

The eight basic shapes (32 conformers) within the outlined area in Figure 3 all gave local conformational minima but the minima associated with conformers of type **3544**, **4444**, and **4435** were relatively high energy. For example, the minimum found for the free base corresponding to **1-3544-A** was 32 kJ/mol less stable than the best free base conformer found. These conformers (**3544**, **4444**, and **4435**), in addition to being of relatively high energy, also appear to be poorly suited for binding to benzenoid substrates because the interior surface of these conformers poorly complements the surface of a benzenoid guest.

COMMENTS ON THE BASIC SHAPES

The twenty lowest energy minima which have been located for host **1** are illustrated in Figure 5. Each unoccupied host is represented by a space filling model and by a Dale projection constructed as detailed above. All conformers are energy minimized structures.

These twenty conformers define roughly rectangular cavities about 6 Å deep. The rectangles are all about 10.5 x 6.5 Å and this is evidently about the right size for accepting a benzenoid guest. The lowest energy shapes are the **2626** conformers. The differences among the free bases are small, however, and the **2626** conformers are only about 1 kcal/mol more stable than the **2617** conformers. All twenty conformers fall in a range of only about 5 kcal/mol, and it's important to emphasize that any one or all of these may well contribute to the host properties of **1** that have been observed. These conclusions were unchanged when the calculations incorporated a mechanical model for the dibenzodiazocene which was created based on crystallographic data [11].

Published crystallographic data for cyclophane hosts are consistent with these results. Presently known crystal structures of cyclophane hosts reveal cavities with parallel sides and rectangular or square cross section very similar to the shapes "discovered" here by calculation [13].

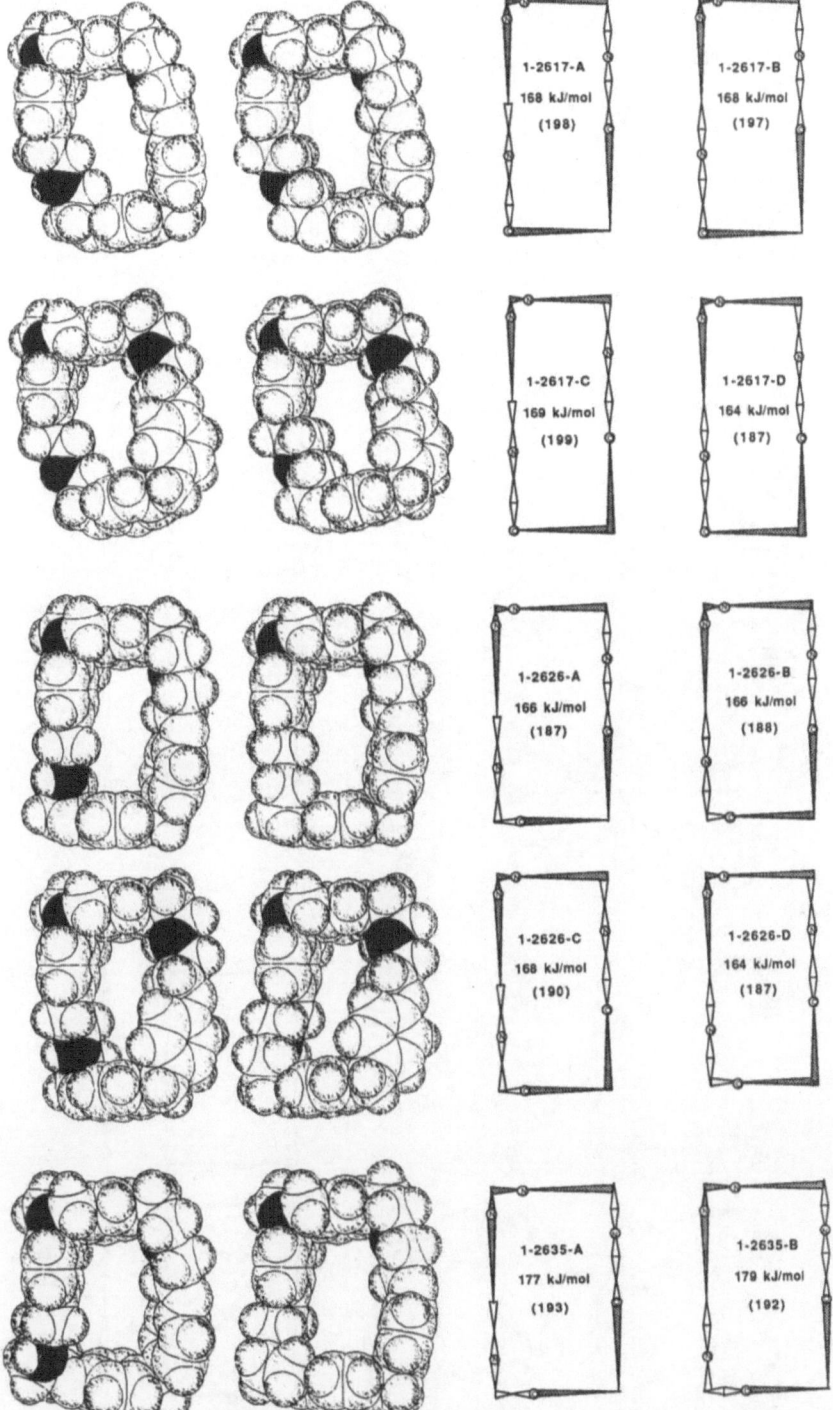

Figure 5. The twenty most stable conformers located for host 1. The conformers are presented in groups of four and each group corresponds to one of the underlined conformers illustrated in the matrix shown in Figure 3. On the left are shown the host conformers (local minima) in space filling format and on the right are the conformational diagrams which represent those conformers.

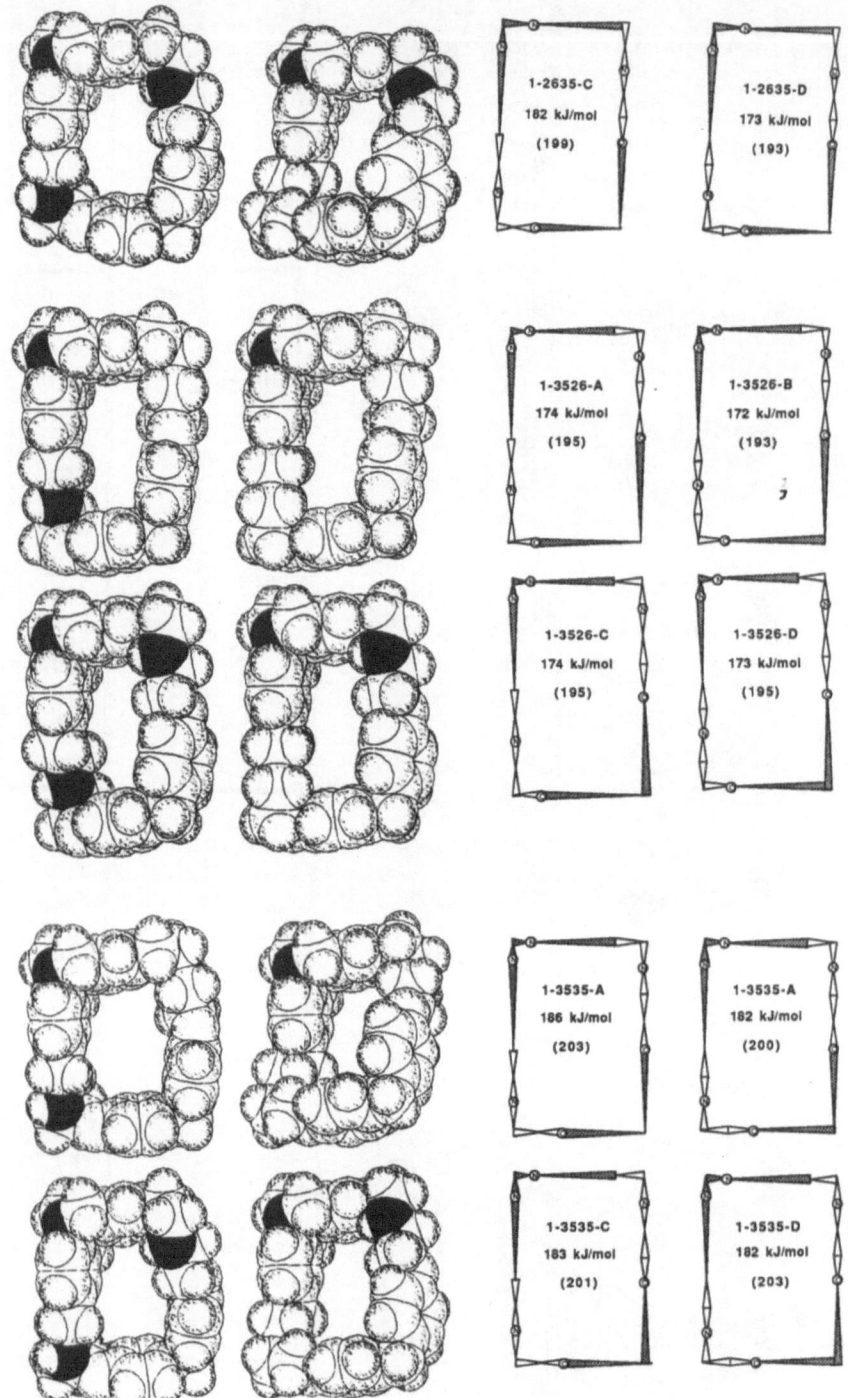

Figure 5 (continued).

A HOST CONTAINING TWO DIBENZODIAZOCENE UNITS GIVES IMPROVED BINDING TO BENZENOID GUESTS

To support these calculations, X-ray crystallographic data for complexes of aromatic substrates with **1** were desired but no suitable crystals have been obtained. A related Tröger's base derived host (**4**) has been crystallized. Figure 6 illustrates the structure of a complex formed between this host and *p*-xylene as determined by analysis of single crystal X-ray diffaction data. The conformation of the host in this complex was illustrated in Figure 2 and it can be seen that this conformer corresponds to a **2626** type of conformer. The general box-like shape of the host is emphatically revealed. The close complementarity of the binding site surface and guest surface is obvious. The fact that the least squares plane defined by the guest is very nearly parallel to the plane of two host aromatic rings reveals that the guest is snugly fitted within the macrocycle. The tight binding of the guest is further established by the small thermal motion observed for the bound guest in comparison with the host. The complex is centrosymmetric and the guest is deeply bound. There is only marginal overlap between the guest and the host aromatic rings that are parallel to the guest. It is interesting that the guest takes part in "edge-to-face" contacts with the two host aromatic rings at the ends of the cavity. It is tempting to suggest, following Burley and Petsko, that these interactions may play an important role in stabilizing the complex [12].

In solution, host **4** binds to 2,4,6-trimethylphenol with a dissociation constant of 1.5 mM, or about 1.3 kcal/mol stronger binding than was observed for host **1** with the same guest.

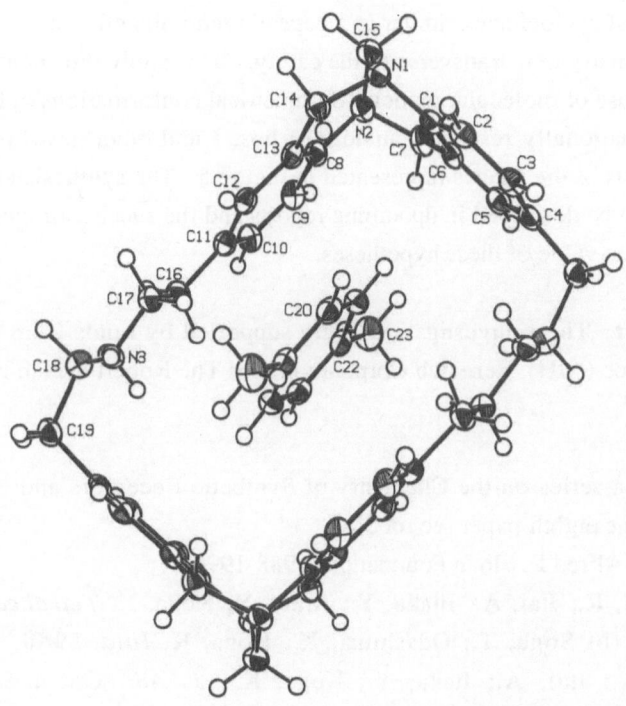

Figure 6. View of a complex formed from *p*-xylene and host **4**. Thermal ellipsoids are scaled to the 50% probability level.

CONCLUSIONS

We suggest that the 20 conformers described in Figure 5 are important conformers available to host **1**. Of these conformers, the **2626** group best complement the shape of a *para*-disubstituted benzene derivative. These results should not be considered as specific for host **1**. Conformational analysis of many cyclophane hosts leads to results comparable to these. The 84 conformers described by Figure 3, and the 20 conformers of Figure 5 are generalizable structures which will be of interest to any group working with cyclophane hosts.

The use of two dibenzodiazocenes in a host macrocycle provides two rigid "corners" which may help to stabilize these hypothetical lipophilic "boxes". Several crystallographic studies support our hypothesis that cyclophane hosts do not bind to guests to afford complexes with a C_2 symmetry axis which traverses the cavity.

Finally, the results of these calculations illuminate an interesting aspect of the many macrocyclic hosts which are designed to have a C_2 axis of symmetry passing across the binding site. A host molecule like **1** may indeed have a connectivity graph which is symmetric and may have a conformation which is symmetric, but if, in solution, the best binding shapes are not symmetrical and/or the symmetrical shapes are of high energy compared to unsymmetrical shapes, then (with the exception of a possibly simplified synthetic effort) nothing has been gained by requiring C_2 symmetry in the host. In fact, this requirement of symmetry may hinder the development of better binding hosts based on preorganized cyclophanes.

On the basis of these considerations and available X-ray structural data we conclude that the host properties of cyclophanes similar to **1** depend predominantly on conformers which do not have a C_2 symmetry axis transverse to the cavity. This study thus confirms conclusions readily reached by use of molecular models. Symmetrical conformations of host are relatively unstable. Conformationally restricted analogs of host **1** and other new cyclophanes can be designed on the basis of the shapes represented in Figure 5. The synthesis and host properties of such analogs will be described in upcoming reports and the success of these designs should be the final test of the value of these hypotheses.

Acknowledgement. These investigations were supported by funds from the United States Public Health Service (NIH), Research Corporation, and The Robert Welch Foundation.

References

1. Number 9 in a series on the Chemistry of Synthetic Receptors and Functional Group Arrays. For the eighth paper see ref 5(a).
2. Fellow of the Alfred P. Sloan Foundation (1988-1990).
3. (a) Odashima, K.; Itai, A.; Iitaka, Y.; Arata, Y; Koga, K. *Tetrahedron Lett.* **1980**, 4347-4350;   (b) Soga, T.; Odashima, K.; Koga, K. *Ibid.* **1980**, 4351-4354;   (c) Odashima, K.; Itai, A.; Itaka, Y.; Koga, K. *J. Am. Chem. Soc.* **1980**, *102*, 2504-2505.   (d) Tabushi, I.; Sasaki, H.; Kuroda, Y. *J. Am. Chem. Soc.* **1976**, *98*, 5727-5728;   (e) Tabushi, I.; Kimura, Y.; Yamamura, K. *Ibid.* **1978**, *100*, 1304-1306;

(f) Tabushi, I.; Kuroda, Y.; Kimura, Y. *Tetrahedron Lett.* **1976**, 3327-3330; (g) Jarvi, E. T.; Whitlock, H. W., Jr. *J. Am. Chem. Soc.* **1980**, *102*, 657-662; (h) Whitlock, B. J.; Jarvi, E. T.; Whitlock, H. W., Jr. *J. Org. Chem.* **1981**, *46*, 1832-1835; (i) Adams, S. P.; Whitlock, H. W., Jr. *Ibid.* **1981**, *46*, 3474-3478; (j) Adams, S. P.; Whitlock, H. W., Jr. *J. Am. Chem. Soc.* **1982**, *104*, 1602-1611; (k) Jarvi, E. T.; Whitlock, H. W., Jr. *Ibid.* **1982**, *104*, 7196-7204; (l) Sheridan, R. E.; Whitlock, H. W., Jr. *Ibid.* **1986**, *108*, 7120-7121; (m) Diederich, F.; Dick, K. *Tetrahedron Lett.* **1982**, *23*, 3167-3170; (n) Diederich, F.; Dick, K. *Angew. Chem., Int. Ed . Engl.* **1983**, *22*, 715-716; (o) Diederich, F.; Dick, K. *J. Am. Chem. Soc.* **1984**, *106* , 8024-8036; (p) Diederich, F.; Griebel, D. *Ibid.* **1984**, *106*, 8037-8046; (q) Diederich, F.; Dick, K.; Griebel, D. *Chem. Ber.* **1985**, *118*, 3588-3619; (r) Diederich, F.; Dick, K. *Ibid.* **1985**, *118*, 3817-3829; (s) Lutter. H.; Diederich, F. *Angew. Chem., Int. Ed. Engl.* **1986**, *25*, 1125-1127; (t) Ferguson, S. B.; Diederich, F. *Ibid.* **1986**, *25*, 1127-1129; (u) Schurmann, G.; Diederich, F. *Tetrahedron Lett.* **1986**, *27*, 4249-4252; (v) Rubin, Y.; Dick, K.; Diederich, F.; Georgiadis, T. M. *J. Org. Chem.* **1986**, *51*, 3270-3278; (w) Sheppodd, T. J.; Petti, M. A.; Dougherty, D. A. *J. Am. Chem. Soc.* **1986**, *108*, 6085-6087; (x) Sheppodd, T. J.; Petti, M. A.; Dougherty, D. A. *J. Am. Chem. Soc.* **1988**, *110*, 1983-1984; (y) Petti, M. A.; Sheppodd, T. J.; Dougherty, D. A. *Tetrahedron Lett.* **1986**, *27*, 807-810.

4.   (a) Odashima, K.; Koga, K. "Cyclophanes in Host-Guest Chemistry" In *Cyclophanes* (Vol. II) P. M. Kheen, S. M. Rosenfeld, Eds.; Academic Press: New York; 1983; (b) Tabushi, I.; Yamamura, K. *Top. Curr. Chem.* **1983**, *113*, 145-182; (c) Tabushi, I.; Kimura, Y.; Yamamura, K. "Facilitated Formation of Tetrahedral Intermediate in Esterase Action by Water Soluble Heterocyclophane". In *Chemical Approaches to Understanding Enzyme Catalysis: Biomimetic Chemistry and Transition State Analogs*; B. S., Y. Ashani, D. Chipman, Eds.; Elsevier: Amsterdam, 1981; (d) Diederich, F. *Angew. Chem., Int. Ed. Engl.* **1988**, 27, 362-386.

5.   (a) The preparation and properties of host 1 were recently described: Cowart, M. D.; Sucholeiki, I. Bukownik, R. R.; Wilcox, C. S. *J. Am. Chem. Soc.* **1988**, *110*, 6204-6210.; (b) For preliminary binding data on a similar host, see Wilcox, C. S.; Cowart, M. D. *Tetrahedron Lett.* **1986**, 5563-5566.

6.   Dale, J. *Top. Stereochem.* **1976**, *9*, 199-270.

7.   Masek, B. B.; Santarsiero, B. D.; Dougherty, D. A. *J. Am. Chem. Soc.* **1987**, *109*, 4373-4379.

8.   The long wedge is counted just once, as a single component. Because one and only one aromatic ring (or long wedge) is found on each of the four sides of these conformers (other possibilities are of impossibly high conformational energy), the lengths of the sides are directly related to the number of wedges counted. Future, more complex, macrocycles may require a more complicated description.

9.   The four isomers differ in regard to the formal sign of the gauche bonds in each connecting chain. It is more informative, however, to describe the four isomers in

terms of the positions of the nitrogen atoms in the connecting chain relative to the Tröger's base unit. This method clearly identifies the four conformers and gives additional useful information in regard to possible host-guest interactions. A nitrogen atom or dialkylammonium group is defined as "facing" the direction from which a solute or solvent can most readily form hydrogen bonds with the nitrogen or ammonium group.

10. For an excellent discussion of the strengths and weaknesses of empirical force field calculations, see: Allinger, N. L. *Adv. Phys. Org. Chem.* **1976**, *13*, 1-82. It has been noted that MM2 calculations underestimate the attractive forces between aromatic rings and heteroatoms: Abe, K.; Hirota, M.; Morokuma K. *Bull. Chem. Soc. Jap.* **1985**, *58*, 2713-2714.

11. Several crystallographic studies of Tröger's base derivatives have been undertaken: (a) Wilcox, C. S. *Tetrahedron Lett.* **1985**, 5749-5752; (b) Wilcox, C. S.; Greer, L. M.; Lynch, V. *J. Am. Chem. Soc.* **1987**, *109*, 1865-1867; (c) Larson, S. B.; Wilcox, C. S. *Acta Crystallogr.* **1986**, *C42*, 224-227; (d) Sucholeiki, I.; Lynch, V.; Phan, L.; Wilcox, C. S. *J. Org. Chem.* **1988**, *53*, 98-104.

12. (a) Burley, S. K.; Petsko, G. A. *Science* **1985**, *229*, 23; (b) Burley, S. K.; Petsko, G. A. *J. Am. Chem. Soc.* **1986**, *108*, 7995.

13. The first crystal structure for a cyclophane host-guest inclusion complex revealed a host in a conformation which can be described as a **6161** conformer [11c]. For an example of a **3333** conformer which forms a conical cavity which partially includes a dioxane guest molecule, see Abbott, A. J.; Barrett, A. G. M.; Godfrey, C. R. A.; Kalindjian, S. B.; Simpson, G. W.; Williams, D. J. *J. Chem. Soc., Chem. Commun.* **1982**, 796-797. In a recent review, Diederich reported crystallographic data which reveal that the same cyclophane host can assume both **6161** and **3434** conformations [4d]. The former conformation is reported for a crystal which also contains disordered toluene molecules. The latter conformation is observed in a crystalline form in which a benzene molecule occupies the central cavity.

COMPLEXATION AND MOLECULAR RECOGNITION OF NEUTRAL AND ANIONIC SUBSTRATES IN THE SOLID AND SOLUTION STATES BY BISPARAQUAT(1,4)CYCLOPHANE

Mark V. Reddington, Neil Spencer, and J. Fraser Stoddart*

Department of Chemistry
The University
Sheffield S3 7HF (UK)

SUMMARY

The bisparaquat(1,4)cyclophane O^{4+}, which can be synthesized in two steps from bipyridine and *para*-xylylene dibromide, is soluble as its tetrakis(hexafluorophosphate) in organic solvents such as acetonitrile and as its tetrachloride in water. In acetonitrile, $O^{4+} \cdot 4PF_6^-$ forms weak 1:1 inclusion complexes with 1,2-, 1,3-, and 1,4-dimethoxybenzene whereas, in water, $O^{4+} \cdot 4Cl^-$ binds hydroquinol as well as its dimethyl ether. X-Ray crystallography shows that the bisparaquat(1,4)cyclophane tetracation O^{4+} forms crystalline 1:1 inclusion complexes with 1,2- and 1,4-dimethoxybenzenes in which the π-electron rich diphenol ethers are sited inside the channels of the π-electron deficient bipyridinium rings of the tetracation O^{4+} separated, as before, by layers of anions and neutral molecules in continuous hollow stacks of a zeolitic nature.

INTRODUCTION

In recent times, we have developed [1-3] cyclophane-like molecular receptors such as bisparaphenylene-34-crown-10 (BPP34C10), which contains π-electron rich hydroquinol units for face-to-face complexation (Figure 1) with Paraquat (PQT^{2+}), a π-electron deficient

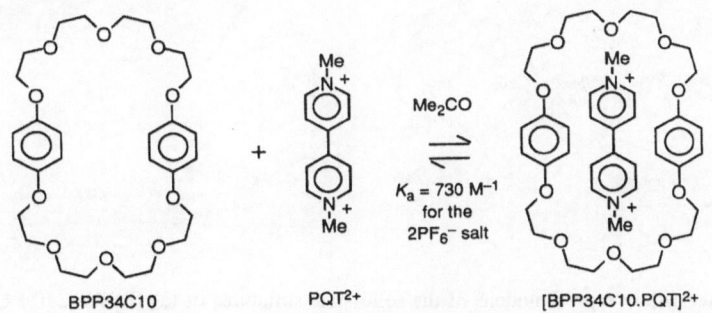

BPP34C10   PQT^{2+} [BPP34C10.PQT]$^{2+}$

Figure 1. The 1:1 complex formation between BPP34C10 and PQT^{2+} in acetone.

Inclusion Phenomena and Molecular Recognition
Edited by J. Atwood
Plenum Press, New York, 1990

bipyridinium dication with herbicidal activity. A charge-transfer absorption band is observed at 435 nm in the visible spectrum of the 1:1 complex ([BPP34C10·PQT][PF$_6$]$_2$) in acetone. Quantitative analysis of the dependence of the absorption intensity upon the concentration of the complex provided evidence for its 1:1 stoichiometry in acetone solution and afforded an association constant (K_a) for its formation of 730 M^{-1} corresponding to a free energy of complexation (-$\Delta G°$) of 16.3 kJ mol^{-1}. The X-ray crystal structure (Figure 2a) of [BPP34C10·PQT]$^{2+}$ establishes that the 1:1 complex is of the inclusion type in the solid state. The binding of the PQT^{2+} by BPP34C10 is believed [2,3] to arise from a combination of: (i) electrostatic interactions including [C-H···O] hydrogen bonding and (ii) dispersive forces including charge-transfer interactions.

BPP(3n+4)Cn	n	j	k
BPP22C6	6	1	1
BPP25C7	7	2	1
BPP28C8	8	2	2
BPP31C9	9	3	2
BPP34C10	10	3	3
BPP37C11	11	4	3
BPP40C12	12	4	4

Decreasing ($n = 6 - 9$) or increasing ($n = 11$ and 12) the macrocyclic ring size within the series of BPP(3n+4)Cn receptors leads to an impairment in the complexation of PQT^{2+}. The X-ray crystal structure (Figure 2b) of BPP34C10 in its uncomplexed state reveals that its conformation is very similar to that observed (Figure 2a) for the receptor in [BPP34C10·PQT]$^{2+}$. Surprisingly, there is a sizable molecular void with a distance of 7.5 Å between the centroids of the hydroquinol rings.

The X-ray crystal structures of BPP28C8, BPP25C7, and BPP22C6 expose [4] an interesting sequence of events as the macrocyclic ring size is decreased from $n = 8$ to $n = 6$.

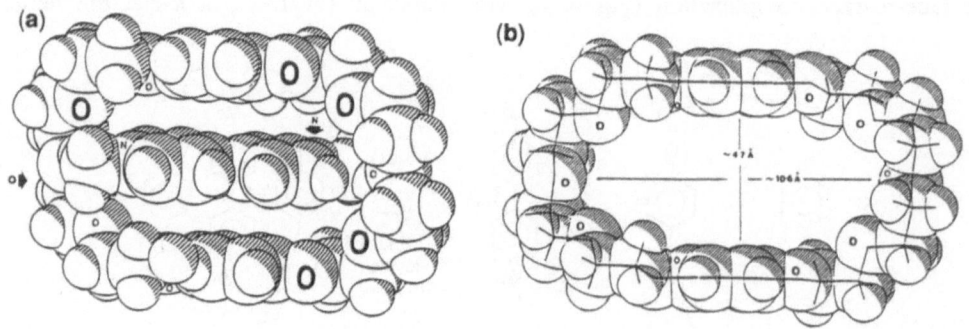

Figure 2. Space-filling representations of the solid state structures of (a) [BPP34C10·PQT]$^{2+}$ and of (b) BPP34C10.

As the centroid-centroid distance between the hydroquinol rings falls from (a) 5.4 Å ($n = 8$) to (b) 5.0 Å ($n = 7$) to (c) 4.3 Å ($n = 6$), the hydroquinol rings go from being face-to-face and displaced sideways (Figure 3a) to being face-to-edge with a hydroquinol ring hydrogen atom directed into the centre of the π-system of the other hydroquinol ring at a distance of 2.8 Å from its centroid (Figure 3b) to being face-to-face and displaced sideways (Figure 3c) again. The T-type interaction between aromatic rings in BPP25C7 (Figure 3b) is one which is well documented [5] in the literature from the structure of benzene in the solid state [6] through small synthetic receptor systems [7] to phenyl-phenyl interactions in proteins [8].

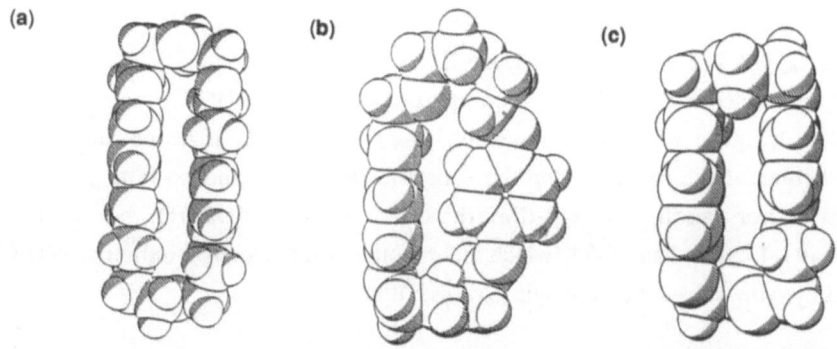

Figure 3. Space-filling representations of the solid state structures of (a) BPP28C8, (b) BPP25C7, and (c) BPP22C6.

The main disadvantage of BPP34C10 as a PQT^{2+} receptor is that it also complexes equally well [2] with Diquat (DQT^{2+}), another π-electron deficient dication with herbicidal activity. In a search for receptors that might complex more selectively with the PQT^{2+} dication, we next turned our attention to cyclophane-like polyethers incorporating two naphtho rings. It was soon established [9] that 1/5DN30C10 forms a crystalline 1:1 inclusion complex (Figure 4a) with PQT^{2+}. However, it was the 2:1 complex (Figure 4b) formed [10] between PQT^{2+} and 1/5DN44C12 that encouraged us to contemplate the design of novel molecular recognition systems (e.g. polymers of polyether-linked PQT^{2+} units interacting with polymers of polyether-linked 1,5-naphtho- or 1,4-benzo-residues forming a double helical arrangement) and new tetracationic molecular receptors incorporating two PQT^{2+} units.

1/5DN(3n+8)Cn	m
1/5DN38C10	1
1/5DN44C12	2

DQT^{2+}

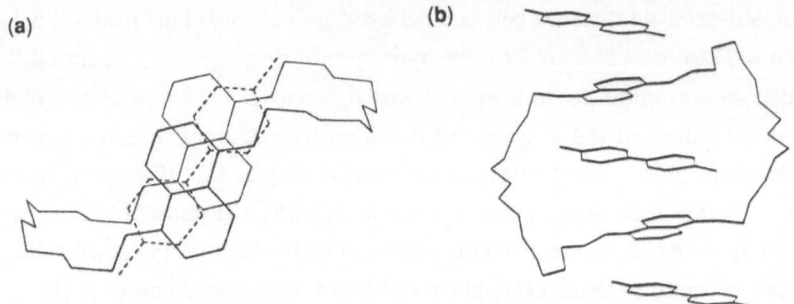

Figure 4. The solid state structure of (a) the 1:1 complex formed between PQT^{2+} and 1/5DN38C10 and (b) the 2:1 complex formed between PQT^{2+} and 1/5DN44C12.

BISPARAQUAT(1,4)CYCLOPHANE

In the knowledge that PQT^{2+} is complexed by BPP($3n+4$)Cn ethers – in particular, by BPP34C10 (Figure 2a) – the question was posed: might it be possible to reverse the structural roles of the receptor and the substrate (Figure 5) and so bind a π-electron rich diphenol ether such as 1,4-dimethoxybenzene (1/4DMB) by a tetracationic bisparaquat(1,4)cyclophane O^{4+}, which incorporates two π-electron deficient PQT^{2+} residues separated by *para*-substituted phenylene rings? It is.

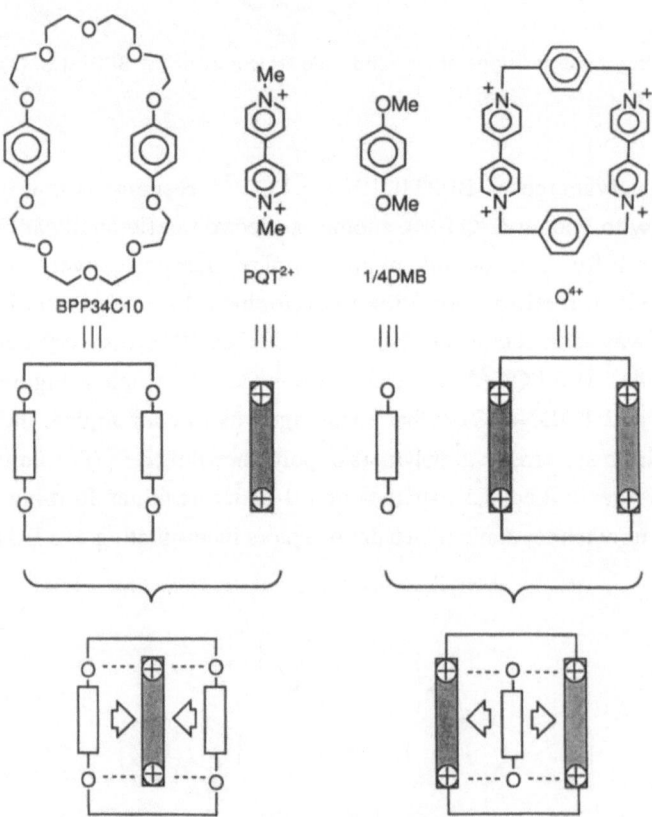

Figure 5. Diagrammatic representation of the reasoning behind the design of the tetracationic bisparaquat(1,4)cyclophane receptor (O^{4+}) for 1,4-dimethoxybenzene (1/4DMB).

44

The tetracationic receptor O^{4+} can be made [11] in 10-20% yield as outlined in Scheme 1. While the tetrakis(hexafluorophosphate) ($O^{4+}\cdot 4PF_6^-$) is soluble in solvents such as acetonitrile and nitromethane, the tetrachloride ($O^{4+}\cdot 4Cl^-$) is water soluble. This means that by simply changing the nature of the counterion, the tetracationic receptor O^{4+} can be rendered soluble in *both aqueous and non-aqueous media*. An X-ray crystal structure analysis of $O^{4+}\cdot 4PF_6^-$,

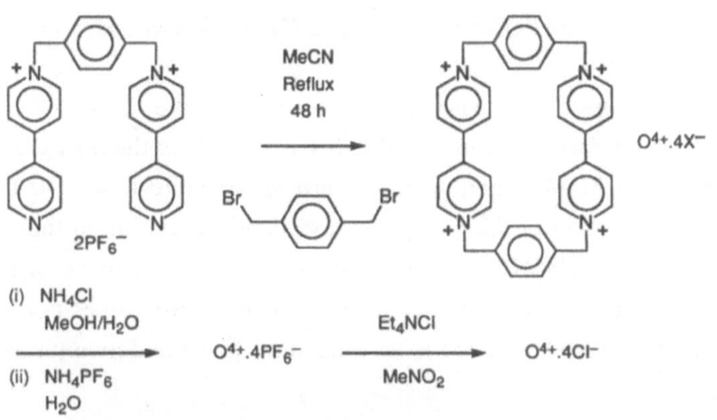

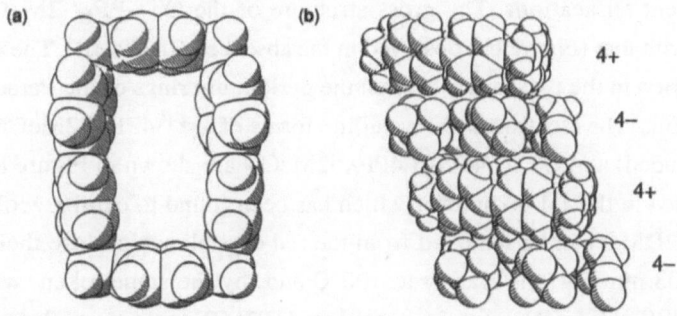

Scheme 1. The synthesis of the tetracationic receptor O^{4+} as its $4PF_6^-$ and $4Cl^-$ salts.

trebly solvated with MeCN, reveals (Figure 6a) that the O^{4+} tetracation adopts a rigid rectangular box-like conformation which is also considerably strained. Molecular mechanics calculations reproduced the solid state structure and semi-empirical calculations suggest that the interiors of the two Paraquat components of the O^{4+} tetracation carry relatively more positive electrostatic potential than their exteriors. Also, each macrocyclic ring is stacked (Figure 6b) almost directly above its neighbour with four PF_6^- counterions and two of the MeCN molecules forming a torus so that the overall solid state structure contains parallel open channels bounded by alternately-charged (4+/4-) entities. The other MeCN molecule is located in part of the void created by these channels. In acetonitrile as solvent, $O_4^+\cdot 4PF_6^-$ forms weak 1:1 complexes with 1/2DMB ($K_a = 8$ M^{-1}/$-\Delta G^\circ = 5.2$ kJ mol^{-1}), 1/3DMB ($K_a = 8$ M^{-1}/$-\Delta G^\circ = 5.2$ kJ mol^{-1}), and 1/4DMB ($K_a = 17$ M^{-1}/$-\Delta G^\circ = 7.0$ kJ mol^{-1}) whereas in

Figure 6. Space-filling representations of (a) the solid state structure of O^{4+} and of (b) the alternately-charged entities of $O^{4+}\cdot 4PF_6^-\cdot 2MeCN$ in the solid state.

45

water $O^{4+} \cdot 4Cl^-$ binds hydroquinol as well as its dimethyl ether. Evidence for inclusion complex formation comes from a number of different experiments:

(i)   Addition of equimolar proportions of any of the isomeric dimethoxybenzenes (DMB's) to $O^{4+} \cdot 4PF_6^-$ in acetonitrile results in the immediate formation of red/orange coloured solutions, indicating that stabilizing charge-transfer (CT) interactions (λ_{max} 478, 484, and 462 nm, respectively, for 1/2DMB, 1/3DMB, and 1/4DMB) are operative between the π-electron rich guests and the π-electron deficient host. (The CT bands were used as quantitative probes to obtain the K_a values.)

(ii)   Although the chemical shift changes ($\Delta\delta$) observed in the ^{1}H NMR spectra recorded in CD_3CN on 1:1 complex formation are small ($\Delta\delta < 0.1$ ppm) for the host protons, those for the protons in 1/2DMB, 1/3DMB, and 1/4DMB are appreciable ($\Delta\delta = -0.10$ to -0.91 ppm), indicating not only inclusion of the guests but also their orientation in the cavity of the host. (Interestingly, when the ^{1}H NMR spectrum of the 1:1 complex formed between $O^{4+} \cdot 4Cl^-$ and hydroquinol is recorded in D_2O, not only were appreciable $\Delta\delta$ values noted for some of the protons in the tetracation O^{4+}, but all the protons in the guest underwent rapid H/D exchange [cf. ref. 12]).

(iii)   As its $4PF_6^-$ salt, doubly solvated with MeCN, X-ray crystallography shows [13] that the O^{4+} tetracation forms crystalline 1:1 inclusion complexes with 1/2DMB and 1/4 DMB (Figure 7) in which the π-electron rich diphenol ethers are located inside the cavities of the

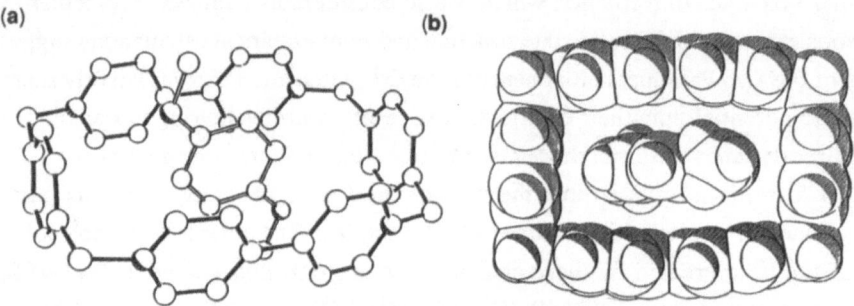

Figure 7. The solid state structure of $O^{4+} \cdot 1/4DMB \cdot 4PF_6^- \cdot 2MeCN$ in (a) ball-and-stick and (b) space-filling representations.

π-electron deficient tetracations. The gross structure of the $O^{4+} \cdot PF_6^- \cdot 2MeCN$ receptor is almost identical with that (Figure 6b) observed in the absence of 1/4DMB. The only significant change is a reduction in the twist angle between the pyridinium rings of the Paraquat residues in the O^{4+} tetracation. The isostructural crystalline forms of $O^{4+} \cdot 4PF_6^- \cdot 2MeCN$ (i.e. the third MeCN is not included) and $O^{4+} \cdot 1/4DMB \cdot 4PF_6^- \cdot 2MeCN$ are shown in Figure 8. The channel structure is retained in the 1:1 complexes which has been found to exhibit zeolitic properties, e.g. the bound 1/2DMB can be removed from the red crystals which lose their colour under high vacuum (0.03 mm Hg) on heating to 100°C and, by the same token, when colourless crystals of $O^{4+} \cdot 4PF_6^- \cdot 3MeCN$ were immersed in 1/2DMB for several days, they gradually assumed a deep red colour and a 1:1 stoichiometry of guest to host. The principal intermolecular interactions in both $O^{4+}\cdot 1/2DMB$ and $O^{4+}\cdot 1/4DMB$ are (i) the dispersive forces involving two Paraquat units which are each separated by ca. 3.45 Å from the central parallelly-aligned DMB molecules and (ii) electrostatic 'T-type' edge-to-face interactions [4,7] involving the *para*-phenylene units and the orthogonally-aligned DMB molecules with centroid-centroid distances of 4.8 and 5.1 Å for $O^{4+}\cdot 1/2DMB$ and $O^{4+}\cdot 1/4DMB$, respectively.

(a) (b)

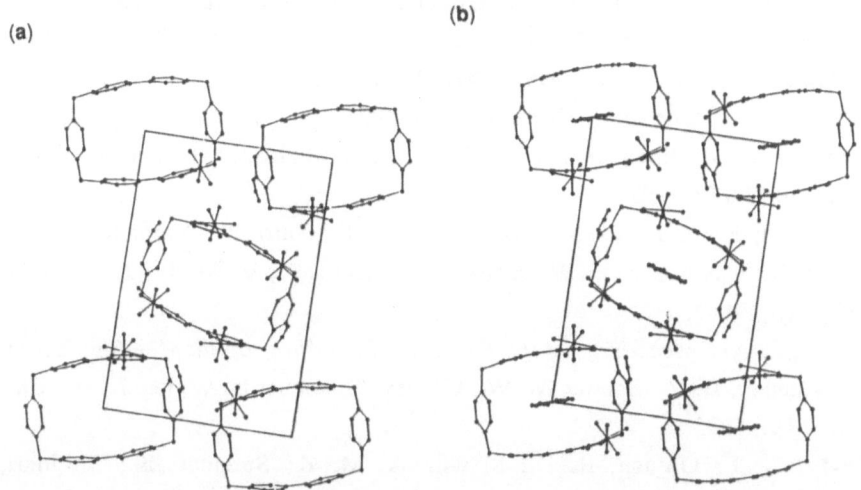

Figure 8.   The isostructural crystalline forms of (a) $O^{4+}\cdot 4PF_6^-\cdot 2MeCN$ and (b) $O^{4+}\cdot 1/4DMB\cdot 4PF_6^-\cdot 2MeCN$.

CONCLUSIONS

Bisparaquat(1,4)cyclophane is a novel tetracationic receptor: it,

- can be made in **two steps** from **inexpensive starting materials** (bipyridine and *para*-xylylene dibromide);
- contains a **rigid rectangular cavity** as shown by X-ray diffraction and molecular mechanics;
- has its **positive charges delocalised** over four pyridinium rings as shown by semi-empirical quantum mechanical calculations;
- is soluble in **water** as its **halides**;
- is soluble in **organic solvents** as its **tetrakis(hexafluorophosphate)**;
- **complexes with neutral aromatic substrates** in both the **solid state** and **in solution**;
- forms **crystalline complexes** which are **isostructural** with **each other** and with the **free crystalline receptor**;
- crystallizes in **continuous hollow stacks** which exhibit **zeolitic properties**;
- catalyzes **H/D exchange** of the aromatic ring protons of **hydroquinol**.

Acknowledgements. We are grateful to Shell Research Limited and the Agriculture and Food, and Science and Engineering, Research Councils in the United Kingdom for financial support. We thank Dr. David Williams and Miss Sandra Slawin for carrying out the X-ray crystallography, Dr. Barbara Odell for performing the calculations, and Dr. Roger Pettman for helpful discussions.

References

1. Colquhoun, H. M.; Stoddart, J. F.; Williams, D. J. *New Scientist* **1986**, *1 May*, 44.

2. Allwood, B. L.; Spencer, N.; Shahriari-Zavareh, H.; Stoddart, J. F.; Williams, D. J. *J. Chem. Soc., Chem. Commun.* **1987**, 1064.

3. Stoddart, J. F. *Pure Appl. Chem.* **1988**, *60*, 467.

4. Slawin, A. M. Z.; Spencer, N.; Stoddart, J. F.; Williams, D. J. *J. Chem. Soc., Chem. Commun.* **1987**, 1070.

5. Ito, M. M.; Kato, J.; Takagi, S.; Nakashiro, E.; Sato, T.; Yamada, Y.; Saito, H.; Namiki, T.; Takamura, I.; Wakatsuki, K.; Suzuki, T.; Endo, T. *J. Am. Chem. Soc.* **1988**, *110*, 5147.

6. Beevers, C. A. *Proceedings of the European Crystallographic Meeting*; Zurich, 1976; 293; Cheney, B. V.; Schulz, M. W.; Cheney, J.; Richards, W. G. *J. Am. Chem. Soc.* **1988**, *110*, 4195.

7. Moody, G. J.; Owusu, R. K.; Slawin, A. M. Z.; Spencer, N.; Stoddart, J. F.; Thomas, J. D. R.; Williams, D. J. *Angew. Chem., Int. Ed. Engl.* **1987**, *26*, 890; Alston, D. R.; Slawin, A. M. Z.; Stoddart, J. F.; Williams, D. J.; Zarzycki, R. *Angew. Chem., Int. Ed. Engl.* **1987**, *26*, 692, 693.

8. Gould, R. O.; Gray. A. M.; Taylor, P.; Walkinshaw, M. D. *J. Am. Chem. Soc.* **1985**, *107*, 5921; Burley, S. K.; Petsko, G. A. *J. Am. Chem. Soc.* **1986**, *108*, 7995.

9. Ashton, P. R.; Chrystal, E. J. T.; Mathias, J. P.; Parry, K. P.; Slawin, A. M. Z.; Spencer, N.; Stoddart, J. F.; Williams, D. J. *Tetrahedron Lett.* **1987**, *28*, 6367.

10. Ortholand, J.-Y.; Slawin, A. M. Z.; Spencer, N.; Stoddart, J. F.; Williams, D. J. *Angew. Chem.*, in press.

11. Geuder, W.; Hünig, S.; Suchy, A. *Angew. Chem., Int. Ed. Engl.* **1983**, *22*, 489; *Tetrahedron* **1986**, *42*, 1665; Odell, B.; Reddington, M. V.; Slawin, A. M. Z.; Spencer, N.; Stoddart, J. F.; Williams, D. J. *Angew. Chem., Int. Ed. Engl.* **1988**, *27*, 1547.

12. Muller, W. M.; Vögtle, F. *Angew. Chem., Int. Ed. Engl.* **1984**, *23*, 712; Franke, J.; Vögtle, F. *Angew. Chem., Int. Ed. Engl.* **1985**, *24*, 219.

13. Ashton, P. R.; Odell, B.; Reddington, M. V.; Slawin, A. M. Z.; Stoddart, J. F.; Williams, D. J. *Angew. Chem., Int. Ed. Engl.* **1988**, *27*, 1550; see also Buhner, M.; Geuder, W.; Gries, W.-K.; Hünig, S.; Koch, M.; Poll, T. *Angew. Chem., Int. Ed. Engl.* **1988**, *27*, 1553.

THE HEXAGONAL LATTICE APPROACH TO MOLECULAR RECEPTORS

Thomas W. Bell*, Albert Firestone, Jia Liu, Richard Ludwig, and Scott D. Rothenberger

Department of Chemistry
State University of New York at Stony Brook
Stony Brook, NY 11794-3400 (USA)

SUMMARY

We have shown that the hexagonal lattice architecture may be used to construct hosts with high affinity for cations and neutral molecules. These planar receptors are easily synthesized, they form complexes rapidly and they show considerable potential for selectivity towards planar guest molecules.

INTRODUCTION

A key problem in the design of selective receptors and catalysts is synthetic access to molecular frameworks that focus binding sites and reactive functional groups on a molecular cavity. In our efforts to prepare relatively rigid complexing agents for metal ions we have

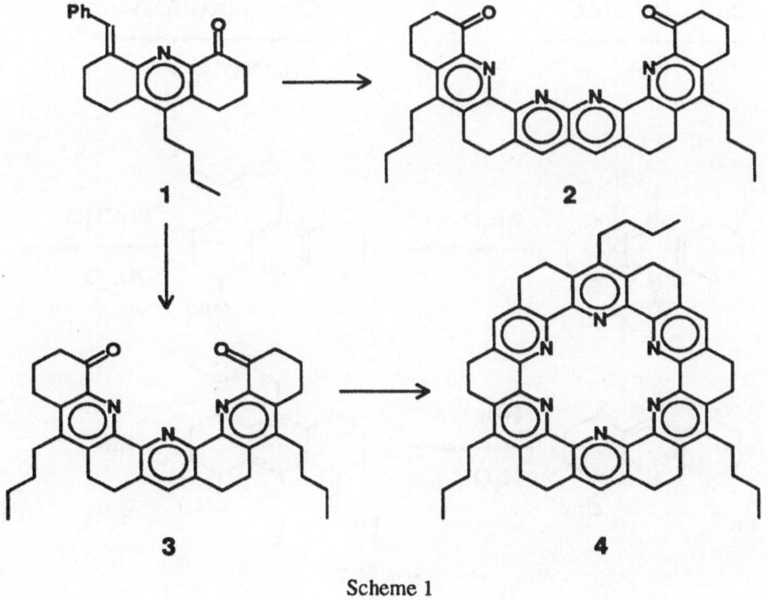

Scheme 1

developed a host architecture consisting of fused carbocyclic and heterocyclic six-membered rings. This "hexagonal lattice" approach to molecular receptors is exemplified in Scheme 1 by three hosts (2-4) [1]. This report describes the improved preparation of key intermediate 1 as well as some of the complexation properties of the three hexagonal lattice hosts.

DISCUSSION

Conventional hosts for alkali metals include crown ethers [2], cryptands [3], cryptaspherands [4], and spherands [4a]. Binding strength and selectivity also generally increase in this order. Unfortunately, exchange rates decline in the same order so that complexes of the most selective hosts equilibrate too slowly for many practical applications. Host 4 represents two extremes: the cavity is extremely rigid but it is nonencapsulating. These considerations led us to devise a synthesis of 4 and to coin the term "torand", which denotes a toroidal ligand composed of mutually fused rings [1a].

New approaches to the quino[8,7-b]1,10-phenanthroline nucleus have been reported by Thummel and Jahng [5] and by Ransohoff and Staab [6], who also reported the parent dodeca-hydrohexaazakekulene. Our synthetic efforts have focussed on n-butyl substituted hosts to

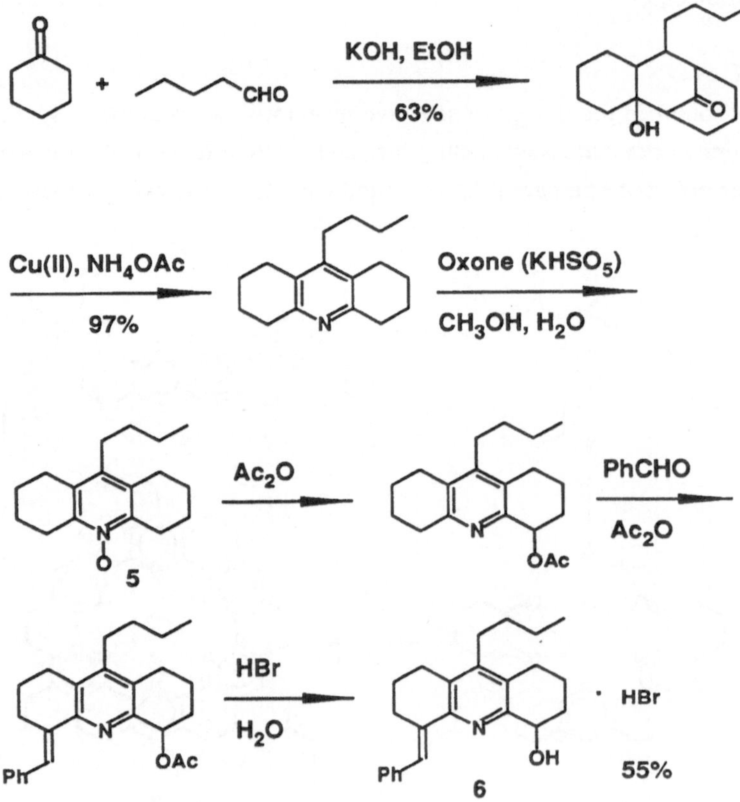

Scheme 2

enhance solubility in nonpolar solvents. Since our initial report on tri-*n*-butyl torand **4** [1a] we have considerably improved the syntheses of 9-*n*-butyloctahydroacridine derivatives that are the building blocks of our hexagonal lattice approach [7]. As shown in Scheme 2, these intermediates are now readily available. Employing our improved synthesis of pyridines from 1,5-diketone precursors [8], N-oxide **5** is now available on a 40 g scale in 59% yield overall from cyclohexanone and valeraldehyde [9]. N-oxide **5** is then converted to benzylidene alcohol **6** by one-pot Katada rearrangement/benzaldehyde condensation, followed by acidic hydrolysis of the resulting ester. Intermediate **6** is easily purified by recrystallization, affording this differentially functionalized acridine in 55% overall yield on a 25-30 g scale. Finally **6** is converted to the free base and oxidized at room temperature by the Albright-Goldman method [10], yielding **1** in 72% yield after recrystallization.

As previously described [1a], benzylideneketone **1** is converted to diketone **3** by Newkome-Fishel dimerization of the trimethylhydrazonium salt, followed by ozonolysis. Diketone **3** is both an intermediate in the synthesis of torand **4** and a hexagonal lattice receptor having interesting complexation properties. This host extracts alkali metal and ammonium picrates from water into chloroform more effectively than crown ethers. The stability constants of Li, Na, K, Cs and ammonium picrate complexes were determined by this method and are compared to those for naphtho-18-crown-6 [11] in Table 1. Even though naphtho-18-crown-6 contains one more donor atom than does **3**, complexes of the latter host are more

Table 1. Stability constants (log K_s) measured by CHCl$_3$ extraction of aqueous picrates.

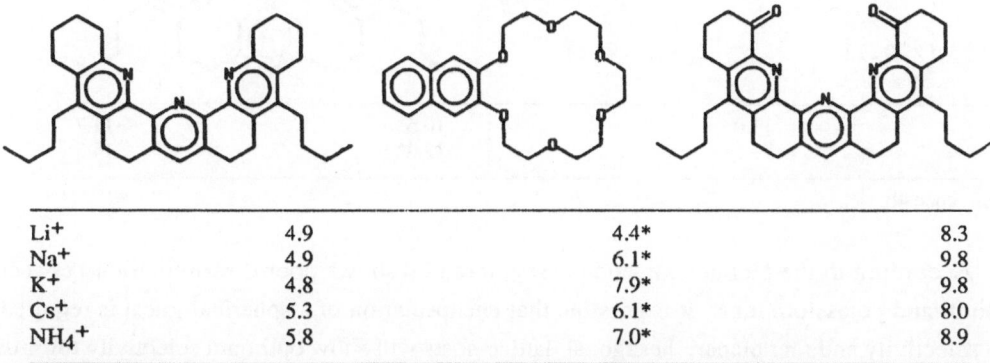

Li$^+$	4.9	4.4*	8.3
Na$^+$	4.9	6.1*	9.8
K$^+$	4.8	7.9*	9.8
Cs$^+$	5.3	6.1*	8.0
NH$_4$$^+$	5.8	7.0*	8.9

* Reference 11.

stable. The metal-binding properties of **3** apparently stem from the conformational organization of three pyridine and two ketone dipoles by the hexagonal lattice framework.

Torand **4** was synthesized by cyclization of **3** as previously described [1a,b]. Initially isolated as its complex with calcium triflate, which was a sequestered impurity, torand **4** is currently prepared under calcium-free conditions and isolated as the monotriflate salt in 30% yield. The overall synthesis from cyclohexanone requires 11 steps and may be used to prepare about a gram of **4** in a few weeks. The product is a versatile complexing agent and 1:1 complexes with Li, Na, K, Rb and Cs triflates are readily prepared by washing a chloroform solution of **4·CF$_3$SO$_3$H** with aqueous solutions of the alkali metal carbonates or hydroxides.

Metal-free **4** may also be obtained by treating **4**·CF$_3$SO$_3$H with tetra-*n*-butylammonium hydroxide in *n*-butanol/acetonitrile.

Chloroform solutions of **4** extract 1 equivalent of each alkali metal picrate salt from water, indicating that the stability constant for each complex exceeds 10^{11}. We have used competition experiments to determine the stabilities of the Na and K picrate complexes. When free torand **4** and the [2.2.1]cryptand complex of sodium picrate or the [2.2.2]cryptand complex of potassium picrate are dissolved in CDCl$_3$ (saturated with D$_2$O), equilibrium is quickly established. When torand complexes and free cryptands are used the same picrate chemical shifts and cryptand/cryptate integrals are obtained. Table 2 shows the log K$_s$ values of the sodium and potassium torand complexes, compared with the values for the cryptands [4b]. Remarkably, the equilibria favor complexes of a planar, nonencapsulating host over those of the encapsulating cryptands.

Table 2. Stability constants (log K$_s$) measured in CHCl$_3$/D$_2$O by NMR competition.

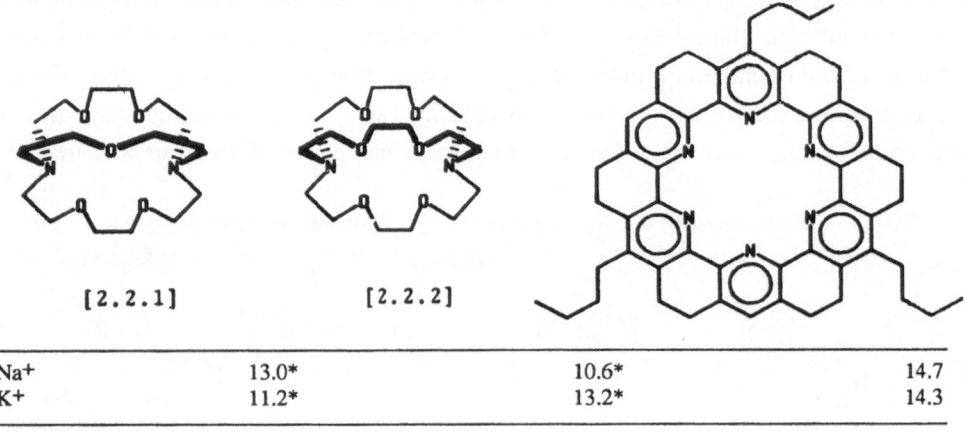

	[2.2.1]	[2.2.2]	
Na+	13.0*	10.6*	14.7
K+	11.2*	13.2*	14.3

* Reference 4b.

According to the picrate extraction assay, torand **4** shows poor discrimination between sodium and potassium ions. It is possible that encapsulation of a spherical guest is required for selectivity and that planar, hexagonal lattice hosts will show optimum selectivity towards planar guest molecules. The hexagonal lattice architecture is particularly well suited to forming

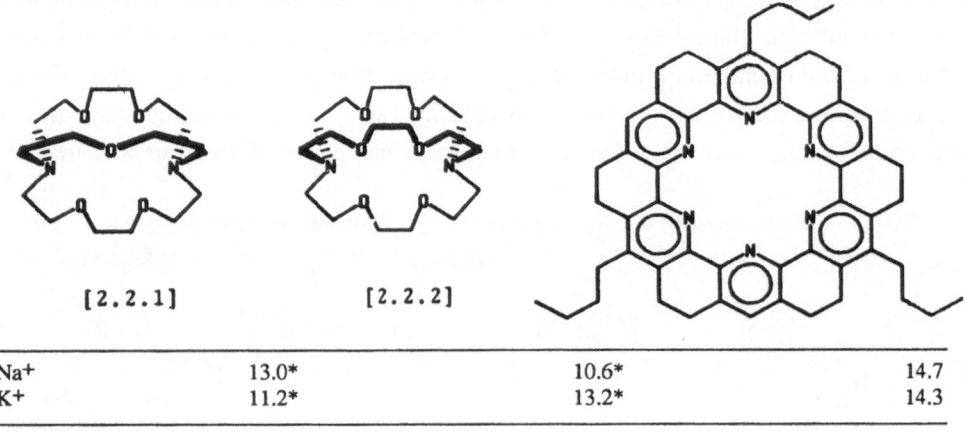

Scheme 3

hydrogen bonding networks with molecules having sp^2-hybridized heteroatoms, as shown in Scheme 3. Thus 8-membered and 10-membered H-bonded chelate rings are nearly ideal because C-C, C-N and C-O bond lengths are approximately half the distance between heteroatoms in N-H⋯N, N-H⋯O, O-H⋯N and O-H⋯O hydrogen bonds. This relationship displaces the guest heteroatoms to the hexagonal centers of the projected host lattice. Both 8-membered and 10-membered chelate rings are combined in Scheme 4 to produce a general design for hydrogen bonding urea receptors.

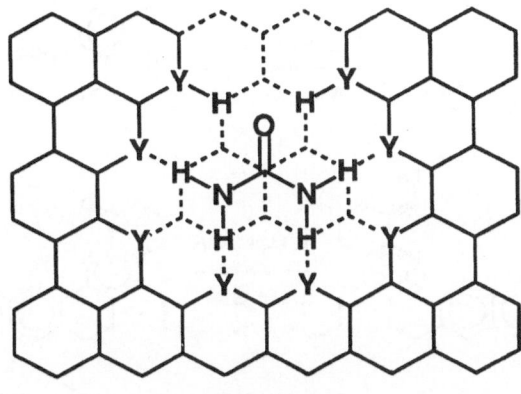

Scheme 4

Diketone 2, a receptor that can form four hydrogen bonds with urea, is synthesized as outlined in Scheme 5 [1c]. Benzylideneketone 1, again a key intermediate, is dimerized in three steps to form a 1,8-naphthyridine ring employing the Friedlander condensation. Ozonolysis is used as before to unmask latent carbonyl functionality, yielding diketone 2 in 46% overall from 1. Crystals of urea rapidly dissolve in chloroform solutions of 2 producing a 1:1 complex. The structure of this complex has not yet been determined crystallographically, but the coplanar host-guest orientation shown in Scheme 5 is predicted by examination of CPK molecular models.

The stability constant of the complex between 2 and urea may be calculated using the solubility of urea in anhydrous chloroform in the absence of host (5 x 10^{-4} M^{-1}) [1c]. According to gravimetric and calorimetric methods, at least 0.95 equivalents of urea are dissolved by a 10^{-2} M solution of 5. Hence the complex/host ratio exceeds 19 and K_s for the urea complex is at least 4 x 10^4 M^{-1}. A 0.02 M solution of diketone 2 also extracts 0.36 equivalents of urea from 2 M aqueous urea. The distribution constant for partition of urea between chloroform and water is approximately 9 x 10^{-5}, hence the K_s for the urea complex in chloroform saturated with H$_2$O may be calculated as 3 x 10^3 M^{-1}.

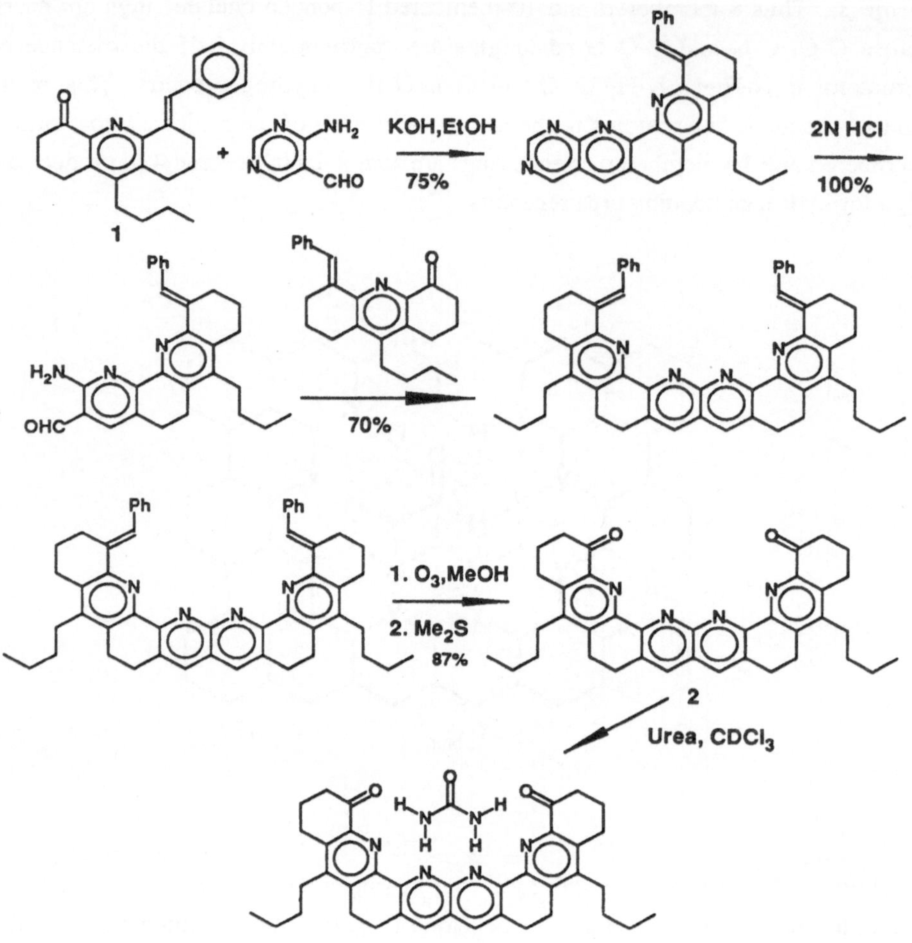

Scheme 5

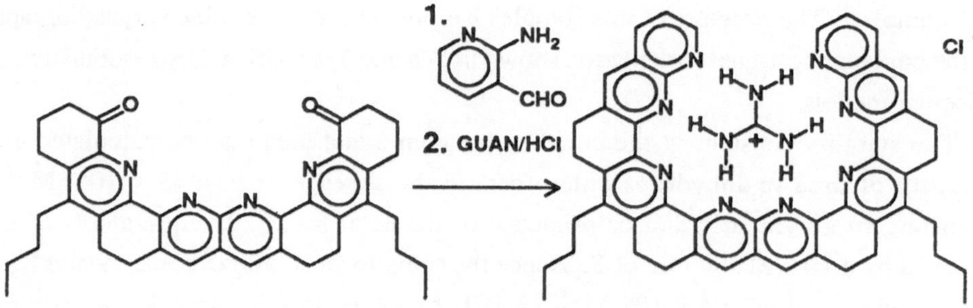

Scheme 6

Table 3. Stability constants of urea complexes.

$$R = \frac{[UREA]}{[HOST]}$$

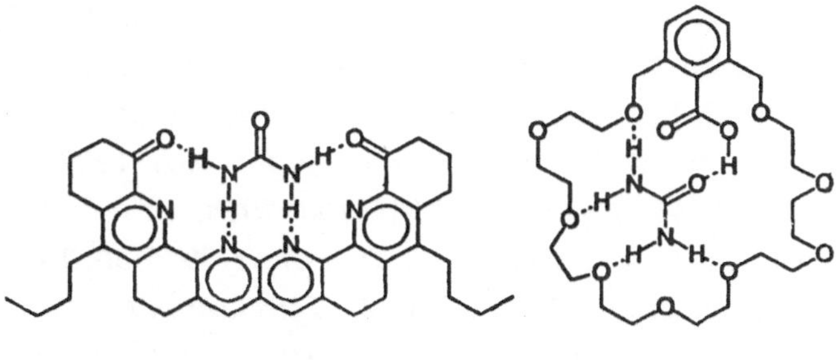

	R	K_s	R	K_s
CHCl₃ or CDCl₃ Solid UREA	> 0.95	>40,000	0.54*	2000
2M UREA in H₂O CHCl₃ or CDCl₃	0.36	3,000	0.1*	600

*Reference 12.

Table 3 compares the stability constants of urea complexes of **2** with those of a benzoic acid crown ether receptor reported by Reinhoudt et al. [12]. Although the latter complex has five hydrogen bonds, one of which is between the urea carbonyl and a good hydrogen bond donor, the hexagonal lattice complex is significantly stronger. This is apparently the consequence of effective conformational organization of four hydrogen bond acceptors by the hexagonal lattice framework. The two spacer pyridines bearing *n*-butyl substituents may also contribute to stabilization of the complex by dipole-dipole interactions. An additional point of interest is that the urea complex of **2** is weakened by the presence of water more than is the urea complex of the benzoic acid crown. This effect is explained by greater stabilization of uncomplexed **2** by hydration than for the benzoic acid crown, which is probably intramolecularly hydrogen bonded in the absence of guest.

We are currently working on the synthesis of hexagonal lattice hosts capable of forming six hydrogen bonds to urea. For this purpose two hydrogen bond donors must be incorporated into the structure (cf. Scheme 4). If two hydrogen bond acceptors are placed in these positions then a guanidinium ion receptor is produced as shown in Scheme 6. The application of guanidinium receptors of this type to ion-selective electrodes for guanidinium derivatives is currently under investigation.

Acknowledgements. We gratefully acknowledge support by the New York Science and Technology Foundation through the center for Biotechnology at Stony Brook and by the National Institutes of Health (PHS Grant GM 32937).

References

1. (a) Bell, T. W; Firestone, A. *J. Am. Chem. Soc.* **1986**, *108*, 8109.;
   (b) Bell, T. W; Firestone, A.; Guzzo, F.; Hu, L.-Y. *J. Incl. Phenom.* **1987**, *5*, 149;
   (c) Bell, T. W; Liu, J. *J. Am. Chem. Soc.* **1988**, *110*, 3673.

2. Pedersen, C. J. *Angew. Chem., Int. Ed. Engl.* **1988**, *27*, 1021.

3. Lehn, J.-M. *Angew. Chem., Int. Ed. Engl.* **1988**, *27*, 89.

4. (a) Cram, D. J. *Angew. Chem., Int. Ed. Engl.* **1988**, *27*, 1009; (b) Cram, D. J.; Ho, S. P. *J. Am. Chem. Soc.* **1986**, *108*, 2998.

5. Thummel, R. P.; Jahng, Y. *J. Org. Chem.* **1985**, *50*, 2407.

6. Ransohoff, J. E. B.; Staab, H. A. *Tetrahedron Lett.* **1985**, *26*, 6179.

7. Bell, T. W; Firestone, A. *J. Org. Chem.* **1986**, *51*, 764.

8. Bell, T. W.; Rothenberger, S. D. *Tetrahedron Lett.* **1987**, *28*, 4817.

9. Bell, T. W.; Cho, Y.-M.; Firestone, A.; Healy, K.; Liu, J.; Ludwig, R.; Rothenberger, S. D. *Org. Syn.*, submitted.

10. Albright, J. D.; Goldman, L. *J. Am. Chem. Soc.* **1967**, *89*, 2416.

11. Helgeson, R. C.; Weisman, G. R.; Toner, J. L.; Tarnowski, T. L.; Chao, Y.; Mayer, J. M.; Cram, D. J. *J. Am. Chem. Soc.* **1979**, *101*, 4928.

12. van Staveren, C. J.; Aarts, V. M. L. J.; Grootenhuis, P. D. J.; Droppers, W. J. H.; van Eerden, J.; Harkema, S.; Reinhoudt, D. N. *J. Am. Chem. Soc.* **1988**, *110*, 8134.

MOLECULAR RECOGNITION BY MACROCYCLIC RECEPTORS

Andrew D. Hamilton

Department of Chemistry
University of Pittsburgh
Pittsburgh, PA 15260 (USA)

Effective molecular recognition requires the precise alignment of binding groups on the receptor with chemical features on the substrate. In the past two years we have initiated a program aimed at the development of artificial receptors containing several recognition sites that are complementary with biologically-interesting molecules. This project was inspired by a lecture at the 4th IPMR in Lancaster by Professor W. Saenger in which he discussed the X-ray structure of the guanine-binding enzyme, ribonuclease T_1 [1]. This shows a guanine substrate bound into the active site by both hydrogen bonding to the peptide backbone and aromatic stacking to a tyrosine aromatic ring (Figure 1). The presence of two recognition interactions increases both the strength and specificity of substrate binding.

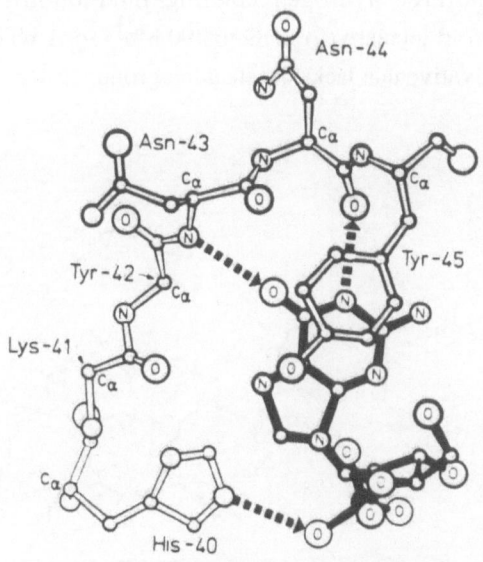

Figure 1. Guanine bound to ribonuclease T_1.

We decided to incorporate this two-site binding strategy into a series of synthetic receptors for the nucleotide bases (shown for thymine in Figure 2). Our continuing goal in this work is to gain insights into the principles or "rules" of nucleotide recognition and then to apply them to the construction of "artificial repressor" molecules capable of binding to specific oligonucleotide sequences.

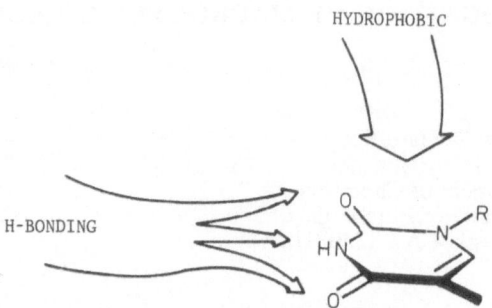

Figure 2. Two-site binding strategy.

Our first receptor **1** (shown with its X-ray structure in Figure 3) combines a 2,7-dialkoxynaphthalene as the π-stacking component with a 2,6-diamidopyridine unit that forms a triple hydrogen bonded complementarity with the imide functional group on thymine [2]. Treatment of **1** with one equivalent of 1-butylthymine **2** in CDCl$_3$ results in large downfield shifts (≈2 ppm) of the amide-NH and imide-NH resonances and small upfield shifts (≈0.3 ppm) of the thymine-CH$_3$, ring-H and N-CH$_2$ resonances. These ^{1}H NMR changes are consistent with the formation of a complex between **1** and **2** that combines simultaneous aromatic stacking with three hydrogen bonding interactions. Association constant measurements show a 3-fold increase (from 89 to 290 M^{-1}) for **1** when compared to a simple 2,6-diamidopyridine derivative that lacks the stacking group.

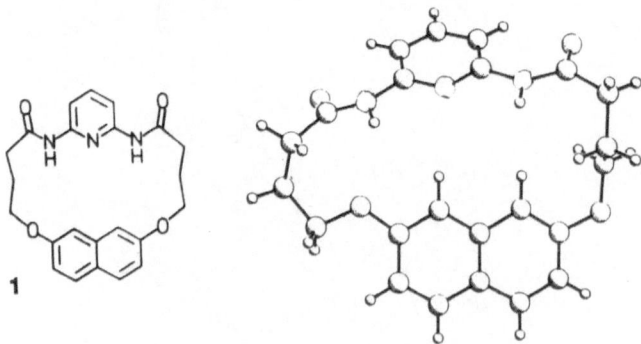

Figure 3. Molecular and crystal structure of **1**.

A side view of the X-ray structure of the complex between **1** and **2** is shown in Figure 4. The naphthalene ring is positioned above the thymine ring with a closest approach of 3.4 Å. Comparison to the structure of **1** shows that the naphthalene swings through a 40° arc to its face-to-face stacking position.

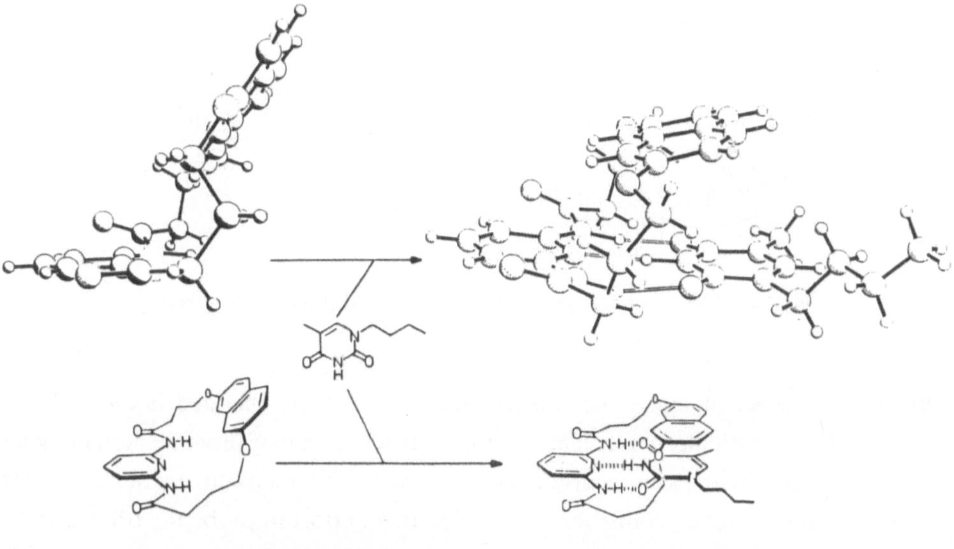

Figure 4. The complex between 1 and 1-butylthymine 2.

The structural and synthetic simplicity of **1** means that many changes can be made to the receptor. Varying the hydrogen bonding region will lead to differences in the binding specificity. 2-Amino-1,8-naphthyridine **3** forms a triple hydrogen bonding complementarity with guanine [3]. Incorporation of **3** into a macrocycle containing a π-stacking unit leads to receptor **4**. ^{1}H NMR experiments show that **4** binds to guanine derivatives by a combination of aromatic stacking and hydrogen bonding interactions as shown in **5**. The participation of the stacking group is confirmed by the increase in K_a (126 to 502 M^{-1}) on going from a simple acyclic aminonaphthyridine to **4**. A similar strategy can be applied to the recognition of adenine. Bis-(2-aminopyridine) derivatives (Figure 5) can form four hydrogen bonds to the periphery of adenine (combining both Watson-Crick and Hoogsteen interactions) [4]. We have prepared a macrocyclic receptor **6** combining both a naphthalene π-stacking unit and the

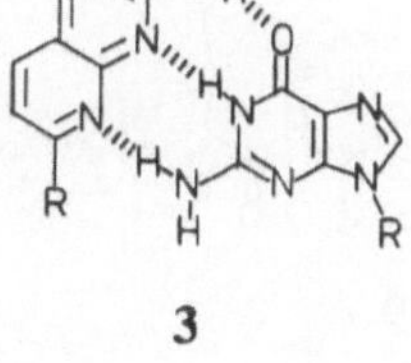

3

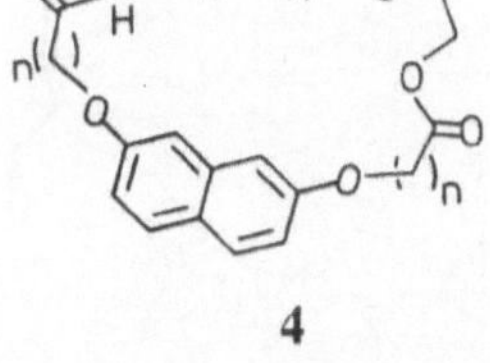

4

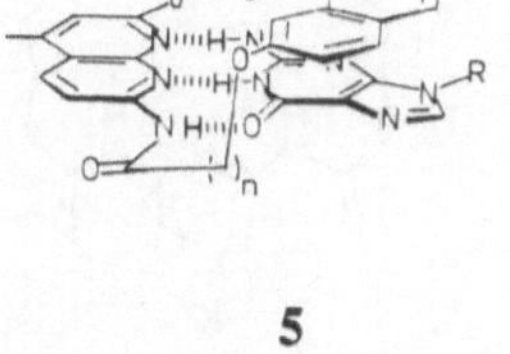

5

potentially tetrahydrogen bonding 1,2-bis-(2-amino-6-pyridyl)ethane group. This forms strong complexes with 9-alkyl- adenine **7** as shown by a K_a value for **6:7** of 3000 M^{-1}. The characteristic upfield and downfield shifts in the NMR spectrum confirm the participation of both hydrogen bonding and aromatic stacking interactions.

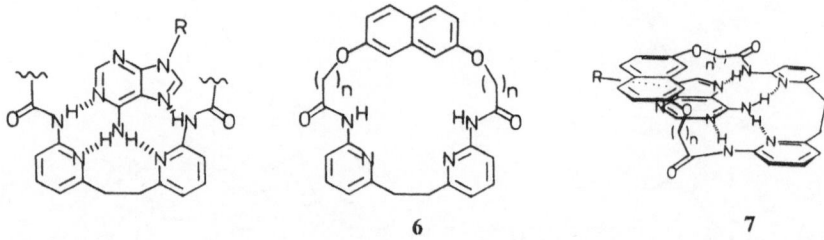

6 **7**

Figure 5. Hydrogen bonding between adenine and bis-(2-aminopyridine).

The second aspect of these two-site receptors that can be changed is the π-stacking component. This is particularly relevant since in a recent survey of protein crystal structures Petsko [5] has identified two important geometries for aromatic-aromatic interactions; face-to-face or edge-to-face geometries (Figure 6). In an effort to probe the origins of these two arrangements we have investigated the importance of the electronic characteristics of the

FACE-FACE **EDGE-FACE**

Figure 6. Possible geometries for aromatic-aromatic interaction.

stacking group on its orientation. Macrocycle **8**, containing two electron withdrawing ester groups on the naphthalene ring, forms stronger complexes (K_a = 570 M^{-1}) than unsubstituted **1** (K_a = 290 M^{-1}) with 1-butylthymine **2**. The X-ray structure of complex **8:2** (Figure 7) shows a parallel, face-to-face interaction with an interplanar distance of 3.5 Å. An insight into

8 **9**

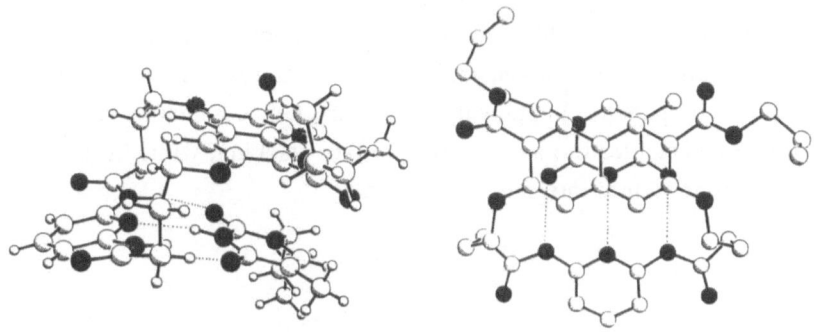

Figure 7. X-ray crystal structure of the **8:2** complex.

the special stabilization involved in stacking comes from MNDO calculations on thymine and 2,7-dimethoxynaphthalene-3,6-dicarboxylate. The resulting charge distributions are superimposed (sign only) on a downward view of structure **8:2** (Figure 8). This shows five points of contact where partially positively charged atoms on the naphthalene precisely align themselves with partially negative regions on the thymine [6]. From this we can conclude that electrostatic interactions between regions of complementary charge distribution on the rings play an important role in π-stacking. If this analysis is correct it should be possible to change the geometry of aromatic-aromatic interactions by varying the electronic characteristics of one ring. Replacing the diester groups in **8** by two ether groups leads to macrocycle **9** which shows substantially weaker binding ($K_a = 138$ M^{-1}) to **2** than **8**. MNDO calculations on 2,3,6,7-tetramethoxynaphthalene show a reversal of sign on the carbons-4 and -5. This would lead to a repulsive electrostatic interaction between receptor and substrate if a face-to-face geometry were to form (Figure 9). The X-ray structure of complex **9:7** confirms that the face-to-face geometry is avoided and that the naphthalene takes up an almost perpendicular, edge-to-face orientation with respect to the substrate. Interestingly, the naphthalene-1,8-protons project towards a region of partial negative charge formed by the thymine imide group. The possible electrostatic stabilization between these regions may account for the small stabilization of **9:2** compared to the simple complex lacking a naphthalene

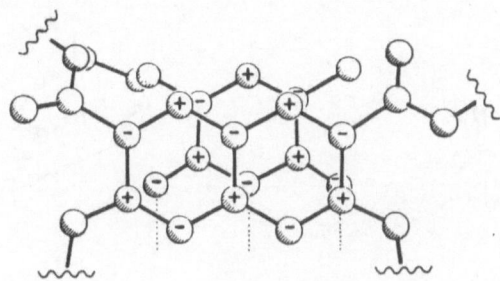

Figure 8. Charge distribution of the **8:2** complex.

(as discussed above). Thus, within a simple series of thymine receptors, we have shown that the geometry of aromatic-aromatic interactions can be controlled by modifying the electronic properties of one component. In particular, an electrostatic complementarity between partial charges on the rings can lead to a strong face-to-face stacking, while in the absence of such effects a weaker edge-to-face interaction is preferred.

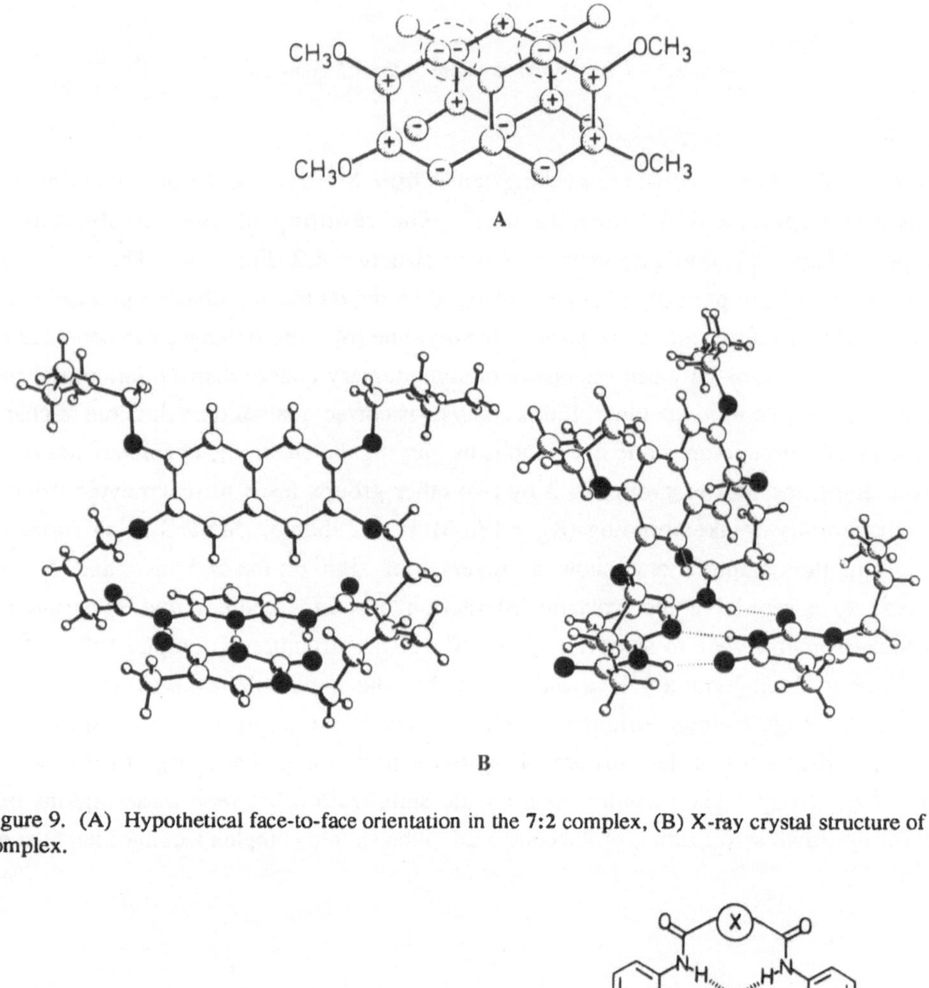

Figure 9. (A) Hypothetical face-to-face orientation in the 7:2 complex, (B) X-ray crystal structure of the 7:2 complex.

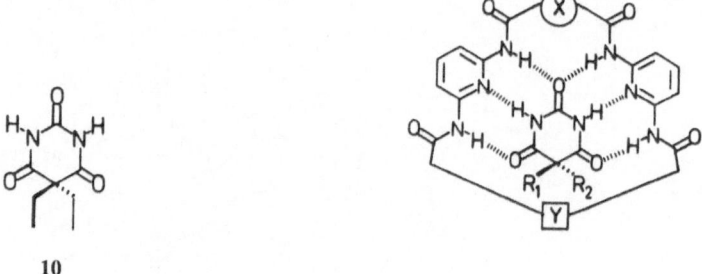

Figure 10. Proposed complex between a macrocycle incorporating two 2,6-diamidopyridine derivatives and a 5,5-dialkylbarbiturate.

The multi-site approach to molecular recognition can be applied to many potential substrates. The barbiturate family of drugs (e.g. barbital **10**) is a particularly attractive target due to their wide-spread use as sedatives and anticonvulsants. Two 2,6-diamidopyridine derivatives incorporated in a macrocyclic framework can provide six complementary hydrogen bonding interactions to the periphery of a 5,5-dialkyl barbiturate (Figure 10). Following a

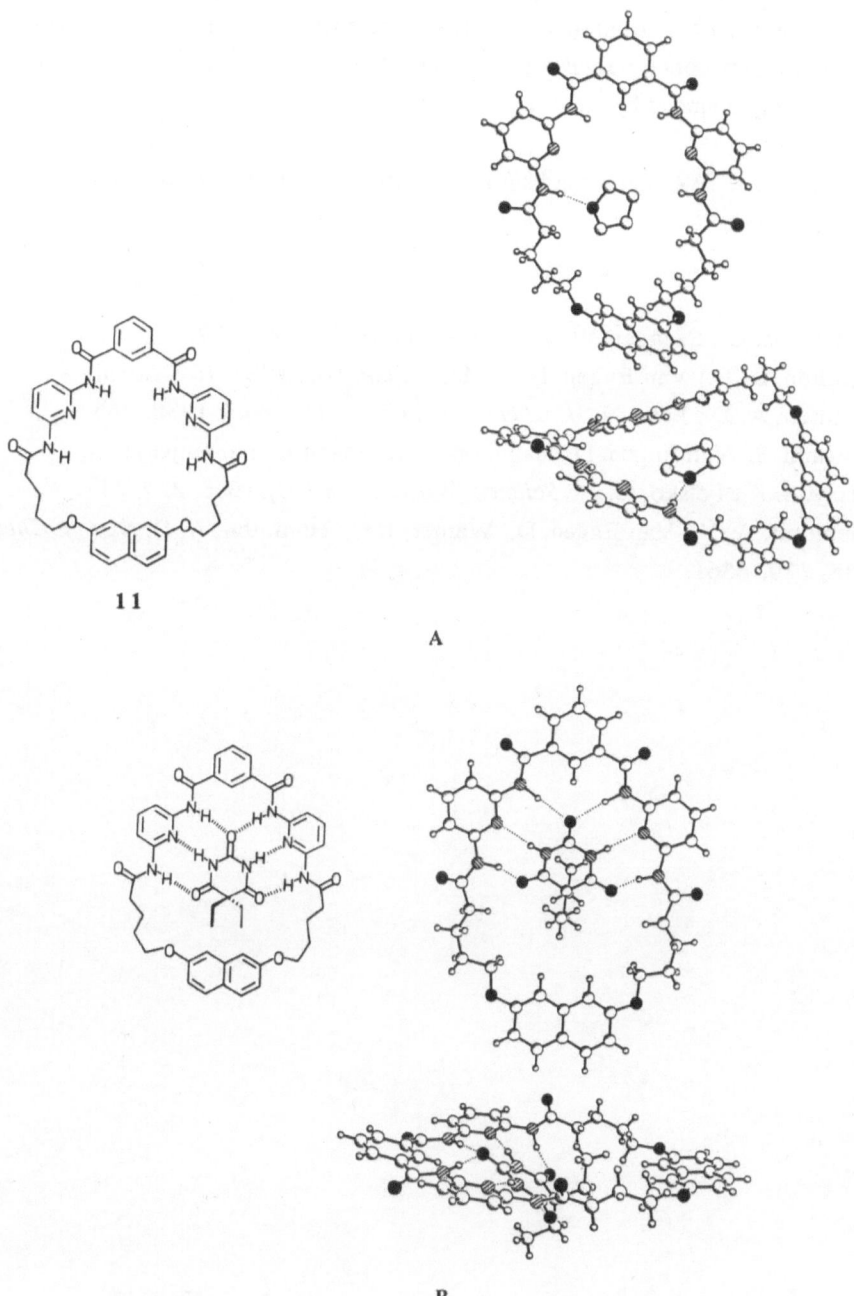

11

A

B

Figure 11. (A) Crystal structure of **11**, (B) Complex between **11** and **10**.

high dilution route, we have prepared macrocycle **11** which contains two 2,6-diaminopyridines separated by an isophthalate spacer. The X-ray structure of **11** (Figure 11a) confirms the preorganization of the binding site with all six H-bonding groups pointing inside the cavity. Indeed, **11** formed strong complexes with barbital **10** ($K_a = 1.35 \times 10^5$ M^{-1}) which can be followed by observing the large downfield shifts in the ^{1}H NMR spectrum of the amide and imide protons involved in H-bonding. The nature of the interaction between **11** and **10** is confirmed by X-ray crystallographic analysis. The structure of the complex (Figure 11b) shows six hydrogen bonds between the receptor and substrate with the two ethyl groups occupying a cavity bounded by the naphthalene ring.

Acknowledgment. We thank the National Institutes of Health (GM 35208) for financial support of this work.

References

1. Heinemann, U.; Saenger, W. *Nature (London)* **1982**, *299*, 27.
2. Hamilton, A. D.; Van Engen, D. *J. Am. Chem. Soc.* **1987**, *109*, 5035.
3. Hamilton, A. D.; Pant, N. *J. Chem. Soc., Chem. Commun.* **1988**, 765.
4. Goswami, S.; Van Engen, D.; Hamilton, A. D., manuscipt in preparation.
5. Burley, S. K.; Petsko, G. A. *Science (Washington DC)* **1985**, *229*, 23.
6. Muehldorf, A. V.; Van Engen, D.; Warner, J. C.; Hamilton, A. D. *J. Am. Chem. Soc.* **1988**, *110*, 6561.

HOST-GUEST BINDING MECHANISMS: EXPERIMENTAL APPROACHES [1]

Hans-Jörg Schneider*, Thomas Blatter, Rüdiger Kramer,
Surat Kumar, Ulrich Schneider, and Isolde Theis

FR Organische Chemie der Universität des Saarlandes
D-6600 Saarbrücken 11 (FRG)

SUMMARY

Non-covalent interactions are analyzed for complex formation in aqueous solution with cyclophanes bearing no positive or negative charges in the cavity, and with cyclodextrins. Geometric fitting as well as substrate solubility is found in several cases to be of lesser significance. The use of the solvent dependence of association for the quantification of hydrophobic binding contributions is exemplified. Coulomb attractions in a large number of systems obey a general correlation with the number of ion-ion interactions, furnishing ≈ 1.3 kcal/mol per salt bridge. Geometric variations of charge locations in host and guest molecules show the expected binding differences. Special van der Waals effects between ammonium ions and aromatic moieties contribute significantly to binding, as demonstrated by comparison to saturated substrates.

INTRODUCTION

In spite of the large number of host-guest systems reported in the literature our understanding of the underlying binding mechanisms is still in a state of infancy. While smaller complexes, e.g. of alkali cations with ammonia, water etc., are quite amenable to MO calculations, the interactions of larger organic systems including crown ethers are difficult to quantify even on the basis of empirical force field potentials [2]. Such potentials also form the basis of many highly sophisticated and commercially available programs aiming at the simulation of e.g. protein structures. It must be noted that, in contrast to the more traditional molecular mechanics calculations with organic molecules, the correctness of non-covalent interaction potentials, which are the heart of host-guest or protein receptor-substrate complexes, are to a large degree experimentally untested. Synthetic host-guest systems, having far less degrees of freedom than proteins, and allowing planned variations of interactions in a well defined environment should provide one of the best ways to explore and to secure non-covalent interactions which have enabled nature for millions of

Inclusion Phenomena and Molecular Recognition
Edited by J. Atwood
Plenum Press, New York, 1990

years, and should enable chemists some day, to operate complex molecular systems with fascinating efficiency.

The discussion of methods necessary for a meaningful investigation of host-guest complexes in solution is not within the scope of the present paper. We have largely used NMR techniques not only to obtain reliable binding constants but also information on the geometries of true intra-cavity inclusion complexes [3]. These methods, which must also include explicit calculations of group anisotropy and electrical field effects besides aromatic ring current effects on NMR shielding constants [4] need further development for rigorous conformational analyses.

HOW IMPORTANT IS A PRECISE GEOMETRIC FIT?

Most researchers in the field of supramolecular chemistry emphasize the importance of a close contact between substrate and cavity, which indeed would be required if dispersion forces, falling off with a steep r^{-6} distance dependence between the molecules, were to dominate here. While such a situation should hold for the binding of e.g. chloroform in lipophilic cavities measured in lipophilic solvents [5], several binding free energies ($\Delta G°$) obtained from equilibrium measurements in aqueous solutions [6,7] (Scheme 1) point more

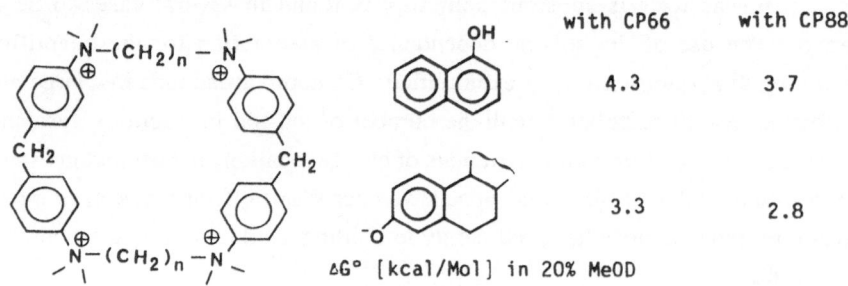

	with CP66	with CP88
OH (naphthol)	4.3	3.7
	3.3	2.8

$\Delta G°$ [kcal/Mol] in 20% MeOD

Scheme 1. n = 6: CP66; n = 8: CP88

towards significant hydrophobic contributions of entropic nature, which do not require a precise surface fitting. Parts of substrate structures which extend out of the cavity, such as in the recently observed complexation of steroids in azoniacyclophanes [8] (Scheme 2) again seem to contribute little to $\Delta G°$; the large contribution of the phenyl ring in the anilinonaphthalenesulfonate (ANS, Scheme 2) compared to naphthalenesulfonate alone points to an additional binding effect which will be discussed below.

HOW SIGNIFICANT ARE SUBSTRATE SOLUBILITY DIFFERENCES IN AQUEOUS SOLUTION?

The lipophilic nature of cavities in cyclodextrins, cyclophanes and other host compounds provide an important driving force for the binding of hydrophobic substrates. However, hydrophilic substituents at the substrates do not lead to a regular binding decrease, as shown by several examples [6,9], e.g. in Scheme 3.

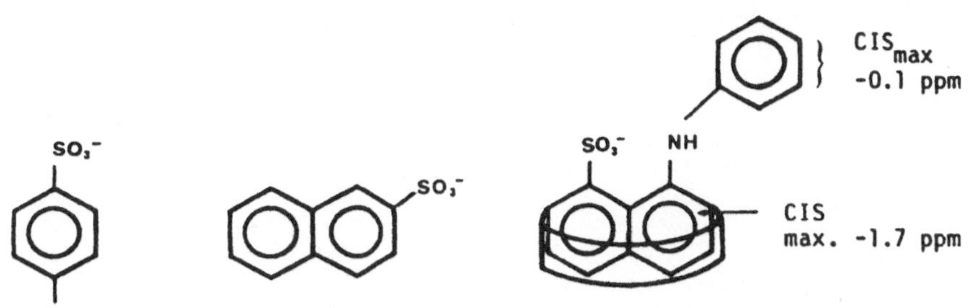

CIS max -0.1 ppm

SO_3^-   SO_3^-   SO_3^- NH

CIS max. -1.7 ppm

3.9   5.5   7.6

ΔG° [kcal/Mol] with CP66^{4+} in 20% MeOD

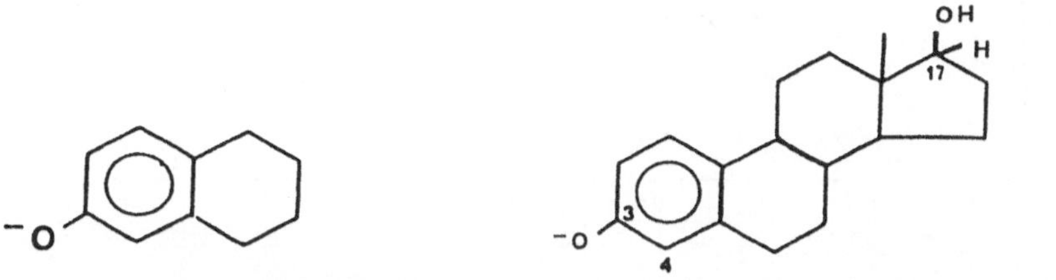

OH
H
17

3

4

2.9 (with CP66: n = 6)

ΔG$_{cplx}$ in 40% MeOD [kcal/Mol]

3.3 (with CP66: n = 6)
⟨ 2.8 (with CP88: n = 8) ⟩

Scheme 2

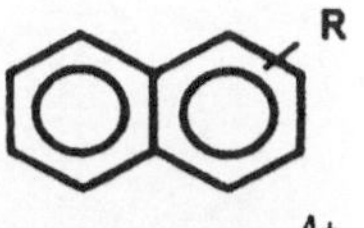

R

R = H   :   4.0 [kcal/Mol]
R = 1-OH :   4.3
R = 2-OH :   3.9

ΔG° with CP66^{4+}
in 20 % MeOD

Scheme 3

SOLVENT EFFECTS ON COMPLEXATION CONSTANTS: A PROBE FOR HYDROPHOBIC BINDING CONTRIBUTIONS

The increase of complexation free energies ΔG° with the water content in the applied solvents provides a quantitative measure for hydrophobic binding contributions; ΔG° correlates well [6,10] with independent parameters for solvent hydrophobicity such as Sp values [11] (Figure 1) derived from free enthalpies of transfer of e.g. tetramethyltin from gas to a given solvent [6]. Such a linear increase of ΔG° of complexation with Sp is found even with a tetraphenolate-ammonium salt complex [12] (Figure 2), which has been shown [12] to be dominated by electrostatic attraction (see below). In the plots of log K vs. Sp the slope measures the sensitivity of the observed binding against solvophobicity changes, and the abscissa log K_0 (for Sp = 0) reflects the complexation energy in a hydrocarbon-like environment,which must be negative (repulsive) if the binding is only of a hydrophobic nature.

67

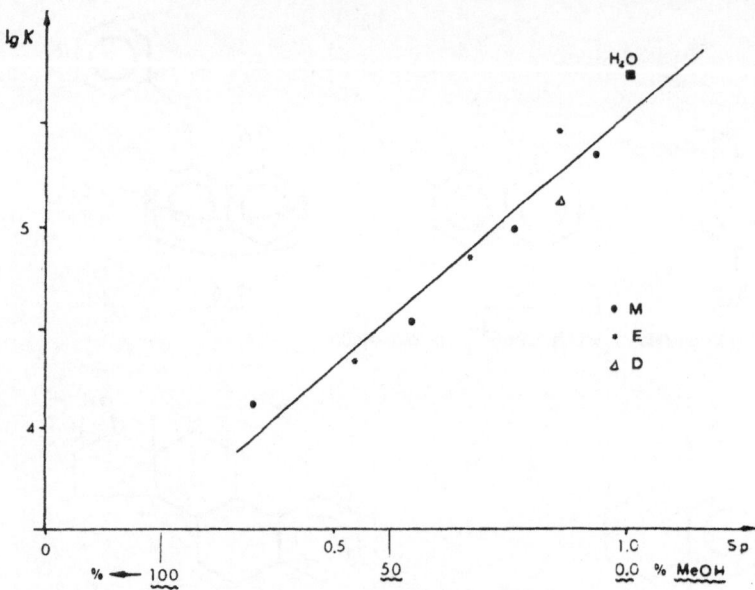

Figure 1. Solvent effects on equilibrium constants K (log K vs. Sp); solvents: M methanol, E ethanol, D dioxane, with different water content; measured with ANS and CP66 [10].

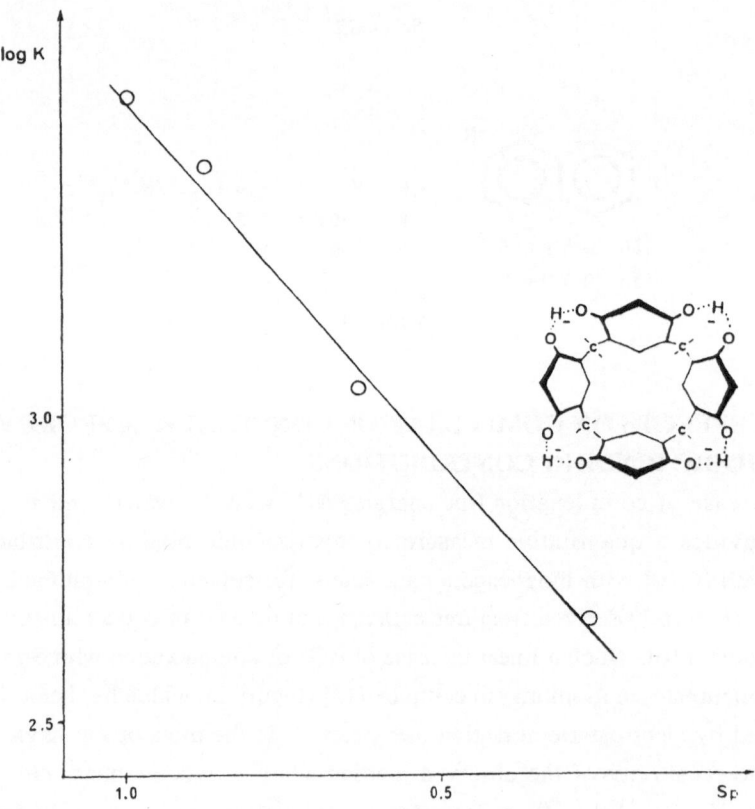

Figure 2. Solvent effects as in Figure 1, measured with tetraethylammonium bromide and the tetraphenolate TP as host [10].

Large sensitivities $\underline{a}$ should correspond to small or negative log K_O values. The almost linear correlation (Figure 3) found between $\underline{a}$ and log K_O indicates that these independent parameters can be used to characterize the hydrophobic binding contributions, which, for the investigated systems, reach a maximum for cyclodextrin complexes [11].

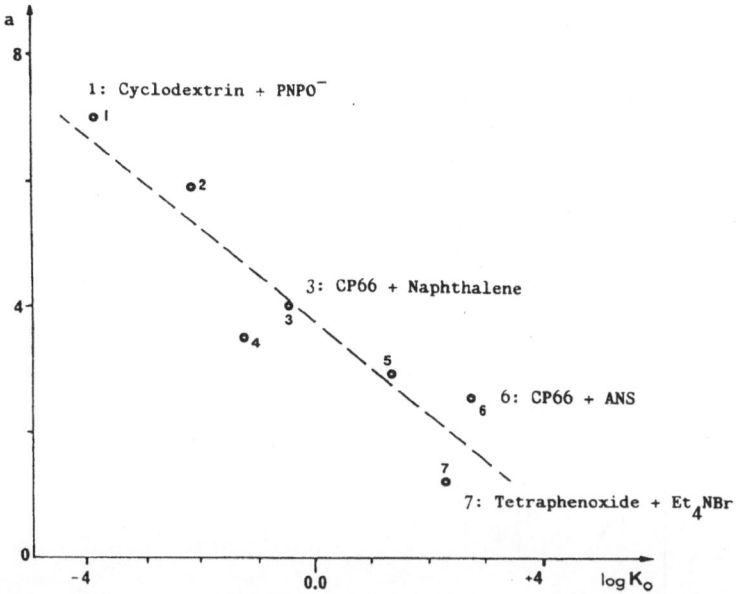

Figure 3. Correlation of sensitivies (slopes $\underline{a}$) vs. abscissas (log K_O) from all measured solvent effect plots (such as Figures 1, 2) [10].

ELECTROSTATIC COULOMB INTERACTIONS

Electrostatic interactions between opposite charges on host and guest molecules have long been used as one of the possible major driving forces for complexation [13,14]. Quantitative analyses of the observed binding constants with the different structures, however, have to our knowledge so far not been carried out, although theoretical approaches to such interactions are available even for proteins [15].

Host-guest complexes with the azoniacyclophanes CPnn such as shown in Schemes 1 and 3 are complicated by the presence of both hydrophobic and electrostatic interactions. If we assume both to be additive, and substract the $\Delta G°$ values observed e.g. for electroneutral naphthalene substrates from derivatives containing 1 or 2 negative charges we obtain 1.3 ± 0.2 kcal/mol for one ion pair formation or so-called salt bridge [16]. This value becomes smaller if the geometry does not allow optimal contact between host and guest charge with simultaneous preservation of an optimal pseudoequatorial inclusion of the naphthalene moiety.

The tetraphenolate TP (Figure 2, Scheme 4) shows very strong binding of cholin-type substrates, almost exclusively due to Coulomb forces. This can be seen from the large $\Delta G°$ difference found between R^+NMe_3 and electroneutral $RCMe_3$ substrates (Scheme 4), from salt effects showing coefficients as expected from Debye-Hückel theory (Figure 4) and from the distance - $\Delta G°$ correlation discussed below. If we again assume additivity, now for

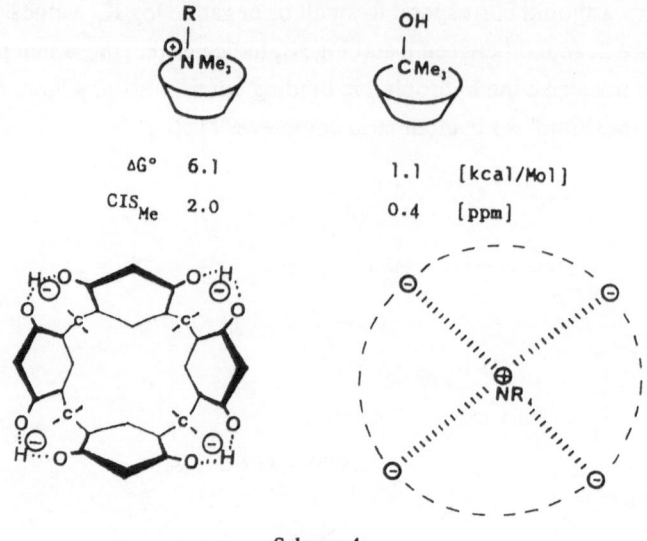

Scheme 4

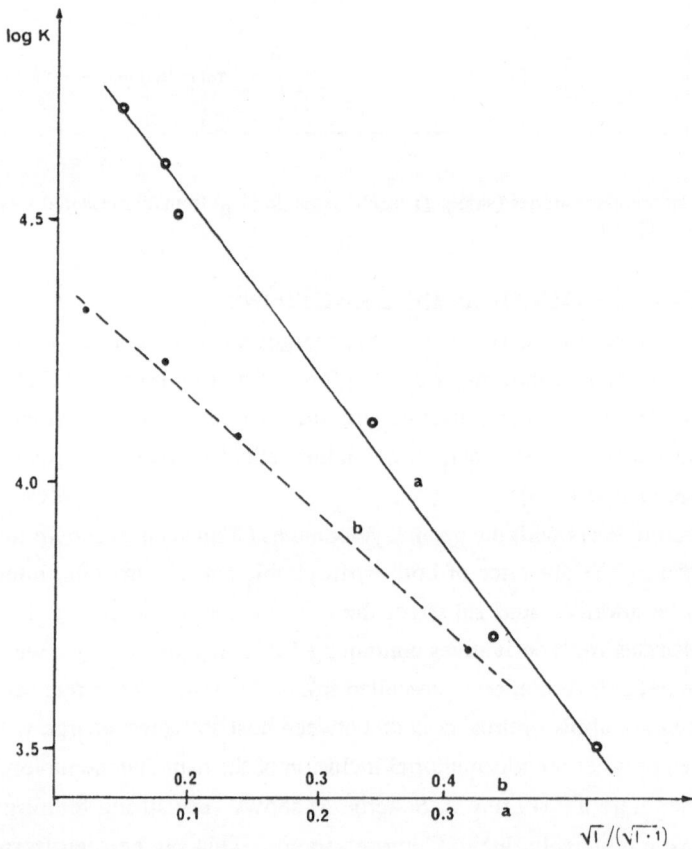

Figure 4. Debye-Hückel correlation of salt effects on equilibrium constants [10]; a: measured with tetraphenolate TP and tetraethylammonium bromide (slope for dilute solutions m = -4.1, theoretically m = -4.07 for z = -4 and z = +1; b: measured with CP66 and 2,4-nitronaphtholate(-1), (in contrast to (a), electrostatic interactions do not dominate; m = -1.0, theoretical value for exclusive electrostatic binding, m = -4.07).

each interaction between the ammonium ion center and four surrounding negative charges, we arrive at 5.0/4 = 1.25 kcal/mol per single salt bridge. Further we analyzed the beautiful series of complexes between carboxylates or phosphates and protonated aza-crown ether derivatives described by Lehn et al. [13] by the same mechanism, counting each ion-ion interaction separately as long as the oppositely charged ion-centers can approach each other. Although this approach neglects any differences in ion-ion distances, as well as in diameters, in polarizabilities, or in desolvation energies for the participating ions one obtains a fairly linear correlation with a slope of again 1.3 kcal/mol (Figure 5), pointing to a quite general applicability of the approach [16].

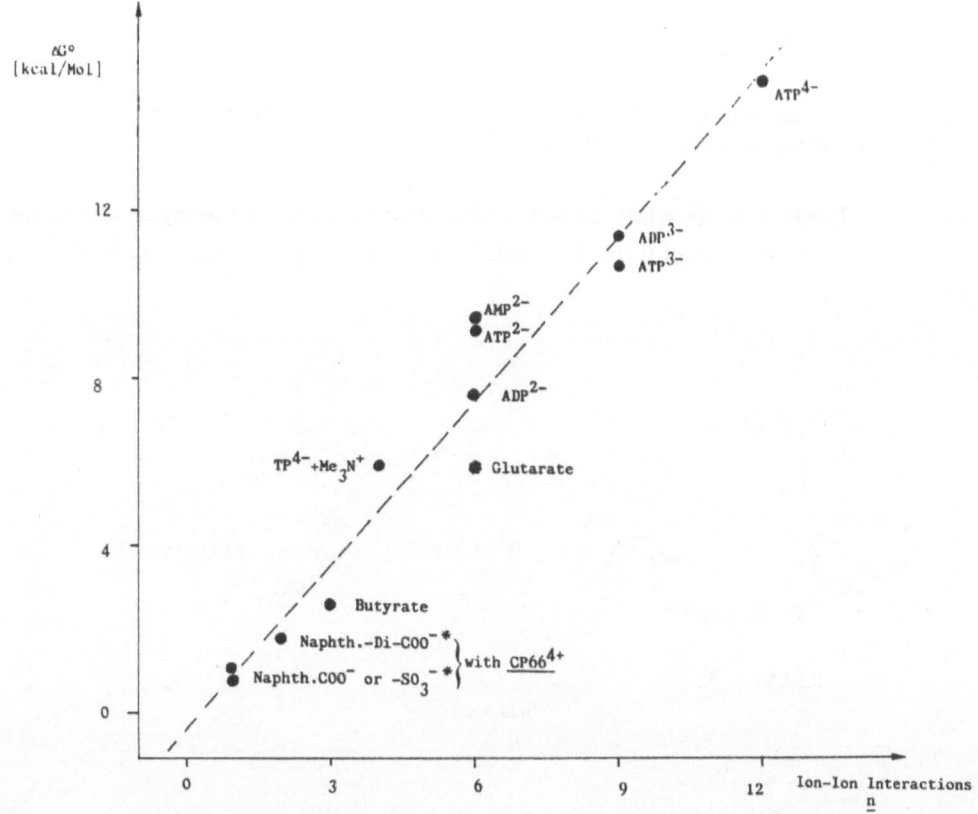

Figure 5. Correlation of complexation $\Delta G°$ values (kcal/mol) vs. the number of ion-ion interactions [16] (see text); r = 0.975, y = 6% (ideal values r = 1.000, y = 0%).

What happens if a close contact between the ions forming a salt bridge is geometrically hindered? For the conformationally less flexible complexes between the tetraphenolate TP and the tetralkylammonium chloride $R_4N^+Cl^-$ (Scheme 4) one can calculate the distance r between N^+ and the negative charge at the oxygens from models; the plot (Figure 6) of the $\Delta G°$ values observed for the compound $R_4N^+Cl^-$ with increasing chain length of R against 1/r shows a Coulomb correlation indicating an effecive dielectric constant $\epsilon = 35$ between typical lipophilic

(kcal/Mol)
ΔG°
7
5
0
-3
Me
Et
nPr
n-Bu
0.25
0.30
1/r (1/Å)

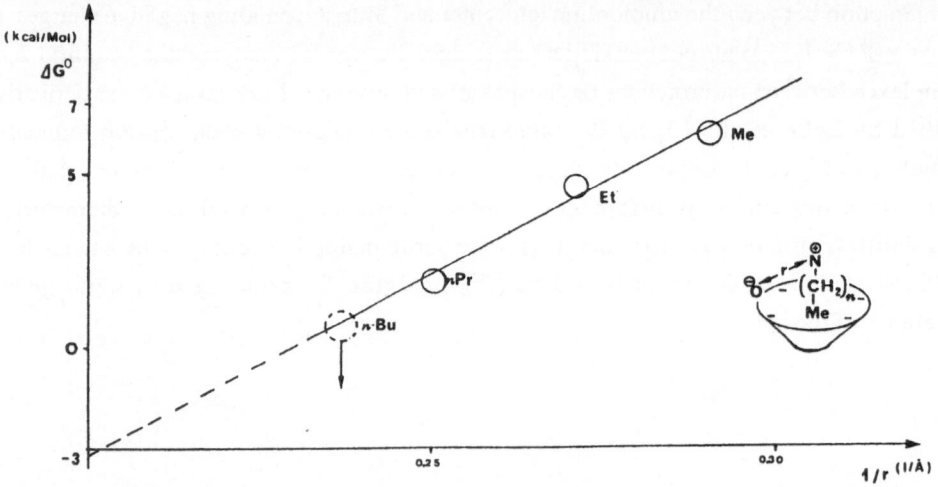

Figure 6. Coulomb correlation of complexation ΔG° values (kcal/mol) as a function of the distance (r^{-1}) separating the charges in host and guest [12].

moieties (ε = 2) and water (ε = 80). Similar ΔG° attenuations are seen in the complexation of positively charged substrates in cyclophanes bearing negative charges outside the cavity (Scheme 5) [17].

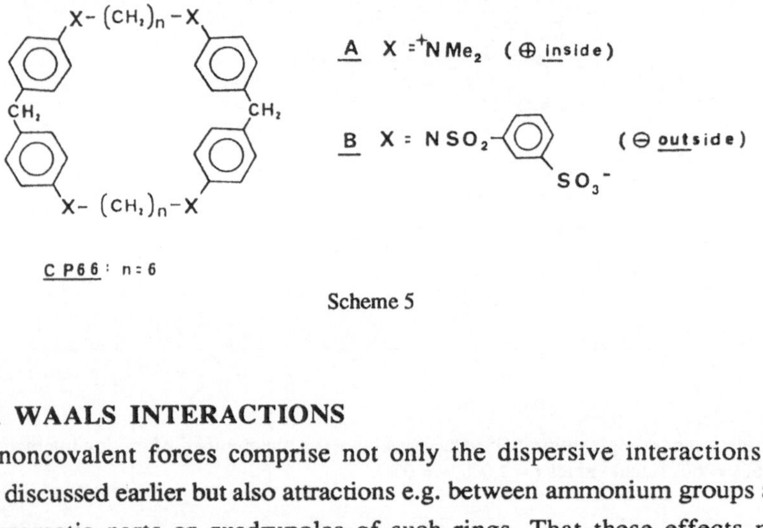

X- (CH2)n -X
CH2
CH2
X- (CH2)n -X
CP66 : n=6
A X = +NMe2 (⊕ inside)
B X = NSO2
SO3-
(⊖ outside)

Scheme 5

VAN DER WAALS INTERACTIONS

These noncovalent forces comprise not only the dispersive interactions with a r^{-6} dependence discussed earlier but also attractions e.g. between ammonium groups and induced dipoles at aromatic parts or quadrupoles of such rings. That these effects must play a significant role is obvious from the association constants of e.g. naphthalene, which are higher with the cyclophane A bearing positive charges inside the cavity (Scheme 6) - an unexpected result in view of the much higher lipophilicity of the corresponding cyclophane B which has no hydrophilic groups in the cavity. It is difficult to exclude e.g. electron donor-acceptor effects as another possible source for these variations as long as both receptor and substrate

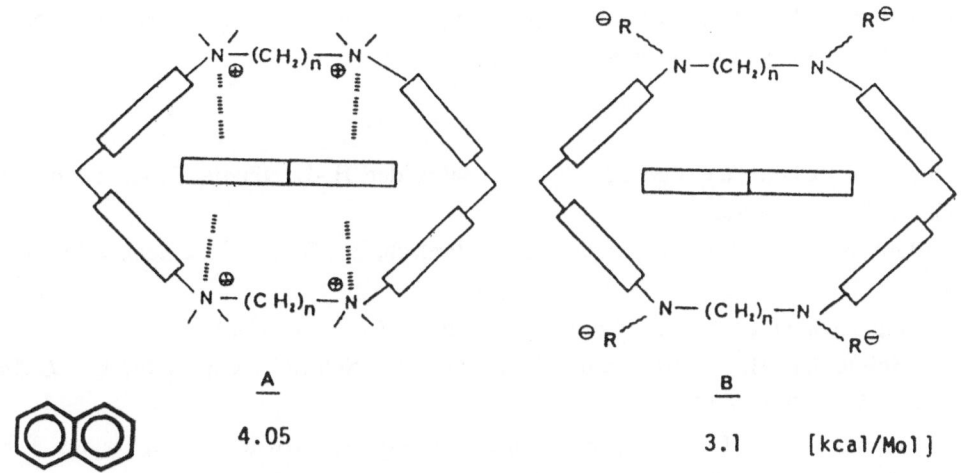

A
4.05

B
3.1   [kcal/Mol]

Scheme 6

contain aromatic parts in close contact [18]. Recent results, however, comparing aromatic with saturated substrates in their association behavior towards the macrocycles A and B (Scheme 7) clearly indicate the above-mentioned van der Waals effects as significant in the binding of aromatic substrates. The unusual binding increase generated by the phenyl ring in ANS (Scheme 2) finds its explanation in the possible approach of this aromatic ring to the N^+ atom of the host. The phenyl group here extends out of the cavity as indicated also by the observed small complexation induced NMR shifts. Noticeably, corresponding potentials are not included in most of the currently used simulation programs, although their presence has been identified both in proteins [19] as well as in some gas-phase measurement and calculations [20].

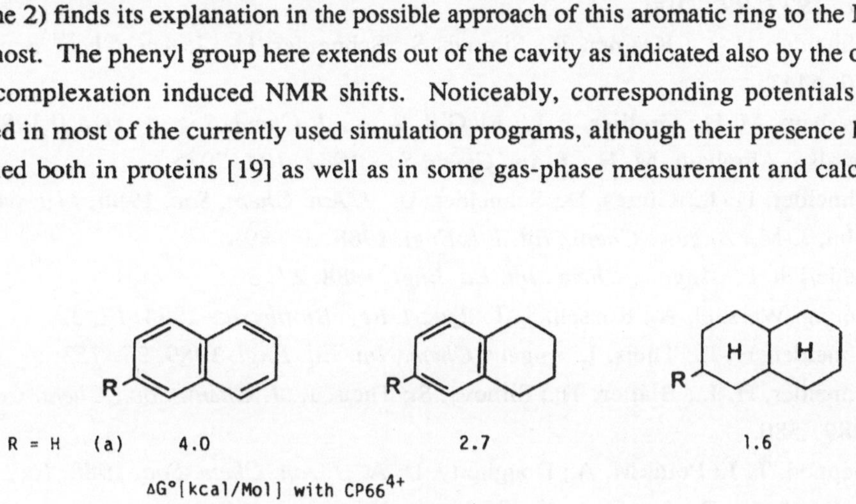

R = H   (a)   4.0   2.7   1.6

$\Delta G°$[kcal/Mol] with CP66^{4+}

Scheme 7

CONCLUSIONS

It is hoped that the application of physical-organic methods and the planned variation of host-guest structures will help us to understand better how biological systems operate on a molecular basis, and to improve calculational techniques. The relatively quite simple quantification of non-covalent forces should also contribute significantly to the rational design of drugs, of efficient host compounds, and of enzyme-analog catalysts.

Acknowledgements. Our work was supported by the Deutsche Forschungsgemeinschaft, Bonn, and the Fonds der Chemischen Industrie, Frankfurt.

References

1. Host-Guest Chemistry, Part 22. Part 21, Schneider, H.-J..; Juneja, R. K.; Simova, S. *Chem. Ber.*, in press.
2. Badertscher, M.; Welti, M.; Portmann, P.; Pretsch, E. *Top. Curr. Chem.* **1986**, *136*, 17.
3. See e.g. Schneider, H.-J..; Pöhlmann, J. *Bioorg. Chem.* **1987** *15*, 183.
4. Cf. Schneider, H.-J..; Buchheit, U.; Becker, N.; Schmidt, G.; Siehl, U. *J. Am. Chem. Soc.* **1985**, *107*, 7287.
5. See e.g. Collet, A. In *Inclusion Compounds*, Vol. 2; Atwood, J. L.; Davies, J. E. D.; MacNicol, D. D. Eds.; Academic Press: London, 1984; Chap. 4.
6. Schneider, H.-J..; Philippi, K.; Pöhlmann, J. *Angew. Chem., Int. Ed. Engl.* **1984**, *23*, 908.
7. Odashima, K.; Koga, K. In *Cyclophanes*, Vol. II; Academic Press: New York: 1983; p. 629.
8. Kumar, S.; Schneider, H.-J.. *J. Chem. Soc., Perkin Trans. II* **1989**, 245.
9. Tabushi, I.; Shimizu, N.; Sugimoto, T.; Shiozuka, M.; Yamamura, K. *J. Am. Chem. Soc.* **1977**, *99*, 7100.
10. Schneider, H.-J..; Kramer, R.; Simova, S.; Schneider, U. *J. Am. Chem. Soc.* **1988**, *110*, 6442.
11. Abraham, M. H.; Grellier, P. L.; McGill, R. A. *J. Chem. Soc., Perkin II* **1989**, 339; see also, Abraham, M. H. *J. Am. Chem. Soc.* **1982**, *104*, 2085.
12. Schneider, H.-J..; Güttes, D.; Schneider, U. *J. Am. Chem. Soc.* **1988**, *110*, 6449.
13. Lehn, J.-M. *Angew. Chem., Int. Ed. Engl.* **1988**, *27*, 89.
14. Diederich, F. *Angew. Chem., Int. Ed. Engl.* **1988**, *27*, 362.
15. See e.g. Warshel, A.; Russell, S. T. *Quart. Rev. Biophysics* **1984**, *17*, 3.
16. Schneider, H.-J..; Theis, I. *Angew. Chem., Int. Ed. Engl.* **1989**, *28*, 753.
17. Schneider, H.-J..; Blatter, Th.; Simova, S.; Theis, I. *J. Chem. Soc., Chem. Commun.* **1989**, 580.
18. Sheppod, T. J.; Petti, M. A.; Dougherty, D. A. *J. Am. Chem. Soc.* **1988**, *100*, 1983.
19. Burley, S. K.; Petsko, G. A. *FEBS Letters* **1986**, *203*, 139.
20. Meot-Ner, M.; Deakyne, C. A. *J. Am. Chem. Soc.* **1985**, *107*, 469, 474.

DESIGNED DNA INTERACTIONS

Kent D. Stewart

Department of Chemistry and Winship Cancer Center
Emory University and Emory School of Medicine
Atlanta, GA 30322 (USA)

This paper is dedicated to the memories of Myron L. Bender (1924-1988) and Emil T. Kaiser (1938-1988).

SUMMARY

Using a fluorescence-detected ethidium displacement array, the complexation properties of several tri- and tetracationic polyamines with natural and synthetic polynucleotides were determined. The compounds exhibited different binding affinities which were dependent upon the structure and charge of the polyamine compound. The GC/AT DNA binding selectivities were measured for several of the polyamines, and two of the tricationic derivatives (compounds **3** and **4**) were observed to exhibit significant GC-selectivity. Computer modelling results were in agreement with a major groove spanning binding mode which possessed site specific interactions between the polyamines and DNA hydrogen bonding sites.

INTRODUCTION

There are approximately 40 compounds that are currently clinically approved for the treatment of cancer [1]. The mechanism of action of over half of these drugs is proposed to be direct interaction with deoxyribonucleic acid (DNA) in the tumor cell. Much more effective anticancer drugs are desperately needed for improved chemotherapeutic treatment of cancer. My research is directed at advancing current understanding of the interaction of organic molecules with DNA in the hope that improved understanding of this process will lead to future design of better chemotherapeutic drugs.

While many examples of DNA intercalating or minor groove binding organic molecules are known, there are few examples of organic molecules which bind selectively in the major groove of DNA [2]. The two Watson-Crick purine/pyrimidine base pairs, adenine-thymine, AT, and guanine-cytosine GC, show different hydrogen bond acceptor and donor patterns in the major groove. DNA-binding proteins have amino acid side chains that provide complementary hydrogen bond donor and acceptor groups which make protein DNA-sequence

Table 1. Polyamines binding to DNA.

POLYAMINE COMPOUND	C50 VALUES			C50 AT / C50 GC
	poly d(AT)	CT-DNA	poly d(GC)	
1 (NH₂...NH₂)	5600	2300	2200	2.5
2 (H₂N...NH...NH₂)	80	41	18	4.4
3 (H₂N...NH...NH...NH...NH₂)	50	17	6.0	8.3
4 (H₂N...NH...NH...NH₂)	55	20	5.3	10.4
5 (H₂N...NH...NH...NH₂)	2.8	1.6	1.2	2.3
6 (NH₂...NH...benzene...NH...NH₂)	2.2	-	0.9	2.4
7 (NH₂...NH...OCH₃/CH₃ benzene...NH...NH₂)	2.4	2.2	2.1	1.1
8 (NH₂...NH...pyridine N...NH...NH₂)	3.8	-	1.5	2.5

All binding C50 values are in micromolar concentration (μM) and were measured in 2 mM HEPES, 8 mM NaCl, 0.05 mM EDTA, pH 7.0.

recognition possible through selective binding of GC or AT base pairs [3]. Although unknown at this time, organic molecules which make use of similar hydrogen bond complementary patterns in the major groove of DNA are in principle possible. For major groove binders to be selective for GC base pairs, this proposed class of compounds should have a relatively high cationic charge. This charge will be complementary to the calculated lower electrostatic potential in the major groove of a GC base pair relative to an AT base pair [4].

The polyamines are a cationic class of molecules known to bind to anionic receptors through primarily electrostatic effects [5]. The common polyamines, putrecine, spermidine, and spermine (compounds 1,2 and 5, respectively, see Table 1) have a number of metabolically important functions and they are believed to play significant roles in normal and malignant cell proliferation [6]. Some combination of groove and phosphate interactions is currently considered to explain polyamine binding to DNA [7]. Given the importance of designing base pair specificity into new DNA binding compounds [8], it is appropriate to more fully evaluate the DNA binding affinities and selectivities shown by the polyamine class of compounds. In this paper, the structure/function relationship of polyamine recognition of DNA base pairs is investigated by use of a fluorescence-detected ethidium displacement assay.

EXPERIMENTAL

All compounds were either purchased (Aldrich) or prepared according to a published procedure [9]. New compounds gave satisfactory nuclear magnetic resonance, mass spectral, and combustion analysis data. Poly d(AT) and poly d(GC) were purchased from Sigma. Calf thymus DNA was the generous gift of Professor W. D. Wilson. The relative binding affinities of the polyamines for DNA were measured by fluorescence spectroscopy according to a literature procedure [10].

The energy change, ΔE, for the complexation of spermine and compound 4 with oligo d(AT)$_5$ and oligo d(GC)$_5$ was estimated by computer simulation according to literature precedent [7b]. The united-atom AMBER force field [11] as implemented in the MacroModel software program [12] was used for all calculations.

RESULTS AND DISCUSSION

The ethidium displacement assay has been shown to be very useful in screening the DNA binding properties of hundreds of synthetic compounds [13]. Table 1 shows the C_{50} ethidium displacement values which were determined for polyamines 1-8 binding to poly d(AT), calf thymus DNA (CT-DNA) and poly d(GC). (Note that in this assay, lower C_{50} values imply tighter DNA binding). The ratio of the C_{50} values for poly d(AT) and poly d(GC) is shown in the final column in these tables. The relative C_{50} values observed for putrecine, spermidine and spermine (compounds 1, 2, and 5, respectively, see Table 1) correlate with the known literature dissociation constants for these compounds [9,10b].

Spermine, compound 5, is tetracationic at neutral pH and has been reported as having either no base pair binding selectivity [14] or GC selectivity [15]. Under the conditions of the ethidium displacement assay of this work, spermine and several spermine analogs, compounds 6, 7 and 8 were observed to have no or slight GC preference in binding to DNA. A crystallographic study of spermine bound to an oligomer of DNA [16] indicates that the polyamine spans the major groove of a GC base pair making both phosphate and groove interactions. The electrostatic interactions between DNA and this series of derivatives must be so dominant over other binding forces (hydrophobic, hydrogen bonding, etc.) that perturbation of the binding character is not observed with the structural changes introduced into the spermine backbone in compounds 6-8. This observation indicated that polyamines possessing hydrogen bonding groups but having total molecular charge less than +4 might exhibit significant DNA-binding selectivity.

According to published pK_a values, spermidine, 2, and compounds 3 and 4 (reference 17, 18, and 19, respectively) have +3 charges in neutral pH aqueous solutions. From studies of spermidine's effects upon the thermal denaturation of various sequences of DNA, it was concluded that any sequence specificity exhibited by spermine was not large [20]. Compounds 3 and 4 are spermidine analogs possessing 1 and 2 uncharged amino groups, respectively. Compound 3 is protonated mainly at the terminal and middle nitrogens (N1, N7 and N13) and compound 4 is protonated at the terminal and one of the central nitrogens (N1, N5 and

Table 2. Compound **4**, nerenil, **9**, and netropsin, **10**, binding to DNA.

COMPOUND	C_{50} VALUES			$\dfrac{C50 \quad AT}{C50 \quad GC}$
	poly d(AT)	CT-DNA	poly d(GC)	
4	55	20	5.3	10.4
9	0.60	1.5	6.6	0.09
10	0.55	-	15	0.04

All binding C_{50} values are in micromolar concentration (μM) and were measured in 2 mM HEPES, 8 mM NaCl, 0.05 mM EDTA, pH 7.0. Data for berenil are from this work and are in agreement with published values (13). Data for netropsin are taken from the literature (13).

N12). The unprotonated amine groups in **3** and **4** increase their affinity for calf thymus DNA relative to spermidine, **2**, as indicated by their lower C_{50} values. The more striking observation about the DNA binding properties of compounds **3** and **4** is their pronounced GC-selectivity in binding (8- and 10-fold, respectively). The uncharged amino groups in these two compounds may act as either hydrogen bond donors or hydrogen bond acceptors for sites in the grooves of DNA. As shown in Table 2, the GC selectivity is opposite to the AT selectivity of other groove binding compounds. The C_{50} values of two well known groove binding compounds, berenil and netropsin are listed in this table. These two compounds show 10- and 25-fold AT selectivities (0.09 and 0.04-fold GC-selective, respectively), respectively, in the ethidium displacement assay used in this work. The opposite direction of the binding selectivities by these classes of compounds is striking and investigations of the specific interactions involved are continuing.

To aid in these future studies, theoretical calculations of the interaction of spermine or compound **4** with a decamer of either alternating GC or alternating AT were carried out using molecular mechanics minimization methods according to literature precedent [7b]. Shown in Table 3 are the changes in energy, ΔE, for these polyamines complexing to DNA in several possible binding modes. In these calculations, three polyamine binding modes were considered:

Table 3. Molecular Mechanics calculations of polyamine-DNA binding energies.

	Major Groove	Minor Groove	Phosphate
d(GC)$_5$:SPM	-344	-261	-290
d(GC)$_5$:cmpd.4	-268	-231	-
d(AT)$_5$:SPM	-335	-322	-259
d(AT)$_5$:cmpd.4	-248	-213	-

All ΔE values in kcal/mole, SPM=spermine.

mode 1: the polyamine spanned the major groove in a manner similar to that observed in the spermine CGCGAATTCGCG crystal structure [16], mode 2: the polyamine bound along the minor groove in a manner similar to that observed for netropsin [21], and mode 3: the polyamine is not bound within either groove but is aligned with the phosphate backbone. The most favored binding position for each compound is an orientation in which the polyamine spans the major groove of a d(GC)$_5$ duplex (mode 1). Less favored binding modes are minor grove and phosphate binding orientation (modes 2 and 3). Specific interaction in the complexes of compound **4** and oligo d(GC)$_5$ predicted by the modelling to be low energy binding arrangements are shown in the Figure. In these arrangements, the uncharged amine of the polyamine either accepts a hydrogen bond from a cytosine 4-NH$_2$ group or donates a hydrogen bond to a guanine N7 position. These specific interactions could lead to a greater affinity of compound **4** for a GC base pair than for an AT base pair, in agreement with the experimental results discussed above.

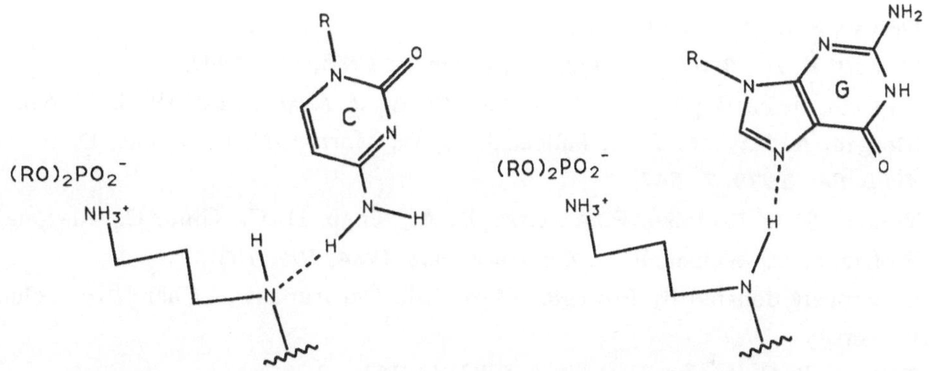

Figure 1. Diagram of the proposed site specific interactions between compound **4** and DNA (only part of compound **4** is shown).

In conclusion, a survey of the DNA binding affinities and selectivities of several polyamines has been presented. As expected, DNA affinity increases with increasing cationic charge of the polyamine. Two of the derivatives, compounds **3** and **4**, exhibited significant GC binding selectivity. With the aid of computer modelling techniques, a binding mode in the DNA major groove is proposed. Although further experiments are necessary, the research reported here indicates that it is possible for organic molecules to selectively recognize GC base pairs solely by groove interactions.

Acknowledgements. The author acknowledges the finanical support of the Donors of the Petroleum Research Fund, administered by the American Chemical Society, the Olin Corporation Charitable Trust Grant of Research Corporation, the Emory University Research program, and the Winship Cancer Clinic of Emory University. The high-field NMR and mass spectrometers used in these studies were made possible through equipment grants from the NSF and NIH.

References

1. Farmer, P. B.; Walker, J. M. Eds. *The Molecular Basis of Cancer*, Wiley: New York, 1985.

2. Saenger, W. *Principles of Nucleic Acid Structure*, Springer-Verlag: New York, 1984.

3. Schleif, R. *Science* **1988**, *241*, 181.

4. Lavery, R.; Pullman, B. *J. Bio. Struct. Dynam.* **1985**, *2*, 1021.

5. (a) Tabor, C. W.; Tabor, H. *Ann. Rev. Bioch.* **1984**, *53*, 749; (b) Morris, D. R.; Marton, L. J. *Polyamines in Biology and Medicine*, Marcel Dekker, Inc.: New York, 1981.

6. (a) Pegg, A. E. *Cancer Res.* **1988**, *48*, 759; (b) Porter, C. W.; Sufrin, J. R. *Anticancer Res.* **1986**, *6*, 525.

7. (a) Feuerstein, B. G.; Marton, L. J. In *The Physiology of Polyamines*, Bachrach, U.; Heimer, Y. Eds., in press; (b) Feuerstein, B. G.; Pattibiraman, N.; Marton, L. J. *Proc. Nat. Acad. Sci. USA* **1986**, *83*, 5948; (c) Zakrzewska, K.; Pullman, B. *Biopolymers* **1986**, *25*, 375.

8. Dervan, P. B. *Science* **1986**, *232*, 464.

9. Stewart, K. D. *Bioch. Biophys. Res. Commun.* **1988**, *152*, 1441.

10. (a) Cain, B. F.; Baguley, B. C.; Denny, W. A. *J. Med. Chem.* **1978**, *21*, 658; (b) Morgan, A. R.; Lee, J. S.; Pullebank, D. E.; Murray, N. L.; Evans, D. H. *Nuc. Acids Res.* **1979**, *7*, 547.

11. Weiner, S. J.; Kollman, P. A.; Case, D. A.; Singh, U. C.; Ghio, C.; Alagona, G.; Profeta, S. Jr.; Weiner, P. *J. Am. Chem. Soc.* **1984**, *106*, 765.

12. Generously donated by Professor Clark Still, Department of Chemistry, Columbia University.

13. Baguley, B. C. *Molec. Cell Bioch.* **1982**, *43*, 167.

14. Hirshman, S. Z.; Leng, M.; Felsenfeld, G. *Biopolymers* **1967**, *5*, 227.

15. Igarashi, K.; Sakamoto, I.; Goto, N.; Kashiwagi, K.; Honma, R.; Hirose, S. *Arch. Bioch. Biophys.* **1982**, *219*, 438.

16. Drew, H. R.; Dickerson, R. E. *J. Mol. Biol.* **1981**, *151*, 535.

17. Kimberly, M. M.; Goldstein, J. H. *Anal. Chem.* **1981**, *53*, 789.

18. Reilly, C. N.; Vavoulis, A. *Anal. Chem.* **1959**, *31*, 243.

19. Paoletti, P.; Fabbrizzi, L.; Barbucci, R. *Inorg. Chem.* **1973**, *12*, 1861.

20. Morgan, J. E.; Blankenship, J. W.; Matthews, H. R. *Arch. Bioch. Biophys.* **1986**, *246*, 225.

21. Goodsell, D.; Dickerson, R. E. *J. Med. Chem.* **1986**, *29*, 727.

A NOVEL APPLICATION OF THE HOST-GUEST PARADIGM: DESIGN OF ORGANIC OPTOELECTRONIC MATERIALS

David M. Walba†*, Noel A. Clark§, Homaune A. Razavi†,
and Devendra S. Parmar§

†Department of Chemistry and Biochemistry
§Department of Physics and Optoelectronic Computing Systems Center
University of Colorado
Boulder, Colorado 80309-0215 (USA)

SUMMARY

A ferroelectric polarization **P** (permanent, spontaneous electric dipole moment) is predicted for certain non-racemic liquid crystals based upon simple symmetry arguments. The magnitude of this polarization is a major factor in determining the electro-optic response times achievable in ferroelectric liquid crystal (FLC) light valves and spatial light modulators – devices with excellent potential utility in devices for optoelectronic computing and large-area, high resolution flat panel displays.

The polarization in FLCs is quite analogous to that of the well known pyroelectric organic crystals, and results from anisotropic, polar orientation of molecular dipoles in the fluid phase. While the symmetry arguments are compelling, insight into the actual molecular origins of **P** is necessary if new materials with high polarization and fast electro-optic response are to be designed in a directed way. Based on the concept that ferroelectricity in FLCs is a manifestation of a novel form of molecular recognition occurring in the FLC phase, we have developed a simple stereochemical model for **P** allowing prediction of the sign (handedness) and magnitude of the polarization for certain compounds. The design, synthesis and ferroelectric properties of several new materials are described in the context of this model.

INTRODUCTION – FERROELECTRIC LIQUID CRYSTALS

Very often, organic materials crystallize such that there is an anisotropic, polar distribution of molecular dipoles, and a crystal possessing a permanent, macroscopic electric dipole moment (polarization) results. Such crystals are often termed pyroelectric, since temperature changes cause changes in the magnitude of the spontaneous polarization. If the direction of the macroscopic polarization is switchable by application of an external field, then the material is termed ferroelectric [1]. While electrically polar organic crystals are very common, no pyroelectric or ferroelectric fluids were known until recently.

Inclusion Phenomena and Molecular Recognition
Edited by J. Atwood
Plenum Press, New York, 1990

Clearly, a ferroelectric material must be anisotropic. In fact, anisotropic fluids have been known since the 19th century. Neat liquids exhibiting anisotropic molecular order are termed thermotropic liquid crystals (LCs). In general, LCs possess rod-shaped structures, although this is not always the case; for example disk-shaped molecules have recently been shown to form ordered liquids called discotics. For the classic rod-shaped LC structures, many types of phases have been characterized, differing in the way the molecular rods are ordered. In all phases, the rods are parallel; the long axis of the rods is termed the LC director.

Interestingly, while polar order is possible in any LC phase, until 1976 no liquid crystal material had ever been shown to be ferroelectric. Thus, even though the individual molecules in the LC phase may possess the symmetry of an arrow, spontaneous orientation of the arrows such that an excess of arrows is pointing in one direction has never been observed. The arrows are parallel (or antiparallel), but they never spontaneously orient in a polar fashion. There is no fundamental reason why such order cannot occur, but empirically, for all LC phases known, all properties of the phase are invarient with the sign of the director.

Molecular orientation in the most ordered of the "monomeric" LC phases, the smectic C phase, is represented schematically in Figure 1. In this phase, the molecules are arranged in layers, with the director ($\underline{n}$) tilted with respect to the layer normal ($\underline{z}$) by the tilt angle (θ). The plane containing $\underline{n}$ and $\underline{z}$ is termed the tilt plane. Within the layers, the centers of mass of the molecules behave as a two-dimensional liquid. The layered structure and director tilt of smectic C phases have been experimentally determined by X-ray scattering experiments and by optical microscopy. Given the layering and the tilt, on the time average the phase can possess a maximum of two symmetry elements: A C_2 axis normal to the tilt plane and a σ plane congruent with the tilt plane. Since no smectic phase has ever been shown to be ferroelectric, it can be stated that all smectic phases studied in fact possess the maximum possible symmetry, since breaking either of these symmetry elements would afford polar order, and ferroelectricity.

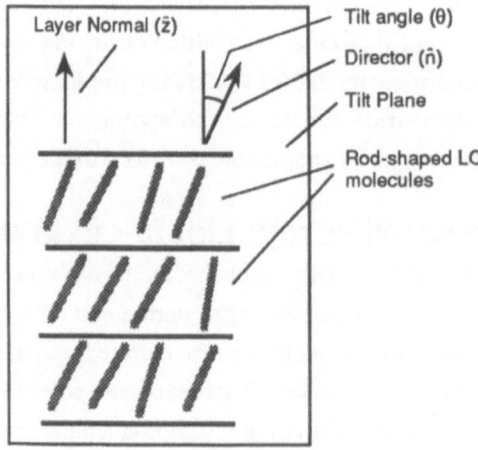

Figure 1. Molecular orientation in the smectic C phase.

In order to achieve a ferroelectric smectic C phase, either the C_2 axis or the σ plane must be removed. Breaking the C_2 symmetry is equivalent to obtaining spontaneous order along the director. This is certainly a worthy and highly interesting goal, but has yet to be achieved in any LC phase. However, it is quite simple to break reflection symmetry in any organic material by simply making the molecules chiral and non-racemic. This concept was first pointed out by Meyer in 1974 and may be stated as follows: The maximum number of time-average symmetry elements possible in a chiral smectic C phase (termed a C* phase) is one – the C_2 axis normal to the tilt plane. Such a phase has polar symmetry, and may possess a ferroelectric polarization P oriented along the C_2 axis (and only along this axis). The polarization is invariant with sign of the director, being oriented normal to the director. Thus, a C* phase must possess polar symmetry and should be a ferroelectric liquid crystal (FLC) [2].

Given the empirical fact that all achiral C phases are non-polar, and spontaneous breaking of the reflection symmetry about the tilt plane and the two-fold axis normal to the tilt plane does not occur with "standard" thermotropic LC structures, the ferroelectric polarization of a C* phase is a "chirality phenomenon" similar to optical activity. The polarization has a sign (positive or negative, which could just as easily be right-handed or left-handed) and if one enantiomer of a C* material possesses, e.g. positive P, the other enantiomer must possess a negative P of exactly equal magnitude.

A 3-D "slice" of C* phase, indicating the spatial relationship between the director, layer normal, and polarization vector is illustrated in Figure 2. By convention, if P for a material is in the direction of $z \times n$ (the cross product of the unit vectors z and n, which is of course normal to both) then that material possesses positive polarization. If P is opposed to $z \times n$, then that material has negative polarization. The material indicated in Figure 2 thus has positive P. It must be noted that the physics convention for the direction of dipoles (from negative to positive) is used here for the direction of P.

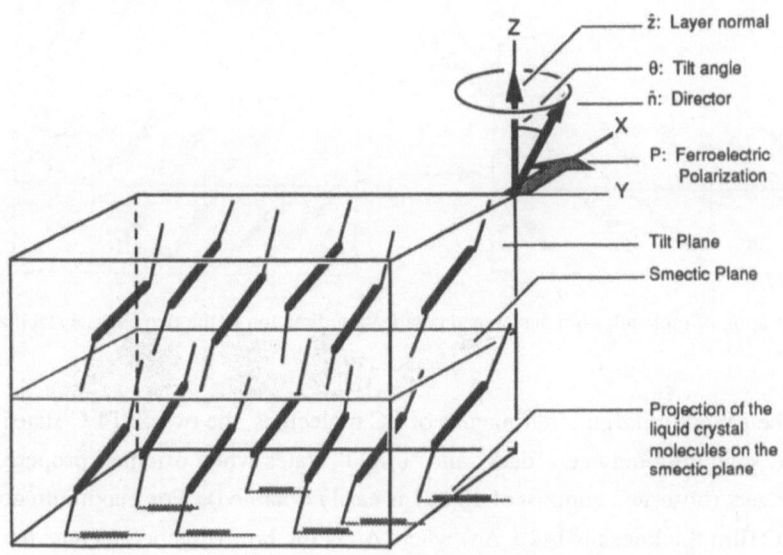

Figure 2. A 3-D "slice" of C* phase.

Immediately following Meyer's prediction, the first demonstrated FLC, DOBAMBC, was synthesized by Patrick Keller [3]. DOBAMBC possesses the phase sequence indicated above, and is ferroelectric between 76° and 95°C. As indicated, the C* phase super cools to 63°C. The measured polarization density at 76° is -3 nC/cm^2, or about -0.009 D/molecule. The DOBAMBC structure exemplifies the typical pattern for thermotropic LCs, with a rigid core (the benzylideneaminocinnamate unit) connecting two floppy tails.

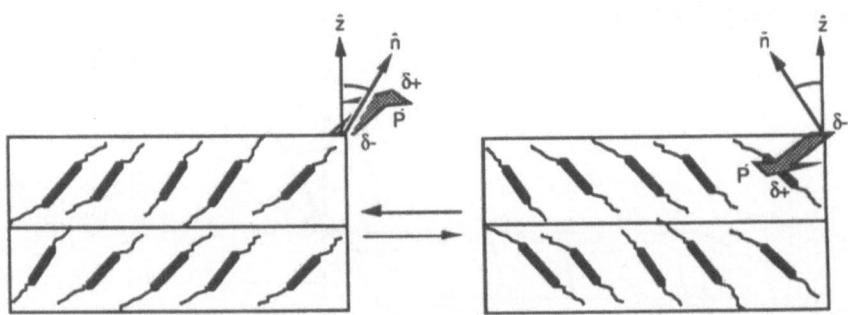

The discovery of FLCs triggered considerable activity among physicists to invent a fast electro-optic device based upon the novel materials. In 1980 Clark and Lagerwall disclosed their solution to this problem: the surface stabilized ferroelectric liquid crystal (SSFLC) light valve [4]. In an SSFLC cell, the molecules are oriented in a thin film between glass plates coated with a transparent conductor. The molecules will often prefer to orient such that the director is parallel to the surface of the plates. In this geometry, when the sample is aligned such that the layers are flat and parallel, then two molecular orientations, or states, are allowed, as indicated in Figure 3. In the absence of applied fields, these states are degenerate. But, upon application of an electric field (by charging the transparent electrodes) the state with its ferroelectric polarization aligned with the field becomes more stable than the other state, and the molecules will switch into the more stable state. Reversal of the sign of the applied field will cause the molecules to switch back.

Figure 3. Illustration of molecular orientation and polarization direction in the two surface stabilized states of an SSFLC cell.

Due to the generally large birefringence of LC molecules, the two SSFLC states can show high contrast, switching between "dark" and "bright" states when oriented properly between crossed polarizers (observed contrast of 1000/1 is easily obtained). For maximum contrast, the optimum FLC film thickness is $l/(2 \times \Delta n)$ where Δn is the birefringence. Thus, for a typical FLC material with a birefringence of 0.15, the optimum thickness for visible light switching is about 1.7 μm. When the applied field is above a certain value (generally about 5 V/μm), then there is no "barrier" for the switching process, and the switching speed is inversely proportional to the "driving force" for switching, given by the polarization density times the applied field strength (**P** x **E**). The switching process is slowed by an orientational viscosity η. The characteristic time constant for the cell, closely related to the observed electro-optic response to an applied voltage step, is thus given by Equation 1:

$$\tau = \frac{\eta}{P \times E} \qquad (1)$$

We estimate that the theoretical limit on switching speed for the SSFLC device is on the order of 10 nsec, or about 10^6 faster than the current generation of nematic liquid crystal light valves used in the fastest watch and computer displays. Materials available today show demonstrated high contrast electro-optic rise times on the order of 10 μsec in response to voltage steps of 15 V/μm at room temperature.

The combination of high speed, high contrast, sharp threshold for switching, low switching energy, bistability, and high achievable spatial resolution (≈ 2 μm) combine to make the Clark-Lagerwall device a unique and potentially very useful solution for the demanding requirements of some optoelectronic processing and information handling systems, especially those requiring high information content (high spatial resolution). These include flat panel display devices, and optically addressed spatial light modulators – a kind of transistor for optical computing.

In order to realize the potential of FLC devices, materials with the fastest possible switching speeds must be obtained. Thus, there is a clear need for new FLC materials with high polarization density and low orientational viscosity. Our recent work has focussed on the former. While the symmetry argument for the existence of a polarization in C* materials is compelling, in order to design in a directed way new materials with high polarization, some understanding of the molecular origins of **P** must be achieved. Our current model for the origins of **P** is presented herein, with recent results testing the model.

A MODEL FOR THE MOLECULAR ORIGINS OF THE POLARIZATION IN FLCS

Experimental measurement of a ferroelectric polarization in FLCs proves that molecular dipoles are anisotropically oriented in the phase in a polar fashion relative to the tilt plane. From the beginning, our goal has been to understand this phenomenon in terms of a kind of molecular recognition occurring in the FLC phase [5]. Thus, one may consider the molecules in a fluid to be oriented with respect to conformation and orientation in space by a "surface of constant molecular mean field" resulting from interactions with neighboring molecules. Instantaneously, this surface in general has a complex shape, and movement of molecules within the fluid is restricted. However, on the time average, the mean field surface, or "binding site", in an isotropic medium takes the shape of a sphere.

In untilted LC phases, the orienting binding site takes the shape of a cylinder. Intuition might suggest that in a tilted phase, the cylinder simply tilts over. However, we propose that the binding site actually changes shape, becomes a bent cylinder, as illustrated in Figure 4.

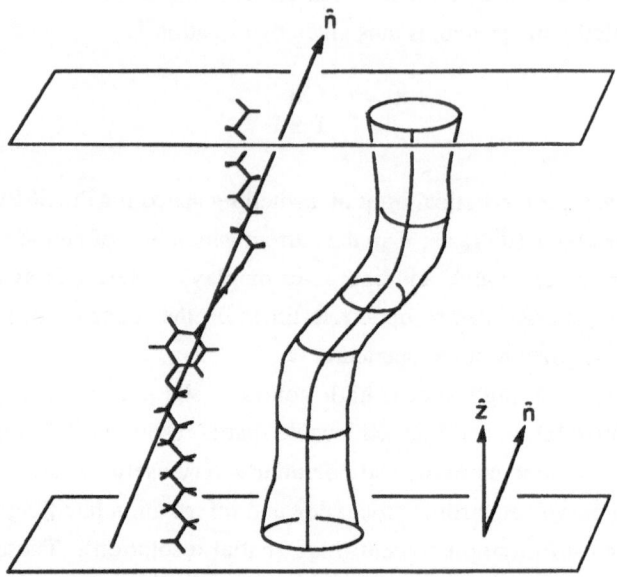

Figure 4. Illustration of the C phase "binding site".

This bent cyclinder binding site possesses a C_2 axis, and is oriented relative to the tilt plane (thus relative to the glass bounding plates) as indicated in the Figure. Thus, our model states that the FLC phase behaves as a "crystal" lattice with cavities taking the shape of the bent cyclinder. LC molecules bound within such cavities are oriented as indicated in the Figure, and rotation of a molecule about the director within the cavity gives rise to rotational states which are diastereomeric on the time average. It is the diastereomeric nature of rotational states relative to the tilt plane that ultimately gives rise to the observed polarization.

Our proposed preferred conformation and rotational orientation of an alkoxyphenyl-nonylbenzoate is shown in Figure 5. In this view, the tilt plane is almost normal to the plane of the page, the layer normal is vertical, and the molecules are tilting back (and a bit to the right). Note that the rotational orientation occurs even for achiral materials. In the Figure, the benzoate carbonyl dipole is pointing to the right (recall that dipoles aim from negative to positive), suggesting a polar orientation of dipoles in that direction. However, for the achiral material there is an exactly equal number density of molecules with the carbonyl pointing in the opposite direction, and no net polarization is observed. Again, this is simply an empirical fact for all known smectic C materials, there is no fundamental reason why the carbonyls cannot spontaneously achieve a polar orientation in the phase.

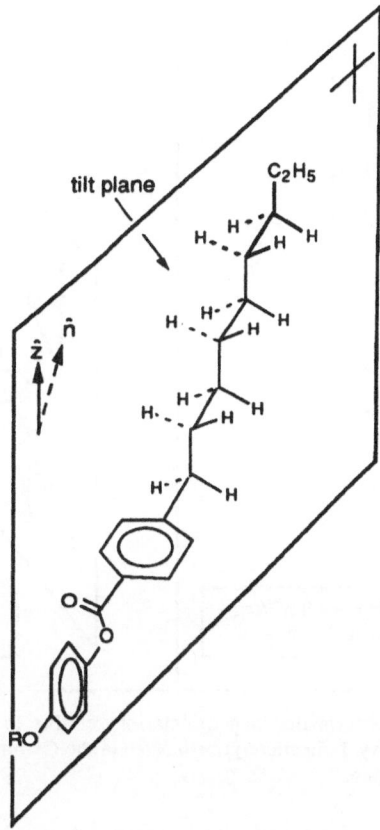

Figure 5. Proposed preferred conformation and rotational orientation of an alkoxyphenyl-nonylbenzoate in the C phase.

A recent test of this simple model for orientation of alkoxyphenyl-alkylbenzoates is presented below.

RECENT RESULTS WITH FLUORINATED FLC MATERIALS

Note that by our model, the carbons of the alkyl tail prefer to lie in the tilt plane, with the tail less tilted than the core. In order to obtain a material with high **P**, it is necessary to place polar functional groups in the molecule such that they will be oriented normal to the tilt plane. To test this idea, and to obtain new materials with large **P**, the disastereomers **1** and **2** of 4'-decyloxyphenyl-4-[(S,S)2,3-epoxy-1-fluorononyl]benzoate (Figure 6) were prepared and characterized [6]. The model suggests that the preferred conformation and rotational orientation for compounds **1** and **2** is as shown in Figure 6, and allows the predictions indicated in the Figure regarding the ferroelectric properties of the stereoisomers: compound **1** should possess small polarization while compound **2** should possess large negative **P**.

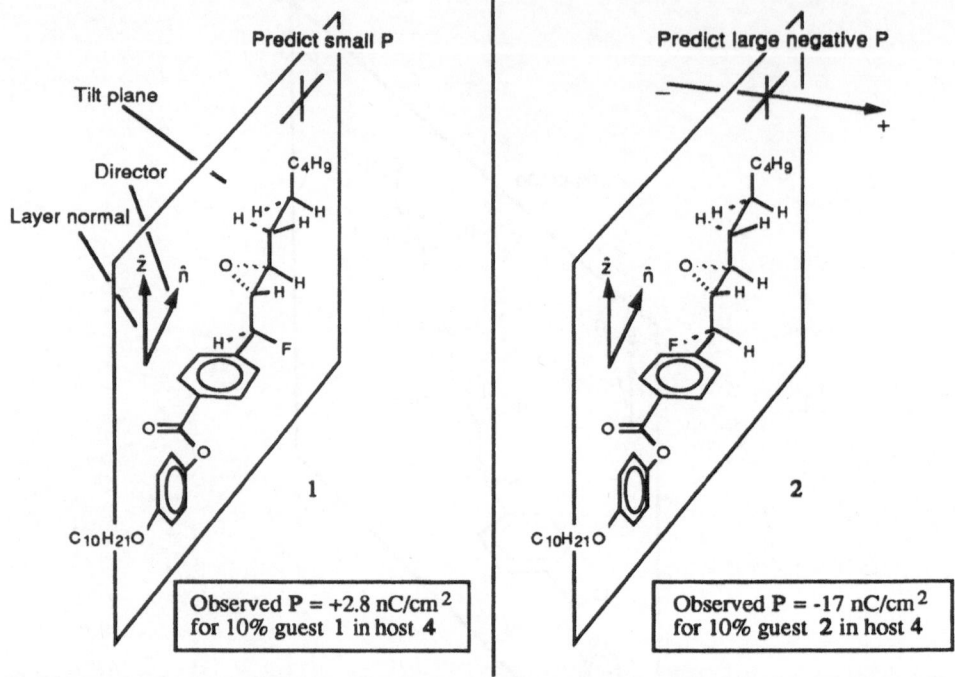

Figure 6. Proposed preferred conformation and orientation relative to the tilt plane for diastereomeric 4'-decyloxyphenyl-4-[(S,S)2,3-epoxy-1-fluorononyl]benzoates in the C* binding site. Predicted and observed polarizations are indicated in the figure.

Fluoroepoxides **1** and **2** were prepared by treatment of epoxyalcohol **3** (from Sharpless kinetic resolution [7] of the racemic allylic alcohol) with diethylaminosulfurtrifluoride (DAST) [8]. When the reaction was run at -78°C, a clean 3:1 mixture of diastereomers was obtained, while at 0°C, a 2:1 mixture of the same products resulted. Separation of the diastereomers by flash chromatography on silica gel gave pure samples of each. The two materials were clearly diastereomers as evidenced by their ^{1}H NMR spectra. Assignment of relative configuration was not possible based upon the spectra, however.

The structures of the diastereomers were in fact established by determination of the ferroelectric polarization of the materials in the C* phase. While neither pure isomer possesses any liquid crystal phases, the binding site model clearly suggests that doping of a chiral compound (the guest) into an achiral (or low-polarization) C phase (the host) will afford orientation of the guest molecules such that measurement of the properties of the mixture can establish the properties of the guest.

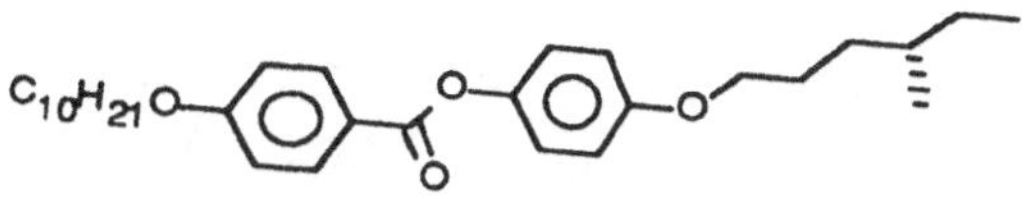

4

We have had excellent results using the FLC phenylbenzoate **4**, first prepared by Keller [9], as a low polarization host [10]. Indeed, when mixed with host **4** (10% by weight of guest), guest compounds **1** and **2** afford stable, well behaved C* phases. Measurement of the polarization of these mixtures in the surface stabilized ferroelectric liquid crystal geometry gave the results shown in the Figure. The major product of the DAST reaction induces small, positive polarization, and may therefore according to the model be assigned structure **1**, while the minor product, affording large, negative polarization, should be assigned structure **2**. It should be noted that the extrapolated polarization of compound **2** (-130 nC/cm^2, or about -0.4 Debye/molecule) is comparable to the measured values for typical "high polarization" materials [11].

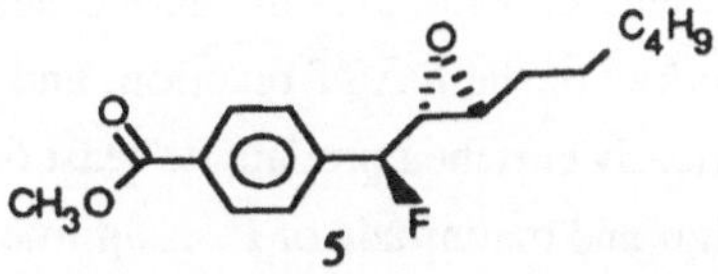

5

Thus, in this system the DAST fluorination apparently proceeds with predominant retention of configuration! This somewhat surprising result [12] was proven unequivocally by single crystal X-ray analysis of the methyl ester **5** deriving from the major product of the DAST reaction (phenyl benzoate **1**) by hydrolysis (LiOH, MeOH, reflux) followed by esterification (DCC, DMAP, MeOH, CH$_2$Cl$_2$).

In this work, the relative configuration of fluoroepoxides **1** and **2** was first established by measurement of the ferroelectric polarization of C* mixtures containing the two compounds as guests, and application of the binding site model as shown in Figure 6. The absolute configuration of both fluoroepoxides was set in a Sharpless epoxidation, though the observed sign of **P** for compound **2** is also indicative of absolute configuration.

In a related test of the model, a synthesis of the decyloxyphenyl 1-fluorononylbenzoate **7** was achieved by treatment of benzylic alcohol **6** (from reduction of a phenyl propargyl ketone with R-Alpine-Borane according to Midland [13]) with DAST. Measurement of the optical rotation of samples of compound **7** prepared at -78°C and 0°C showed that at least at the higher temperature, considerable racemization was occurring. Which was the major enantiomer, **7(R)** or **7(S)**, was not known, since the process could occur with predominant inversion or retention.

Measurement of the properties of a mixture of compound **7** with host **4** shows that the material formed in the fluorination has a moderately large positive polarization (P_{obs} = +21 nC/cm^2 in a 1:1 mixture by weight with host **4**), in agreement with that expected for the fluoride formed with predominant inversion of configuration (**7(R)**).

In order to more fully establish the stereochemistry of the fluorination of alcohol **6**, the mesylate derived from **6** was treated with CsF/18-Crown-6 in refluxing THF. This process was quite inefficient, but gave a low yield (1%) of pure fluoride, which must be **7(R)**, since this process must proceed with inversion of configuration. The measured optical rotation of fluoride **7(R)** prepared in this way corroborates the above described assignment of absolute configuration of the major product of the DAST reaction, and indicates the DAST process at -78°C gives highly enantiomerically enriched product (at least 69% ee) with inversion.

By measurement of the sign and magnitude of **P**, assignment of absolute configuration for compound **7** is accomplished independently of any optical activity or crystallography measurements, and is therefore to our knowledge completely unique. While prediction of the sign and magnitude of **P** in FLCs is in general still a very difficult task, the success achieved in this particular system serves to illustrate how beautifully direct the connection between a macroscopic property of an organic thin film (the ferroelectric polarization) and the microscopic structure of the molecules in the film can be.

Acknowledgement. This work was supported by the National Science Foundation (Grant # DMR-8611192), the NSF through the Engineering Research Center Program with funding for the Optoelectronic Computing Systems Center (Grant # CDR-8622236), the Office of Naval Research, IBM, and Displaytech, Inc.

References

1. Paul, I. C.; Curtin, D. Y. "Gas-Solid Reactions and Polar Crystals" in *Organic Solid State Chemistry*; Desiraju, G. R. Ed.; Elsevier: Amsterdam, 1987; pp 331-370.
2. Meyer, R. B., presented at the *Vth International Liquid Crystal Conference*, Stockholm, 1974.

3. Meyer, R. B.; Liébert, L.; Strzelecki, L.; Keller, P. *J. Phys. (Les Ulis, Fr.)* **1975**, *36*, L-39.

4. Clark, N. A.; Lagerwall, S. T. *Appl. Phys. Lett.* **1980**, *36*, 899.

5. (a) Walba, D. M.; Slater, S. C.; Thurmes, W. N.; Clark, N. A.; Handschy, M. A.; Supon, F. *J. Am. Chem. Soc.* **1986**, *108*, 5210-5221;   (b) Walba, D. M.; Vohra, R. T.; Clark, N. A.; Handschy, M. A.; Xue, J.; Parmar, D. S.; Lagerwall, S. T.; Skarp, K. *J. Am. Chem. Soc.* **1986**, *108*, 7424-7425;   (c) Walba, D. M.; Clark, N. A. "Model for the Molecular Origins of the Polarization in Ferroelectric Liquid Crystals" *in Spatial Light Modulators and Applications II*; Uzi Efron Ed., Proc. SPIE, 1988; Vol. 825, pp. 81-87.

6. A preliminary account of this work appears in the *Proceedings of the First International Symposium on Ferroelectric Liquid Crystals*, Bordeaux-Arcachon, France, September 21-23, 1987; Walba, D. M.; Clark, N. A. *Ferroelectrics* **1988**, *84*, 65-72. A full paper describing the details of the work is given in: Walba, D. M.; Razavi, H. A.; Clark, N. A.; Parmar, D. S. *J. Am. Chem. Soc.* **1988**, *110*, 8686-8691.

7. Martin, V. S.; Woodard, S. S.; Katsuki, T.; Yamada, Y.; Ikeda, M.; Sharpless, K. B. *J. Am. Chem. Soc.* **1981**, *103*, 6237-6240.

8. Middleton, W. J. *J. Org. Chem.* **1975**, *40*, 574.

9. Kerller, P. *Ferroelectrics* **1984**, *58*, 3-7. Compound **4**, 4-[(S)-(+)-(4-methylhexyl)-oxy]phenyl 4-(decloxy)benzoate, is available from Aldrich.

10. The ferroelectric polarization of host **4** is too small to measure directly. Indirect measurements done in our laboratories indicate that for neat compound **4**, **P** $\approx$ -1 nC/cm^2 at 30°C.

11. The classic high polarization FLC materials are α-haloesters exemplified by 4'-((S)2-chloro-3-methylbutyryloxy)phenyl-4-decyloxybenzoate (i) (Mohr, K.; Köhler, S.; Worm, K.; Pelzl, G.; Diele, S.; Zaschke, H.; Demus, D.; Andersson, G.; Dahl, I.; Lagerwall, S. T.; Skarp, K.; Stebler, B. *Mol. Cryst. Liq. Cryst.* **1987**, *146*, 151-171. See also (a) Yoshino, K.; Ozaki, M.; Kishio, S. I.; Sakurai, T.; Mikami, N.; Higuchi, R. I.; Masao, H. *Mol. Cryst. Liq. Cryst.* **1987**, *144*, 87-103.;   (b) Bahr, C. H.; Heppke, G. *Mol. Cryst. Liq. Cryst. Letters* **1986**, *4*, 31-37.) Both enantiomers of compound **i** are available from Aldrich. As measured in

i

our laboratories, the observed polarization of neat chloroester **i** is -125 nC/cm^2, and the observed polarization of a mixture containing 11% by weight of guest **i** in host **4** is -11 nC/cm^2.

12. We know of only one other example in the literature where DAST fluorination proceeds with retention in the absence of standard neighboring group participation: (a) Bird, T. G. C.; Felsky, G.; Fredericks, P. M.; Jones, E. R. H.; Meakins, G. D. *J. Chem. Res. (S)* **1979**, 388; (b) Bird, T. G. C.; Fredericks, P. M.; Jones, E. R. H.; Meakins, G. D. *J. Chem. Soc., Chem. Commun.* **1979**, 65. In this case, it is reported that 12ß-hydroxyandrostan-17-one gives 12ß-fluoroandrostan-17-one upon DAST fluorination.

13. Midland, M. M.; McDowell, D. C.; Hatch, L. H.; Tramontano, A. *J. Am. Chem. Soc.* **1980**, *102*, 867. The alcohol was >95% enantiomerically enriched as judged by NMR using tris[3-(heptafluoropropylhydroxymethylene)-(+)-camphorato]-europium(III) derivative (Eu(hfc)$_3$).

COMPLEXES OF NEUTRAL MOLECULES AND CYCLOPHANE HOSTS

François Diederich

Department of Chemistry and Biochemistry
University of California
Los Angeles, CA 90024-1569 (USA)

SUMMARY

Synthetic cyclophane hosts with apolar binding sites form highly structured complexes with aromatic substrates in aqueous and organic solutions. In this report, the stabilities of the complexes of a variety of neutral and charged arenes are reviewed. Complexation in aqueous solution is largely driven by a strong enthalpic hydrophobic effect, and the nature of this effect is discussed. Multiple host-guest interaction modes can lead to enzyme-like selectivity in binding. The progress in the development of optically active cyclophane hosts for chiral molecular recognition in aqueous solution is described. The selective functionalization of cyclophane hosts has generated water-soluble, enzyme-like catalysts for redox processes, benzoin condensations, and the cleavage of activated carboxylic esters.

INTRODUCTION

After the early work on water-soluble cyclophane hosts in the research groups of Murakami [1], Tabushi [2], and Whitlock [3] during the seventies, Koga et al. in 1980 [4] provided the first unambiguous evidence for inclusion complexation between a protonated tetraazaparacyclophane and neutral aromatic guests in acidic aqueous solution and in the solid state. Their work triggered vigorous research efforts by an increasing number of researchers [5-12] and, within a short period of time, the development of efficient cyclophane receptors for neutral guests advanced into the center of interest in chemical molecular recognition.

We started in 1981 the synthesis of cyclophane hosts, e.g. **1**, with the objective to study in detail the special driving force, known as the hydrophobic effect, which promotes supramolecular complexation of apolar binding partners in aqueous solution. In our receptor systems, binding sites of pronounced apolar character are generated by the specific location of the water-solubility providing ammonium ions remote from the cavity [13]. X-ray crystallographic studies demonstrate that the binding sites of these cyclophanes are highly preorganized in the solid state [5,14]. The crystal structure analysis of a benzene complex (2) shows the perfect inclusion of the aromatic guest in the intramolecular cavity of the host and strongly suggests that π-π-stacking interactions as well as edge-to-face aryl-H_{guest}-π-aryl$_{host}$ interactions represent major stabilizing forces in the complex. Such interactions are also effective in the solution complexes of cyclophane hosts and arene guests that are discussed in the following sections.

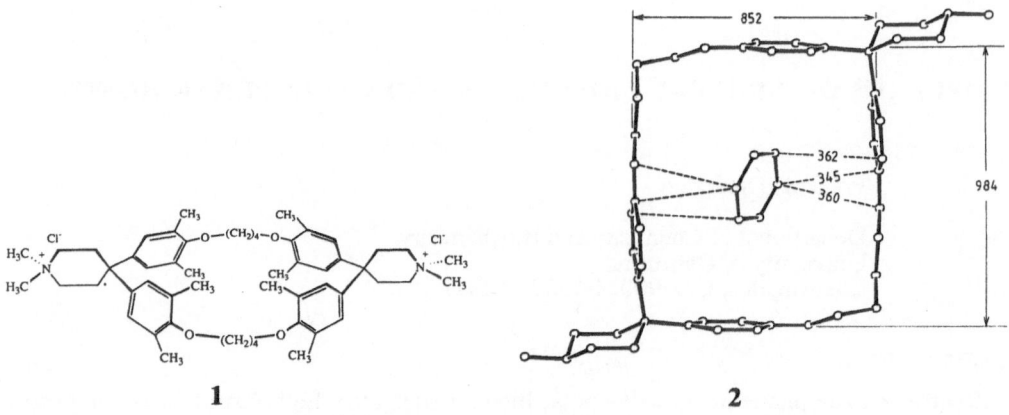

1   **2**

Table 1. Association constants (K_a) and free enthalpies of complexation (-ΔG^o) for the 1:1 complexes of host **3** in aqueous solution (T = 293-295 K) [15].

Guest	K_a [L mol^{-1}]	-ΔG^o [kcal mol^{-1}]
Perylene	1.6×10^7	9.6
Fluoranthene	1.8×10^6	8.4
Pyrene	1.8×10^6	8.4
Biphenyl	2.2×10^4	5.8
Azulene	2.1×10^4	5.8
Naphthalene	1.2×10^4	5.5
Durene	1.9×10^3	4.4

COMPLEXES OF AROMATIC GUESTS IN AQUEOUS AND ORGANIC SOLUTION

In aqueous solutions, our cyclophane hosts form complexes with arene guests which are comparable in stability to enzyme-substrate complexes. Table 1 shows the association constants (K_a) and the free enthalpies of complexation (-ΔG^o) for the highly structured 1:1 complexes (Figure 1) of host **3** and aromatic hydrocarbons in water (T = 293-295 K) [15,16]. Complexes of similar stability form between **3** and guest derivatives bearing non-ionic substituents, e.g. methyl, hydroxy, or dimethylamino groups. Additional charge-charge interactions can lead to extraordinary, enzyme-like selectivity in the complexation of a series of guests which all fit well in the binding site of **3** [15,17]. Table 2 shows that **3** forms complexes with naphthalene mono- and disulfonates that are considerably more stable than those of neutral naphthalene derivatives (Table 1). These complexes are stabilized by apolar binding interactions in the cavity in addition to attractive Coulombic interactions between the

Table 2. Association constants (K_a) and free enthalpies of complexation ($-\Delta G^o$) for the 1:1 complexes of host 3 and charged naphthalene derivatives in aqueous solution (T = 293-295 K) [15,17].

Guest	K_a [L mol^{-1}]	$-\Delta G^o$ [kcal mol^{-1}]
2,6-Naphthalenedisulfonate	$> 10^6$	> 8.0
1,5-Naphthalenedisulfonate	4.4×10^5	7.6
2-Naphthalenesulfonate	4.0×10^5	7.5
1-Naphthalenesulfonate	3.5×10^5	7.4
1-(Trimethylammonium)-naphthalenefluorosulfonate	1.7×10^3	4.3
1,5-Bis(dimethylammonium)-naphthalenebis(deuterium chloride)	$\approx\, < 10$	$\approx\, < 1.3$

anionic residues of the guests and the quaternary nitrogen atoms of the piperidinium rings attached to the aliphatic bridges of the host. This is schematically shown in Figure 1 for the 3·2,6-naphthalenedisulfonate complex. Coulombic interactions severely destabilize the complexes of **3** and cationic naphthalene derivatives, and Table 2 shows that the additional charge-charge interactions can lead to differences in the free enthalpy of complexation, $\Delta(\Delta G^o)$, of more than 6.5 kcal mol^{-1}.

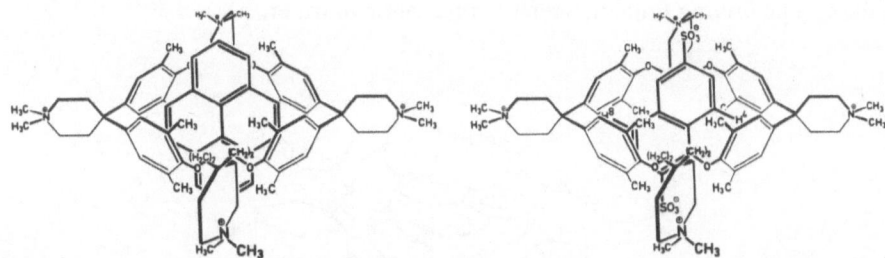

Figure 1. Favored geometries of the 3·pyrene and 3·2,6-naphthalenedisulfonate complexes in water.

We studied the complexation between the macrobicyclic host **4** and polycyclic arenes in a wide range of organic solvents [18]. An impressive dependency of the complexation strength on the nature of the solvent was observed, as shown in Table 3 for the **4**·pyrene complex. Since the geometry of the **4**·pyrene complex is very similar in all solvents (Figure 2), the large decrease in binding strength upon changing from methanol ($\Delta G^o = -6.4$ kcal mol^{-1}) via the dipolar-aprotic solvents to carbon disulfide ($\Delta G^o = -1.3$ kcal mol^{-1}) does not primarily result

4

Table 3. Association constants (K_a) and free enthalpies of complexation (-ΔG^O) for the **4**·pyrene complex in various organic solvents (T = 303 K)[18].

Solvent	K_a [L mol^{-1}]	-ΔG^O [kcal mol^{-1}]
Methanol	4.4 x 10^4	6.4
Ethanol	2.5 x 10^4	6.1
Acetone	1.2 x 10^3	4.3
Me$_2$SO	6.9 x 10^2	3.9
THF	84	2.7
Benzene	12	1.5
Carbon Disulfide	9	1.3

from differences in attractive van der Waals interactions between host and guest in the complex. Rather, solvation effects are responsible for the differences in complexation strength. Host-guest complexation is strongest in polar-protic solvents which compete less efficiently with the guest for the cavity binding site. The strong cohesive interactions in these solvents make the transfer of a solvent molecule for solvation from the bulk into the host cavity energetically unfavorable. Also, polar-protic solvent molecules possess less affinity for the apolar host cavity than aromatic or highly polarizable solvent molecules such as benzene and carbon disulfide. Very similar solvation effects are also considered in the following chapter to explain the special driving force for arene complexation in water.

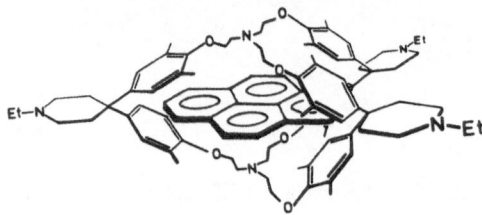

Figure 2. Geometry of the **4**·pyrene complex in organic solvents [18].

A STRONG ENTHALPIC HYDROPHOBIC EFFECT IN AQUEOUS SOLUTION

With the octamethoxy-substituted cyclophanes **5** and **6**, we analyzed the thermo-dynamic characteristics for the complexation of neutral benzene derivatives in aqueous solution [19]. The X-ray crystal structure of **5** (Figure 3) shows a very deep, pre-organized cavity occupied by two highly disordered water molecules. Both cyclophanes **5** and **6** form tight and stable complexes with neutral *para*-substituted benzene derivatives in aqueous solution. The association constants K_a (T = 293 K) for selected complexes of **6** in water are included in Table 4. These K_a values are extraordinarily large, and a comparison shows that host **6** ranges between the generally poorer binding enzymes and the better binding antibodies in its affinity for neutral, non-hydrogen bonding benzene guests [20].

Table 4. Association constants (K_a), enthalpic (ΔH^O), and entropic ($T\Delta S^O$) contributions to the free enthalpies of complexation (ΔG^O) at 293.4 K for the complexes of host 6 with 1,4-disubstituted benzene guests in D_2O and methanol-d_4 [19].

	$K_a{}^a$ (L mol^{-1})	ΔG^{Ob}	ΔH^O (kcal mol^{-1})	$T\Delta S^O$
Complexes in D_2O				
Dimethyl-*p*- benzenedicarboxylate	1.17×10^5	- 6.81	-10.7±1.0	-4.0±1.0
p-Nitrotoluene	3.00×10^4	- 6.01	- 9.6±3.0	-3.6±3.0
p-Dimethoxybenzene	1.02×10^4	- 5.38	-10.2±2.5	-4.8±2.5
p-Xylene	9.33×10^3	- 5.33	- 7.4±1.0	-2.1±1.0
p-Dicyanobenzene	7.83×10^3	- 5.23	- 9.5±1.0	-4.3±1.0
Complexes in methanol-d_4				
p-Dicyanobenzene	2.4×10^1	- 1.86	- 4.2±1.5	-2.4±1.5
p-Dimethoxybenzene	8×10^0	- 1.20	- 4.4±1.5	-3.2±1.5

a Uncertainty: 10 % in D_2O and 25% in methanol-d_4.
b Uncertainty: 0.07 kcal mol^{-1} in D_2O and 0.17 kcal mol^{-1} in methanol-d_4.

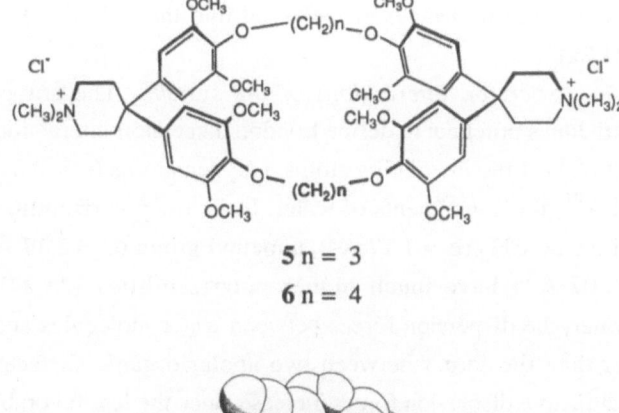

5 n = 3
6 n = 4

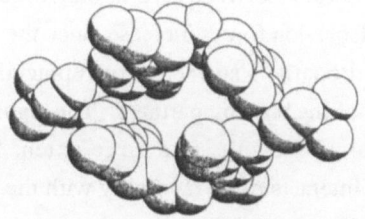

Figure 3. X-ray structure of 5.

Table 4 reveals that the complexation in aqueous solution is predominantly enthalpically driven. It should, however, be noted that, although the $T\Delta S^O$-terms in water are very negative, complexation is also promoted by a positive entropy contribution resulting from the desolvation of the host cavity and the guest surfaces upon complexation. In the absence of such an entropically favorable desolvation process, the $T\Delta S^O$-term would be far more negative.

Attractive host-guest interactions in the very tight complexes of 6 and benzene derivatives certainly contribute significantly to the favorable large ΔH^o-term in water. By comparing the thermodynamic characteristics for the complexation processes in water and methanol (Table 4), another major enthalpic driving force, provided by water as the solvent, becomes apparent.

Complexes of 6 in methanol-d_4 are 3 - 4 kcal mol^{-1} less stable than the corresponding complexes in water (Table 4) [19]. This difference mainly results from an enhanced enthalpic driving force for binding in water. The analysis of the ^{1}H NMR complexation shifts at saturation binding suggests that the geometry of the complexes in water and methanol is very similar; therefore, the observed differences in enthalpic contributions cannot be explained with large differences in attractive host-guest interactions in the two solvents. Rather, a large part of the favorable complexation enthalpy in water results from specific contributions of the solvent. We have identified two major components of this *strong enthalpic hydrophobic effect* :

a) *Changes in cohesive interactions of water.* Water molecules that solvate the free guest and especially the deep cavity of the free host have reduced cohesive interactions and are enthalpically higher in energy than water molecules in the bulk solvent. They participate in fewer strong hydrogen bonds than solvent molecules in the bulk. Upon inclusion complexation, these solvent molecules are released into the bulk and become enthalpically lower in energy [21,22].

b) *Changes in dispersion interactions of the system.* The attractive B-term in the "Ar^{-12}-Br^{-6} Lennard-Jones potential to define London dispersion interactions is proportional to the polarizability α ($Å^3$) of the interacting atoms. Oxygen atoms ($\alpha = 0.84$ $Å^3$) and hydroxyl residues ($\alpha = 1.20$ $Å^3$), the constituents of water, have low polarizabilities whereas organic residues, e.g. an aliphatic CH_2 ($\alpha = 1.77$ $Å^3$), a methyl group ($\alpha = 2.17$ $Å^3$), or an aromatic CH group ($\alpha = 2.07$ $Å^3$) have much higher polarizabilities [23,24]. Due to the low polarizability of water, the dispersion forces between water molecules and an apolar organic surface are weaker than the forces between two apolar organic surfaces. Upon inclusion complexation, the attractive dispersion forces increase since the less favorable contacts between water molecules and the hydrocarbon surfaces are replaced by the more favorable close contacts between the surfaces of the binding partners. The significant decrease in the enthalpic driving force in methanol solution results to a large extent because methanol possesses a polarizable methyl group that interacts more favorably with the host and guest than water does. Therefore, the differences in dispersion interactions between the nonbinding and the binding state of host and guest in methanol are significantly reduced.

The enthalpic hydrophobic effect represents an important, generally occuring driving force for complexation in aqueous solution [25]. A literature survey shows that cyclodextrin [26], enzyme, or antibody [20] binding to aromatic substrates in water is defined by thermodynamic characteristics (large negative ΔH^o and $T\Delta S^o$) that are similar to those measured for the complexation by 6 (Table 4). We predict that the enthalpic hydrophobic effect will also be recognized as a major driving force in water for the complexation of polar, hydrogen-bonding organic molecules, e.g. oligosaccharides, by natural receptors [27] as well as by unnatural ones that await to be prepared.

WATER-SOLUBLE OPTICALLY-ACTIVE CYCLOPHANE HOSTS

In no other area of our research, computer modeling proved to be as effective as in the design of water-soluble, optically active cyclophane hosts and of the chiral spacers that shape their binding sites. Our first target macrocycle **7** with a 2,2',6,6'-tetrasubstituted biphenyl

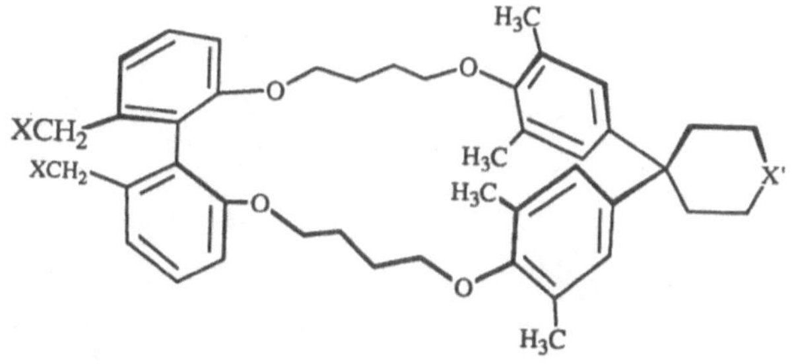

7 X = Me$_3$N$^+$ X' = $^+$NMe$_2$
8 X = Me$_2$N X' = NMe

unit as a chiral spacer was designed based on CPK model examinations. Binding studies with **8** in acidic aqueous solution, however, clearly revealed that this cyclophane does not possess an efficient binding site and, hence, does not act as a host [28]. According to force-field calculations, the biphenyl unit is not a suitable chiral spacer group [29]. With an O···O distance of 8.91 Å, the diphenylmethane unit of **8** favorably shapes and widens binding sites for arenes and flat cycloalkanes. At an O···O distance of only 4.10 Å, the biphenyl unit of **8** does not sufficiently widen the cavity. In addition, the bridging of two very differently sized spacers in **8** leads to an extensively twisted helical geometry (**9**) which further reduces the size of a possible binding site. The helical conformation seems to be imposed on the macrocycle by the tendency of the bridging alkane chains to maximize the number of *anti* torsional angles [29].

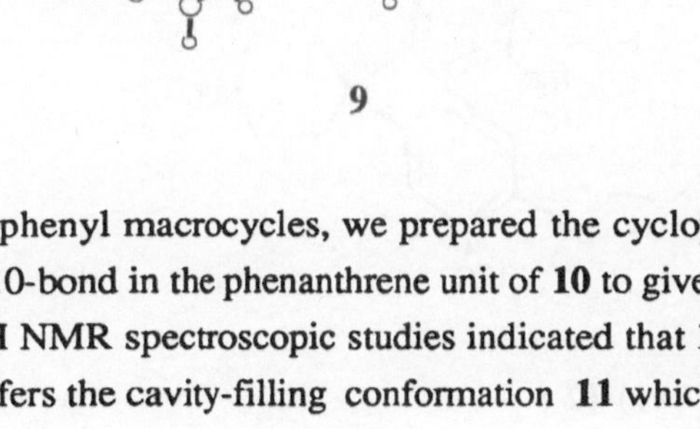

9

On the way to biphenyl macrocycles, we prepared the cyclophane **10** [28]. However, the ozonation of the 9,10-bond in the phenanthrene unit of **10** to give the chiral biphenyl spacer was not successful. ^{1}H NMR spectroscopic studies indicated that **10**, under all experimental conditions, highly prefers the cavity-filling conformation **11** which protects the 9,10-double bond of the phenanthrene moiety from ozonation. In collaboration with Professor K. N. Houk, a rigorous conformational analysis was undertaken which included force-field minimizations using MM2 and AMBER as well as molecular dynamics simulations [30].

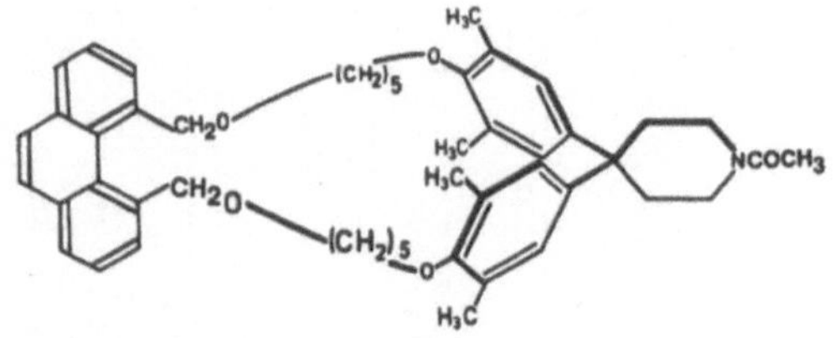

10 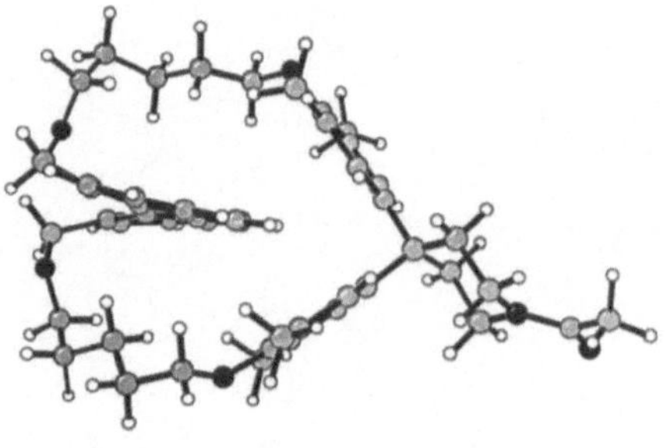   **11**

These computational studies fully supported the ^{1}H NMR results and indicated that **10** prefers by 4 - 5 kcal mol^{-1} the cavity-filling conformation **11**. This conformation is specifically stabilized by enhanced van der Waals interactions. The minimizations performed with the two different force-fields AMBER and MM2 were in good agreement [30].

The computer-modeling work led to the conclusion that efficient hosts for aromatic and flat alicyclic guests should form when an achiral or chiral aromatic spacer with an O···O distance of $\geq$ 6.5 Å and a diphenylmethane unit are bridged by suitably sized aliphatic chains [29]. We prepared the optically active host (+)-**12** which incorporates a 4-phenyl-1,2,3,4-tetrahydroisoquinoline unit as a chiral spacer with an O···O distance of 7.58 Å and found its binding properties in agreement with the force-field predictions. In water/methanol (60:40), (+)-**12** preferentially binds 2,6-disubstituted naphthalenes which take a pseudo-axial position in the cavity. By ^{1}H NMR, we could demonstrate the formation of diastereomeric complexes between (+)-**12** and (*R,S*)-naproxen or its methyl ester (**13**) [29].

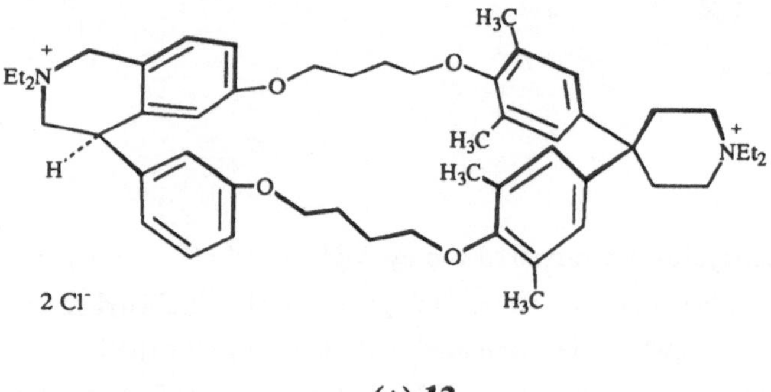

(+)-**12**

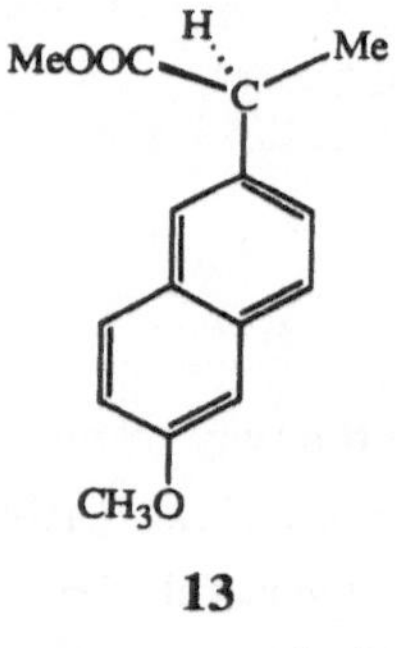

13

(S)-naproxen methyl ester

The absence of symmetry in (+)-12, however, was of disadvantage in quantitative chiral molecular recognition studies. This host generates a complex ^{1}H NMR spectrum which severely limits the spectral window needed for monitoring the shifting guest resonances during quantitative ^{1}H NMR titrations. The difference in stability of the diastereomeric complexes formed by (+)-12 and the naproxen enantiomers could not be determined. Further disadvantages of (+)-12 were the long synthesis of the chiral spacer as well as its considerable conformational flexibility which led to a less preorganized and, hence, less efficient binding site [29].

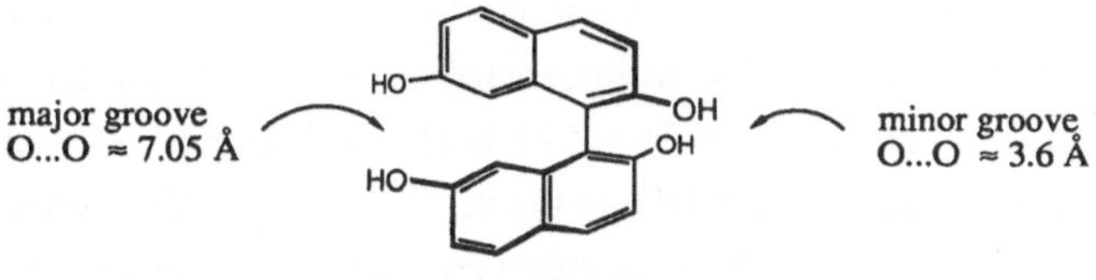

major groove
O...O ≈ 7.05 Å

minor groove
O...O ≈ 3.6 Å

14

With the 2,2',7,7'-tetrahydroxy-1,1'-binaphthyl unit [31] we now have identified the desired efficient, readily available, and versatile chiral spacer for our cyclophane hosts [32]. This ditopic component has C_2-symmetry and possesses two distinctive geometric clefts. Having a dihedral angle about the chirality axis of 88°, the O···O distance at the minor groove is ≈ 3.60 Å, whereas the major groove, with an O···O distance of 7.05 Å, is almost twice as wide. The minor groove has been most successfully incorporated into chiral cation binding receptors by Cram et al. [33]. The major groove of 14 is ideal to shape an aromatic binding site in chiral cyclophane hosts. In 15, the functionalities at the minor groove are used to introduce water-solubility providing residues remote of the apolar binding site. In the ditopic host 16, the minor groove shapes an additional crown ether binding site.

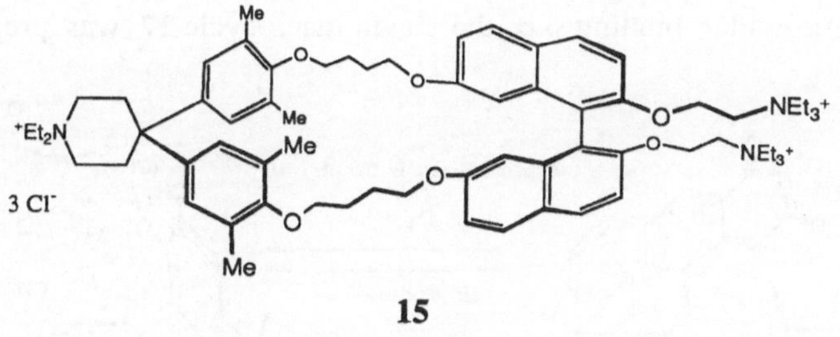

15

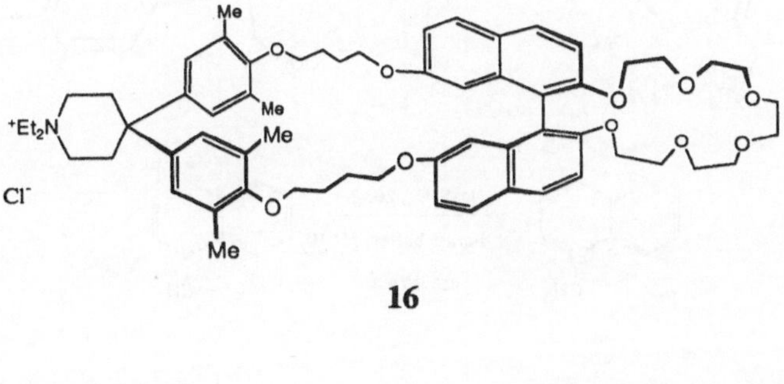

16

Both hosts **15** and **16** form stable and highly structured complexes with neutral naphthalene guests in aqueous solution [32]. A comparison shows that they are much better binders than host (+)-**12**. At T = 293 K in D_2O/methanol-d_4 (60:40), the association constant for the complex of 6-methoxy-2-naphthonitrile and (+)-**12** (K_a = 336 L mol^{-1}) [29] is considerably smaller than the K_a values for the complexes of both **15** (K_a = 1990 L mol^{-1}) and **16** (K_a = 4510 L mol^{-1}). Cyclophane **16** is a ditopic receptor which can be switched from an efficient binder of neutral aromatics to a good host for potassium cations by changing the water content of a water-methanol mixture. The stability of the 6-methoxy-2-naphthonitrile complex decreases with increasing methanol content from -ΔG^o = 4.90 kcal mol^{-1} in 40% aqueous methanol to -ΔG^o = 1.85 kcal mol^{-1} in pure methanol. In contrast, the stability of the potassium cation complex increases with increasing methanol content from -$\Delta G^o \approx$ 1.5 kcal mol^{-1} in 40% aqueous methanol to -ΔG^o = 4.35 kcal mol^{-1} in pure methanol. Preliminary studies with the racemic hosts **15** and **16** and the methyl ester of (*S*)-naproxen (**13**) provided clear evidence for the formation of diastereomeric complexes with different geometries and/or stabilities. Significant differential complexation shifts are observed for all aromatic host resonances in the two formed diastereomeric complexes [32]. The stabilities of these diastereomeric complexes are now being determined quantitatively in separate ^{1}H NMR titrations with optically pure host and each guest enantiomer. With their C_2-symmetry, hosts **15** and **16** provide large open spectral windows to monitor the guest resonances in these titrations.

SUPRAMOLECULAR CATALYSIS BY WATER-SOLUBLE CYCLOPHANE HOSTS [5]

The geometry of the isoalloxazine unit of flavin coenzymes is highly dependent on its redox state [34]. The oxidized unit is planar while the 2e$^-$-reduced dihydroisoalloxazine unit is butterfly shaped with an angle of approximately 30° about the two nitrogens N-5 and N-10 of the central ring. To explore how the different geometry in the two redox states influences the properties of a molecular binding site, the flavin-macrocycle **17** was prepared [35]. Both

17a and **17b** form complexes in aqueous solution with 2,6-disubstituted naphthalene derivatives. In the complexes of **17a**, these guests assume a near cofacial π-π-stacking orientation to the isoalloxazine unit and presumably bind outside the cavity [36]. In the complexes of **17b**, they are included into the cavity and take a similar position as in the complexes of bis(diphenylmethane)-hosts (e.g. see Figure 1). The complexes of very different geometry, formed by the oxidized and the reduced flavinophane, are of similar stability, and the driving force for complexation in deuterated borate buffer at pH 10 is 3-4 kcal mol^{-1}. Electron donor-acceptor interactions are presumably responsible for the strongly redox-dependent complexation observed with acceptor-acceptor substituted guests, e.g. 2,6-naphthalenedicarbonitrile in methanol-d_4/D_2O (1:1). These guests only bind to the reduced host **17b**. The absence of any significant binding to the oxidized host **17a** is best explained by electrostatic repulsion between the electron-accepting isoalloxazine unit of **17a** and the cofacially oriented acceptor guest. The complexation properties of flavinophane **17** show the correct direction for the development of mediators with switchable binding properties to catalyze the "active transport" of arenes across aqueous solutions. Flavinophanes resembling **17** but with an externally bridged and, hence, shielded isoalloxazine should exhibit complete redox-dependent complexation behavior. The bridging of the alloxazine will prevent external π-π-stacking and binding will only occur in the large cavity of the reduced flavin-host.

Our work aimed at the development of efficient synthetic esterases which follow the covalent catalysis mechanism of α-chymotrypsin in their mode of action has provided a convincing example for the concept of productive versus non-productive binding as a selectivity-generating mechanism [37]. The two closely related macrocycles **18** and **19** form complexes of similar stability with 4-nitro-1-naphthol in aqueous solution [38]. However, supramolecular catalysis is only observed with **19**, the host with the larger binding site. For the acylation of **19** by complexed 4-nitro-1-naphthylacetate in an aqueous phosphate buffer at pH 8.0, T = 293 K, Michaelis-Menten-type saturation kinetics were observed whereas the acylation of **18**, despite similarly strong substrate binding, follows strict second-order kinetics. A comparison of second-order rate constants shows that the supramolecular reaction proceeds 22 times faster than the bimolecular reaction. Steric considerations explain the differences in reactivity between the two similar hosts. In the supramolecular complex of **18**, the phenol ring completely blocks one side of the cavity. Hence, the ester residue of the complexed substrate extends out of the cavity on the side opposite to the nucleophile and reacts in an intermolecular

18 n = 6

19 n = 6

reaction with a second macrobicyclic phenol. In the larger cavity of **19**, productive binding is possible. The ester residue of the bound substrate extends out of the cavity on the phenol side and takes a favorable orientation for intra-complex nucleophilic attack. Host **19** not only acts as a transacylation reagent, but the deacylation step is fast enough to allow for modest catalytic turnover in the hydrolysis of 4-nitro-1-naphthylacetate [38].

With the macrobicyclic system **20**, we have prepared a very complete model system for thiamine pyrophosphate-dependent ligases [39,40]. In protic solvents, where complexation takes place, **20** is a better turnover catalyst for the benzoin condensation than non-macrocyclic thiazolium derivatives. In aqueous buffers, we observed a specific activation of the thiazolium ring by the binding site of **20** which is of lower polarity than the surrounding solution. Large enhancements of the H/D exchange rate at C-2 of the macrobicyclic thiazolium ring are best explained by a micropolarity effect of the cavity on the kinetic acidity of the proton at this position [41]. A similar effect was not detectable with non-macrocyclic thiazolium derivatives.

20

CONCLUSIONS

Cyclophanes provide powerful, preorganized binding sites in aqueous and organic solvents for aromatic guests of stereoelectronic complementarity. With these hosts, the driving forces for molecular complexation in biological and chemical systems can be studied in great detail. Efficient chiral spacers for optically active cyclophanes are now available and future systematic studies will advance the understanding of multiple binding interactions at the origin of chiral recognition in biological and chemical systems. The ready functionalization of tetraoxa[n.1.n.1]paracyclophanes has generated catalytically active receptors which exhibit many of the characteristic properties of enzymatic systems. These designed, structurally well defined catalysts allow to analyze in detail mechanistic aspects that have been advanced to explain properties of the natural systems.

Within a short period of time, cyclophanes have gained recognition as one of the most versatile and most efficient classes of molecular receptors for neutral guests. Undoubtedly, research with future generations of cyclophane hosts will continue to contribute to some of the most fascinating developments in the field of molecular recognition.

Acknowledgement. We thank the National Science Foundation and the Office of Naval Research for their support of this work

References

1. Murakami, Y. *Top. Curr. Chem.* **1983**, *115*, 107-159.
2. Tabushi, I.; Yamamura, K. *Top. Curr. Chem.* **1983**, *113*, 145-182.
3. Jarvi, E. T.; Whitlock, H. W. *J. Am. Chem. Soc.* **1980**, *102*, 657-662.
4. (a) Odashima, K.; Itai, A.; Iitaka, Y.; Koga, K. *J. Am. Chem. Soc.* **1980**, *102*, 2504-2505; (b) Odashima, K.; Koga, K. In *Cyclophanes*; Keehn, P. M.; Rosenfeld, S. M., Eds.; Academic: New York, 1983; Vol. 2, pp 629-678.
5. Diederich, F. *Angew. Chem.,* **1988**, *100*, 372-396; *Angew. Chem., Int. Ed. Engl.* **1988**, *27*, 362-386.
6. Winkler, J.; Coutouli-Argyropoulou, E.; Leppkes, R.; Breslow, R. *J. Am. Chem. Soc.* **1983**, *105*, 7198-7199.
7. Schneider, H.-J.; Kramer, R.; Simova, S.; Schneider, U. *J. Am. Chem. Soc.* **1988**, *110*, 6442-6448.
8. Merz, T.; Wirtz, H.; Vögtle, F. *Angew. Chem.* **1986**, *98*, 549-550; *Angew. Chem., Int. Ed. Engl.* **1986**, *25*, 567-569.
9. Cowart, M. D.; Sucholeiki, I.; Bukownik, R. R.; Wilcox, C. S. *J. Am. Chem. Soc.* **1988**, *110*, 6204-6210.
10. Collet, A. *Tetrahedron* **1987**, *43*, 5725-5729.
11. Sheridan, R. E.; Whitlock, Jr., H. W. *J. Am. Chem. Soc.* **1988**, *110*, 4071-4073.
12. Fornasier, R.; Reniero, F.; Scrimin, P.; Tonellato, U. *J. Incl. Phenom.* **1988**, *6*, 175-181.
13. (a) Diederich, F.; Dick, K. *Tetrahedron Lett.* **1982**, *23*, 3167-3170; (b) Diederich, F.; Dick, K.; Griebel, D. *Chem. Ber.* **1985**, *118*, 3588-3619.
14. Krieger, C.; Diederich, F. *Chem. Ber.* **1985**, *118*, 3620-3631.
15. Diederich, F.; Dick, K. *J. Am. Chem. Soc.* **1984**, *106*, 8024-8036.
16. Diederich, F.; Griebel, D. *J. Am. Chem. Soc.* **1984**, *106*, 8037-8046.
17. Diederich, F.; Dick, K. *Chem. Ber.* **1985**, *118*, 3817-3829.
18. Diederich, F.; Dick, K.; Griebel, D. *J. Am. Chem. Soc.* **1986**, *108*, 2273-2286.
19. Ferguson, S. B.; Seward, E. M.; Diederich, F.; Sanford, E. M.; Chou, A.; Inocencio-Szweda, P.; Knobler, C. B. *J. Org. Chem.* **1988**, *53*, 5593-5595.
20. Biltonen, R. L.; Langerman, N. *Methods Enzymol.* **1979**, *61*, 287-318.
21. Tabushi, I.; Kiyosuke, Y.; Sugimoto, T.; Yamamura, K. *J. Am. Chem. Soc.* **1978**, *100*, 916-919.
22. Saenger, W. *Angew. Chem.* **1980**, *92*, 343-361; *Angew. Chem., Int. Ed. Engl.* **1980**, *19*, 344-362.
23. Fersht, A. *Enzyme Structure and Mechanism*; Freeman: New York, 1985; pp 293-310.
24. McCammon, J. A.; Wolynes, P. G.; Karplus, M. *Biochemistry*, **1979**, *18*, 927-942.
25. Abraham, M. H. *J. Am. Chem. Soc.* **1982**, *104*, 2085-2094.
26. Harata, K.; Tsuda, K.; Uekama, K.; Otagiri, M.; Hirayama, F. *J. Incl. Phenom.* **1988**, *6*, 135-142.

27. Lemieux, R. U.; Venot, A. P.; Spohr, U.; Bird, P.; Mandal, G.; Morishima, N.; Hindsgaul, O.; Bundle, D. R. *Can. J. Chem.* **1985**, *63*, 2664-2668 and preceeding papers in that journal.

28. Rubin, Y.; Dick, K.; Diederich, F.; Georgiadis, T. M. *J. Org. Chem.* **1986**, *51*, 3270-3278.

29. Dharanipragada, R.; Ferguson, S. B.; Diederich, F. *J. Am. Chem. Soc.* **1988**, *110*, 1679-1690.

30. Loncharich, R. J.; Seward, E.; Ferguson, S. B.; Brown, F. K.; Diederich, F.; Houk, K. N. *J. Org. Chem.* **1988**, *53*, 3479-3491.

31. (a) Brass, K.; Patzelt, R. *Chem. Ber.* **1937**, *70*, 1349-1353; (b)Feringa, B.; Wynberg, H. *Tetrahedron Lett.* **1977**, 4447-4450; (c) Pirkle, W, H.; Schreiner, J. L. *J. Org. Chem.* **1981**, *46*, 4988-4991; (d) Tisler, M. *Org. Prep. Proc. Int.* **1986**, *18*, 17-78.

32. Diederich, F.; Hester, M. R.; Uyeki, M. A.; *Angew. Chem.* **1988**, *100*, 1775-1777; *Angew. Chem., Int. Ed. Engl.* **1988**, *27*, 1705-1707.

33. (a) Cram, D. J.; Cram, J. M. *Acc. Chem. Res.* **1978**, *11*, 8-14; (b) Knobler, C. B.; Gaeta, F. C. A.; Cram, D. J. *J. Chem. Soc., Chem. Commun.* **1988**, 330-333.

34. Walsh, C. *Enzymatic Reaction Mechanisms*, Freeman: San Francisco, 1979.

35. Seward, E.; Diederich, F. *Tetrahedron Lett.* **1987**, *28*, 5111-5114.

36. Seward, E.; Hopkins, R. B.; Sauerer, W.; Tam, S.-W.; Diederich, F. *J. Am. Chem. Soc.*, submitted.

37. See ref. 23), Chapter 12, pp 311-346.

38. Diederich, F.; Schürmann, G.; Chao, I. *J. Org. Chem.* **1988**, *53*, 2744-2757.

39. Lutter, H.-D.; Diederich, F. *Angew. Chem.* **1986**, *98*, 1125-1127; *Angew. Chem., Int. Ed. Engl.* **1986**, *25*, 1125-1127.

40. Breslow, R.; Kool, E. *Tetrahedron Lett.* **1988**, *29*, 1635-1638.

41. (a) Haake, P.; Bausher, L. P.; Miller, W. B. *J. Am. Chem. Soc.* **1969**, *91*, 1113-1120; (b) Crosby, J.; Stone, R.; Lienhard, G. E. *J. Am. Chem. Soc.* **1970**, *92*, 2891-2900.

CATALYTIC FUNCTIONS OF PARACYCLOPHANES BASED ON MOLECULAR RECOGNITION

Yukito Murakami

Department of Organic Synthesis
Faculty of Engineering
Kyushu University
Fukuoka 812 (Japan)

SUMMARY

A cubic cyclophane surrounded by six faces, each being constructed with the [3.3.3.3]paracyclophane ring, acted as a polycationic host in acidic aqueous media and exhibited pH-dependent guest-binding behavior. Hydrophobicity of its internal cavity was markedly enhanced as medium pH was raised. The cubic cyclophane exercised the size-sensitive molecular discrimination that originates from restricted and rigid geometry of its hydrophobic cavity and perfectly protected an incorporated guest from oxygen attack. An octopus cyclophane, constructed with the [3.3.3.3]paracyclophane skeleton and eight hydrocarbon chains, provided a cavity that is deep and hydrophobic enough to incorporate hydrophobic guests of various bulkiness through an induced-fit mechanism. Unique biphasic guest-binding behavior was observed for an octopus cyclophane having L-aspartate residues as connector units interposed between the macrocyclic skeleton and the alkyl branches. An artificial holoenzyme system, composed of a hydrophobic vitamin B_{12} and the octopus cyclophane, could simulate catalytic functions of methylmalonyl-CoA mutase and glutamate mutase in the presence of vanadium trichloride under aerobic irradiation conditions.

INTRODUCTION

Cyclophanes with a sizable internal cavity of hydrophobic character have been widely employed as simplified and effective models capable of simulating enzymatic functions [1]. Although moderate substrate discrimination in host-guest complexation has been exercised by various cyclophanes composed of a single macrocyclic skeleton, more specific molecular recognition can be achieved by modified cyclophane hosts which are capable of providing a three-dimensionally extended internal cavity. We have now prepared two different types of hydrophobic host molecules by utilizing a tetraaza[3.3.3.3]paracyclophane ring as a basic molecular skeleton. The one type is the cubic cyclophane (1) composed of six faces, each being constructed with the [3.3.3.3]paracyclophane ring [2], and the other are the

Inclusion Phenomena and Molecular Recognition
Edited by J. Atwood
Plenum Press, New York, 1990

octopus cyclophanes (2-4) which have eight flexible hydrocarbon branches introduced into the macrocyclic skeleton [3]. This article describes characteristic substrate-binding behavior of the cyclophanes in aqueous media and catalytic functions of a host-guest complex formed with octopus cyclophane **2** and a hydrophobic vitamin B_{12} derivative as a vitamin B_{12}-dependent artificial holoenzyme.

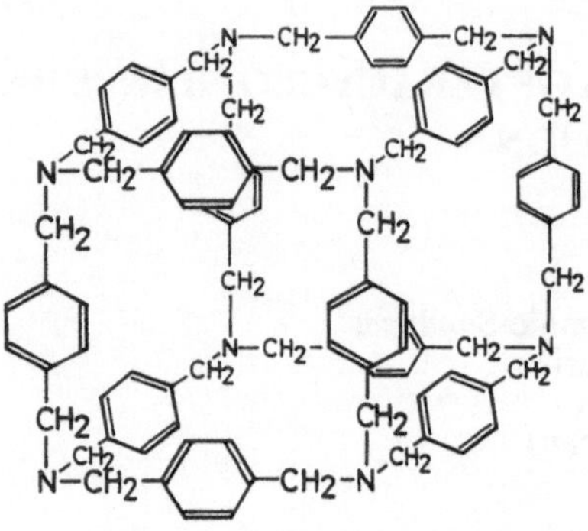

1

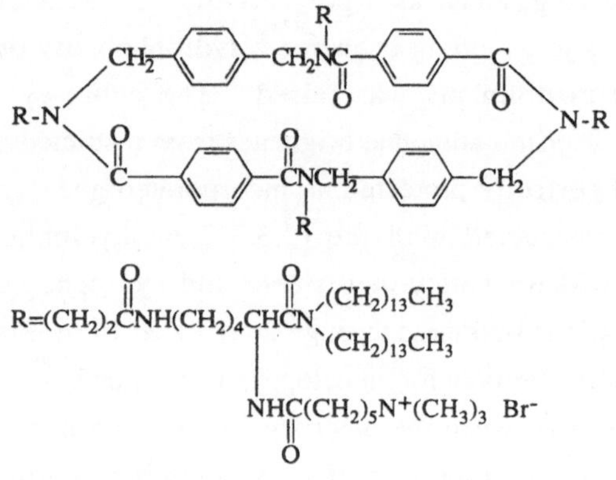

2

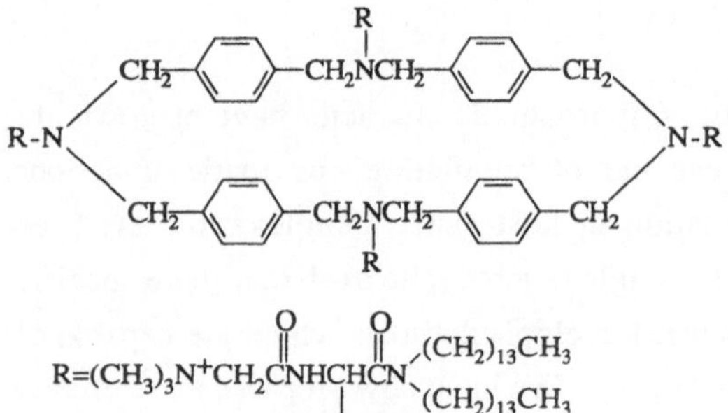

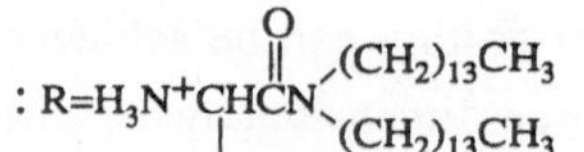

3

4

MOLECULAR RECOGNITION BY CUBIC CYCLOPHANE [4]

In the light of a structural study using CPK models, it became apparent that cyclophane **1** may provide a relatively large and globular hydrophobic cavity with a maximum internal diameter of ca. 9 Å (Figure 1) and that guest molecules capable of passing through the 2,11,20,29-tetraaza[3.3.3.3]paracyclophane ring (hole size, 5.5 - 7 Å) can be incorporated into the internal cavity of the host. The cubic cyclophane possesses eight tertiary amino moieties, and all the amino nitrogens were completely protonated below pH 2.5 as confirmed by ^{1}H NMR spectroscopy. Thus, cyclophane **1** acts as a polycationic host in acidic aqueous media, and its pH-dependent guest-binding behavior was examined by fluorescence spectroscopy at 30.0°C.

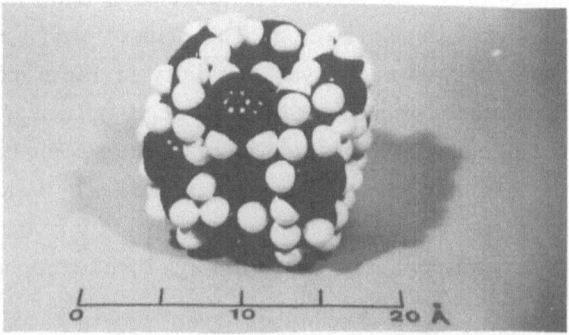

Figure 1. CPK molecular model of **1**.

Figure 2 shows fluorescence spectra of ANS incorporated into **1** under various pH conditions. The fluorescence intensity increased with a concomitant blue shift of the

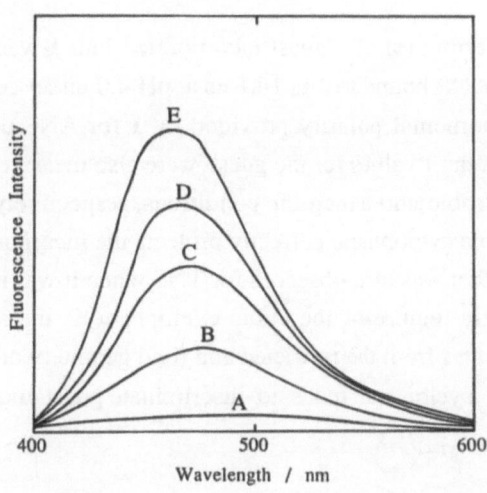

Figure 2. Fluorescence spectra of ANS (1.0×10^{-6} mol dm^{-3}) upon addition of **1** (1.0×10^{-5} mol dm^{-3}) in an aqueous acetate buffer [0.01 mol dm^{-3}, μ 0.1 (KCl)] at 30.0°C under various pH conditions: A, 2.5; B, 2.9; C, 3.2; D, 3.4; E, 3.9.

Table 1. Binding constants for host-guest complexes of 1 with hydrophobic guests in aqueous acetate buffer [0.01 mol dm^{-3}, μ 0.10 (KCl)] at 30.0°C.

Guest	K (dm^3 mol^{-1})		
	pH 2.6	pH 3.0	pH 4.0
TNS	4600	30000	420000
ANS	3800	10000	160000
α-PNA		890	88000

fluorescence maximum as the the medium pH was raised. This spectral behavior indicated that the hydrophobicity of the internal cavity of 1 was markedly enhanced as the extent of deprotonation of the host increases. The CPK molecular model of 1 clearly demonstrates that all the lone-pair orbitals of the amino-nitrogens are directed towards the interior cavity rather than sticking out from the molecular skeleton. This state of orbital orientation provides a reasonable explanation for weakened hydrophobicity at lower pH's. Four stepwise proton-dissociation equilibria for the host-guest complexes composed of 1 with ANS and TNS were detected in a pH range of 2.5-4.0 by fluorescence spectroscopic titrations. The binding constants for inclusion of anionic and nonionic guest molecules by 1 are very sensitive to pH change (Table 1). On the other hand, 1 exhibited no capability of binding cationic guests. Thus, the cubic cyclophane discriminates guests through both hydrophobic and electrostatic interactions.

ANS   TNS

The fluorescence lifetime (τ) of a guest incorporated into 1 was greatly increased. For example, the τ value for ANS bound to 1 is 16.1 ns at pH 4.0 under aerobic conditions. Since an effective microenvironmental polarity provided by 1 for ANS under such conditions is equivalent to that of THF, the τ values for the guest were also measured in the organic solvent: 9.2 and 16.1 ns under aerobic and anaerobic conditions, respectively. This implies that the molecular cage of the cubic cyclophane perfectly protects the incorporated guest from oxygen attack. A similar cage effect was also observed for TNS when it was included in 1.

Another characteristic feature of the cubic cyclophane is the size-selective molecular discrimination that originates from the restricted and rigid geometry of the hydrophobic cavity. In other words, the cubic cyclophane tends to discriminate guest molecules on the basis of a

α-PNA   β-PNA

Table 2. Binding constants for host-guest complexes of 1 with hydrophobic guests in an aqueous acetate buffer [0.01 mol dm^{-3}, μ 0.10 (KCl)] at pH 4.0 and 30.0°C.

Guest	K (dm^3 mol^{-1})
Naphthalene	3000
β-PNA	69000
α–PNA	88000
Pyrene	5600
Perylene	710

lock-and-key mechanism. Thus, α-PNA fits the cavity most tightly among the guest molecules listed in Table 2, even though pyrene and perylene are more hydrophobic than α-PNA.

MOLECULAR RECOGNITION BY OCTOPUS CYCLOPHANES

Octopus cyclophane **2**, constructed with a rigid macrocylic skeleton and eight flexible hydrocarbon chains, provides a cavity that is deep and hydrophobic enough to incorporate hydrophobic guests of various bulkiness through an induced-fit mechanism originating from the flexible character of the alkyl branches [3]. We have now prepared other octopus cyclophanes (**3** and **4**), both of which have L-aspartate residues as connector units interposed between the macrocyclic ring and the alkyl branches. When an aqueous stock solution of **3** was injected into an aqueous buffer prepared with 2-[4-(2-hydroxyethyl)piperazinyl]ethane-sulfonate (HEPES) [0.01 mol dm^{-3}, pH 8.0, μ 0.10 (KCl)] containing a fluorescent guest, the host-guest complexation took place immediately without any complicated process (Table 3) [5]. In an analogous fashion to the guest-binding behavior exercised by **2**, host **3** can provide a highly apolar and viscous binding site for hydrophobic guests of various bulkiness and recognize them through hydrophobic and electrostatic interactions.

Table 3. Binding constants and microenvironmental parameters for host-guest complexes of octopus cyclophane **3** in an aqueous HEPES buffer [0.01 mol dm^{-3}, μ 0.10 (KCl)] at pH 8.0 and 30.0°C.

Guest	Binding Constant K (dm^3 mol^{-1})	Polarity $E_T(30)$ (kcal mol^{-1})	Flourescence polarization (P)
ANS	530000	49	0.33
TNS	1400000	51	0.32
α-PNA	100000	34	0.25
β-PNA	230000	44	0.13
DASP		No complex formation	

On the other hand, unique biphasic inclusion behavior was observed by using organic stock solutions of **3** in place of its aqueous stock solution under otherwise identical conditions. As shown in Figure 3, the fluorescence maximum and intensity originating from

ANS showed biphasic time courses upon addition of a methanol solution of **3** into an aqueous buffer. This behavior is consistent with fast incorporation of the guest molecule into the hydrophobic host cavity followed by slow and long-range conformational change of the host, as induced by the incorporated guest, to provide a highly desolvated microenvironment for the tight host-guest interaction. This behavior was also reflected in other fluorescence parameters such as lifetime and rotational correlation time. The biphasic behavior is schematically illustrated in Figure 4. However, octopus cyclophane **4** did not show such biphasic complexation behavior. On the basis of circular dichroism measurements, the biphasic host-guest characteristic of **3** must come from a rigid conformation of the L-aspartate moiety presumably caused by its intramolecular hydrogen bonding in organic media. This seems to be the primary reason why the complexation behavior of **3** is so different from that of **4** [5].

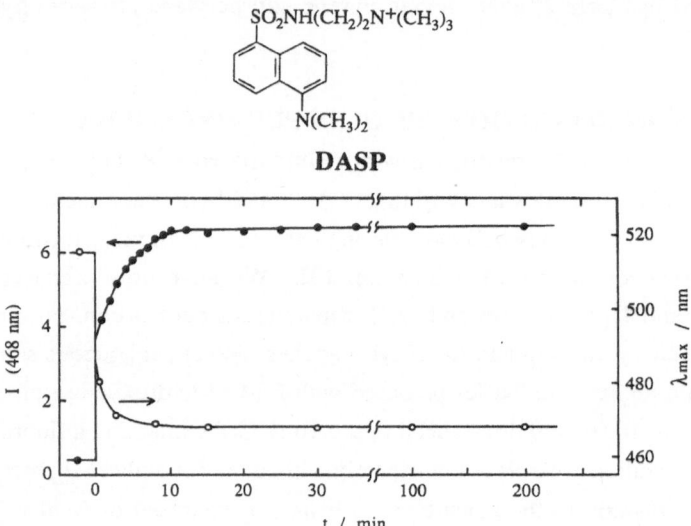

DASP

Figure 3. Time courses for changes of fluorescence intensity (I) and its maximum wavelength (λ_{max}) for ANS (2.5×10^{-7} mol dm^{-3}) upon addition of a methanol solution of **3** (2.5×10^{-6} mol dm^{-3}) in an aqueous HEPES buffer [0.01 mol dm^{-3}, pH 8.0, μ 0.1 (KCl)] at 30.0 °C.

CATALYTIC FUNCTIONS OF ARTIFICIAL HOLOENZYME COMPOSED OF OCTOPUS CYCLOPHANE AND HYDROPHOBIC VITAMIN B$_{12}$

The isomerization reactions catalyzed by vitamin B$_{12}$-dependent enzymes, which result in carbon-skeleton rearrangements, are of particular interest from the viewpoints of synthetic organic and organometallic chemistry [6]. The naturally occurring apoproteins, which provide relevant reaction sites for vitamin B$_{12}$, are considered to play crucial roles in the isomerization reactions [7]. However, non-enzymatic reactions so far studied have been exclusively related to the catalytic functions of the coenzyme, and relevant apoprotein models have been little explored. We have been interested in the catalytic activity of vitamin B$_{12}$ in a hydrophobic microenvironment, to simulate the catalytic functions of methylmalonyl-CoA mutase and glutamate mutase (Reactions 1 and 2, respectively) and have previously investigated various reactions of hydrophobic vitamin B$_{12}$ derivatives in single-compartment vesicles of peptide amphiphiles [8-10].

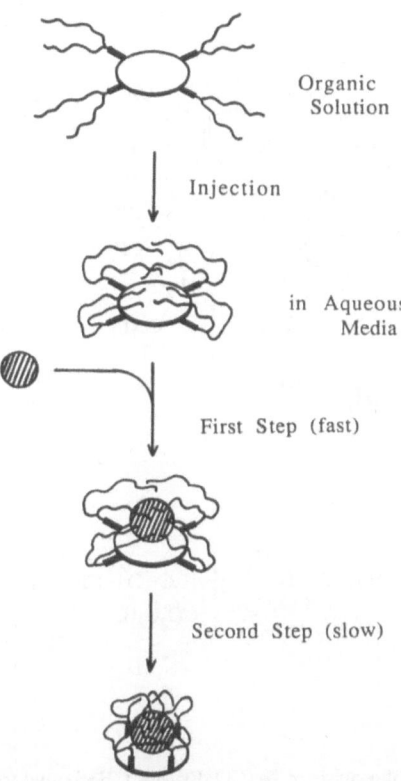

Figure 4. Schematic illustration of biphasic inclusion behavior.

$$\begin{array}{ccc}
\text{H} & \text{COOH} & \\
| & | & \\
\text{H-C-C-H} & \rightleftharpoons & \text{H-C---C-H} \\
| & | & \\
\text{H} & \text{CO} & \\
| & & \\
\text{SCoA} & & \\
\end{array}$$

methylmalonyl-CoA   Succinyl-CoA   (1)

$$\text{HOOC-C-CH}_2\text{-CH-COOH} \rightleftharpoons \text{HOOC-CH-CH-COOH}$$

L-glutamic acid (S)   L-threo-β-methylaspartic acid
(2S, 3S)   (2)

We have examined the catalytic activity of an artificial holoenzyme composed of hydrophobic vitamin B_{12} **5** and octopus cyclophane **2** in aqueous media. Hydrophobic vitamin B_{12} derivatives bearing various alkyl ligands at one axial site of the nuclear cobalt were incorporated into the hydrophobic cavity of **2** in aqueous media and then irradiated with visible light under anaerobic conditions. As an example, product analyses for the photolysis of a hydrophobic vitamin B_{12} derivative bearing the $Ac(CO_2Et)MeCCH_2$ moiety in the octopus cyclophane, methanol, and benzene are listed in Table 4 (see Reaction 3 for an explanation).

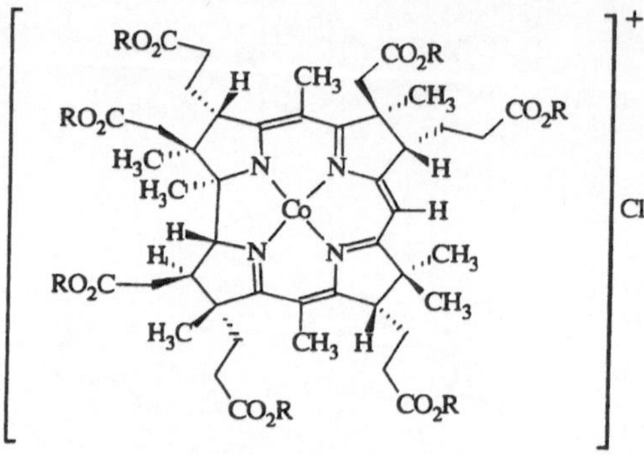

$R=C_3H_7$

5

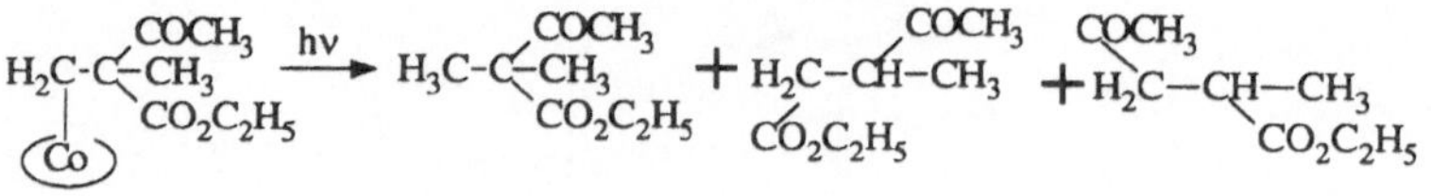

(3)

Table 4. Product analyses for the photolysis of Ac(CO$_2$Et)MeCCH$_2$ bound to **5** in various media at 20.0°C.[a]

Medium	Yield (%)		
	A	B	C
Methanol	70	trace	8.0
Benzene	65	trace	10
2[b]	23	trace	65

[a] A solution containing Ac(CO$_2$Et)MeCCH$_2$ bound to **5** (5.0 x 10^{-5} mol dm^{-3}) was irradiated with a 500 W tungsten lamp for 1 h from a distance of 30 cm.
[b] **2** (5.0 x 10^{-5} mol dm^{-3}), Ac(CO$_2$Et)MeCCH$_2$ bound to **5** (5.0 x 10^{-5} mol dm^{-3}) in an aqueous phosphate-borate buffer (0.05 mol dm^{-3}, pH 9.2).

The yield of acetyl-migration product C was significantly increased in the octopus cyclophane compared to those in methanol and benzene. Thus the octopus cyclophane is effective as an apoenzyme model for functional simulation of vitamin B$_{12}$-dependent enzymes. The 1,2-migration of electron-withdrawing groups was found to arise from both repression of motion and desolvation effects operating on the alkylated cobalt complexes placed in the cyclophane. Furthermore, the nuclear bivalent cobalt of the hydrophobic vitamin B$_{12}$ promotes the isomerization reaction via formation of a tight pair with the radical intermediate as shown in Scheme 1 [11].

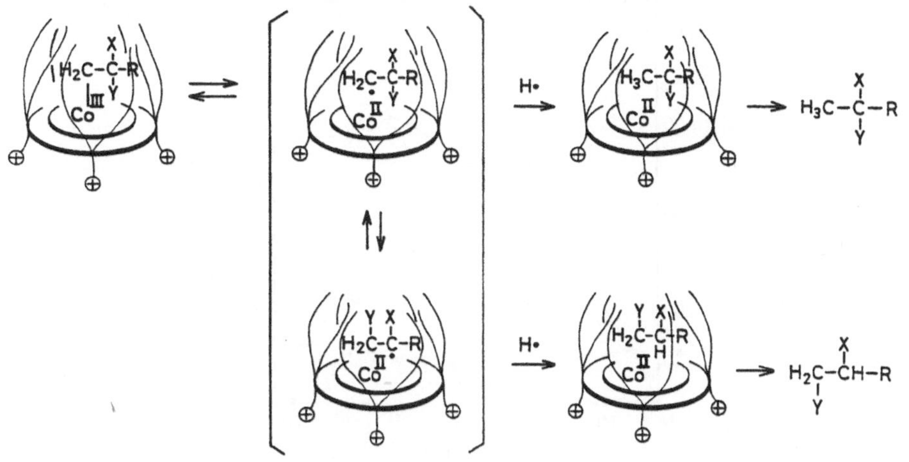

Scheme 1

The reaction given in Scheme 1 is a stoichiometric one even though the carbon-skeleton rearrangement takes place in such a manner. In order to set up a real catalytic system, we adopted a substrate-activation process, as reported by Schrauzer et al. [12], by employing vanadium trichloride and molecular oxygen. A substrate radical generated by the activation process reacted with the hydrophobic vitamin B_{12} to form the corresponding alkylated complex as an intermediate, which subsequently underwent photolysis to afford the rearrangement and reduction products, the latter being recycled as substrate (Scheme 2). The artificial holoenzyme system composed of the hydrophobic vitamin B_{12} and the octopus cyclophane is now capable of catalyzing the methylmalonyl-CoA mutase-like and glutamate mutase-like reactions with turnover of the catalyst as shown, for example, in Figure 5 [13].

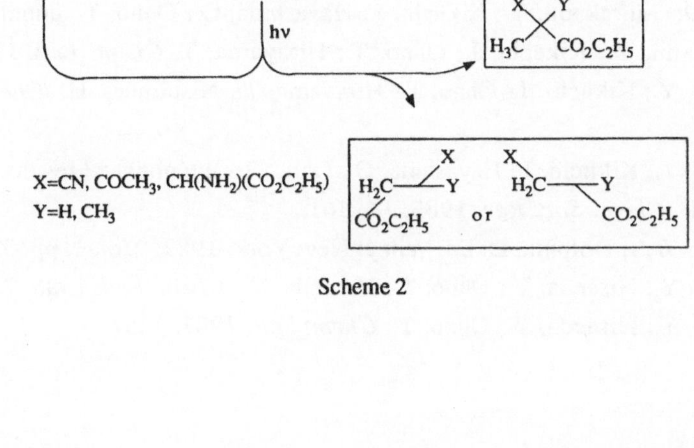

Scheme 2

115

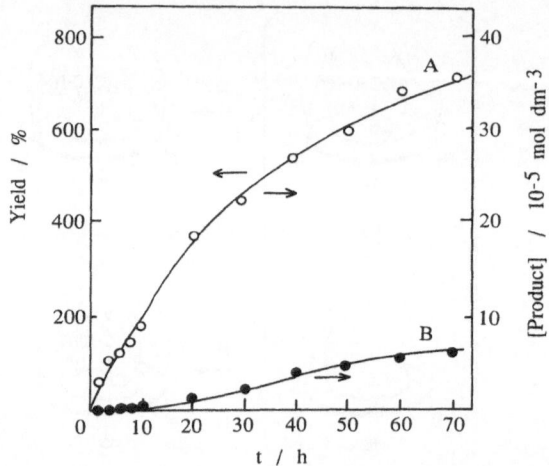

Figure 5. Conversion of diethyl-2-acetyl-2-ethoxycarbonylpropane (3.0×10^{-3} mol dm^{-3}) into 1-acetyl-2-ethoxycarbonylpropane in the presence of **2** (5.0×10^{-5} mol dm^{-3}) and VCl$_3$ (0.1 mol dm^{-3} every 12h) in an aqueous phosphate-borate buffer [0.05 mol dm^{-3}, pH 1.0] at 20 ± 1 °C under aerobic irradiation conditions: A, the Co^{II} species of **5** (5.0×10^{-5} mol dm^{-3}); B, without the Co^{II} complex. Yield was based on the hydrophobic vitamin B$_{12}$.

CONCLUSION

The rigid cubic cyclophane and the flexible octopus cyclophane provide three-dimensionally extended hydrophobic cavities and recognize guest molecules through biomimetic lock-and-key and induced-fit mechanisms, respectively. An artificial holoenzyme system, which can simulate catalytic functions of methylmalonyl-CoA mutase and glutamate mutase, is constructed by the combination of the hydrophobic vitamin B$_{12}$ and the octopus cyclophane, and its turnover is achieved by adopting a specific substrate-activation process.

References

1. Tabushi, I; Yamamura, K. *Top. Curr. Chem.* **1983**, *113*, 145; Murakami, Y. *Top. Curr. Chem.* **1983**, *115*, 107; Diederich, F. *Angew. Chem., Int. Ed. Engl.* **1988**, *27*, 362.
2. Murakami, Y.; Kikuchi, J.; Hirayama, T. *Chem. Lett.* **1987**, 161.
3. Murakami, Y.; Kikuchi, J.; Suzuki, M.; Matsuura, T. *J. Chem. Soc., Perkin Trans. 1* **1988**, 1289; Murakami, Y.; Kikuchi, J.; Hayashida, O.; Ohno, T., unpublished results.
4. (a) Murakami, Y.; Kikuchi, J.; Ohno, T.; Hirayama, T. *Chem. Lett.* **1989**, 881; (b) Murakami, Y.; Kikuchi, J.; Ohno, T.; Hirayama, T.; Nishimura, H. *Chem. Lett.* **1989**, 1199.
5. Murakami, Y.; Kikuchi, J.; Hayashida, O.; Ohno, T., unpublished results.
6. Pratt, J. M. *Chem. Soc. Rev.* **1985**, *14*, 161.
7. Pratt, J. M. *B$_{12}$*; Dolphin, D. Ed; Wiley: New York, 1982; Vol. 1, pp. 325-392.
8. Murakami, Y.; Hisaeda, Y.; Ohno, T.; Matsuda, Y. *Chem. Lett.* **1986**, 731.
9. Murakami, Y.; Hisaeda, Y.; Ohno, T. *Chem. Lett.* **1987**, 1357.

10. Murakami, Y.; Hisaeda, Y.; Ohno, T. *J. Chem. Soc., Chem. Commun.* **1988**, 856.

11. Murakami, Y.; Hisaeda, Y.; Kikuchi, J.; Ohno, T.; Suzuki, M.; Matsuda, Y.; Matsuura, T. *J. Chem. Soc., Perkin Trans. 2* **1988**, 1237.

12. Maihub, A.; Grate, J. W.; Xu, H. B.; Schrauzer, G. N. *Z. Naturforsch., Teil B* **1983**, *38*, 643.

13. Murakami, Y.; Hisaeda, Y.; Ohno, T., unpublished results.

SYNTHESIS AND COMPLEXING PROPERTIES OF CHIRAL GUANIDINIUM RECEPTORS DESIGNED FOR MOLECULAR RECOGNITION OF ANIONS

Antonio Echavarren,† Amalia Galán,§ Jean-Marie Lehn,#
and Javier de Mendoza*§

†Instituto de Química Orgànica
CSIC, Juan de la Cierva 3
28006-Madrid (Spain)
§Departamento de Química
Universidad Autónoma de Madrid
Cantoblanco
28049-Madrid (Spain)
#Institut Le Bel
Université Louis Pasteur
4 rue Blaise Pascal
67000-Strasbourg (France)

In sharp contrast to the well established host-guest chemistry of cations, anion-binding hosts have been developed only recently, and less than 1% of the current host-guest literature is devoted to their coordination, despite the obvious importance of anions in chemistry and biology [1].

One of the most useful ways to complex a negative substrate is by means of positively charged binding sites. This may be accomplished either by quaternary ammonium salts [2] or by protonated receptor molecules. In addition to the resulting charge to charge stabilization, protonation offers the possibility of further stabilizing the complex by hydrogen bonding. Therefore, most anion receptors belong to the class of polyprotonated cyclic polyamines [3]. Driving forces for such complexation are electrostatic interactions and hydrogen bonding. However, due to the generally low pK_a values of most amines, acidic solutions are usually required to ensure a full protonation. This problem could be obviated if a strong basic group is employed instead of the amine group.

Guanidinium units may serve as basic building blocks for anionic functional groups. Such is the role of this moiety in the arginine residues of biological receptor sites. The guanidinium group ($pK_a \approx 13.5$) remains protonated over a much wider range of pH than does the ammonium group. In addition, it can form characteristic pairs of zwitterionic hydrogen bonds $N-H^+ \cdots X^-$, which provide binding strength by their charge and structural organization by their arrangement. This can be observed in the crystal structures of many guanidinium salts, which involve a bidentate hydrogen-bonding pattern of the general type shown in Scheme 1. A similar arrangement is present in carboxylate salts. In this context, an important step in the development of anion coordination chemistry was the synthesis and study of some acyclic and cyclic guanidinium receptors in 1978 [4,5]. Potentiometric data showed that anion complexes of definite stoichiometries were formed with phosphate guests, and the receptors displayed some structural selectivity features, like macrocyclic and chelate effects. However, stability constants were generally lower for guanidinium than for ammonium receptors of similar shapes and geometries. This fact emphasizes the importance of the electrostatic factors in anion complexation, since the guanidinium group has a lower charge density than ammonium.

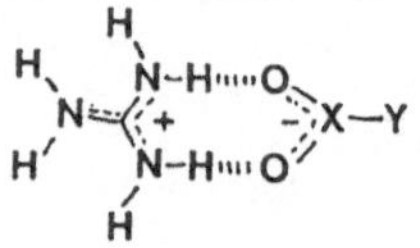

Scheme 1

In order to maintain the structure of the guanidinium group and to enhance its binding abilities, one may incorporate it into a rigid bicyclic framework. Some natural products contain the guanidine functionality as part of a cyclic or bicyclic system [6], and some abiotic bicyclic amidinium [7] and guanidinium [8-11] compounds have been prepared synthetically, but none are endowed with appropriate functional groups so as to allow their subsequent introduction into macrocyclic and macropolycyclic anion receptors. This could be accomplished by means of the bicyclic guanidine targets **1-SS**, **1-RR**, and **1-R*-S***, which can be assembled from chiral amino acids. Moreover, the optically active subunits **1-SS** and **1-RR** would be of interest for chiral recognition of substrates bearing appropriate anionic functional groups, such as carboxylic acids or phosphates.

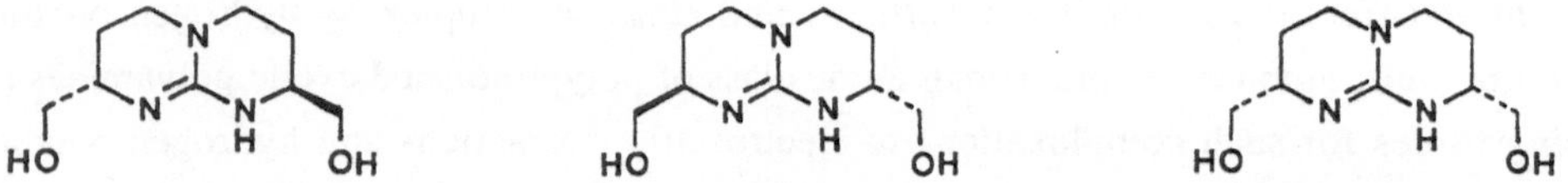

1-SS   **1-RR**   **1-R*-S***

Asparagine was selected as the precursor of choice, since it contains all of the atoms present in **1** except for the quaternary carbon. Asparagine was transformed in a few steps (Scheme 2) into the reduced and protected derivative of cyanoalanine **2**. Intermediate **2** was easily transformed into the secondary amine **3-SS**, by catalytic hydrogenation with rhodium-alumina in acetic acid, in a 60% yield. It is well known that catalytic hydrogenation of nitriles affords variable amounts of secondary amines, but this process has always been considered to be an undesirable side reaction and no examples are known of its use as a synthetic tool [12]. An alternative stepwise synthesis of amine **3** is also represented in Scheme 2. Removal of the tosyl groups from **3-SS** to give triamine **4-SS** was easily performed by sodium/ammonia, in a 70% yield, and the synthesis was completed by introduction of the last central carbon atom of the guanidinium target. This was achieved in about 60% yield with

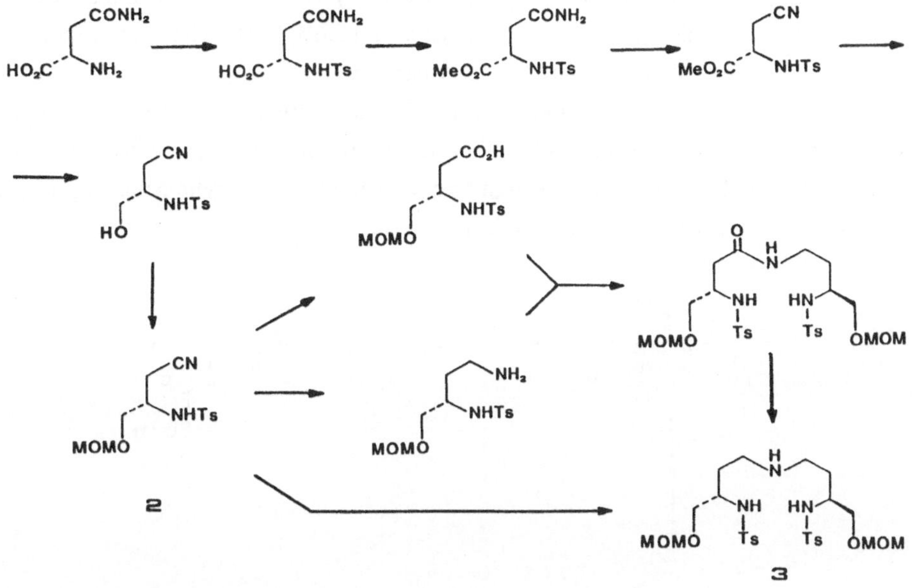

Scheme 2

thiocarbonyldiimidazole. Presumably, the reaction takes place via a thiourea intermediate, but no activation was necessary to achieve cyclization (Scheme 3). This represents, to our knowledge, one of the few examples of guanidine synthesis from thioureas not requiring activation to thiouronium derivatives [13], a fact that could be explained by the intramolecular character of the reaction.

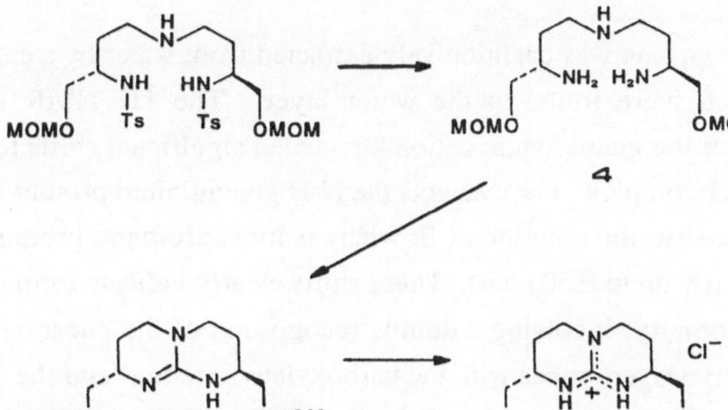

Scheme 3

Direct introduction of the guanidinium carbon by orthocarbonate esters was an attractive alternative. Our first attempts, following the Schmidtchen conditions (acid catalysis, dimethyl sulfoxide) [10] were unsuccessful, but after removal of the catalyst and use of an excess of reagent, guanidine 5-SS was obtained, now in a 40% yield. The final elimination of the methoxymethyl protecting groups was performed under acidic conditions, with formation of the corresponding salt of 1-SS in 50% yield. The overall yield of this nine-step synthesis of a chiral, rigid, bicyclic guanidine from L-asparagine was 17.6% [14]. The same sequence was also performed with the low-cost enantiomer, D-asparagine.

Ester derivatives 6-8 were prepared by reaction of 1 with excess of the corresponding acid chloride (Scheme 4). Compound 6 was designed for anion transport across model membranes, and 7 and 8 for molecular recognition of aromatic carboxylic acids or phosphates. The following preliminary results were obtained from 8.

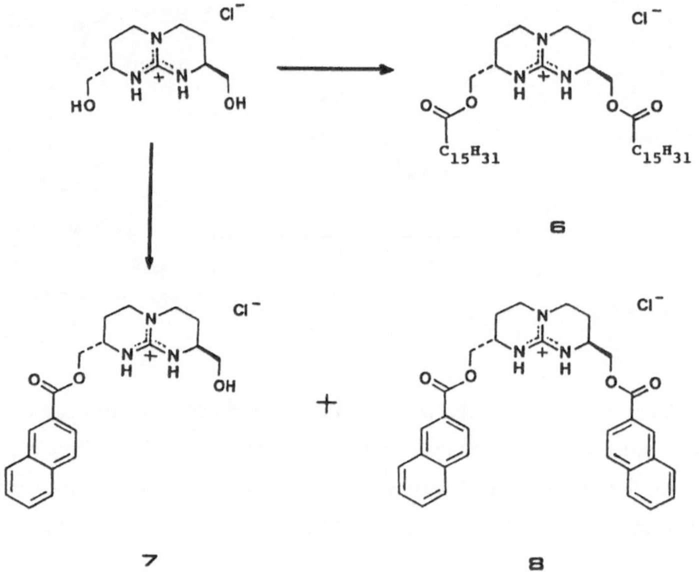

Scheme 4

Sodium p-nitrobenzoate was quantitatively extracted from water by a chloroform solution of 8. No traces of 8 were found in the water layer. The ^{1}H NMR spectrum of the p-nitrobenzoate salt of the guanidinium cation 8 revealed significant shifts for most signals of both species in the 1:1 complex. For instance, the N-H guanidinium protons shifted downfield by 1.78 ppm compared to the chloride of 8, whereas most aromatic protons of the host and guest shifted upfield (by up to 0.50 ppm). These shifts clearly indicate formation of a complex of a well defined geometry, involving a double recognition of the guest by the guanidinium cation (zwitterionic hydrogen-bonds with the carboxylate function) and the napthoyl residues at the side arms (stacking with the p-nitrophenyl moiety). NMR titration of 8 chloride with triethylammonium p-nitrobenzoate in CDCl$_3$ gave $K_S = 1609$ M^{-1} (calculated by the program 'SALS' from the chemical shift of the 1-napthoyl proton; standard deviation 12%).

Similar results were obtained by addition of triethylammonium *p*-methoxybenzoate or (+)-6-methoxy-α-methyl-2-naphthalene-acetate (naproxenate) to **8**.

The chirality of our host should force any substrate to bind in a asymmetric environment, allowing for an enantioselective recognition of chiral carboxylic acids. *N*-acetyl- and *N*-*tert*-butoxycarbonyl derivatives of sodium tryptophan were selected, for they contain all the features necessary for our goal, namely two recognition functions (the carboxylate and the well known π-donor indole ring) and a bulk substituent to interact sterically with the host aromatic side arm not involved in the stacking (see Scheme 5). Indeed, extraction of an excess of the racemic salt with **8** chloride afforded two diastereomeric salts, with an enantiomeric excess of ca. 30-40% for the L-tryptophan derivative. NMR titration of the triethylammonium salts of *N*-acetyltryptophan in CDCl$_3$ gave K_S = 1051 M^{-1} and K_S = 534 M^{-1} (standard deviations, 19.3% and 15.6%) for the L- and D-enantiomers, respectively. This represents, to our knowledge, the first example of enantioselective recognition of amino acid carboxylate salts by an abiotic receptor.

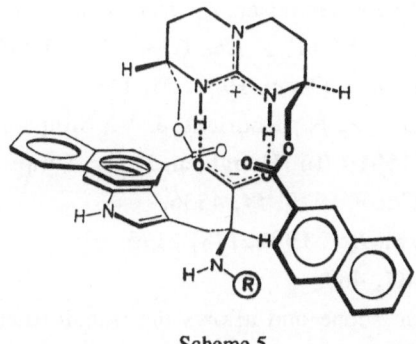

Scheme 5

Acknowledgements. We gratefully acknowledge support of this research by the "Comisión Asesora de Investigación Científica y Técnica" (CAICYT Grant 84-0410).

References

1. Some recent reviews on anion complexation have been published: Pierre, J.-L.; Baret, P. *Bull. Soc. Chim. Fr.* **1983**, II-367; Schmidtchen, F. P. *Nachr. Chem. Tech. Lab.* **1988**, *36*, 8.

2. For quaternary ammonium containing macropolycycles see: (a) Schmidtchen, F. P. *Angew. Chem., Int. Ed. Engl.* **1977**, *20*, 720; (b) Schmidtchen, F. P. *Chem. Ber.* **1980**, *113*, 864; (c) Schmidtchen, F. P. *Chem. Ber.* **1981**, *114*, 597; (d) Schmidtchen, F. P. *Chem. Ber.* **1984**, *117*, 725 and 1287; (e) Schmidtchen, F. P. *J. Chem. Soc., Perkin Trans. 2* **1986**, 135.

3. (a) Park, C. H.; Simmon, H. E. *J. Am. Chem. Soc.* **1968**, *90*, 2431; (b) Dietrich, B.; Hosseini, M. W.; Lehn, J.-M. *J. Am. Chem. Soc.* **1981**, *103*, 1282; (c) Kimura, E.; Sakonaka, A.; Yatsunami, T.; Kodama, M. *J. Am. Chem. Soc.* **1981**, *103*, 1041;

      (d) Kimura, E.; Kodama, M.; Yatsunami, T. *J. Am. Chem. Soc.* **1982**, *104*, 3182.

4.   Dietrich, B.; Fyles, D. L.; Fyles, T. M.; Lehn, J.-M. *Helv. Chim. Acta* **1979**, *62*, 2763.

5.   Dietrich, B.; Fyles, T. M.; Lehn, J.-M.; Pease, L. G.; Fyles, D. L. *J. Chem. Soc., Chem. Commun.* **1978**, 934.

6.   (a) Furusaki, A.; Tomije, Y.; Nitta, I. *Bull. Chem. Soc. Jap.* **1970**, *45*, 3332; (b) Wong, J.; Rapoport, H. *J. Am. Chem. Soc.* **1971**, *93*, 4633; (c) Bycroft, B. W.; Cameron, D.; Johnson, A. W. *J. Chem. Soc. C* **1971**, 3040; (d) Bycroft, B. W.; Croft, L. R.; Johnson, A. W.; Webb, T. *J. Chem. Soc., Perkin Trans.1* **1972**, 820; (e) Yoshioka, H.; Aoki, T.; Goko, H.; Nakatsu, K.; Noda, T.; Sakakibara, H.; Take, T.; Nagata, A.; Abe, J.; Wakamiya, T.; Shiba, T.; Kaneko, T. *Tetrahedron Lett.* **1971**, 2043; (f) Bodanszki, M.; Izdebski, J.; Muramatsu, I. *J. Am. Chem. Soc.* **1969**, *91*, 2351; (g) Hart, N. K.; Johns, S. R.; Lamberton, J. A.; Willing, R. I. *Aust. J. Chem.* **1970**, *23*, 1679; (h) Khuong-Huu, F.; Le Forestier, J.-P.; Goutarel, R. *Tetrahedron* **1972**, *28*, 5207.

7.   Heinzer, F.; Soukup, M.; Eschenmoser, A. *Helv. Chim. Acta* **1978**, *61*, 2851.

8.   (a) McKay, A. F.; Hatton, W. G. *J. Am. Chem. Soc.* **1956**, *78*, 1618; (b) McKay, A. F.; Kreling, M. E. *Can. J. Chem.* **1957**, *35,* 1438.

9.   (a) Bosin, T. R.; Hanson, R. N.; Rodricks, J. V.; Simpson, R. A.; Rapoport, H. *J. Org. Chem.* **1973**, *38*, 1591; (b) Houghten, R. A.; Simpson, R. A.; Hanson, R. N.; Rapoport, H. *J. Org. Chem.* **1979**, *44*, 4536.

10.   Schmidtchen, F. P. *Chem. Ber.* **1980**, *113*, 2175.

11.   Esser, F. *Synthesis* **1987**, 460.

12.   This reaction is of wide scope and allows the transformation of many nitriles into secondary amines. Echavarren, A.; Galán, A.; de Mendoza. J, to be published.

13.   For other examples of direct guanidine formation from thiourease in the presence of a metal oxide (PbO) see: (a) Bhargara, P. N.; Singh, H. *J. Med. Chem.* **1969**, *12*, 558; (b) Macholdt-Erdniss, J. *Chem. Ber.* **1958**, *91*, 1992; (c) Malik, W. V.; Srivastara, P. K.; Mehra, S. *J. Med. Chem.* **1968**, *11*, 126.

14.   Echavarren, A.; Galán, J.; de Mendoza, J.; Salmerón, A.; Lehn, J.-M. *Helv. Chim. Acta* **1988**, *71*, 685.

MOLECULAR DESIGN OF CALIXARENE-BASED HOST MOLECULES

Seiji Shinkai

Department of Organic Synthesis
Faculty of Engineering
Kyushu University, Fukuoka 812 (Japan)

SUMMARY

Calixarenes are applied as novel ionophores, host molecules, chiral receptors, and superuranophiles.

INTRODUCTION

The chemistry of cyclodextrins and crown ethers has been a focus of interest in host-guest chemistry for the last two decades. More recently, Gutsche and co-workers [1] have reported on a series of new cyclic oligomers called "calixarenes". We have currently been interested in the functionalization of calixarenes which may act as host molecules in solution [2]. We here report new applications of calixarenes as (i) ionophores, (ii) host molecules, (iii) chiral receptors, and (iv) superuranophiles.

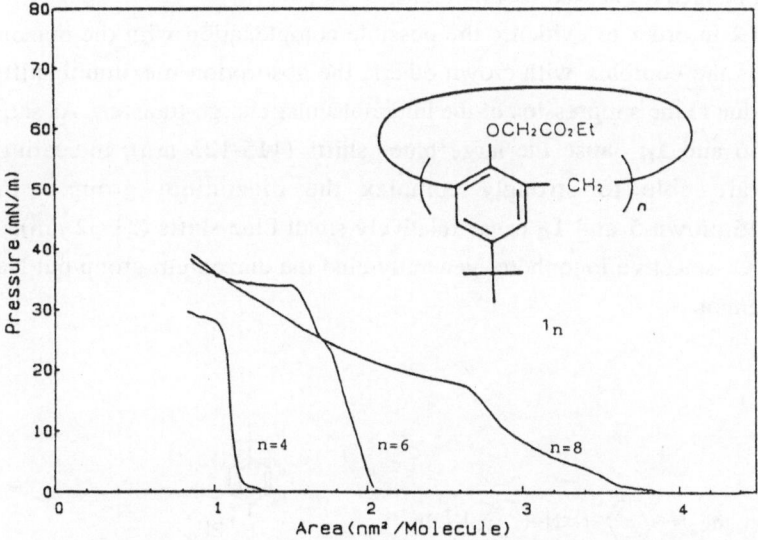

Figure 1. Area-pressure curves of 1_n on pure water at 20°C.

Inclusion Phenomena and Molecular Recognition
Edited by J. Atwood
Plenum Press, New York, 1990

CALIXARENES AS IONOPHORES

It is known that calixarene esters (1_n) act as ionophores for alkali and alkaline earth metal cations [3,4]. The tetramer (1_4) shows the selectivity order of $Na^+ > K^+ > Rb^+$ while the hexamer (1_6) shows the selectivity order of $K^+ > Rb^+ > Na^+$. However, the sharp metal selectivity is observed only for 1_4 which can selectively bind Na^+. The low metal selectivity of 1_6 and 1_8 is related to the flexibility of these calixarene cavities. In contrast, 1_4, the lower rim of which is sterically crowded, can provide the conformationally rigid cavity resulting in a high selectivity toward Na^+.

Figure 1 shows the area-pressure relationship of 1_n for the monolayer formation at the air-water interface. It is clearly seen from Figure 1 that 1_4 and 1_6 form the condensed phase at 1.1 and 1.9 nm^2, respectively, which exactly corresponds to the area of the larger side of the calixarene cavity (i.e., the upper rim). Very interestingly, the area-pressure curve for 1_6 is more gently-sloping than that for 1_4, and 1_8 has the most gently-sloping curve among three calixarenes. The difference is well accounted for by the flexibility of the calixarene cavity: that is, the conformation of 1_8 having the most flexible cavity is reorganized several times with increasing surface pressure and finally fixed in a condensed phase. We found that the area-pressure relationship is sensitively affected by the added MCl. The condensed phase of 1_4 was expanded to 2 nm^2 only when NaCl (1.0 M) was added to the subphase. Similarly, the condensed phase of 1_6 was expanded to 4 nm^2 only when KCl (0.1 M) was added to the subphase. The results are well complementary to the ion selectivity in the homogeneous solution and suggest the potential for receptors or ion-sensors.

Gokel and Cram [5] found that crown ethers of proper dimensions can solubilize several arenediazonium salts in nonpolar solvents. Of additional interest is the finding that the thermal decomposition of arenediazonium salts is slowed down when the salts are complexed by crown ethers [6]. As 1_n can complex alkali metal cations as crown ethers do, it is of great interest to study how 1_n behave for arenediazonium ions.

We used **2** in order to evidence the possible complexation with the diazonium group. When **2** forms the complex with crown ethers, the absorption maximum shifts to shorter wavelengths due to the suppression of the intramolecular charge-transfer. As shown in Table 1, 18-crown-6 and 1_6 cause the large blue shifts (115-123 nm), indicating that these ionophores are able to strongly complex the diazonium group. In contrast, monobenzo-15-crown-5 and 1_4 cause relatively small blue shifts (21-32 nm). The results suggest that K^+-selective ionophores generally bind the diazonium group but Na^+-selective ionophores cannot.

2

3

Table 1. Absorption maxima and association constants (40°C, tetrachloroethane).

Host molecule	λ_{max} (nm)	$\Delta\lambda$ (nm)	K (M^{-1})
None	680 (= λ_0)		
Monobenzo-15-crown-5	648	32	230
18-Crown-6	565	115	178,000
1_4	659	21	96
1_6	557	123	16,500
1_8	637	43	210

We measured the dediazoniation rate of **3** in the presence of these additives. As reported previously [6], 18-crown-6 suppressed the rate efficiently whereas the inhibition effect was scarcely observed for 1_n. Two opposing mechanisms have been proposed for the inhibition effect of crown ethers: that is, the thermal decomposition is suppressed because of (i) the reduction of positive charge at the diazo group or (ii) the macrocyclic effect which sterically inhibits the reaction going from the linear diazo group to the bulky π-diazo-intermediate. The fact that 1_n having the flexible ionophoric cavities showed no effect supports the view that the mechanism (ii) is more likely.

CALIXARENES AS HOST MOLECULES

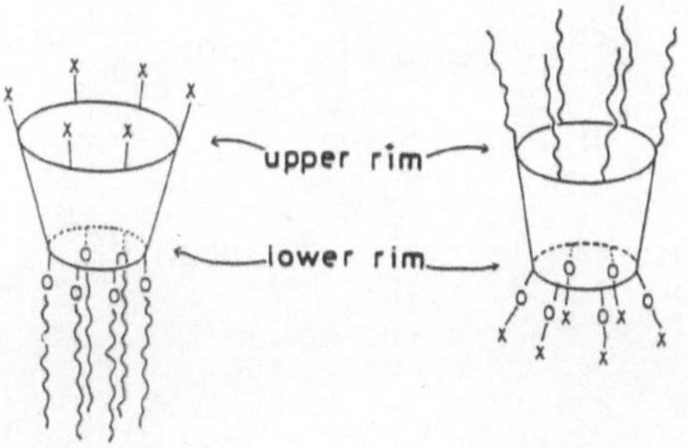

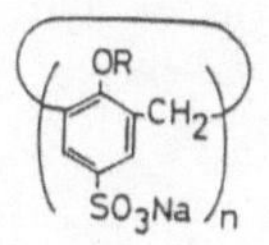

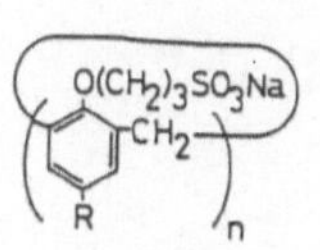

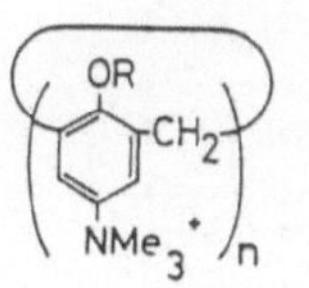

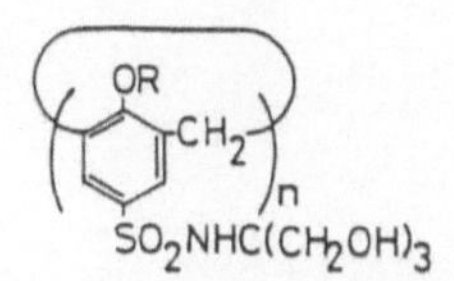

Gutsche described in his review article [1] that there are no published data in support of solution complexation by calixarenes. In order to find evidence for solution complexation, we have synthesized several water-soluble calixarenes because one can generally expect the "hydrophobic interaction" in an aqueous system. They are anionic ($4_n, 5_n$), [7,8] cationic (6_n) [2,9], and neutral calixarenes (7_n) [10]. In 5_n the alkyl groups are incorporated into the p-position, so that they can act cooperatively with the cavity as a binding site.

We are interested in the diazo-coupling with calix[4]arene (8_4) because the product $10(x)$ has the extended cavity and shows the chromogenic nature. The reaction was carried out at 5°C in THF in the presence of pyridine. Interestingly, we found that the reaction shows an all-or-nothing aspect: that is, regardless of the molar ratio of 8_n vs. 9 the main product is always the fully-substituted $10(x=4)$ and the yields for $10(x=1)$, $10(x=2)$, and $10(x=3)$ are always low [11]. This means that the diazo-coupling reaction with 8_4 occurs autoacceleratively. The unusual finding is rationalized in terms of strong intramolecular hydrogen-bonding among the hydroxyl groups. Introduction of the electron-withdrawing azo group into the p-position lowers not only the pK_a of the OH group in the substituted phenol unit but also those in the neighboring phenol units through strong hydrogen bonding. Thus, the substitution facilitates the dissociation of the OH groups acceleratively and the diazo-coupling occurs autoacceleratively.

What happens when equimolar amounts of anionic and cationic calixarenes are mixed in water? We synthesized cationic calixarenes according to Scheme 1 and mixed these with anionic 4_n. We found that when calixarenes with the same ring size (e.g. 4_4 and 6_4) are mixed, they give a white precipitate that is insoluble in any solvent (Figure 2). Even more interesting is the finding that this capsule is capable of entraping small guest molecules. They did not leak out of the capsule by solvent extraction or by reduced pressure. This is a new class of compartmentalization effected by only two molecules.

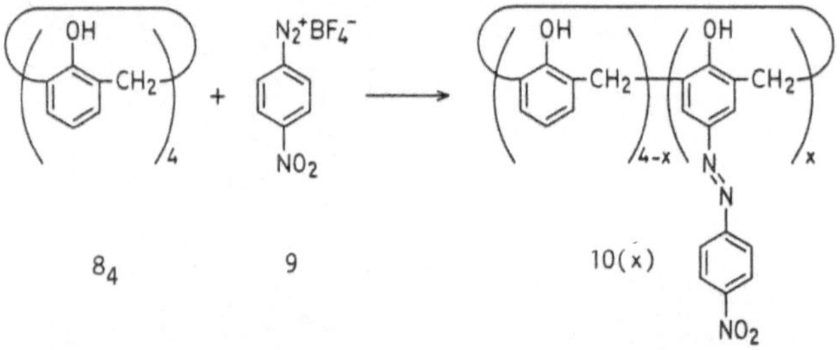

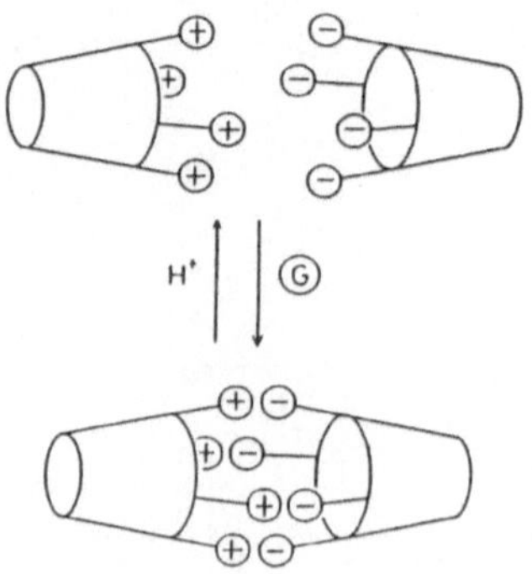

Figure 2. Formation of a molecular capsule from anionic and cationic calixarenes.

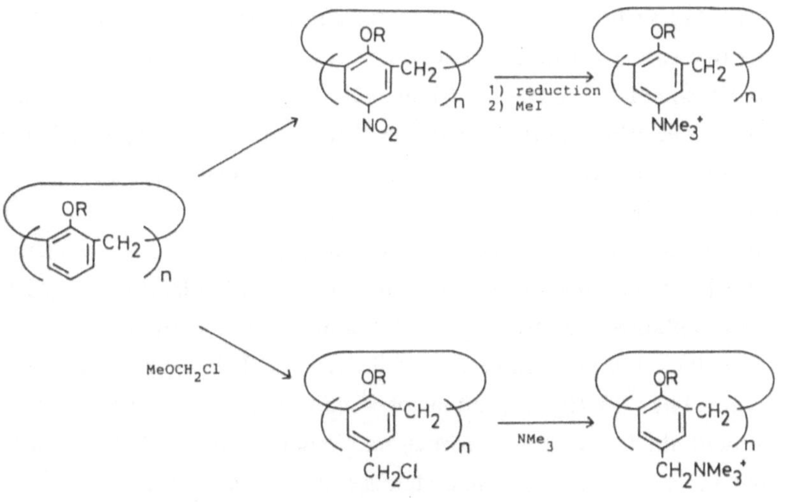

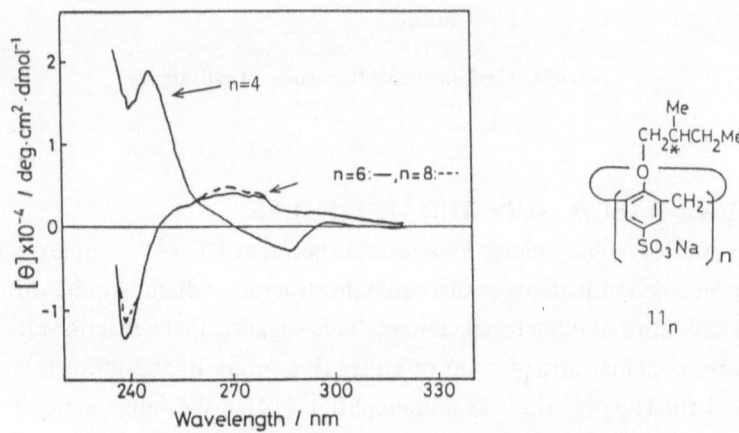

Scheme 1

CHIRAL CALIXARENES

When comparing calixarenes with cyclodextrins, one can raise several essential differences. Among them, the particular difference is the ability of cyclodextrins to catalyze certain reactions asymmetrically owing to the presence of the chiral cavity made up of D-glucose units. It thus occurred to us that introduction of chiral substituents into calixarenes would be of great value for development of a new class of chiral host molecules. We thus synthesized chiral calixarenes (11_n) with the (S)-2-methylbutyl groups [12,13].

In an aqueous solution 11_n gave the absorption maximum ($^1\lambda a$) at 249 nm. As shown in Figure 3, the CD maximum for 11_4 appeared at the same wavelength. In contrast, the CD spectra for 11_6 and 11_8 showed the presence of exciton coupling with (R)-chirality. Furthermore, the CD spectrum of 11_4 was totally unaffected by the added guest molecules whereas

Figure 3. CD spectra of 11_n (20°C, [11_n] = (2.5-5.0) x 10^{-4} M).

those of 11_6 and 11_8 were weakened with increasing guest molecule concentrations. From the CD change one can estimate the association constants (K): for 11_6, for example, K = 1.4 x 10^2 for 1-hexanol, 1.2 x 10^3 for 1-heptanol, 7.8 x 10^3 for 1-octanol, and ca. 80 M^{-1} for cyclohexanol [12]. What is the origin of the difference in the CD spectra between 11_4 and others? As mentioned above, calix[4]arene features conformational rigidity. Thus, the ring cannot fluctuate enough for each exciton to interact at the excited state. In contrast, the ring flexibility still remains in calix[6]arene and calix[8]arene and each exciton interacts when the ring adopts the deformed conformation (Figure 4). The CD change induced by the guest addition is thus explained by the fact that the conformational fluctuation is considerably suppressed by inclusion of guest molecules in the cavity. This view is also supported by the ^{1}H-NMR data. The ArCH$_2$Ar protons of vacant 11_6 gave a sharp singlet which is commensurate with the "alternate" conformation. When the guest (1-dodecanol) was added, the peak changed to a few sets of double doublets which is commensurate with the "cone"-like conformation. The finding clarifies an interesting facet of the calixarene conformation: the benzene units in vacant 11_6 can freely rotate in solution, but when the guest is added they orient to one direction in order to construct a hydrophobic cavity to bind the guest.

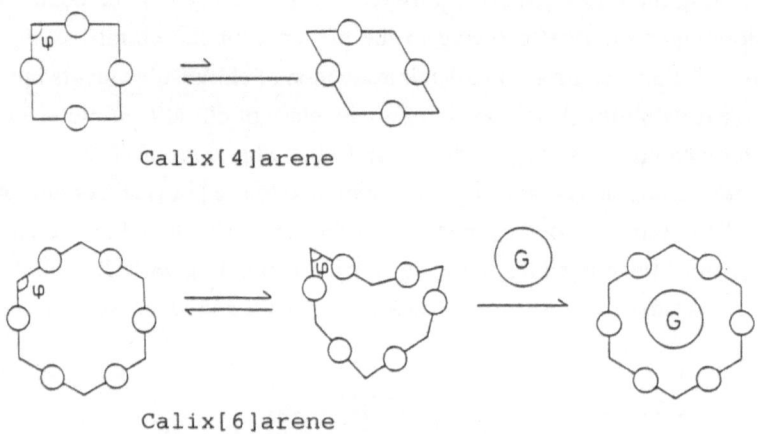

Calix[4]arene

Calix[6]arene

Figure 4. Conformational fluctuation of calixarenes.

CALIXARENE-BASED SUPERURANOPHILES

X-ray crystallographic studies have established that UO_2^{2+} complexes adopt either a pseudoplanar pentacoordinate or hexacoordinate structure, which is quite different from the coordination structures of other metal cations. This suggests that a macrocyclic host molecule having a nearly coplanar arrangement of either five or six ligand groups would serve as a specific ligand for UO_2^{2+} (i.e., as a uranophile) [14]. We thus applied water-soluble calixarenes $12_5(R)$ and $12_6(R)$ as a new class of uranophiles [15,16].

Table 2. Stability constants for $12_n(R)$ (25°C).

Calixarene	log K_{uranyl}
12_4 (R=H)	3.2
12_4 (R=CH$_2$COOH)	3.1
12_5 (R=H)	18.9
12_5 (R=CH$_2$COOH)	18.4
12_6 (R=H)	19.2
12_6 (R=CH$_2$COOH)	18.7

As shown in Table 2, $12_5(R)$ and $12_6(R)$ have remarkably large stability constants ($K_{uranyl} = 10^{18.4-19.2}$ M^{-1}), whereas 12_4 have very small stability constants ($K_{uranyl} = 10^{3.1-3.2}$ M^{-1}). The trend is very compatible with the X-ray data but cannot be rationalized by the so-called "hole-size selectivity". Hence, the high stability is better explained by "coordination-geometry selectivity". The selectivity factors ($K_{uranyl}/K_{M^{n+}}$) for (R=H) are recorded in Table 3. It was found that the selectivity factors for 12_6 (R=H) are surprisingly large, $10^{10.6-17}$, as compared with competing Ni^{2+}, Zn^{2+}, and Cu^{2+}. The remarkably high selectivity is attributed to the moderately rigid skeleton of calix[6]arene which can provide the preorganized hexacoordination geometry for the binding of UO$_2{}^{2+}$ but cannot accommodate to the square-planar or tetrahedral coordination geometry for other metal cations in an induced-fit manner.

Table 3. Selectivity factors of 12_6 (R=H) for UO$_2{}^{2+}$ (25°C).

Metal (M^{n+})	log $K_{M^{n+}}$	$K_{uranyl}/K_{M^{n+}}$
UO$_2{}^{2+}$	(19.2)	1.0
Mg^{2+}	2	10^{17}
Ni^{2+}	2.2	10^{17}
Zn^{2+}	5.5	10$^{13.7}$
Cu^{2+}	8.6	10$^{10.6}$

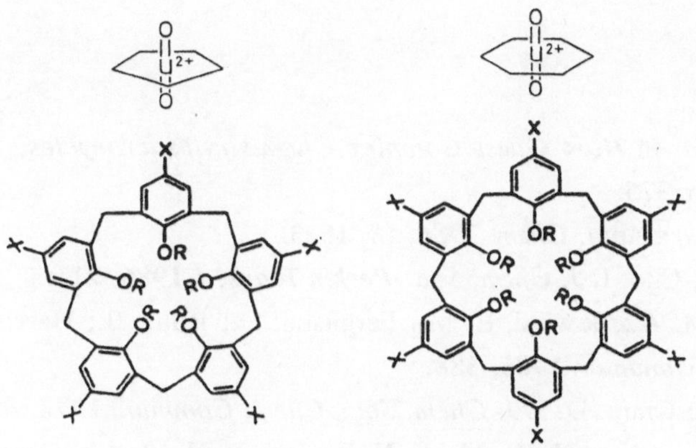

$12_5(R)$ $12_6(R)$

X = SO$_3$Na, R = H or CH$_2$COOH

In general, there are two possible strategies for improving the metal selectivity of macrocyclic ligands: the first one is to enhance the stability constant for the target cation and the second one is to lower the stability constants for competing metal cations (Figure 5). If the first strategy is employed, one should design some very rigid macrocycles. Hence, this approach is frequently accompanied by the disadvantage that the dynamic process becomes very slow and is not necessarily recommended for the design of uranophilies. In contrast, the second strategy does not have this disadvantage and the uranophiles would be applicable as carriers in dynamic processes such as solvent extraction and membrane transport. Table 2 shows that calixarenes provide an ideal basic skeleton for the second strategy.

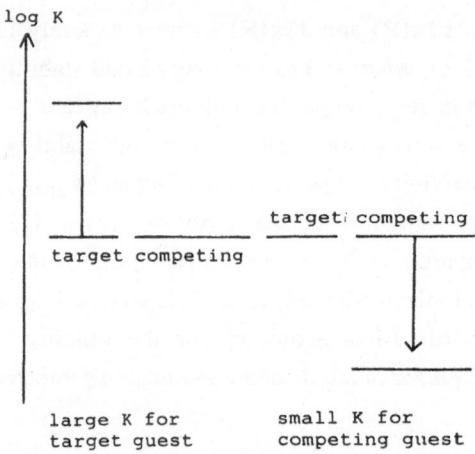

Figure 5. Strategies to enhance the guest selectivity.

CONCLUSION

It is thus demonstrated that properly functionalized calixarenes behave as a new class of host molecules.

Acknowledgment. We are indebted to Professors T. Kunitake and Y. Ishikawa for the monolayer studies.

References

1. Gutsche, C. D. In *Host Guest Complex Chemistry/Macrocycles*; Springer-Verlag: Berlin, 1985; p 375.
2. Shinkai, S. *Pure Appl. Chem.* **1986,** *58,* 1523.
3 Chang, S.-K.; Cho, I. *J. Chem. Soc., Perkin Trans. 1* **1986,** 211.
4. McKervey, M. A.; Seward, E. M.; Ferguson, G.; Ruhl, B.; Harris, S. J. *J. Chem. Soc., Chem. Commun.* **1985,** 388.
5. Gokel, G. W.; Cram, D. J. *J. Chem. Soc., Chem. Commun.* **1973,** 482.
6. Bartsch, R. A.; Chen, H.; Haddock, N. F.; Juri, P. N. *J. Am. Chem. Soc.* **1976,** *98,* 6753.

7. Shinkai, S.; Mori, S.; Tsubaki, T.; Sone, T.; Manabe, O. *Tetrahedron Lett.* **1984**, *25*, 5315.

8. Shinkai, S.; Mori, S.; Koreishi, H.; Tsubaki, T.; Manabe, O. *J. Am. Chem. Soc.* **1986**, *108*, 2409.

9. Shinkai, S.; Tsubaki, T.; Manabe, O. to be submitted.

10. Shinkai, S.; Kawaguchi, H.; Manabe, O. to be submitted.

11. Shinkai, S.; Araki, K.; Shibata, J.; Manabe, O. *J. Chem. Soc., Perkin Trans. I* **1989**, 195.

12. Shinkai, S.; Arimura, T.; Satoh, H.; Manabe, O. *J. Chem. Soc., Chem. Commun.* **1987**, 1495.

13. Arimura, T.; Edamitsu, S.; Shinkai, S.; Manabe, O.; Muramatsu, T.; Tashiro, M. *Chem. Lett.* **1987**, 2269.

14. Tabushi, I.; Kobuke, Y.; Ando, K.; Kishimoto, M.; Ohara, E. *J. Am. Chem. Soc.* **1980**, *102*, 5948.

15. Shinkai, S.; Koreishi, H.; Ueda, K.; Manabe, O. *J. Chem. Soc., Chem. Commun.* **1986**, 233.

16. Shinkai, S.; Koreishi, H.; Ueda, K.; Arimura, T.; Manabe, O. *J. Am. Chem. Soc.* **1987**, *109*, 6371.

COMPLEXATION OF IONS AND NEUTRAL MOLECULES BY FUNCTIONALIZED CALIXARENES

Rocco Ungaro*, Andrea Pochini, and Arturo Arduini

Istituto di Chimica Organica dell'Universita
Viale delle Scienze I
43100 Parma (Italy)

INTRODUCTION

After the pioneering work of C. D. Gutsche [1] and ourselves [2], which was mainly devoted to establishing the structures, both in solution and in the solid state, of the already [3] known cyclic products of the base catalyzed condensation of phenols and formaldehyde, the Calixarenes 1 became rather popular among the chemical community as potential hosts for Molecular Recognition [4].

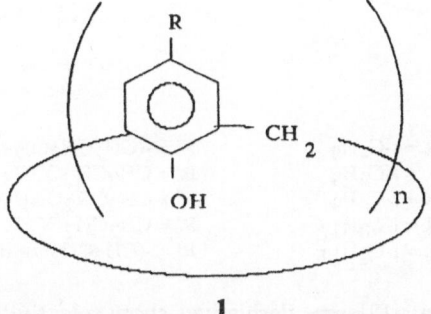

1

CALIXARENES AS BUILDING BLOCKS FOR CATION RECEPTORS AND CARRIERS

Calixarene Polyether Derivatives

Our first interest in calixarenes arose when we recognized in these macrocycles the potential to act, after functionalization of the phenolic oxygen atoms with suitable binding groups, as cation complexing agents [5].

The first series of cation ligands which were synthesized in good yield was that of the calixarene podands 2 obtained by introducing oligoethylene-glycol units of different length on calixarenes.

Table 1. Extraction ($H_2O \rightarrow CH_2Cl_2$) of alkali metal and ammonium picrates by calixarene podands, $T = 22°C$.

$$M^+(aq) + Pi^-(aq) + L_{org} \Leftrightarrow (L,M^+,Pi^-)_{org}$$

Ligand		K_e x 10^{-3} M^{-2}				
R	n	Na^+	K^+	Cs^+	NH_4^+	$C(NH_2)_3^+$
2b t-C$_4$H$_9$	6	0.20	0.21	1.70	0.32	28.30
2c t-C$_4$H$_9$	8	0.10	0.25	1.60	0.40	15.30
2d t-C$_8$H$_{17}$	8	0.20	0.36	1.86	1.00	25.04
2e t-C$_8$H$_{17}$	8	1.70	1.91	5.15	2.90	52.10
Dibenzo-18-crown-6	8.90	137.00	33.20	23.10	<1.00	

Although less powerful than normal crown ethers or other calixarene cation ligands reported recently by us [6] and other authors [7], the ligands 2b-e show the basic features of the complexing properties of the overall class of calixarene-based ionophores (Table 1).

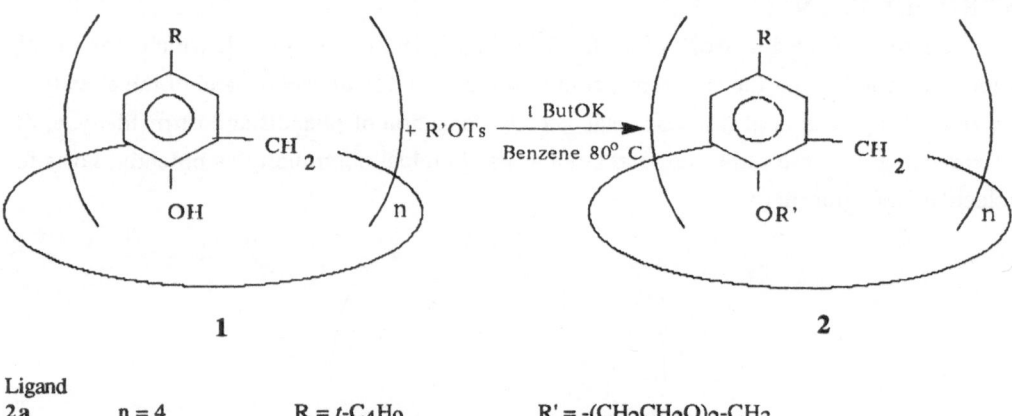

Ligand			
2a	n = 4	R = t-C$_4$H$_9$	R' = -(CH$_2$CH$_2$O)$_2$-CH$_3$
2b	n = 6	R = t-C$_4$H$_9$	R' = CH$_2$CH$_2$OCH$_3$
2c	n = 8	R = t-C$_4$H$_9$	R' = CH$_2$CH$_2$OCH$_3$
2d	n = 8	R = t-C$_8$H$_{17}$	R' = CH$_2$CH$_2$OCH$_3$
2e	n = 8	R = t-C$_8$H$_{17}$	R' = -(CH$_2$CH$_2$O)$_2$-CH$_3$

In fact, calix[6]- and calix[8]arene derivatives show selectivity toward large cations such as cesium among the alkali metals and guanidinium among the ammonium cations.

Moreover Table 1 shows that calix[6]arene podands are more efficient and selective than the corresponding calix[8]arene derivatives and that the extraction ability of the compounds can be improved either by increasing the lipophilicity of the backbone (compare 2c and 2d) or increasing the length of the binding chain (compare 2d and 2e). The selectivity towards sodium which has been evidenced in other calix[4]arene derivatives [6,7] could not be observed clearly with ligand 2a because this compound does not exist in the "cone" conformation [5] and shows very little extraction ability. However, Nakamoto and co-workers have been able recently to synthesize a calix[4]arene podand analogous to 2a but in the fixed cone structure, and this compound shows selectivity toward sodium over other alkali metal cations [8].

X-ray crystallography allowed us to establish the conformational preferences of ligands **2** in the solid state [9], and solution studies showed that calix[6]- and calix[8]arene derivatives **2b-e** are conformationally mobile in solution [5].

Calixarene Esters and Amides

Conformationally immobile calix[4]arene derivatives with fixed "cone" structures and convergent binding chains have been synthesized [6,7]. In particular we have obtained the ester **3** and the amide **4** in good yields and established their fixed cone structure both in solution and in the solid state [6].

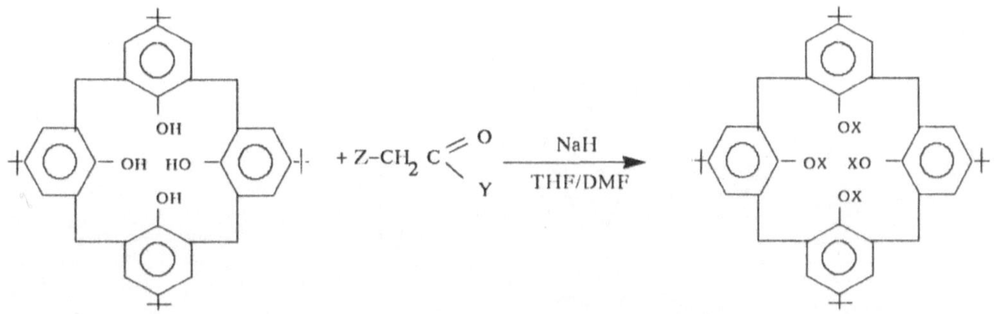

1a

Z = Cl or Br

Y = N(C₂H₅)₂ or OBuᵗ

3: X = CH₂COOBuᵗ

4: X = CH₂CON(C₂H₅)₂

Table 2 reports the association constants and the binding free energies of complexes of compounds **3** and **4** with alkali metal cations established via the extraction method reported by D. J. Cram [10].

Table 2. Association constants (K_a) and binding free energies ($-\Delta G°$) of complexes of hosts **3** and **4** with alkali metal picrates in $CDCl_3$ saturated with H_2O at 22°C[a].

			K_a M^{-1} [$-\Delta G°$ (kcal mol^{-1})]		
Compound	Li⁺	Na⁺	K⁺	Rb⁺	Cs⁺
3	8.5×10^5 [8.0]	7.8×10^9 [13.3]	4.2×10^6 [8.9]	4.2×10^5 [7.6]	1.7×10^6 [8.4]
4	1.5×10^9 [12.3]	8.3×10^{11} [16.1]	2.3×10^9 [12.6]	8.5×10^7 [10.7]	1.2×10^5 [6.8]

[a]Determined according to Cram et al. [10].

These two ligands show very high efficiency and good selectivity towards sodium cations. The Na⁺/Li⁺ (≈ 10⁴ for **3** and ≈ 5 x 10² for **4**) and the Na⁺/K⁺ selectivity (≈ 2 x 10³ for **3** and 4 x 10² for **4**) makes these two ligands and all the family of calix[4]arene ester and amides attractive for sodium-selective electrodes or other sensors. Some results on compounds very similar to **3** and **4** have been recently reported [11,12].

Another interesting aspect of ligands **3** and **4** is that they completely encapsulate the complexed cation and efficiently deshield it from the counterion and from the solvent. The X-ray crystal structure of the potassium complex of ligand **4** (Figure 1) shows the cation surrounded by four amide and four ether oxygen atoms in the form of an antiprism.

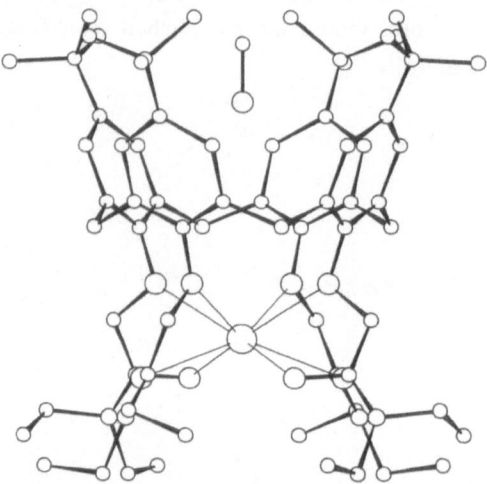

Figure 1. K^+ complex of **4**.

This last property prompted us to investigate, together with V. Balzani and N. Sabbatini, the binding properties of ligand **4** toward lanthanide ions. A 1:1 complex between Eu^{3+} and ligand **4** was prepared and the luminescence properties studied in water solution [13].

By analyzing the lifetimes of the luminescent $^5D_0$ excited state of Eu^{3+}, it was possible to establish that only one molecule of water is coordinated to the complexed cation, a result which encourages further synthetic efforts towards new calixarene amide ligands with higher stability and better energy transfer toward Eu^{3+}.

IONIZABLE CALIXARENE LIGANDS

Besides neutral ionophores we have also synthesized calix[4]arene derivatives having ionizable phenolic or carboxylic groups which could be used in the active transport of cations through liquid membranes.

As expected ligands **5** and **6** which have ionizable phenolic OH groups are effective in the transport of monovalent cations from basic conditions (pH 12-13) to acidic water solution.

The presence of a crown bridge not only increases the binding ability of the parent *p-tert*-butylcalixarene (**1a**) (which shows no cation transport under our conditions), but introduces a double control of selectivity. The first one is due to ion-pair formation and to its lipophilicity and the second one to the complementarity between the size of the crown and the ionic radii of the complexed cations.

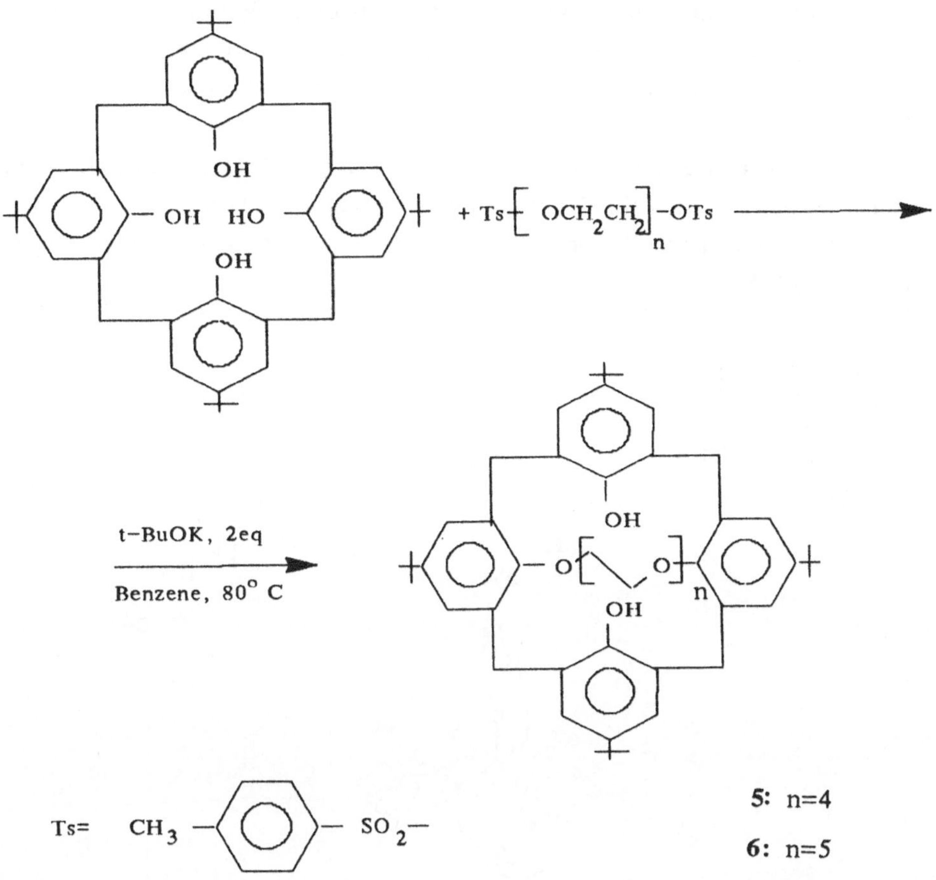

$Ts = CH_3 \!-\!\!\langle O \rangle\!\!- SO_2\!-$

5: n=4

6: n=5

In fact, when calix[4]-, calix[6]- and calix[8]arenes, without functionalization, have been used for the active transport of alkali metals through liquid membranes [14], the observed selectivity was always towards the less hydrated cesium cation which gives the most lipophilic ion pair with the ionized ligands.

Table 3. Single ion active transport of alkali metal cations through a liquid membrane (CH_2Cl_2) containing calixarene macrocycles.[a]

Ionophore	Transport rate 10^{-6} mol/24 hr			Selectivity
	Na^+	K^+	Cs^+	
p-tert-butylcalix[4]arene **1a**	No transport			
calixarene-crown-5 **5**	2.2	5.4	1.1	$K^+ > Na^+ > Cs^+$
calixarene-crown-6 **6**	3.0	2.6	16.1	$Cs^+ > Na^+ > K^+$
dibenzo-18-crown-6	No transport			

[a]U tube experiments: Source phase: 0.1 M M^+OH^-; Receiving phase = 0.1 M HCl; [Ionophore] = 10^{-3} M in CH_2Cl_2; transfer area = 2.2 cm^2.

Table 3 shows that, under the conditions used, *p-tert*-butylcalix[4]arene-crown-5 (**5**) is selective toward potassium and *p-tert*-butylcalix[4]arene-crown-6 (**6**) transports cesium more efficiently, according to the complementarity between ring size of the crown ether and the ionic radii of the cations.

The introduction on the "bottom rim" of calixarenes of carboxylic groups as in ligands **7**, **8** and **9** enhances the extraction abilities of these macrocycles toward divalent cations, ligand **7** being the most efficient and the crown-5-diacid (**9**) the most selective, especially for calcium and uranyl cations [15].

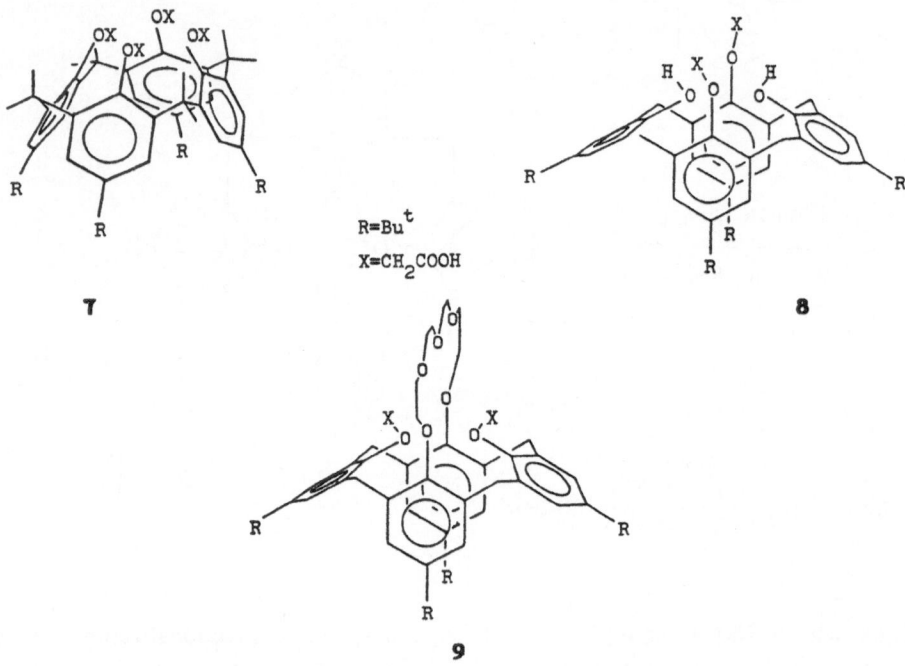

R=Bu^t

X=CH$_2$COOH

The single ion transport of alkaline earth cations through a CH$_2$Cl$_2$ liquid membrane shows (Table 4) these ligands to be effective at low pH, where calixarene derivatives having only phenolic ionizable OH groups are inactive. As in the extraction studies [15], the transport rates also confirm the higher selectivity of compound **9** towards calcium, compared with ligands **7** and **8**.

Table 4. Single ion active transport of alkaline earth cations through a liquid membrane (CH$_2$Cl$_2$) containing calixarene carboxylic-acid ionophores[a].

Ionophore	Transport rate 10^{-6} mol/24 hr			Selectivity
	Ca^{2+}	Sr^{2+}	Ba^{2+}	
Tetraacid **7**	2.6	0.6	1.5	Ca^{2+}>Ba^{2+}>Sr^{2+}
Diacid **8**	5.25	5.1	0.1	Ca^{2+}≥Sr^{2+}>Ba^{2+}
Diacid-crown **9**	7.6	0.7	1.7	Ca^{2+}>Ba^{2+}>Sr^{2+}

[a]U tube experiments: Source phase: [MCl$_2$] = 0.1 M, pH = 9.1 (TRIS Buffer); [Ionophore] = 1.5 x 10^{-4} M in CH$_2$Cl$_2$; receiving phase = 0.1 M H$_3$PO$_4$; transfer area = 2.2 cm^2.

INCLUSION OF NEUTRAL MOLECULES IN CALIXARENES

Calix[4]arenes exist in the solid state in the cone conformation which is mainly determined by four intramolecular hydrogen bonds [1,2]. In this conformation, a hydrophobic cavity is created which eventually includes neutral molecules of suitable size.

After our first report on the X-ray crystal structure of the 1:1 complex between p-tert-butylcalix[4]arene and toluene, which showed the endo-calix character of the complex, several other molecular complexes between calix[4]arenes and organic molecules have been isolated and characterized.

Especially interesting have been the complexes between p-tert-butyl- (**1a**) or p-tert-octylcalix[4]arene (**1**, n = 4, R = t-C$_8$H$_{17}$) with simple aromatic guest molecules, which always show a well defined stoichiometry (Table 5).

Table 5. Isolated complexes of p-tert-butylcalix[4]arene (**1a**) and p-tert-octylcalix[4]arene (**Ib**).

Host	Guest	Host/guest stoichiometry
1a	Benzene	1:1
	Toluene	1:1
	Phenol	1:1
	Anisole	2:1
	o-Xylene	2:1
	m-Xylene	2:1
	p-Xylene	2:1
1b	Toluene	1:1
	p-Xylene	1:1

X-ray crystal structure determinations of both the 1:1 and 2:1 complexes of host **1a** show that the inclusion is always of an endo-calix character [2,16] (Figure 2), whereas macrocycle **1b** forms clathrates in which the guest is held inside the intermolecular voids of the crystal lattice [17].

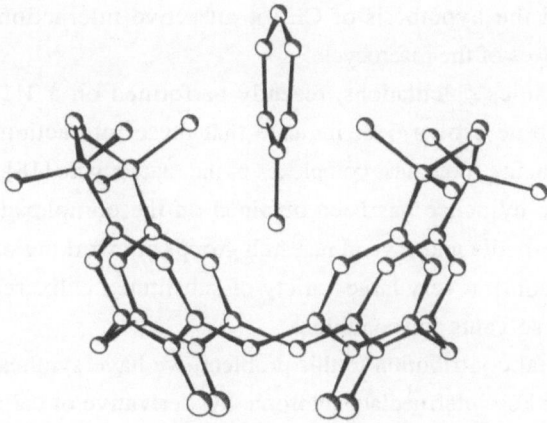

Figure 2. X-ray structure of the 1:1 complex between toluene and p-tert-butylcalix[4]arene.

Table 6. Guest selectivity properties of calix[4]arenes toward aromatic molecules.

Host	Recrystallization solvent mixture (50/50 v/v)	Respective percentage of guest included[a]
1a	p-Xylene / m-Xylene	80 / 20
	p-Xylene / o-Xylene	100 / 0
	p-Xylene / Toluene	90 / 10
	p-Xylene / Anisole	75 / 25
	Anisole / Toluene	95 / 5
	Benzene / Toluene	80 / 20
	Benzene / p-Xylene	75 / 25
	Benzene / Anisole	90 / 10
1b	Benzene / Toluene	50 / 50
	p-Xylene / o-Xylene	45 / 55
	p-Xylene / m-Xylene	55 / 45

[a] by GLC analyses or ^{1}H NMR.

Table 6 shows the guest selectivity properties of the two macrocycles **1a** and **1b** towards aromatic molecules as inferred from competitive crystallization experiments in the presence of equal volumes of two competing guests.

The results show that host **1a** has molecular recognition properties whereas **1b** is not able to discriminate between, for example, xylene isomers. This indicates that the *endo* calix mode of binding of **1a** can give rise to molecular recognition whereas clathrate formation in crystal cavities does not allow high discrimination between guest species.

Benzene seems to be the preferred guest for the cavity of *p-tert*-butylcalix[4]arene although also noteworthy is the high discrimination of *p*-xylene over the *ortho* and *meta* isomers. The X-ray crystal structures of all complexes of *p-tert*-butylcalix[4]arene with aromatic molecules (see for example Figure 2) always show a close proximity between the methyls of the *p-tert*-butyl groups and the aromatic nucleus of the guest (CH$_3$···$\pi \approx 3$ Å) and this observation led us to put forward the hypothesis of CH$_3$/π attractive interactions which stabilize the intramolecular complexes of the macrocycle.

Molecular Mechanics calculations, recently performed on a 1:1 pyridine complex of *p-tert*-butylcalix[4]arene-crown-6 **6**, indicate that these interactions can indeed play an important role in stabilizing molecular complexes of the macrocycle [18].

So far, very little evidence has been obtained on the complexation characteristics of calixarenes in organic media and several research groups are working on this subject. In order to firmly assess this point, a very large variety of substituted calixarenes should be studied extensively in various solvents and conditions.

As an experimental contribution to this problem, we have synthesized [19] several new host molecules via the key intermediate chloromethyl derivative of calixarene **1c** (**1**, R = H, n = 4).

The inclusion abilities of these new calix[4]arene derivatives are presently under study.

Acknowledgements. We thank Professor G. D. Andreetti and Dr. F. Ugozzoli for the X-ray structures and Dr. Eleonora Ghidini and Dr. Alessandro Casnati for their cooperation in part of the work described. We also thank Bruno Bottarelli for the drawings and for typing the manuscript. This work was partially supported by the CNR (Consiglio Nazionale delle Ricerche) and MPI (Ministero Pubblica Istruzione).

References

1. Gutsche, C. D.; Muthukrishnan, R. *J. Org. Chem.* **1978**, *43*, 4905; Gutsche, C. D.; Dhawan, B.; No, K. H.; Muthukrishnan, R. *J. Am. Chem. Soc.* **1981**, *103*, 3782.

2. Andreetti, G. D.; Ungaro, R.; Pochini, A. *J. Chem. Soc., Chem. Commun.* **1979**, 1005; **1981**, 533.

3. Zinke, A.; Kretz, R.; Leggewa, E.; Hössinger, K. *Monatsch. Chem.* **1952**, *83*, 1213.; Hayes, B.; Hunter, R. *Chem. Ind.* **1956**, 193.; Hayes, B.; Hunter, R. *J. Applied Chem.* **1958**, *8*, 743; Cornforth, J. W.; D'Arcy-Hart, P.; Nicholls, G. A.; Rees, R. J. W.; Stock, J. A. *Brit. J. Pharmacol.* **1955**, *10*, 73.

4. For a recent review article on calixarenes see : Gutsche, C. D. In *Synthesis of Macrocycles: Design of Selective Complexing Agents*; Izatt, R. M.; Christensen, J. J., Eds.; Wiley: New York, 1987; pp. 97-165.

5. Bocchi, V.; Foina, D.; Pochini, A.; Ungaro, R.; Andreetti, G. D. *Tetrahedron* **1982**, *38*, 373.

6. Arduini, A.; Pochini, A.; Reverberi, S.; Ungaro, U. *J. Chem. Soc., Chem. Commun.* **1984**, 981.; Arduini, A.; Pochini, A.; Reverberi, R.; Ungaro, R; Andreetti, G. D.; Ugozzoli, F. *Tetrahedron* **1986**, *42*, 2089; Arduini, A.; Ghidini, E.; Pochini, A.; Ungaro, R.; Andreetti, G. D.; Calestani, G.; Ugozzoli, F. *J. Incl. Phenom.* **1988**, *6*, 119.

7. McKervey, M. A.; Seward, E. M.; Ferguson, G.; Ruhl, B.; Harris, S. J. *J. Chem. Soc., Chem. Commun.* **1985**, 388.; Chang, S. K.; Cho, I. *J. Chem. Soc., Perkin Trans. 1* **1986**, 211.

8. Nakamoto, Y., personal communication.

9. Ungaro, R.; Pochini, A.; Andreetti, G. D.; Domiano, P. *J. Incl. Phenom.* **1985**, *3*, 35; Ungaro, R.; Pochini, A.; Andreetti, G. D.; Ugozzoli, F. *J. Incl. Phenom.* **1985**, *3*, 409.

10. Cram, D. J.; Lein, G. M. *J. Am. Chem. Soc.* **1985**, *107*, 3657.

11. Diamond, D.; Svetla, G.; Seward, E. M.; McKervey, M. A. *Analitica Chemical Acta* **1988**, *204*, 223.

12. Kimura, E.; Matsuo, M.; Shono, T. *Chem. Lett.* **1988**, 615.

13. Sabbatini, N.; Guardigli, N.; Balzani, V.; Ghidini, E.; Pochini, A.; Ungaro, R. *Abstract Book of the 13th International Symposium on Macrocyclic Chemistry*; Hamburg, 1988; p. 184.

14. Izatt, R. M.; Lamb, J. D.; Hawkins, R. T.; Brown, P. R.; Izatt, S. R.; Christensen, J. J. *J. Am. Chem. Soc.* **1983**, *105*, 1782; **1985**, *107*, 63.

15. Ungaro, R.; Pochini, A.; Andreetti, G. D. *J. Incl. Phenom.* **1984**, *2*, 199.

16. Ungaro, R.; Pochini, A.; Andreetti, G. D.; Domiano, P. *J. Chem. Soc., Perkin Trans. 2* **1984**, 197.

17. Ungaro, R.; Pochini, A.; Andreetti, G. D. *J. Chem. Soc., Perkin Trans. 2* **1983**, 1773.

18. Andreetti, G. D.; Ori, O.; Ugozzoli, F.; Alfieri, C.; Pochini, A.; Ungaro, R. *J. Incl. Phenom.* **1988**, *6*, 523.

19. Arduini, A.; Casnati, A.; Pochini, A.; Ungaro, R. *Abstract Book of the 5th International Symposium on Inclusion Phenomena and Molecular Recognition*; Orange Beach, Alabama, 1988; H-13.

BIOMIMETIC ION TRANSPORT: PORES AND CHANNELS IN VESICLE MEMBRANES

V. E. Carmichael, P. J. Dutton, T. M. Fyles, T. D. James, C. McKim,
J. A. Swan, and M. Zojaji

Department of Chemistry
University of Victoria,
Victoria, British Columbia, V8W 2Y2 (Canada)

SUMMARY

Two approaches to biomimetic membrane transport systems for vesicle membranes are discussed. The first seeks to mimic the pore-forming antibiotic amphotericin, while the second involves the construction of a unimolecular ion channel.

Pore-forming antibiotics are proposed to form aggregates in the membrane by self association. The aggregate is proposed to present a hydrophobic exterior to the membrane, yet contain a hydrophilic interior for transport of ions and small molecules. The mimic system is based on side-discriminated bola form amphiphiles possessing hydrophilic and hydrophobic edges. Transport across the vesicle wall is followed by the pH dependent fluorescence of a fluorescein bolaamphile embedded in the vesicle wall. Even though the vesicles are leaky, specific structural effects can be discerned.

The channel forming structures are constructed on a central crown ether framework with functionalized arms projecting towards the two faces of a bilayer membrane. In response to a pH gradient, the channel mediates the collapse of the gradient by mediating the cation-proton antiport. Structural effects, and the effects of carrier additives are consistent with the channel providing a mechanism for cation translocation.

INTRODUCTION

Biomimetic ion transport studies seek to mimic the structures and functions of natural transport systems in greatly simplified model systems. This immediately poses a serious problem in that the structural details of natural systems are virtually unknown on a molecular level. In some cases, natural transporters can be directly imaged by electron microscopy and the overall shape and size of the active protein assembly can be deduced [1,2]. Moreover, the functional activity of many transporters are well understood: a particular energetic input results in an output expressed as an ion flux for example [3]. Despite these types of accessible information, there is little structural guidance for aspiring biomimetic chemists to act upon.

Inclusion Phenomena and Molecular Recognition
Edited by J. Atwood
Plenum Press, New York, 1990

We must then content ourselves with an examination of ion transporting antibiotics, which themselves mimic some of the functions of natural transporters. These fall into several classes: (i) Neutral carrier antibiotics such as valinomycin [3], (ii) carboxylate carrier ionophores such as monensin [4], (iii) oligopeptide toxins such as melittin which act via helical aggregates [5], (iv) oligopeptide antibiotics such as the gramacidins in which ion transport occurs down the axis of a transmembrane helical dimer [6] and (v) self-aggregating pore-forming antibiotics such as amphotericin [7]. Since the structural details of the latter two classes are relatively well characterized, and since both these classes involve structures which span the width of a bilayer membrane (like natutal transport proteins), it is these two classes which have attracted our recent attention.

AMPHOTERICIN MIMICS - PORES BY AGGREGATION OF SUBUNITS

Extrapolation from the structure of the amphotericin pore [7] suggests that an amphotericin mimic should possess the following structural characteristics:

a)   It should be the same size and shape as the membrane lipids to permit free diffusion within the membrane.

b)   It should possess both hydrophobic and hydrophilic edges. The hydrophobic edges will contact the lipid core of the membrane while the hydrophilic edges will line the pore.

c)   The hydrophilic edges should present sites for self-aggregation and stabilization, by hydrogen bonding as one example.

The above presciption is very broad, and alternate approaches have been reported by Kunitake [8] and Furhop [9]. Our approach follows on Furhop's use of a bolaamphilic membrane system as illustrated schematically in Figure 1. Since the structural requirements are

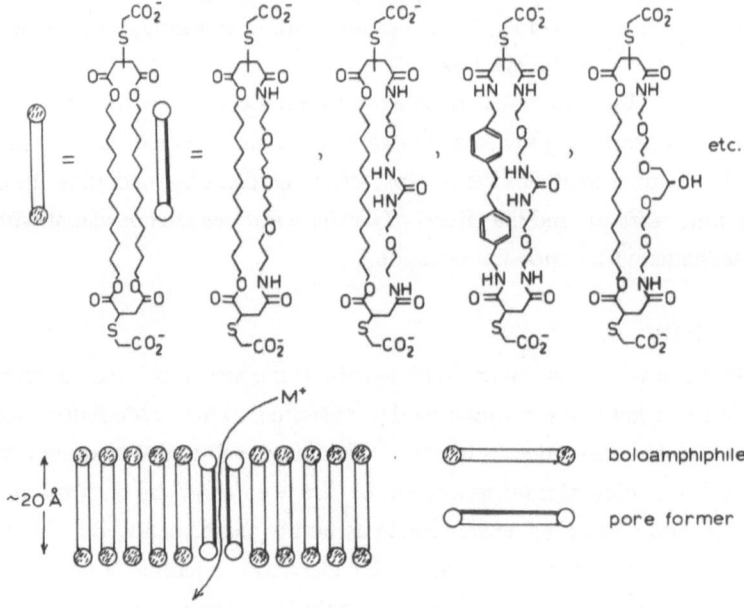

Figure 1

146

so loosely defined, we sought a system which would permit simple assembly of candidates for a survey of structural influences on pore forming function. The synthesis of pore formers such as **2** could be achieved using Furhop's procedure based on acid catalysed dehydration of mixtures of maleic anhydride, pentaethylene glycol and dodecanediol. A statistical mixture of products was separated to give the diene precursor to **2** in poor yield. This brute force approach failed for **3-5** due to the low reactivity of the polar diols. Fortunately, the diacid from two moles of maleic anhydride and one mole of dodecanediol could be converted to its bis cesium salt and alkylated with a polar arm di-iodide to give **3-5** diene precursors. Michael addition of sulfur nucleophiles (thioacetate, thioethanol, thioglucose) gave the systems sketched in Figure 1.

Transport function was assessed using vesicles of **1** containing small amounts (0.1 wt %) of pore formers **2-5**. The vesicles formed by sonication in water were small (60 Å) [9] hence entrapment of indicators was very inefficient. A bis-fluorescein version of **1** was constructed and used to label the vesicle walls. In acidic (pH 5.5) solution fluorescence is low, but increases with pH as the anionic form of the fluorescein appears. Thus the labelled **1** reports the trans membrane pH gradient when the external heads are bleached in pH 5.5 buffer. As the pH gradient collapses, the internal fluorescein fluorescence decreases, providing a probe to monitor transport.

The vesicles without pore former are leaky: with a half-life of only 100 seconds this effectively limits the dynamic range of the transport experiment. Nonetheless, pore formers accelerate the pH gradient collapse in reliable and reasonable ways. For a given pore former with varying head groups, uncharged ones (thioethanol, thioglucose) are more active than charged ones (thioacetate, sulphonate). Pore formers with hydrogen bonding potential (**3-5**) are more active than polyether pore formers (**2**). Despite the experimental flaws, this behavior is consistent with pores formed by aggregation.

GRAMACIDIN DIMER MIMIC - UNIMOLECULAR ION CHANNELS

Gramacidin D forms a Π_{DL} helix in non-polar solvents. In a bilayer membrane, two helices dimerize end-to-end to form a 28 Å structure with inward turning carbonyl groups. Ions traverse the channel down the axis fo the helix [6]. Extrapolation to a generalized gramacidin dimer mimic suggests that the following features are essential:

a)   The mimic should have an overall length close to the membrane thickness (40 Å for bilayers).

b)   The mimic should present a hydrophobic exterior to the membrane yet have inward turning polar functionality to create a hydrophilic core.

c)   In view of the size of object required (3000-5000 g/mole as a minimum estimate) the synthesis should be feasible from well behaved subunits by simple assembly steps.

As with the amphotericin mimics, this prescription is hardly restrictive and functional channels of similar conception have been reported by Notte [10] and Tabushi [11]. Our approach is sketched in Figure 2. Diene tetraesters from octanediol were prepared as discussed above and monofunctionalized with thioethanol. Six of these combined with a crown ether hexaacid chloride derived from three tartaric acid units [12] to yield a hexaene channel precursor. The crown ether core exhibits an extremely marked preference for hexaaxial disposition of substituents [13], thus the core unit provides a structure to maintain the channel. Addition of thiol nucleophiles provides polar head groups to complete the channel. An artist's conception of the complete channel in relation to phosphatidylcholine membrane components is shown in Figure 3. The crown ether lies near the midplane of the bilayer with the macrocyclic tetraester walls extending to the bilayer surfaces.

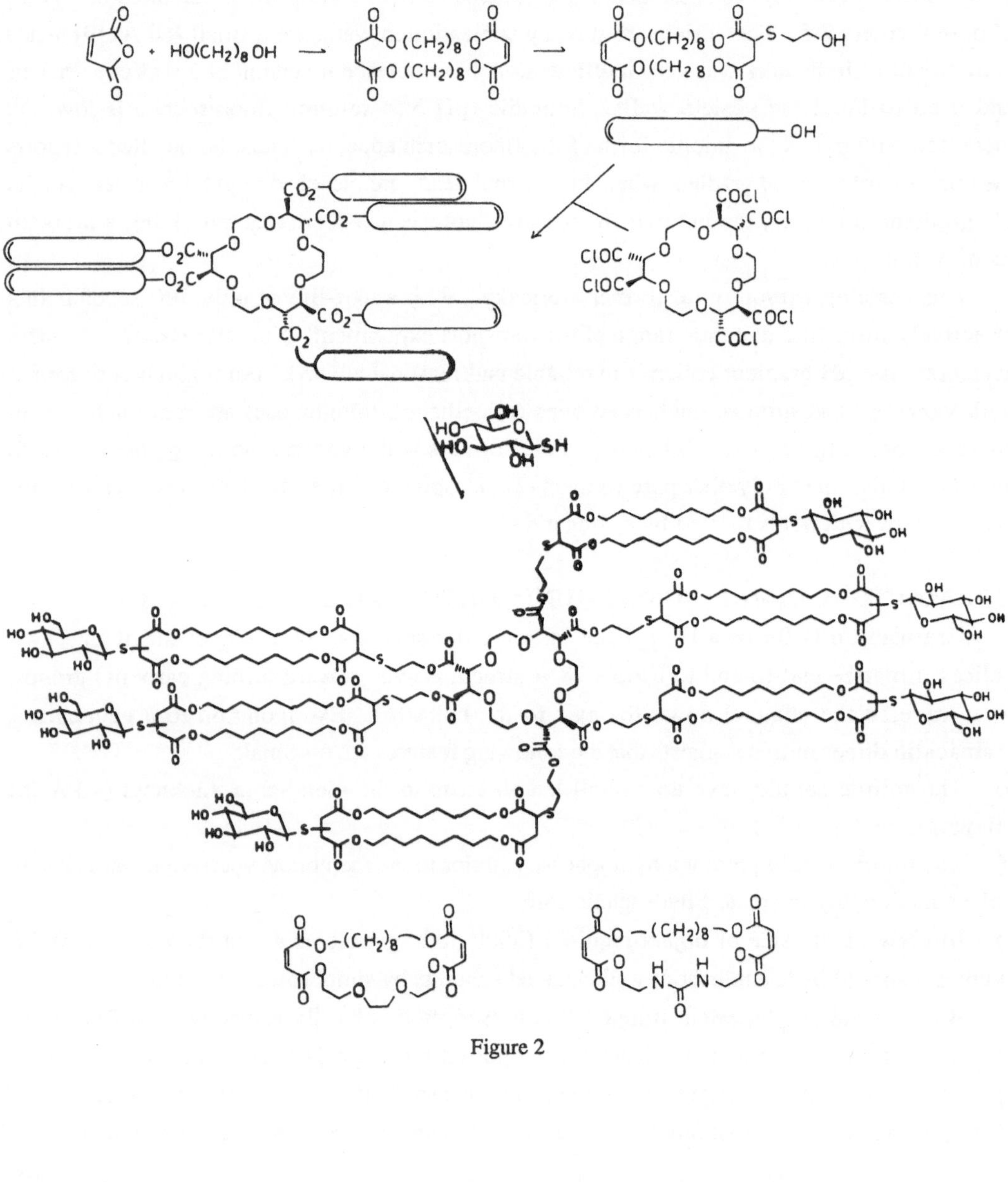

Figure 2

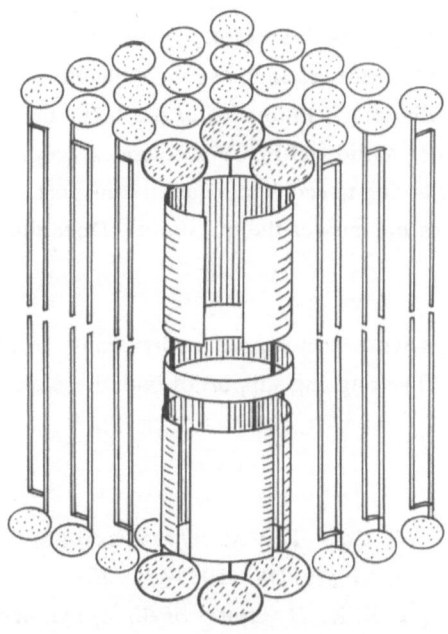

Figure 3

Large unilamellar vesicles with entrapped fluorescein were prepared by reverse evaporation. A pH gradient (5.2 outside, 6.5 inside) collapsed very slowly in the absence of added channel ($t_{1/2} > 30,000$ sec). In the presence of the ion channel mimic, rapid proton gradient collapse occurs as shown for increasing doses of the hexaglucose channel in Figure 4. At the lowest concentration, illustrated, there are about 10^2 - 10^3 copies of the channel per vesicle. At similar molar concentrations the natural toxin melittin, gramacidin D and amphotericin exhibit comparable activities.

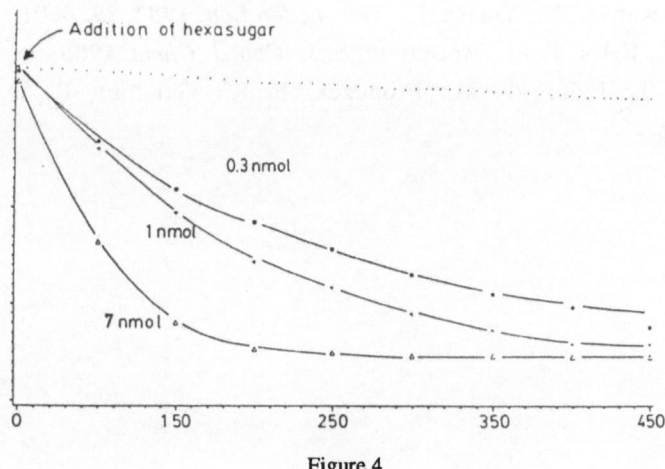

Figure 4

The base rate of leakage across this vesicle system is controlled by the rate of cation permeation. Addition of proton carriers such as FCCP has little effect on the rate of leakage, but cation carriers such as valinomycin dramatically accelerate the pH gradient collapse. In the presence of added ion channel, additional FCCP accelerates the proton gradient collapse. This suggests that the channel is acting to accelerate cation transport. Further mechanistic studies are ongoing with the goal to the answer the question: "Does the channel act by transfer of cation through the crown?"

Acknowledgements. This work was supported by grants from the Petroleum Research Fund and NSERC-Canada. The ongoing support of both these granting agencies is gratefully acknowledged.

References

1. Barry, P. H.; Gage, P. W. *Curr. Top. Membr. Transpt.* **1984**, *21*, 34.
2. Inouye, M. *Proc. Nat. Acad. Sci. USA* **1974**, *71*, 2396.
3. Houselay, M. D.; Stanley, K. K. *Dynamics of Biological Membranes*; John Wiley and Sons: New York, 1982.
4. Westley. J. N. *Polyether Antibiotics*; Marcel Dekker: New York, 1982.
5. Bernheimer, A. W.; Ruby, B. *Biochem. Biophy. Acta* **1986**, *864*, 123.
6. Läuger, P. *Angew. Chem., Int. Ed. Engl.* **1985**, *24*, 905.
7. Finkelstein, A.; Holz, R. In *Membranes: Lipid Bilayers and Antibiotics*, Vol. 2; G.; Eisenman ed.; Marcel Dekker: New York, 1973; p. 357.
8. Kunitake, T. *Ann. N.Y. Acad. Sci.* **1986**, *471*, 70.
9. Fuhrhop, J.-H.; David, H.-H.; Mathieu, J.; Liman, U.; Winter, H. J.; Boekema, E. *J. Am. Chem. Soc.* **1986**, *108*, 1785.
10. Neevel, J. C.; Nolte, R. J. M. *Tetrahedron Lett.* **1984**, *25*, 2263.
11. Tabushi, I.; Kuroda, Y.; Yokota, K. *Tetrahedon Lett.* **1982**, *23*, 4601.
12. Dutton, P. J.; Fyles, T. M.; McDermid, S. J. *Can. J. Chem.* **1988**, *66*, 1097.
13. Dutton, P. J.; Fyles, T. M.; Fronczek, F. R.; Gandour, R. D., unpublished observations.

THERMODYNAMICS OF MICELLIZATION FOR THE K+ AND Ba^{2+} COMPLEXES OF A LIPOPHILIC CRYPTAND

Lourdes E. Echegoyen, Steve R. Miller, George W. Gokel, and Luis Echegoyen*

Department of Chemistry
University of Miami
Coral Gables, FL 33124 (USA)

SUMMARY

Temperature variation of the critical micelle concentration (cmc) has been used to obtain the thermodynamic parameters for micellization of $2\text{-}n\text{-}C_{14}H_{29}\text{-}[2.2.2]$cryptand **1**, in the presence of one equivalent of K+ or Ba^{2+}. ΔH_{mic} for the K+ complex is 5.5 kJ/mol and 12.6 kJ/mol for the Ba^{2+} complex. ΔS_{mic} was 0.1 kJ/molK for both complexes, K+ and Ba^{2+}. The process of micellization is entropy driven, but the major thermodynamic difference is of enthalpic origin. This result is consistent with increased charge repulsion on the micellar surface in the case of Ba^{2+} when compared to K+.

INTRODUCTION

Polyoxyethylene groups have often been employed as the polar heads of a variety of non-ionic amphiphiles [1]. Some of these molecules are capable of associating to form liposomes [2], while the majority of them associate spontaneously to form micelles [3]. Just as these acyclic polyoxyethylene groups (podands) have been used as polar head groups [4], so have their cyclic versions, the crown ethers [5-7]. Noteworthy in the latter category is the work of Okahara [5], Turro [6], and others [7]. They all report the formation of micellar aggregates for the uncomplexed as well as for the complexed states. Complexation leads to micellar property modification and provides interesting possibilities for a variety of practical applications. For example, Gratzel et al. have shown that it is possible to use a micelle formed from a [tetraaza-12-C-4]-C$_{14}$H$_{29}$ system as an electron storage system when it is complexed to Cu^{2+} [8]. Many other applications have been reported, most of them employing the crown ether-cation complexation reaction as a way to modify the micellar properties.

Surprisingly, it was not until recently that we reported the formation of the first non-ionic liposome, niosome, based on cyclic crown ether polar head groups [9]. These amphiphiles contain cholesterol-type lipophilic tails and result in extremely rigid bilayer-like membranes [10]. But perhaps more surprising is the fact that bicyclic versions of the amphiphilic crown ethers, the cryptands, were not employed as polar head groups in micellar aggregates until our very recent report [11]. This is even more surprising in view of the fact that such systems were synthesized by Montanari et al. several years ago [12], and used as phase transfer agents at that time.

Since cation complexation by crown ethers in aqueous media is usually relatively weak, micellar aggregates incorporating crown ethers on their surface have poorly defined complexation properties. The problem must be further aggravated by the decreased cation affinity resulting from ion-ion repulsion interactions on the micellar surface. Since the bicyclic cryptands are much stronger complexing agents than their monocyclic crown counterparts, it seemed interesting to study the micellar aggregates of $C_{14}H_{29}$-[2.2.2], **1**, in the absence and

1

presence of binding cations. The work described here follows closely the methods described by D'Aprano and Sesta in their recent publication [13]. K^+ and Ba^{2+} were selected as the binding cations due to the fact that they have essentially identical ionic radii (1.33 and 1.35 Å for K^+ and Ba^{2+}, respectively) [14]. In addition, the stability constants for their complexation in water by [2.2.2] are very large, $10^{5.4}$ for K^+ and $10^{9.5}$ for Ba^{2+} [15]. Compared to typical values for crown ethers in water, $\approx 10^2$ [13], these high stability constants should provide stronger complexation of the cations, hopefully even after micellization occurs. While D'Aprano and Sesta compared the properties of a neutral crown ether amphiphile with those of the partially formed Ba^{2+} complex, our work directly compares the controlled effect of charge on the micellization properties. Furthermore, the cation complexes formed are stronger and thus better characterized.

EXPERIMENTAL

The 2-*n*-tetradecyl-[2.2.2]cryptand was prepared as described previously [11,12]. Barium chloride ($BaCl_2.H_2O$) (Certified ACS Fisher product) was used as received. Potassium chloride (KCl) (USP-FCC J.T.Baker product) was also used as received. Water was distilled and deionized through a Barnstead Nanopure II system.

Critical micelle concentrations (cmc's) were determined by surface tension measurement as a function of concentration at various temperatures [16]. Typical plots are shown in Figure 1.

10 mM stock solutions of 1:1 [ligand]:[metal] were prepared. Dilutions were made to cover a range above and below the cmc for a total of 14 solutions for each cation.

Surface tension was measured with a Fisher model 20 tensiometer (based on the du Nouy ring method). Temperature was maintained to within ±0.1°C using a jacketed beaker which contained the solution, and through which thermostated water was pumped. Five measurements were obtained for each solution at each temperature. Reproducibility was typically within 0.3 dynes cm^{-1}.

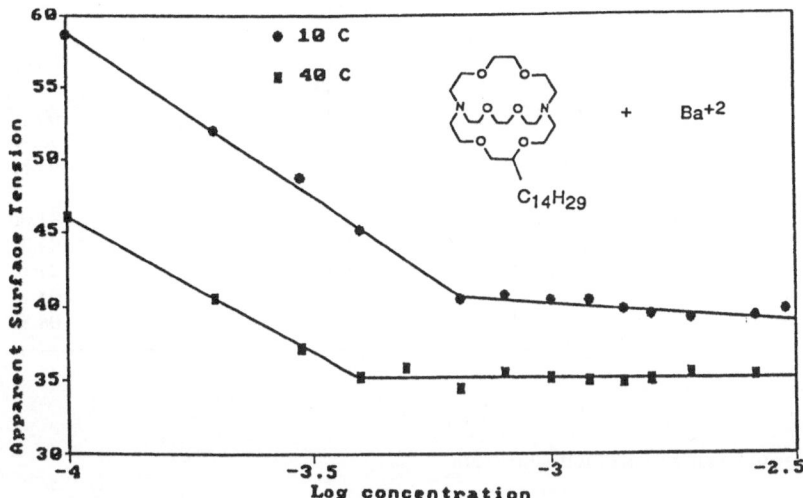

Figure 1. Typical surface tension (g) vs. log c plots at two temperatures, 10° and 40°C, for the Ba^{2+} complex of **1**.

RESULTS

The cmc's obtained from these plots are presented in Table 1. Several general trends can be noted from these values. Cmc's decrease with increasing temperature for the K$^+$ and Ba^{2+} complexes. This behavior is not unexpected and has been observed for other surfactants [17], particularly for the case of a lipophilic crown complex with Ba^{2+} [13]. For a given temperature, the cmc for the K$^+$ complex is always smaller than for the corresponding Ba^{2+} complex. Comparison of these values with those for the neutral amphiphile in pure water was only possible at 10°C. The reason for not being able to determine variable temperature cmc's for the uncomplexed compound is that its cloud point occurs at 12.5°C. The available temperature interval was thus extremely reduced and precluded such a study.

Other calculated values presented in Table 1 include the surface excess concentrations, Γ_{max}, and the minimum area required by a surfactant head group at the air/water interface, A_{min}. These quantities were calculated from the Gibbs adsorption equation [18] as described elsewhere [13]. The activity coefficient was assumed to be unity at the relatively low concentrations used. Thus,

$$\Gamma_{max} = [d\gamma/d \log c]/2.3 \, RTB$$

where dγ/d log c is the slope of the line of the surface tension vs. log c plot (Figure 1) below the cmc's, and B = 2.

Table 1. Critical micelle concentrations (cmc's), surface excess concentrations (Γ_{max}), and minimum areas per surface head group (A_{min}) as a function of temperature for **1** and its complexes.

Medium	T, °C	10^4 (cmc), M	$10^{10}\Gamma_{max}$, mol/cm^2	$100\,A_{min}$, nm^2
H$_2$O	10	1.4	3.10	53.57
KCl	10	0.93±0.04	3.77	44.05
	18	0.82±0.04	3.30	50.32
	25	0.82±0.06	2.76	60.17
	32	0.82±0.06	2.26	73.48
	40	0.71±0.02	1.77	93.82
BaCl$_2$	10	6.52±0.41	2.05	81.00
	18	6.13±0.37	1.80	92.25
	25	5.52±0.34	1.71	97.11
	32	4.74±0.43	1.58	105.10
	40	3.93±0.21	1.53	108.53

The minimum area was calculated from

$$A_{min} = 10^{14}/N\Gamma_{max}$$

where N is Avogadro's number. For all cases Γ_{max} values were observed to decrease with, while A_{min} values increase with, increasing temperatures.

Thermodynamic parameters for the micellization process were determined using the standard equations and the cmc value [19]. Thus,

$$\Delta G°_{mic} = RT \ln(cmc)$$

Plots of $\Delta G°_{mic}$ vs. T were used to determine $\Delta S°_{mic}$ (from the negative of the slopes). These are shown in Figure 2 for the K$^+$ and Ba^{2+} complexes.

$\Delta H°_{mic}$ values were calculated from these using

$$\Delta G°_{mic} = \Delta H°_{mic} - T\Delta S°_{mic}$$

Table 2. Thermodynamic parameters for micellization of **1·K$^+$** and **1·Ba^{2+}**.

	$\Delta G°_{mic}$, 25°C (kJ/mol)	$\Delta H°_{mic}$ (kJ/mol)	$\Delta S°_{mic}$ (J/mol°K)
1·K$^+$	-23.3	5.5	100
1·Ba^{2+}	-18.6	12.6	97

All the thermodynamic parameters are presented in Table 2.The most interesting observation that can be made from Figure 2 is that the slopes of the two complexes are essentially identical. Indeed the analytical values derived from these nearly parallel lines confirm that the ΔS°_{mic} values are the same for both complexes, ≈ 100 J/molK.

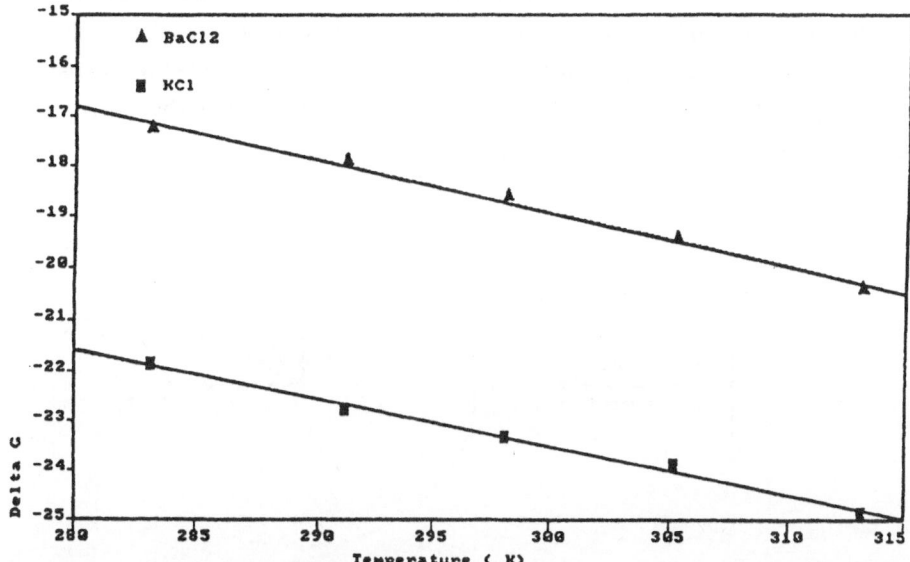

Figure 2. Plots of ΔG°_{mic} (kJ/mol) vs. T (°K) for $1 \cdot K^+$ and $1 \cdot Ba^{2+}$.

DISCUSSION

There are two basic equilibria which have to be considered in order to explain the results. The first one is the complexation equilibrium involving the surfactant and the corresponding cation, Equation 1, with its corresponding stability constant, K_S. Although direct measurements were not performed in this study to determine K_S, the cation binding ability of the cryptand head group was assumed to be similar to that of the parent compound, [2.2.2]cryptand. Although this is admittedly an oversimplification, the high stability constants exhibited by [2.2.2] with K^+ ($10^{5.4}$) and Ba^{2+} ($10^{9.5}$) in water suggest that the lipophilic tail should not result in appreciable changes, probably by less than a power of 10. Indirect support for this assumption comes from the observed decrease in stability constant upon substitution of diaza-18-C-6 with an n-dodecyl chain [20,21]. The stability constant of diaza-18-C-6 with Ba^{2+} decreases from 955 to 158 when the lipophilic substitution is made at one of the nitrogens. If this 6-fold decrease is used as a guide for the case of 1, stability constants of

Equation 1

155

$10^{\approx 4.7}$ and $10^{\approx 8.8}$ would be anticipated for K^+ and Ba^{2+}, respectively. In any case, the stability constants are expected to be much higher than those reported for the lipophilic diaza-18-C-6 by D'Aprano and Sesta [13].

The second important equilibrium is that for formation of the micelle from the monomeric surfactant molecules (Equation 2) whose equilibrium constant is denoted as K_M. There is a potentially serious difficulty with this reaction, as presented in Equation 2. Formation of the micelle must bring the complexed cryptates closer to each other and result in a destabilizing effect. A possible result of this is partial decomplexation by the micelle to liberate metal ions to the water medium. The arguments presented here assume that the extent of cation decomposition is relatively low.

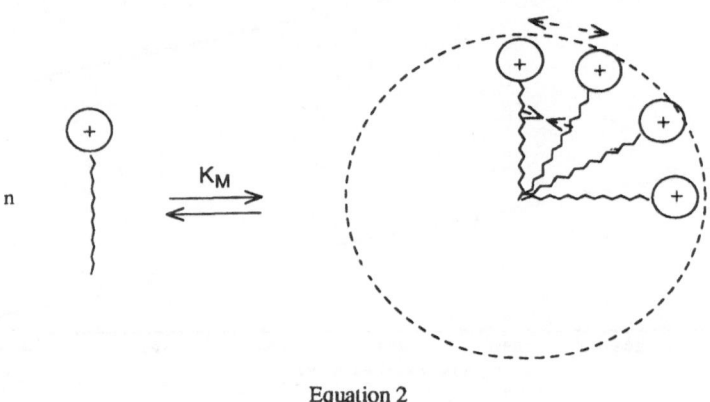

Equation 2

Values of cmc's, particularly those for the Ba^{2+} complex, are in close agreement with others reported for analogous systems [22]. The value for the K^+ complex is, on the other hand, drastically reduced. Although it could be argued that this reduced cmc is a consequence of salting out from the water structuring Cl^- anion [23], such an effect would necessarily have to be more pronounced for the divalent salt case. A competition between salting out by the Cl^- anion and salting in by formation of the cation complex has been used to explain the behavior of crown-substituted surfactants [13]. Recent work has compared the cmc of n-dodecyl-diaza-18-C-6 with those for the same system in the presence of increasing $BaCl_2$ concentrations. While at low salt concentrations the cmc's are lower than for the pure compound, at high salt concentrations they are actually higher. The explanation is that salting out by Cl^- dominates at low concentrations, where the salting in due to complex formation is very low. As the concentration of salt is increased, so does the concentration of the complex and salting in results in increased cmc's. While this explanation is consistent with the observations, it seems insufficient to explain our results.

Our experiments directly compare the effect of a monovalent vs. a divalent cation on the micellization of a strong complexing agent. Both cations must result in pronounced salting in effects due to the strong complexation while both cases should exhibit similar salting out due to the presence of Cl^-. If anything, salting out by Cl^- should be more pronounced in the case of $BaCl_2$ based on simple stoichiometric considerations. Since the observed cmc with $BaCl_2$ is much larger than with KCl, salting out by Cl^- cannot be the determinant factor controlling micellization. The inevitable conclusion seems to be that salting in must be the determinant factor. We find it difficult to believe that Ba^{2+} would have such a strong salting in effect that it will offset the unfavorable Cl^- effect and still exhibit an almost ten-fold increase in the cmc when compared to K^+. We think the answer lies in the thermodynamic parameters. Before presenting the thermodynamic arguments, a potential difficulty due to equilibrium 1 is now discussed.

If we assume that the lipophilic tail does not affect the complexation reaction much, Equation 1, it is possible to use the stability constant for [2.2.2] to determine the fraction of ligand which is complexed at the cmc. Thus for Ba^{2+} at 25°C, with a cmc = 5.5 x 10^{-4} M and $K_s = 10^{9.5}$, the fraction complexed is calculated to be 99.92%. For K^+ the value is only 78.08%. This incomplete complexation could result in a decreased salting in effect and thus be the cause of the relatively low cmc. However, the relative amount of uncomplexed ligand does not seem to have the right order of magnitude to explain the observed differences.

We feel that the explanation can be found in the thermodynamic parameters for Equation 2. The case of $1 \cdot Ba^{2+}$ exhibits a $\Delta G°_{mic}$ of -18.6 kJ/mol. This value is very similar to that of the Ba^{2+} complex of n-dodecyl-diaza-18-C-6, -17.3 kJ/mol [13]. The corresponding cmc's are 5.52 x 10^{-4} M and 9.33 x 10^{-4} M, respectively. As expected, the more negative the free energy change, the more likely the system is to aggregate according to Equation 2. The most negative $\Delta G°_{mic}$ is exhibited by the K^+ complex, -23.3 kJ/mol, resulting from a very small cmc, 0.82 x 10^{-4} M. Since all of these systems exhibit exactly the same $\Delta S°_{mic}$, 100 J/mol°K, differences in micellization properties must be the result of enthalpic effects. The overall drive for Reaction 2 at the studied temperatures comes from the favorable entropy change due to destruction of the water structure around the lipophilic tails. This effect is essentially constant for these systems since it does not involve the head groups, where complexation is used to vary the amphiphilic properties.

These arguments indicate that $\Delta H°_{mic}$ is the discriminating factor. A more favorable (less positive) $\Delta H°_{mic}$ for $1 \cdot K^+$, 5.5 kJ/mol, results in easier micellar formation and lower cmc value. This is in comparison with $1 \cdot Ba^{2+}$, which exhibits a $\Delta H°_{mic}$ of 12.6 kJ/mol. The reason for this difference is clearly the charge difference between the complexed head groups. Complexed Ba^{2+} causes very strong repulsion between the head groups on the micellar surface.

CONCLUSION AND FURTHER WORK

When [2.2.2]cryptand was used as the polar head group in an amphiphile **1**, a controlled and well characterized thermodynamic charge effect was studied for the process of micellization. Micellar properties were found to vary significantly with the charge of the complexed cation, with cmc's being lower for the lower charged cation. Entropy changes upon micellization were found, as anticipated, to be essentially identical and positive. Overall micellization is entropy driven while enthalpy differences account for the charge effect.

Work is in progress using $C_{14}H_{25}$-[2.2.1] and Na^+ and Ca^{2+} as the complexing anions. Size and charge effects will be explored in a similar manner as described here. Characterization of possible cation decomplexation upon micellar formation is currently being probed using conductance measurements for $1 \cdot M^{n+}$ and $C_{14}H_{25}$-[2.2.1]$\cdot M^{n+}$.

Acknowledgements. We thank the National Institutes of Health for a grant, GM 33940, that supported this work.

References

1. Doscher, T. M.; Myers, G. E.; Atkins, D. C. *J. Colloid Sci.* **1951**, *6*, 223.

2. (a) Okahata, Y.; Tanamachi, S.; Nagai, M.; Kunitake, T. *J. Colloid Interface Sci.* **1981**, *82*, 401; (b) Baillie, A. J.; Florence, A. T.; Hume, L. R.; Muirhead, G. T.; Rogerson, A. *J. Pharm. Pharmacol.* **1985**, *37*, 863.

3. (a) Maclay, W. N. *J. Colloid Sci.* **1956**, *11*, 272; (b) Schick, M. *J. Colloid Sci.* **1962**, *17*, 801; (c) Schott, H. *J. Colloid Interface Sci.* **1973**, *43*, 150.

4. Fendler, J. H. *Membrane Mimetic Chemistry*; Wiley: New York, 1982; p. 9.

5. Okahara, M.; Kuo, P.-L.; Yamamura, S.; Ikeda, I. *J. Chem. Soc., Chem. Commun.* **1980**, 586.

6. Gould, I. R.; Kuo, P.-L.; Turro, N. J. *J. Phys. Chem.* **1985**, *89*, 3030.

7. (a) Shinkai, S.; Nakamura, S.; Manabe, O.; Yamada, T.; Nakashima, N.; Kunitake, T. *Chem. Lett.* **1986**, 49; (b) Moroi, Y.; Pramauro, E; Gratzel, M. *J. Colloid Interface Sci.* **1979**, *69*, 341; (c) LeMoigne, J.; Gramain, P. H. *J. Colloid Interface Sci.* **1977**, *60*, 565.

8. Monserrat, K.; Gratzel, M.; Tundo, P. *J. Am. Chem. Soc.* **1980**, *102*, 5527.

9. Echegoyen, L. E.; Hernandez, J. C.; Kaifer, A. E.; Gokel, G. W.; Echegoyen, L. *J. Chem. Soc., Chem. Commun.* **1988**, 836.

10. Fasoli, H.; Echegoyen, L. E.; Hernandez, J. C.; Gokel, G. W.; Echegoyen, L. *J. Chem. Soc., Chem. Commun.* **1989**, 578.

11. Echegoyen, L. E.; Portugal, L.; Miller, S. R.; Hernandez, J. C.; Echegoyen, L.; Gokel, G. W. *Tetrahedron Lett.* **1988**, *29*, 4065.

12. Cinquini, M.; Montanari, F. *J. Chem. Soc., Chem. Commun.* **1975**, 393.

13. Sesta, B.; D'Aprano, A. *J. Phys. Chem.* **1988**, *92*, 2992.

14. Purcell, K. F.; Kotz, J. C. *Inorganic Chemistry*; Saunders: Philadelphia, 1977; p. 284.

15. Döbler, M. *Ionophores and their Structures*; Wiley: New York, 1981; pp. 16-17.

16. Mukerjee, P.; Mysels, K. J. *Critical Micelle Concentrations of Aqueous Surfactant Systems*; U.S. Department of Commerce. National Bureau of Standards. U.S. Government Printing Office: Washington, D. C.,1970; p. 8.

17. Schick, M. J. *J. Phys. Chem.* **1963**, *67*, 1976.

18. Gibbs, J. W. *Collected Work*; Longmans Green: New York, 1928; Vol. 1, pp. 218-237.

19. Mukerjee, P.; Korematzu, K.; Okawauchi, M.; Sugihara, G. *J. Phys. Chem.* **1985**, *89*, 5308.

20. Von Anderegg, G. *Helv. Chim. Acta* **1975**, *58*, 1218.

21. Le Moigne, J.; Simon, J. *J. Phys. Chem.* **1980**, *84*, 170.

22. Berchiesi, G.; Berchiesi, M. A.; La Mesa, C.; Sesta, B. *J. Phys. Chem.* **1984**, *88*, 3665.

23. Mukerjee, P. *Adv. Colloid Interface Sci.* **1967**, *1*, 241.

REDOX ACTIVE CROWNS: TOWARDS THE SECOND GENERATION

Stephen R. Cooper

Inorganic Chemistry Laboratory
University of Oxford
Oxford OX1 3QR (UK)

INTRODUCTION

Redox-active crown ethers arise from a simple idea. In an equimolar solution of benzoquinone and 18-crown-6 no relationship would exist between the redox state of the quinone moiety on one hand and the ion binding state of the crowns on the other. If the quinone group is incorporated into the crown ether, however, the reactivity of the quinone would influence that of the crown, and *vice versa*. The two reactions - electron transfer and ion binding - are now coupled.

Thus, by suitable design of a quinone-containing crown ether (Figure 1) we can intentionally couple these formerly unrelated reactivities: oxidation/reduction and ion-binding. Reduction of the quinone to the unprotonated semiquinone yields an anionic ligand that should exhibit greater affinity for cations (alkali metal ions, for example).

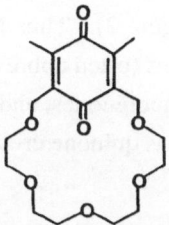

Figure 1. Generic structure of the quinone crown ethers.

This paper extends our earlier work [1] to three types of investigations on redox-active crown ethers. The first part examines their electrochemical properties and why these properties are of interest. The second part touches briefly upon EPR studies of the semiquinone crowns and their complexes. The third part discusses the structural chemistry of the redox-active crown ethers with emphasis on its implications for complexation in solution. This work also complements the beautiful investigations of Misumi and his coworkers earlier in the same system [2]. It also complements the elegant work of George Gokel and Luis Echegoyen [3] on a variety of other redox-active crowns, and that of Fritz Vögtle [4] on a different type of quinone crown.

Motivation for the redox-active crowns comes from both conceptual implications and practical applications. Conceptually, redox-active crowns bear a functional analogy to the alkali metal cation transport of, e.g., neurons. This analogy arises from the ability of redox-active crowns to mediate redox-driven ion transport across a membrane. Nerve cells of course perform the same function, for which they use energy ultimately derived from electrons flowing down an electrochemical potential gradient in oxidative metabolism. (Quinone crown ethers are not, of course, involved in the biological process.) Nevertheless, by using a membrane to separate an oxidizing compartment from a reducing one we could use a quinone crown ether to accomplish the same task: transduction of energy from the form of a redox potential into an ion gradient. Hence quinone crown ethers provide a model for active transport of alkali metal ions, as well as neutral substrates such as amino acids and neurotransmitters.

Potential practical applications also motivate this work. The most obvious one involves sensors. Since redox state influences ion binding, the converse must also be true: ion binding must affect redox properties. Thus the quinone/semiquinone redox potential must reflect the extent to which a quinone crown interacts with alkali metal cations. Electrochemical measurements therefore provide a means to determine the concentration of an alkali metal cation in solution. This would be the basis of a simple sensor system for M^+ cations; at present the greatest interest attaches to Li^+ because of its clinical use in treatment of manic depression.

Another possible application arises from chiral redox-active crowns. Transport of chiral species across membranes mediated by chiral redox-active crowns could offer an energy-efficient means of resolving racemic mixtures. In addition, redox-active crowns (although probably not quinone crowns per se) might find use in hydrometallurgical applications, but that remains a glimmer in the eye at this point.

We use an abbreviated nomenclature based upon the number of oxygen atoms available for complexation to a given metal ion (Figure 2). Thus the molecules in Figure 2 are termed, respectively, the 5- and 6-quinone crowns (often abbreviated 5- and 6-QC) [5]. Although the work described deals with quinones, semiquinones, and hydroquinone dimethyl crown ethers, no confusion arises if all are referred to as quinone crown ethers - the context makes clear the oxidation state involved.

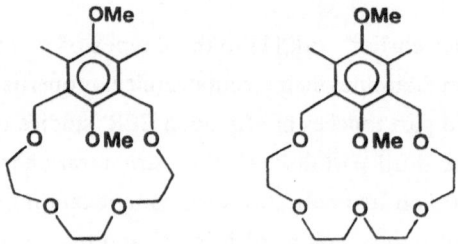

Figure 2. The 5- and 6-quinone crowns (abbreviated to 5- and 6-QC).

ELECTROCHEMICAL STUDIES

Electrochemical properties provide a central focus for interest in redox-active crowns. As indicated above, coupling of redox reactions with ion binding implies that measurement of redox potentials in the presence of alkali metal cations reflects the strength of coupling of the two phenomena. Quantitatively the thermodynamics of complexation shifts redox potential according to Equation 1, where $E_f^{complex}$ and E_f^{free} are the formal potentials of the complexed and free ligands, respectively, and K_{ox} and K_{red} are the stability constants of the oxidized (quinone) and reduced (semiquinone) forms, respectively. Thus electrochemical measurements give direct insight into the relative stability constants of the oxidized and reduced ligands.

$$E_f^{complex} - E_f^{free} = -(RT/nF)\ln\{K_{ox}/K_{red}\} \qquad (1)$$

Cyclic voltammetric measurements of quinone crowns show that the presence of group IA cations shifts E_f to more positive values. This implies (from Equation 1) that, as expected, the semiquinone form of the crown binds M^+ more strongly than does the uncharged quinone state (i.e., K_{red} exceeds K_{ox}). Put another way, addition of M^+ makes the quinone crown easier to reduce, and thereby shifts E_f to more positive values.

Shifts in E_f arise principally from two sources: complexation and ion pairing. Shifts arising from ion pairing of M^+ with SQ^- are not of interest here. Accordingly, all measurements have been performed in the highly polar DMF to minimize ion pairing. In addition, all potential shifts have been corrected for the ion pairing shift obtained for the control compound (2,6-dimethoxymethyl-1,4-benzoquinone). This compound should ion pair to a comparable extent, but it lacks a crown moiety for complexation.

Studies on the quinone crowns show that addition of M^+ (M = Li - Cs) yields potential shifts (E_f) well beyond those attributable to ion pairing (Figure 3). After correction for the ion pairing contribution, the shifts for 5-QC reveal a broad maximum at K^+ and Rb^+, with sharper dropoff toward Li^+ than toward Cs^+. Those for 6-QC show a clear-cut maximum at K^+, again with sharper dropoff toward Li^+ than toward Cs^+. For 7-QC and 8-QC the maximum in E_f moves from K^+ to Cs^+ (or possibly Rb^+; unfortunately, reliable E_f values have not yet been obtainable for Rb^+). The results in hand show that E_f increases monotonically from Li^+ to Cs^+. The electrochemical behavior of 8-QC differs only quantitatively from that of 7-QC, with smaller shifts for Cs^+ and especially Li^+.

Analysis of the electrochemical results shows that the ratios of stability constants K_Q/K_{SQ} (i.e., the potential shifts) with group IA cations parallel the absolute stability constants of corresponding anisole crowns with the same ions. Hence the differences in metal ion affinity (i.e., the selectivities of the quinone crowns) derive solely from the crown ring, not from the $M^+\cdots O(quinoid)$ interaction, because the quinoid moiety pivots with respect to the plane of the crown (vide infra). In this fashion it easily accommodates ions of different size. This constitutes a major conclusion of this work: *any attempt to improve selectivity/affinity must concentrate solely on the crown moiety.*

Combination of these resuts with those of Misumi et al. suggests some conclusions about the absolute stability constants of the semiquinone crown complexes. Misumi's NMR work [2] shows that in the oxidized form, the quinone crown ethers bind M^+ weakly, if at all, The low values for K_{ox} probably change little as a function of crown ring size. If so, the observed shifts in E_f arise largely from changes in K_{red}. Hence the observed pattern of shifts reflects the changes in binding affinity of the semiquinone form. Structural studies (vide infra) buttress this tentative conclusion.

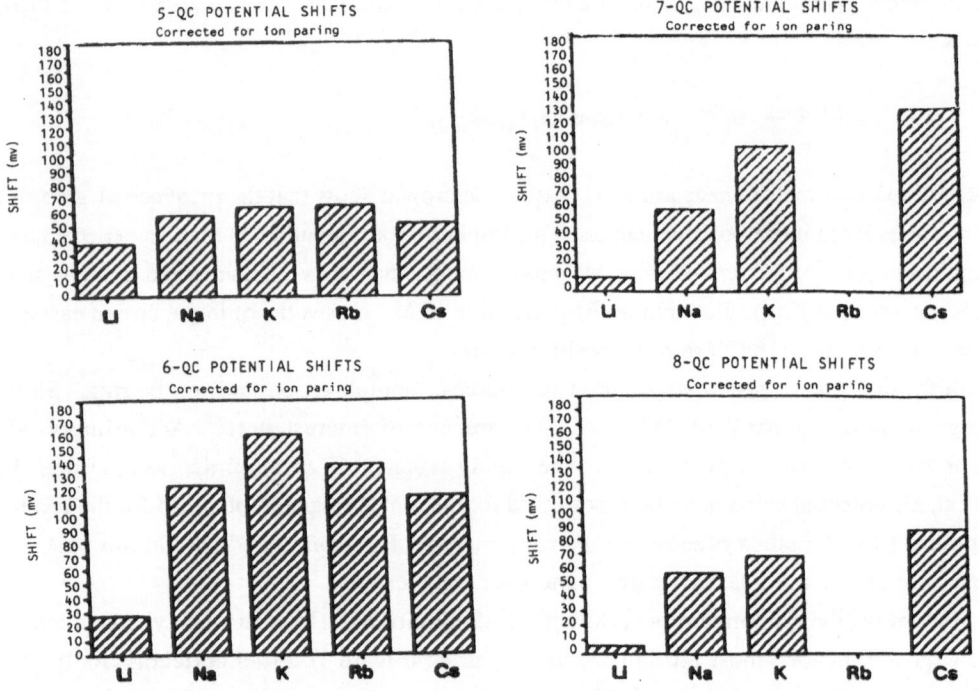

Figure 3. Potential shifts (corrected for ion pairing) of the quinone crowns in the presence of 0.1 M M^+OTs^-. Upper left: 5-QC; lower left: 6-QC; upper right: 7-QC; lower right: 8-QC.

EPR STUDIES

EPR studies of electrochemically generated semiquinone crowns provide further evidence for strong M^+-ligand interaction in solution. For simplicity we will confine discussion to the most impressive example. Introduction of Na^+ to a solution of the 6-QC in the semiquinone form dramatically changes the EPR spectrum (Figure 4). In the unbound form the semiquinone crown shows ^{1}H hyperfine splitting (HFS) from the methylene and methyl protons. In the presence of Na^+ the large [23]Na superhyperfine splitting (SHFS) largely obscures the HFS arising from ^{1}H coupling. The large [23]Na nuclear magnetic moment ([23]Na, I = 3/2, 100%) generates the quartet-like structure that dominates the spectrum. This alkali metal SHFS indicates that Na^+ binds strongly to the semiquinone because simple semiquinones do not afford similar SHFS (in DMF).

These results show that the anisole O atom in the 1-position is the one that binds Na$^+$, and that it does so strongly. Detailed analysis indicates that in the complex the hyperfine splitting at the methyl positions increases and that in the methylene positions decreases. In simple handwaving terms association of Na$^+$ with the O-atom in the 1-position forces electron density to concentrate at that end of the molecule (to balance the positive charge). This results in spin density residing on the opposite (4-position) end of the molecule.

Because of these electronic rearrangements the hyperfine splitting at the 1-position end decreases, while that at the opposite end increases. A more rigorous argument based upon MO theory and the symmetry of the MOs gives the same result. Thus the HFS pattern shows that these compounds bind Na$^+$ (and by extension other alkali metal ions) at the 1-position; the existence of ^{23}Na SHFS indicates that they do so strongly. That agrees with the results of the electrochemical studies that there is an appreciable coupling between redox reaction on one hand and ion binding on the other. Now it remains to take a look at the geometrical properties that influence the strength of coupling in order to optimize the M$^+$-SQ interaction.

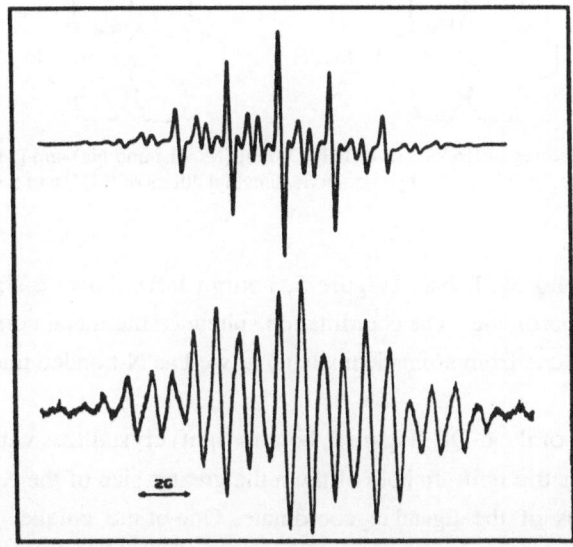

Figure 4. EPR spectra of the 6-QC electrochemically reduced to the semiquinone state. Top: free; bottom: after addition of Na$^+$.

STRUCTURAL STUDIES

Structural studies of the complexes provide this geometrical insight into the mechanism of coupling. Complexes of 5-QC (as the hydroquinone dimethyl ether) with Li$^+$ and Na$^+$ (shown schematically in Figure 5) both exhibit incomplete coordination of the ligand to the cations. For Li$^+$ only four oxygen donor atoms (three from the crown, one from the hydroquinone dimethyl ether) coordinate. An N-bonded thiocyanate group completes the square-pyramidal coordination sphere. The Na$^+$ complex behaves similarly except that the

[NaL] units dimerize to yield a di-μ-NCS bridged unit. Apparently the crown loops lack sufficient size to permit complete coordination without undue ring strain. Consequently both complexes feature an unbound benzylic ether O atom.

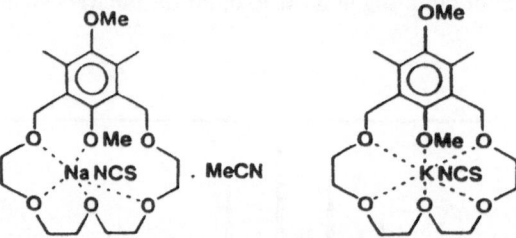

5-QUINONE CROWN HQDME STRUCTURES

DIMER

6-QUINONE CROWN HQDME STRUCTURES

Figure 5. Schematic structures of [MNCS·(5-QC-HQDME)] (M = Li and Na) and [MNCS·(6-QC-HQDME)] (M = Na and K), where the ligands are the hydroquinone dimethyl ethers of 5-QC and 6-QC.

The 6-QC complex with Na⁺ (Figure 5, bottom left) shows a similar failure of one benzylic O atom to coordinate. The coordination sphere of the metal comprises five O atoms (four from the crown, one from aromatic nucleus) as well as N-bonded thiocyanate and MeCN groups.

The K⁺ complex of the 6-QC (Figure 5, bottom right) crystallizes with two quite different molecules per asymmetric unit. In both of them the greater size of the K⁺ ion permits all six potential donor atoms of the ligand to coordinate. One of the unique molecules (Figure 6)

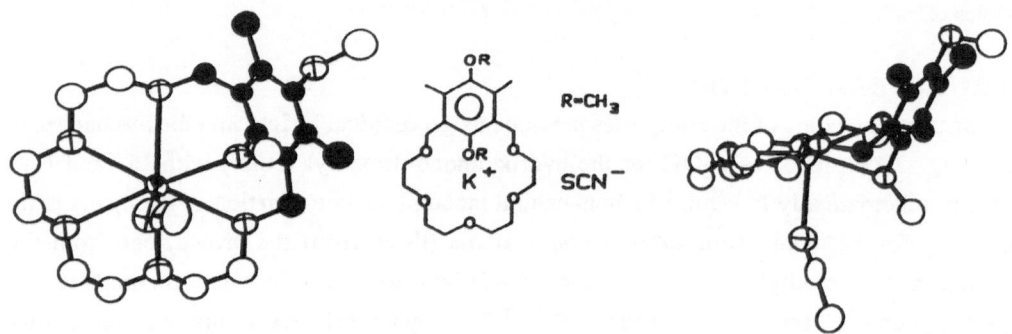

Figure 6. Structure of [KNCS·(6-QC)] in a) top and b) side views. Note the K⁺-methoxy interaction.

exists as a monomer with a seven-coordinate K$^+$ cation in the plane of the crown. An N-bonded NCS ion completes the coordination offered by the six ligand O atoms. The other molecule per asymmetric unit (not shown) is a dimer analogous to [Na(5-QC)]$_2$. A top view of the first molecule shows clearly the quinoid moiety and the conformation of the crown loop. The side view, while complicated, more clearly reveals the coordination of the methoxy O atom to the K$^+$ ion. In effect, this ORTEP constitutes a snap shot of the interaction that gives rise to the coupling between redox reactivity and ion binding.

Thus the greater shift found for [KNCS·(6-QC)] presumably derives in part from the greater propensity for six-coordination observed by X-ray diffraction. This conclusion must be tempered by the usual caveat regarding reasoning from the solid to the solution state. Nevertheless, it is likely - although not proven - that behavior in solution parallels that in the solid state.

Examination of the less than optimal cases (e.g., Na$^+$ with 6-QC) shows the result of a metal-ligand mismatch. This information is important, for it provides the basis on which to design the next generation of ligands. Such information is clearly necessary for attempts to optimize M$^+$ affinity, but it can also be used to build in poor interactions for metal ions we wish to select against. Thus it is also important for improving selectivity.

Comparison of the structure of 6-QC elucidates the role - or more accurately, the lack of role - of the quinoid moiety in determining selectivity. The angle between the quinoid group and the mean plane of the crown pivots through a wide range (29°) on replacement of Na$^+$ with K$^+$. (Analogous changes occur in the 5-QC series.) As a consequence of its ability to pivot the quinoid group readily accommodates ions of differing size. For this reason *the quinoid group makes no contribution to selectivity.*

Pivotting of the quinoid functionality is a mixed blessing. Because of it the quinoid group does not generate steric problems that diminish affinity. At the same time, however, it prevents the M$^+$-quinoid interaction from contributing to selectivity. Selectivity suffers most severely, however, from the flexibility of the crown loop. The ability of the loop to accommodate ions of differing size (in a fashion well-documented for simple crowns) results in necessarily low selectivity.

The work discussed above therefore establishes the successful coupling of ion binding with redox reactivity. It also reveals that metal ion selectivity comes only from the crown loop. Second generation crowns, therefore, must contain stiffer ion binding loops to improve selectivity. Last, the recurring failure of the benzylic O atoms to coordinate indicates the need for redesign of the quinone-crown linkage in second generation molecules.

FUTURE DIRECTIONS

A prototype for future work in this direction comes from a molecule that was reported in 1955 (Figure 7) [6]. More recently Kobuke and coworkers [7] examined the lithium coordination chemistry of this ligand and found remarkably strong binding. This strong binding probably arises in large measure from the conformationally "stiffened" nature of the ionophore. As a consequence this tetrakis(THF) crown lacks the considerable conformational freedom that characterizes most crowns. The connectivity of the ligand constrains the donor atoms to positions commensurate only with the coordination of lithium. Consequently this molecule has not only high affinity but also high selectivity.

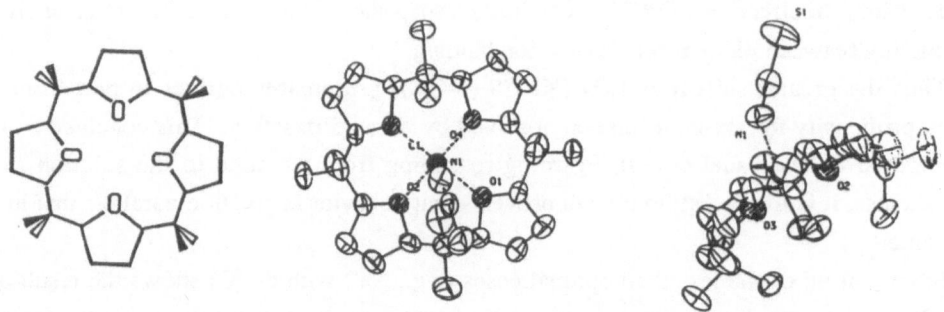

Figure 7. A conformationally "stiffened" crown and two views of its LiNCS complex, [LiNCS·(L)].

X-ray diffraction of the Li^+ complex reveals a square-pyramidal coordination sphere with four oxygen donor atoms disposed on one side of the mean plane of the macrocycle (Figure 7). The *gem*-dimethyl groups add to the selectivity: they contribute four methyl groups that block approach to the pocket where the metal ion binds. This situation, which resembles that of the picket fence porphyrins, doubtless disfavors replacement of Li^+ with any larger ion.

These results indicate that second generation redox-active crowns for practical applications must rely upon conformationally stiffened crown loops. Tetrahydrofuran and tetrahydropyran units are ideal for this purpose. Both the well-worked out synthetic routes to THF-containing fragments, and the ubiquity of THF and THP units in the naturally occurring macrolides encourage this development.

CONCLUSIONS

Electrochemical, EPR and X-ray diffraction studies of quinone crown ethers show that these molecules successfully couple ion binding with a redox reaction. Solution EPR and crystallographic studies indicate that the quinoid O atom in the 1-position (not the 4-position) interacts with alkali metal ions. Analysis of structural and stability constant data suggests that all M^+ selectivity derives from the crown loop; pivoting of the quinoid group promotes affinity for ions but not selectivity. Future improvements in redox-active crown ethers should incorporate conformationally stiffened loops to increase selectivity.

Acknowledgements. I would like to thank my coworkers Bob Wolf, Judy Hartman, Simon Rawle, Gill McCafferty and David Watkin, as well as Milly Delgado. Thanks are also due to the Petroleum Research Fund (administered by the American Chemical Society), the Science and Engineering Council (U.K.), and Genetics International (U.K.) for support of this research.

References

1. Wolf, R. E., Jr.; Cooper, S. R. *J. Am. Chem. Soc.* **1984**, *106*, 4646.

2. Sugihara, K.; Kamiya, H.; Yamaguchi, M.; Kaneda, T.; Misumi, S. *Tetrahedron Lett.* **1981**, *22*, 1619.

3. Delgado, M.; Gustowski, D. A.; Yoo, H. K.; Gatto, V. J.; Gokel, G. W.; Echegoyen, L. *J. Am. Chem. Soc.* **1988**, *110*, 119 and references therein.

4. Bock, H.; Hierholzer, B.; Vögtle, F.; Hollmann, G. *Angew. Chem., Int. Ed. Engl.* **1984**, *23*, 57.

5. To date there is no evidence for interaction of the O atom in the 4 position with any alkali metal ion.

6. Ackman, R. G.; Brown, W. H.; Wright, G. F. *J. Org. Chem.* **1955**, *20*, 1147; see also: Healy, M. de Sousa; Rest, A. J. *J. Chem. Soc., Perkin Trans.* **1985**, 973.

7. Kobuke, Y.; Hanji, K.; Horiguchi, K.; Asada, M.; Nakayama, Y.; Furukawa, J. *J. Am. Chem. Soc.* **1976**, *98*, 7414.

HEAVY METAL CHEMISTRY OF MIXED DONOR MACROCYCLIC LIGANDS: STRATEGIES FOR OBTAINING METAL ION RECOGNITION

Leonard F. Lindoy

Department of Chemistry and Biochemistry
James Cook University of North Queensland
Townsville, 4811 (Australia)

INTRODUCTION

The metal ion preferences of mixed donor coordination sites are often difficult to predict, especially when transition and other heavy metal ions are involved. For example, it can be quite difficult to assign binding preferences to many metal ion binding sites in biological systems even though a knowledge of the metal ion chemistry of such a site is often crucial to an understanding of its biochemical function. In one sense, it is strange that this is still the case since classical coordination chemistry has now been studied for more than a century. Of course, the HSAB theory [1] or the a and b classification [2] provide useful guidelines for predicting and interpreting binding preferences but their usefulness becomes less when mixed donor sites of the type just mentioned are being considered. Alternatively, it is sometimes possible to predict the metal ion binding characteristics of such sites by analogy with the behavior of well studied "simple" ligand systems but this lacks the virtue of wide applicability and hence is less than satisfactory.

An aim of our research over recent years has been to investigate the relationship between mixed donor ligand design and their coordination preferences towards metal ions such as Co(II), Ni(II), and Cu(II); Zn(II) and Cd(II); as well as Ag(I) and Pb(II). Strategies have been developed for achieving discrimination between these series of ions but more importantly, for understanding the nature of such discrimination when it is achieved [3].

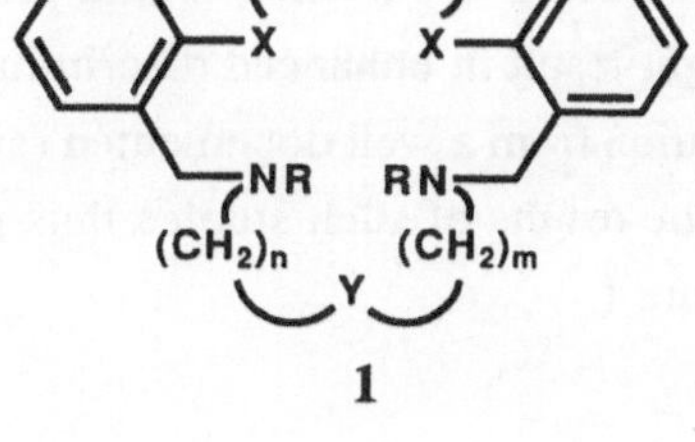

1

Inclusion Phenomena and Molecular Recognition
Edited by J. Atwood
Plenum Press, New York, 1990

Our studies in this area have involved mixed donor, dibenzo macrocyclic ligand systems of general type 1. Such cyclic species tend to give metal complexes which exhibit 1:1 stoichiometries – a factor which aids the interpretation of the solution properties of such complexes. There is a further advantage in using mixed donor species: these usually yield complexes which do not show the very large thermodynamic (and kinetic) stabilities sometimes found for all nitrogen donor ring systems. The absence of extreme stability facilitates the measurement of stability constants. The presence of fused aromatic rings is also advantageous. The benzo groups lead to macrocyclic systems of intermediate flexibility and this is desirable in studies of the type to be described. Further, the aromatic substituents usually result in cyclic products which are crystallizable solids rather than oils; this tends to aid the purification of these large rings.

STRATEGIES FOR OBTAINING METAL ION RECOGNITION

As discussed elsewhere [3], strategies for achieving metal ion recognition (when one metal ion is involved) or discrimination (when more than one ion is present) have resulted from a number of systematic investigations. In a representative study, a macrocyclic ligand is selected which might be expected to show some selectivity for the particular ion(s) of interest. The ligand is then synthesized and an investigation of its solution and solid state metal ion chemistry with the above metal ions undertaken. Typically, the properties of the corresponding complexes are then investigated in some depth. Where appropriate, attempts are then made to relate the solid state and solution structures (often NMR is useful for this). In particular cases, the next step has involved full thermodynamic (calorimetric) investigations but more usually the studies have been limited to the classical determination of stability constants. For some systems, parallel kinetic investigations are also performed at this stage to obtain the rates of formation and/or dissociation associated with complexation.

Taken together, the above investigations give rise to a more complete understanding of the individual systems and provide an important background against which any observed discrimination may be interpreted. That is, the results may be used to assess the factors influencing any observed discrimination; the assessment then forms the basis for modifying the original ligand system in an attempt to achieve enhanced discrimination. The entire process is then repeated and such "tuning" is continued until discrimination reaches a peak (see Figure 1). Typically, several cycles of tuning have been necessary to achieve this goal. In essence, the assessment step acts as both a monitor and a control for the synthetic program.

Molecular mechanics calculations have also been used in such studies [4,5]. In particular instances, it has been possible to simulate features underlying observed complexation behavior and, more recently, we have been using this technique in a predictive way for the design of ligand modifications which might result in enhanced discrimination. Provided this process involves a rather small extrapolation from a well documented (and modelled) system, we have found it to be quite reliable. The results of such studies thus provide the second "feedback loop" in the plan outlined in Figure 1.

A feature of the strategy summarized in Figure 1 is that numbers of ligands are produced which vary in small structural detail from one another; for example, in past studies ranges of macrocycles have been synthesized in which there is a gradual but regular change in such parameters as the macrocyclic ring size, the donor atom set and/or the type of ring substituent present. If three structural parameters are altered, then the process can be thought of as producing a three dimensional ligand matrix (in which each axis is represented by one of the parameters). During a given investigation not all members of such a matrix will need to be synthesized; however, it is has proved useful to picture the tuning process as a series of stepwise movements within such a structural matrix.

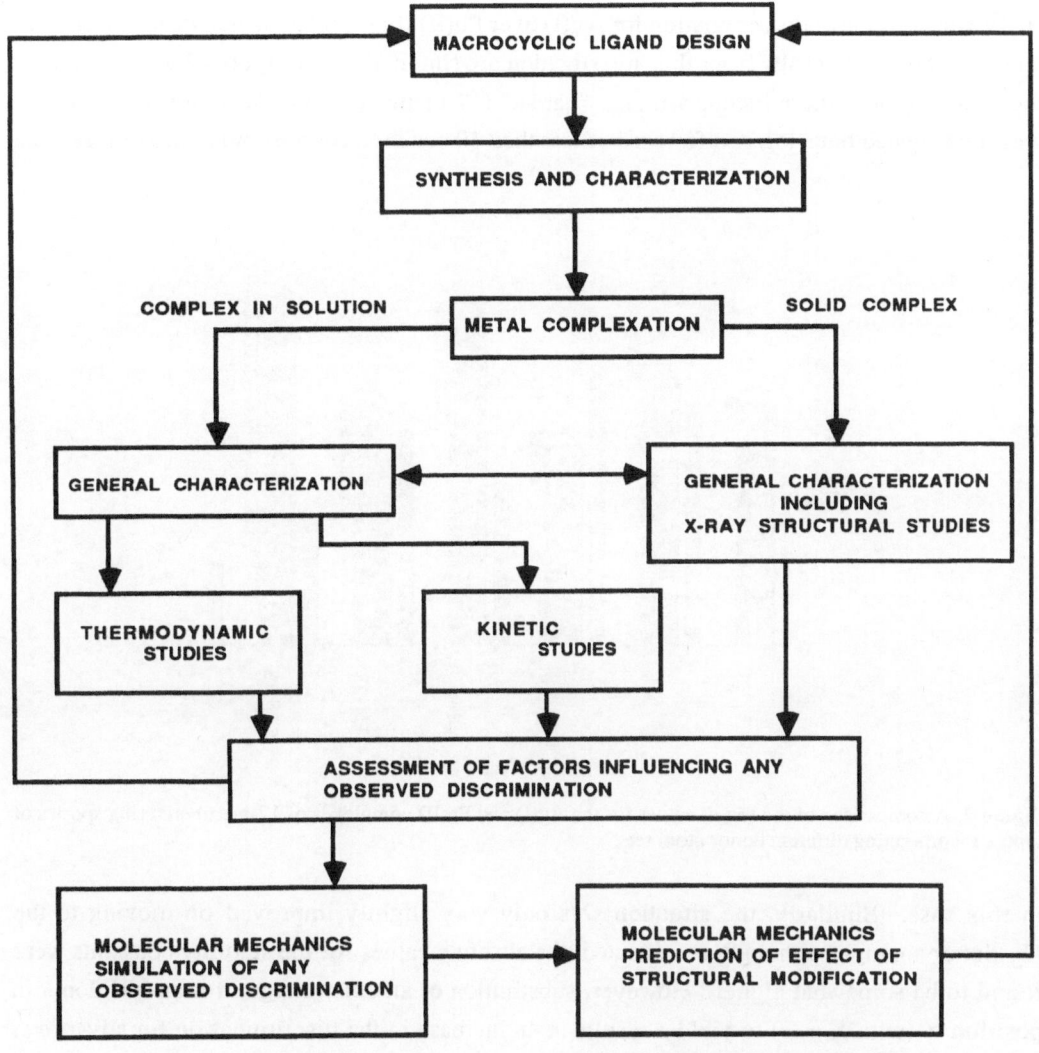

Figure 1. Typical research plan for achieving metal-ion recognition (from ref. 3).

An aim of the discussion which follows is to illustrate, within the overall strategy just discussed, the manner by which particular ligand (structural) aspects may influence discrimination. Examples involving donor atom set variation, macrocyclic ring size variation, and the effect of substitutents will all be discussed. Although treated separately, it is noted that, for a given system, all three structural parameters are available for maximizing metal ion discrimination.

DONOR ATOM SET VARIATION

Donor atom set variation has long been used to alter the affinities of ligand systems towards particular metal ions. We have employed such a procedure for the present systems of general type **1** in which the macrocyclic ring size was maintained at 17-membered and the donor atoms at positions X and Y were varied in a regular manner. The aim of one such study was to achieve discrimination for Ag(I) over Pb(II) [6] – metals which occur together in nature. The log K results from this investigation are summarized in Figures 2 and 3. Initially the interaction of these metals with the "parent" (17-membered) O_2N_3-donor macrocycle **2** was investigated but a log K difference of less than 10 occurred for the complexes of these ions

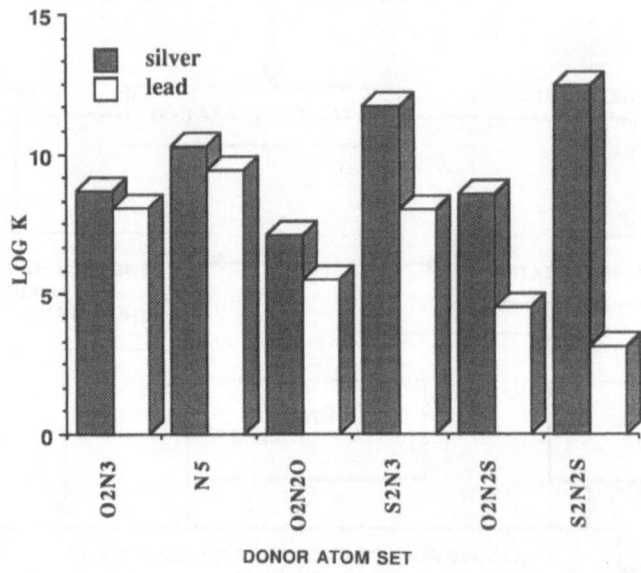

Figure 2. A comparison of the log K values for the Ag(I) and Pb(II) complexes of 17-membered ring species of type **1** incorporating different donor atom sets.

in this case. Similarly, the situation was only very slightly improved on moving to the N_5-donor analogue although, as expected, the absolute values for the stability constants were found to be somewhat higher. However, substitution of an ether oxygen for an NH donor in position Y (with X = O) to yield **3** results in an increase in the discrimination for silver over lead. The effect appears to be a consequence of the known affinity of Ag(I) for a "linear diammine" arrangement relative to Pb(II). That is, although the ether oxygen donors in positions X and Y [see **1**] may coordinate in each case, the inherent affinity of silver for *trans* amine donors "lifts" the stability of this complex relative to that of its Pb(II) analogue.

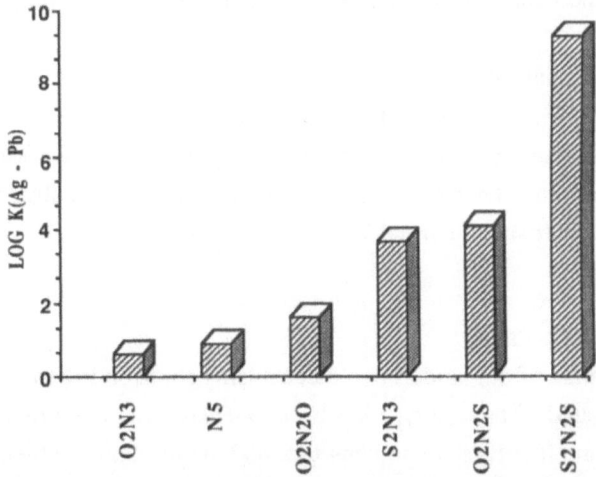

Figure 3. Log K differences between the Ag(I) and Pb(II) complexes of various 17-membered rings of type 1 incorporating different donor atom sets.

At this point the effect of introducing a sulfur donor to yield **4** was investigated. Since, Ag(I) is a b class (or soft) metal ion while Pb(II) is border line, it was hoped that this might also aid enhancement of the log K value for Ag(I) relative to that of Pb(II). This was found to be the case with the stability difference now being approximately 10^4 in favor of Ag(I) – see Figure 3.

The rationale discussed above is supported by the behavior of the next ligand **5** in the series which contains an S_2N_3-donor set. Restoration of the N_3-donor string results in the stability of the Pb(II) species being raised relative to that of Ag(I) and lower discrimination

(2); OenNdienH$_4$

(3); OenNdien(O)H$_4$

(4); OenNdien(S)H$_4$

(5); SenNdienH$_4$

(6); SenNdien(O)H$_4$

(7); SenNdien(S)H$_4$

results.   Nevertheless, the inclusion of two sulfur donors in the ring still leads to an enhancement of the absolute log K value for the Ag(I) complex.   Unfortunately, because of precipitation, it was not possible to obtain the value for lead with the next (S_2N_2O-donor) macrocycle **6** but the value for the silver species is in accordance with the trends discussed so far.

Finally, exceptional discrimination was achieved with the final member of the series **7** incorporating an S_2N_2S-donor set. In this case the discrimination is approximately 10^9 in favor of silver! This result provides a dramatic illustration of the successful use of the "tuning" strategy outlined in Figure 1 – a strategy which should also enable substantial discrimination to be achieved within other suitable metal ion/ligand systems.

RING SIZE VARIATION

During the course of our studies two mechanisms contributing to ring size discrimination have been investigated. The first type has been extensively studied by numerous workers [7] (including us) [8] and involves maximizing thermodynamic stability by matching the radius of the metal ion to the hole size of the coordinated ligand. Discrimination involving this mechanism will not be further discussed here. Rather, the discussion will be restricted to another mechanism – "dislocation" discrimination [3]. In general terms a dislocation occurs when the gradual change of properties along a ligand series results in a sudden change in coordination behavior for the complexes of adjacent ligands. When such a dislocation occurs at different points along the ligand series for different metal ions then this may form the basis for discriminating between these ions (see Figure 4).

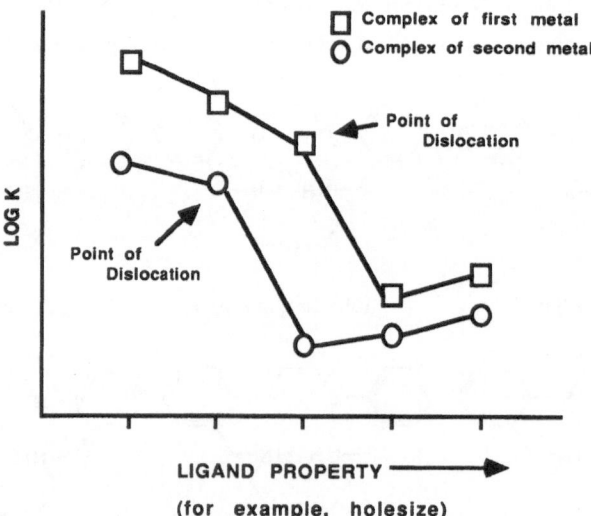

Figure 4. An illustration of the manner by which dislocation behavior may give rise to enhanced metal ion discrimination – in this (hypothetical) case, the log K difference is enhanced for the third ligand along the series.

Dislocation behavior has now been observed for a number of cyclic ligand systems [9-12] and, in most cases, the subtle factors underlying such discrimination have been successfully elucidated using a combination of solution equilibrium, NMR, molecular mechanics, and X-ray diffraction techniques.

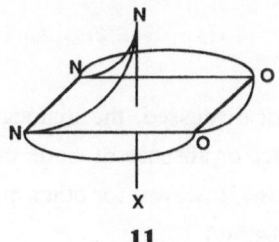

(8); OenNentnH$_4$ (9); OenNditnH$_4$ (10); OenNenbnH$_4$

An early study in the author's laboratory involved the interaction of Ni(II) with the O$_2$N$_3$-donor macrocycles **1, 8-10** [11]. Table 1 lists the log K values for the 1:1 nickel complexes of these ligands. The respective values provide evidence that the complexes fall into two categories. First there are the 17- and 18-membered ring complexes of **1** and **8** which yield log K values of 10.0 and 9.8, respectively. The similarity of these values suggests that both complexes have related geometries (the 18-membered ring complex would be expected to be slightly less stable because it incorporates an additional 6-membered chelate ring). In contrast, the two 19-membered ring ligands **9** and **10** yield complexes whose log K values are substantially lower.

Table 1. Log K values (ML^{2+}) for the Ni(II) complexes of the O$_2$N$_3$-donor macrocycles at 25°C [0.1 M (CH$_3$)$_4$NCl; 95% methanol].

macrocycle	ring size	log K
2; OendienH$_4$	17	10.0
8; OenNentnH$_4$	18	9.8
9; OenNditnH$_4$	19	6.4
10; OenNenbnH$_4$	19	5.1
12; OenNdipnH$_4$	17	6.9

In agreement with the stability data, visible spectral evidence clearly indicated that a structural change occurs along this series between the 18- and the 19-membered ring complexes [11]. Indeed, inspection of molecular models suggested that the 17- and 18-membered rings will adopt the configuration shown in **11** in which the N$_3$-portion of

11

macrocycle is arranged facially. A subsequent X-ray structural investigation of the 17-membered ring complex confirmed that the postulated structure does occur for this complex in the solid (Figure 5) [4]. In constrast, inspection of a model of the complex of **9** suggested that facial coordination will be destabilized in the case of this 19-membered ring species because of steric crowding in the N3-portion of the coordinated macrocycle. Instead, it appeared likely that meridional coordination will occur in this case. Molecular mechanics investigations on both the facial and meridional isomers of both of the 17- and 19-membered ring complexes support the relative stabilities just discussed [4,5]. That is, for the 17-membered ring complex the facial form is more stable than the meridional form whereas the reverse order was found for the (19-membered ring) complex of **9**.

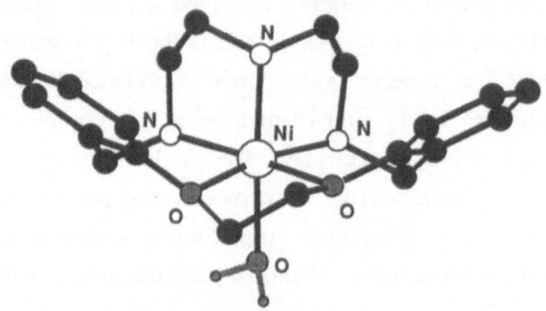

Figure 5. The X-ray structure of [Ni(OenNdienH$_4$)(H$_2$O)]$^{2+}$ illustrating the facial coordination of the N$_3$-fragment of the macrocycle.

It is noted that the complex of the dimethylated derivative **12** also shows a substantial drop in stability relative to the (unsubstituted) parent complex of **2** – see Table 1. Once again, the evidence suggests that the presence of the methyl groups results in steric crowding in the case of the facial form such that formation of the meridional complex is favored. Molecular mechanics calculations again support this postulate [5].

(12) OenNdipnH$_4$ (13); OenN(CH$_3$)$_3$dienH$_4$ (14) OenN(CH$_3$)ditnH$_4$

For the Ni(II) complexes just discussed, the dislocation behavior thus appears to be largely associated with the presence or absence of steric crowding in the N$_3$-portions of the respective coordinated macrocycles. However, for other metal ion systems, the origins of the dislocations may be different (see below).

Table 2. Log K values (ML^{2+}) for the Zn(II) and Cd(II) complexes of the O_2N_3-donor macrocycles at 25°C [0.1 M $(C_2H_5)_4NClO_4$; 95% methanol].

Macrocycle		ring size	log K values	
			Zn(II)	Cd(II)
2;	OenNdienH$_4$	17	7.5	8.5
8;	OenHentnH$_4$	18	7.1	7.9
9;	OenNditnH$_4$	19	6.6	5.3
10;	OenNenbnH$_4$	19	6.0	5.0
12;	OenNdipnH$_4$	17	5.6	8.9
13;	OenN(CH$_3$)$_3$dienH$_2$	17	5.1	6.1
14;	OenN(CH$_3$)ditnH$_4$	19	5.9	4.7

The interaction of Zn(II) and Cd(II) with the unsubstituted O_2N_3-donor ligands 2, 8, 9, and 10 as well as with their substituted derivatives, 12 - 14, has been investigated [12]; the relative stabilities of the various species are summarized in Table 2. Initially, it is convenient to consider the relative values for the unsubstituted 17-, 18-, and 19-membered systems mentioned above. In the case of zinc complexes, the log K values for the 17- to 19-membered rings fall in a regular manner as the number of six-membered chelate rings is increased along the series. However, in the case of the cadmium analogues, a dislocation appears to occur on passing from the 18- to the 19-membered ring species (Figure 6). For the 17- and 18-membered rings, the cadmium complexes are more stable than their zinc analogues whereas the order is reversed for the 19-membered ring species. Indeed, this latter order is that expected for simple linear aliphatic amines. [1]

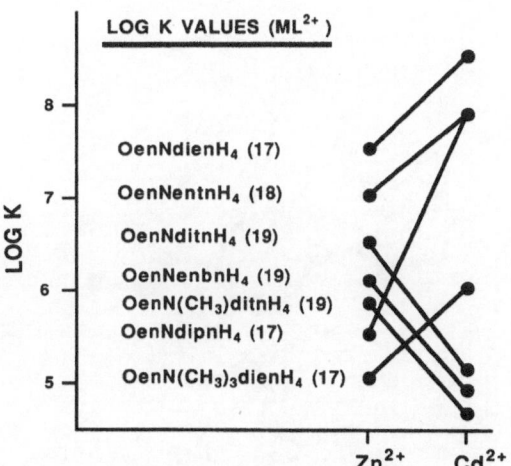

Figure 6. Comparison of the log K values [0.1 M, $(C_2H_5)_4NClO_4$; 95% methanol] at 25°C for the Zn(II) and Cd(II) complexes of the macrocycles shown. The number in parenthesis refers to the ring size of the respective macrocycles.

A combination of NMR and X-ray studies provides an explanation for the above behavior. The X-ray structure of the zinc complex of the 17-membered ring 2 clearly shows that the ether oxygens are not coordinated (Figure 7a) and the NMR evidence suggests that a similar situation occurs in solution. In view of this, it appears likely that similar non-coordination of the ethers is maintained down the series as the macrocyclic ring size is increased. In constrast, the X-ray structure of the Cd(II) complex of the 17-membered ring **2** shows that the macrocylic ligand now coordinates via all five of its donor atoms (Figure 7b). In part, this may be a consequence of the larger radius of Cd(II) relative to Zn(II) although electronic factors are also likely to be important. The enhanced stability arising from such full coordination is presumably the reason that the log K value for this species is higher than that for the Zn(II) analogue. From the log K data, a similar situation appears to occur for the respective 18-membered ring complexes of **8**. In contrast, the X-ray structure of the Cd(II) complex of the 19-membered ring **9** shows that the ether oxygens are now outside bonding distance (Figure 7c). That is, a related situation to that postulated for the corresponding zinc complex occurs. Other evidence once again suggests that a similar situation is present in solution. Hence, the relative stabilities of the complexes of this 19-membered ring now fall in the expected order of zinc > cadmium.

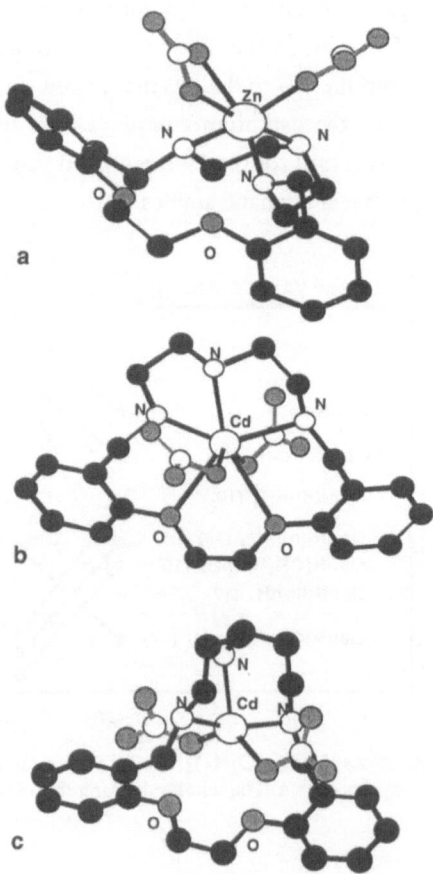

Figure 7. The X-ray structures of (a) [Zn(OenNdienH$_4$)(NO$_3$)$_2$], (b)Cd(OenNdienH$_4$)(NO$_3$)$_2$], and (c) [Cd(OenNditnH$_4$)(NO$_3$)$_2$].

Since the dislocation behavior just discussed appears to be largely a ring size phenomenon, the appending of substituents to the rings would be expected not to alter the stability patterns observed. Inspection of the data in Table 2 and Figure 6 confirms that this is so. In each case the 17-membered rings once again yield complexes in which the stabilities fall in the order zinc < cadmium while, for the 19-membered ring species, this order is reversed.

MACROCYCLE SUBSTITUTION AND DISCRIMINATION

The effect of appending N-benzyl substituents to the parent 17-membered O_2N_3-macrocycle 2 [to yield 15] on metal ion discrimination has been investigated [14]. It might be anticipated that the presence of bulky N-substitutents on all three nitrogen donors would lead to the formation of weaker complexes than those obtained with the unsubstituted analogue 2. Indeed, the stabilities of the Co(II), Ni(II), Zn(II), Cd(II), and Pb(II) complexes of the tribenzylated derivative are all significantly reduced relative to the corresponding

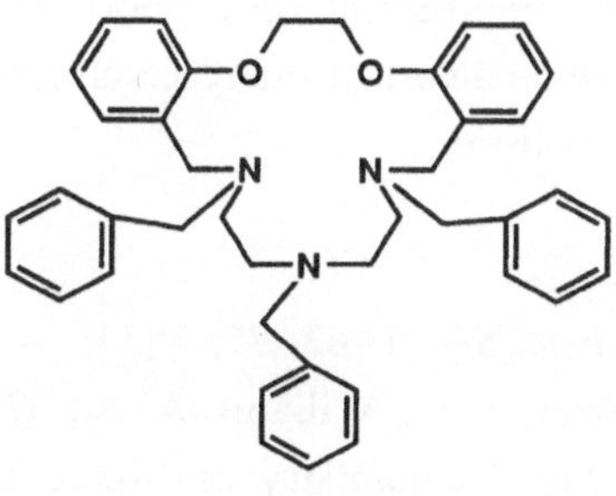

15

complexes of 2; in all of these cases log K values of 4.3 or lower were observed (Table 3). In contrast, the log K value for the Ag(I) species is 9.3 and this is even slightly higher than that obtained for the parent complex. In the absence of further data, the reason for the high selectivity for the latter ion cannot be assigned with certainty. However, molecular modelling suggests that this ion may trigger a major conformational change in the ligand to produce a relatively inflexible conformation which is ideal for Ag(I).

Table 3. Comparison of the log K values for the 1:1 complexes of 2 and its tribenzylated derivative 15[a].

	macrocycle	
	2	15
Co(II)	7.6[b]	<3.5
Ni(II)	10.0[b]	<3.5
Cu(II)	14.4[b]	_[c]
Zn(II)	7.5	~3.5
Cd(II)	8.7	~3.5
Ag(I)	8.7	9.3
Pb(II)	8.1	~4.3

[a]Log K values determined in 95% methanol at 25°C with I = 0.1; $(C_2H_5)_4NClO_4$.
[b]I = 0.1, $(CH_3)_4NCl$. [c]Precipitation.

However, it is emphasized that confirmation of this postulate must await the results of further studies. Clearly, the appending of the bulky benzyl groups to the nitrogen donors effectively detunes the affinity of this ligand system towards the first six metal ions mentioned previously but not towards Ag(I).

CONCLUSIONS

In the present discussion a number of individual studies have been described in order to illustrate the various means by which ion discrimination may be achieved within a ligand matrix of the type discussed. In each of these cases, the discrimination was achieved by variation of a single (but different) structural parameter. However, for a given study, the opportunity may exist to vary simultaneously several such parameters during the tuning process – in such cases the prospect of achieving enhanced discrimination will be substantially greater.

Acknowledgment. The author thanks the Australian Research Council and the Australian Institute of Nuclear Science and Engineering for support. He also thanks his collaborators (see reference list) as well as the members of his research group, both past and present, for their contributions to the work described.

References

1. Pearson, R. G. *J. Am. Chem. Soc.* **1963**, *85*, 3533.
2. Ahrland, S.; Chatt, J.; Davies, N. R.; Wiliams, A. A.; *Quart. Rev.* **1958**, *12* , 264.
3. Lindoy, L. F. "Heavy Metal Chemistry of Mixed Donor Macrocyclic Ligands: Strategies for Obaining Metal Ion Recognition" in *Progress in Macrocylic Chemistry*; Izatt, R. M.; Christensen, J. J. Eds.; Wiley: New York, 1987; Vol. 3, pp. 53-92.
4. Adam, K. R.; Brigden, L. G.; Henrick, K.; Lindoy, L. F.; McPartlin, M.; Mimnagh, B.; Tasker, P. A. *J. Chem. Soc., Chem. Commun.* **1985**, 710.
5. Adam, K. R.; Brigden, L. G.; Lindoy, L. F., unpublished results.
6. Baldwin, D.; Leong, A. J.; Lindoy, L. F., unpublished results; Baldwin, D.; Lindoy, L. F.; Graddon, D. P. *Aust. J. Chem.* **1988**, *41*, 1347.
7. Lindoy, L. F. *The Chemistry of Macrocyclic Ligand Complexes*; Cambridge University Press: Cambridge, 1989; Henrick. K.; Tasker, P. A.; Lindoy, L. F. *Prog. Inorg. Chem.* **1985**, *33*, 1.
8. Anderegg, G.; Ekstrom, A.; Lindoy, L. F.; Smith, R. J. *J. Am. Chem. Soc.* **1980**, *102*, 2670; Goodwin, H. J.; Henrick, K.; Lindoy, L. F.; McPartlin, M.; Tasker, P. A. *Inorg. Chem.* **1982**, *21*, 3261; Drummond, L. A.; Henrick, K.; Kanagasundaram, M. J. L.; Lindoy, L. F.; McPartlin, M.; Tasker, P. A. *Inorg. Chem.* **1982**, *21*, 3923; Henrick, K.; Lindoy, L. F.; McPartlin, M.; Tasker, P. A.; Wood, M. P. *J. Am. Chem. Soc.* **1984**, *106*, 1641; Adam, K. R.; Leong, A. J.; Lindoy, L. F.; Anderegg, G. *J. Chem. Soc., Dalton Trans.* **1988**, 1733.
9. Adam, K. R.; Dancey, K. P.; Harrison, B. A.; Leong, A. J.; Lindoy, L. F.; McPartlin, M.; Tasker, P. A. *J. Chem. Soc., Chem. Commun.* **1983**, 1351.

10. Adam, K. R.; Leong, A. J.; Lindoy, L. F.; Lip, H. C.; Skelton, B. W.; White, A. H. *J. Am. Chem. Soc.* **1983**, *105*, 4645.

11. Adam, K. R.; Ansell, C. W. G.; Dancey, K. P.; Drummond, L. A.; Leong, A. J.; Lindoy, L. F.; Tasker, P. A. *J. Chem. Soc., Chem. Commun.* **1986**, 1011.

12. Adam, K. R.; Dancey, K. P.; Leong, A. J.; Lindoy, L. F.; McCool, B. J.; McPartlin, M.; Tasker, P. A. *J. Am. Chem. Soc.* **1988**, *110*, 8471.

13. Smith, R. M.; Martell, A. E. *Critical Stability Constants, Vol. 2: Amines*; Plenum: New York, 1975.

14. Leong, A. J.; Strixner, T.; Lindoy, L. F., unpublished work.

LANTHANIDE CHELATES AS LUMINESCENT PROBES

John L. Toner

Exploratory Sciences Division
Life Sciences Research Laboratories
Eastman Kodak Company
Rochester, NY 14650-2115 (USA)

SUMMARY

A series of molecules is described which efficiently complex Eu^{+3} and Tb^{+3} and sensitize their luminescence in water. The general approach has been the incorporation of convergent binding sites on molecules which are efficient triplet sensitizers for lanthanides in aprotic solvents. One application for these chelators involves immunoluminescence assays in which a biologically relevant molecule present at low concentration could be quantitatively detected. Progress towards this goal is described.

INTRODUCTION

In 1978, we began searching for potential methods of replacing radioimmunoassay (RIA) with a luminescence based approach [1]. A representative immunoassay consists of competition for a known, limited number of antibody binding sites by labelled and native hapten. The hapten is usually a biologically relevant molecule such as a hormone or a drug. A separation between material bound to the antibody and unbound material is conducted followed by quantification according to the type of label present. Initial calibration with known concentrations of analyte allows subsequent quantification of unknowns.

Fluoroimmunoassays (FIA's) using conventional fluorescent dyes such as fluorescein and the rhodamines have some advantages over RIA methods [2]. FIA techniques can be done without the use of short-lived radioisotopes, no special licensing is needed, long counting times are not necessary, the hazards due to isotope handling are removed, and they can be performed on reasonably priced equipment. However, the sensitivity of RIA, which is its great virtue, is lessened significantly with conventional FIA methods. This is due in part to the intrinsic fluorescence of the medium, especially serum, which can yield high background from 400-600 nm. In addition, scattering in biological media is a significant problem due to high concentrations of macromolecules such as fats and proteins. Typical fluorochromes have detection limits in serum that are 50 to 1000 times those in buffer [3].

Inclusion Phenomena and Molecular Recognition
Edited by J. Atwood
Plenum Press, New York, 1990

The use of luminescence of certain lanthanides, notably Tb^{+3} and Eu^{+3}, appeared to us to be potentially ideal as a method for signal generation in immunoassays. The predominant emission line for Tb^{+3} is the $^5D_4 \rightarrow {}^7F_5$ transition around 545 nm, while the principal

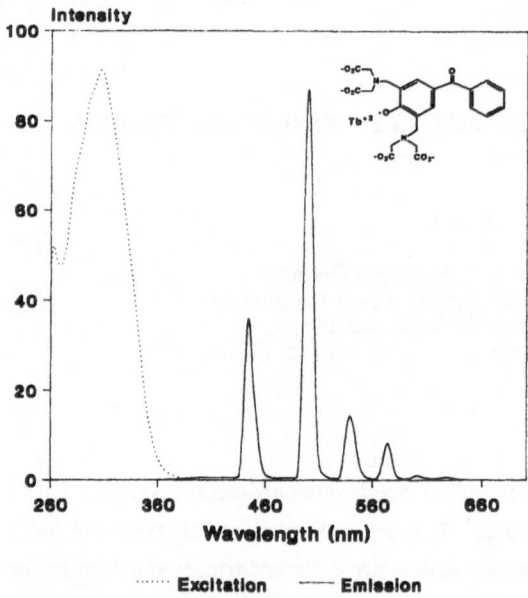

Figure 1. Excitation and emission spectra for the Tb^{3+} complex of benzophenone tetraacid **3**. Ex = 307 nm, Em = 544 nm; [**3**] = 10^{-5} M, [Tb^{3+}] = 10^{-4} M; borate buffer, pH 8.5.

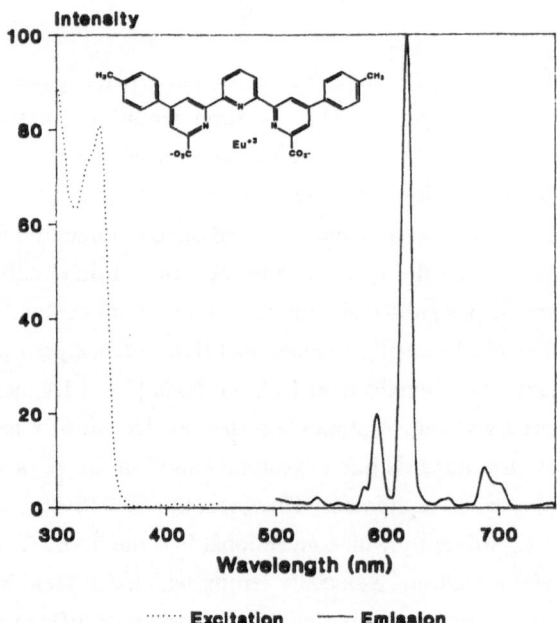

Figure 2. Excitation and emission spectra for the Eu^{3+} complex of terpyridine diacid **10**. Ex = 340 nm, Em = 616 nm; [**10**] = 5 x 10^{-6} M, [Tb^{3+}] = 10^{-4} M; borate buffer, pH 8.5.

emission line for Eu^{+3} is the $^5D_0 \rightarrow {}^7F_2$ transition around 618 nm (Figures 1 and 2). In particular, the luminescence of Eu^{+3} is spectrally distinct from the fluorescence of most biological systems. In addition, the luminescence of chelates of both metals is quite long-lived, with lifetimes typically from 0.1-1 ms. This feature makes simple time resolution schemes easy to accomplish because the background fluorescence has a typical lifetime of about 10-20 ns.

Previous studies, primarily with acetylacetonate complexes, have demonstrated that efficient triplet sensitization of Tb^{+3} and Eu^{+3} can be practical in nonaqueous media [4]. The excitation spectrum of a lanthanide trisacetylacetonate complex is essentially the absorption spectrum of the chelator since Tb^{+3} and Eu^{+3} have extinction coefficients of 0.3 and 3 M^{-1} cm^{-1}, respectively [5]. Consequently, the Stoke's shifts for a sensitizer with its absorption maximum at 400 nm would be 145 nm with Tb^{+3} and 218 nm with Eu^{+3}. The large Stoke's shifts make scattering by macromolecules insignificant.

Unfortunately, there are drawbacks to the use of Tb^{+3} and Eu^{+3} as luminescent probes. As has been mentioned, the lanthanides must be sensitized for effective emission to occur. With water acting as a competitive ligand, the sensitizer must be capable of strongly complexing the lanthanide. However, a worse problem in aqueous media is that the lanthanide luminescence is strongly quenched by interaction with water [4,6]. Chelated water allows nonradiative energy transfer from the excited states of the lanthanide to the OH vibrational manifold. Since both Tb^{+3} and Eu^{+3} are nonahydrates in aqueous media, a great deal of water needs to be removed to ensure effective luminescence. Even if the lanthanide is complexed to the sensitizer and not lost to the medium, the addition of extra waters as ligands severely quenches luminescence. Efficient sensitization in water is a severe enough restraint to have forced some researchers to separate the lanthanide from an initial complex in water and form a secondary organic soluble chelate which luminesces under much less stringent conditions [7].

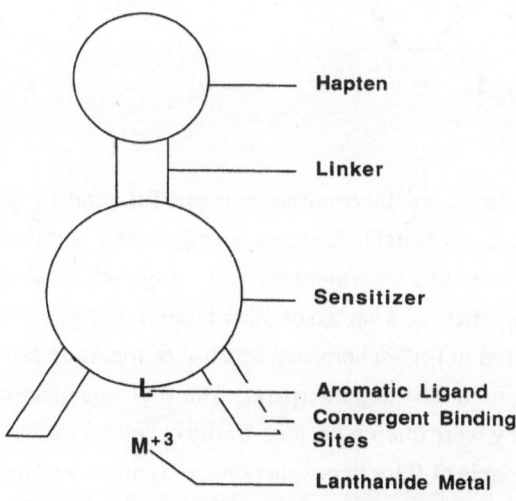

Figure 3. Generalized lanthanide luminescent probe.

The drawbacks to the use of lanthanides in immunoluminescence assays govern the design of effective sensitizers. The objective of using lanthanides as luminescent probes demands that the intensity of emitted light be maximized. To accomplish this goal several criteria must be met. The sensitizer must bind the lanthanides powerfully at low concentrations in water, at or above room temperature. In addition, the sensitizer must have as high an extinction coefficient as possible ($>10,000$ M^{-1} cm^{-1}) and a quantum efficiency for lanthanide luminescence as close to unity as possible. Ideally, it should also absorb above 350 nm. Finally, the sensitizer must have a site for attachment of a hapten. Figure 3 is a schematic showing some of the features we believe to be important in effectively chelating lanthanides and maximizing the resulting luminescence.

Recent work by Balzani, Lehn and coworkers [5,8] highlights the difficulties of accomplishing the above requirements in a given sensitizer host. Choosing bipyridines or phenanthrolines as the lanthanide sensitizers led to the synthesis of the model cryptands 1and 2. Even with the lanthanide completely encapsulated, luminescence quantum efficiencies for the cryptates of 1 were quite low in water (0.03 for Tb^{+3} and 0.02 for Eu^{+3}). From the relative luminescence lifetimes in H$_2$O and D$_2$O the number of water molecules still coordinated to the Tb^{+3} and Eu^{+3} cryptates of host 1 was calculated (3.0 and 2.5 $\pm$ 0.5, respectively) [5].

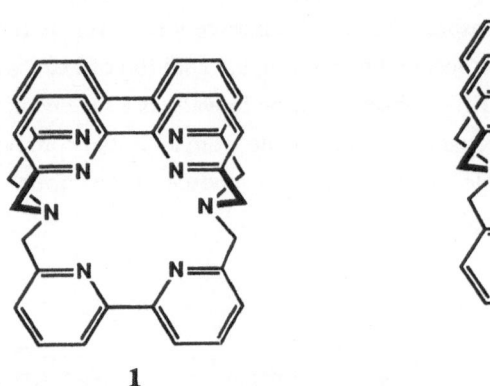

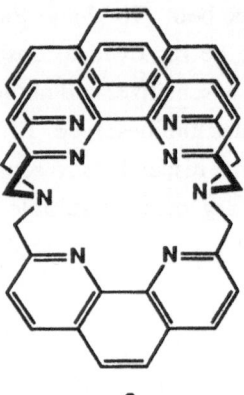

1 2

MODEL SYSTEMS

Our approach has involved incorporating powerful binding groups, usually imino-diacetates or carboxylates, on suitable lanthanide sensitizers. As model studies progressed, appropriate bifunctional chelators were synthesized having both the ability to bind lanthanides, and a remote site for attachment of a hapten or other biologically interesting molecule. We have been particularly interested in Eu^{+3} chemistry because of the more favorable emission spectra of Eu^{+3} chelates versus their Tb^{+3} counterparts. The first sensitizer class we examined was the benzophenones. They were chosen because of the excellent intersystem crossing quantum efficiencies (Φ_{isc}) of models (1.0 for benzophenone). Synthesis of model host 3 was readily accomplished by treatment of p-hydroxybenzophenone with iminodiacetic acid and CH$_2$O under basic conditions. The resulting liquid has a reasonable extinction coefficient and quantum efficiency with Tb^{+3} ($\Phi_{Yb} = 0.10$), but with Eu^{+3} the quantum efficiency was quite low (Table 1). Benzophenone has a triplet energy of 68.5 kcal/mol while Tb^{+3} emitting at 545 nm corresponds to an energy of 52.5 kcal/mol and Eu^{+3} emitting at 614 nm corresponds to an energy of 46.6 kcal/mol. A better match between the sensitizer triplet level and the lanthanide might result in more efficient energy transfer [4].

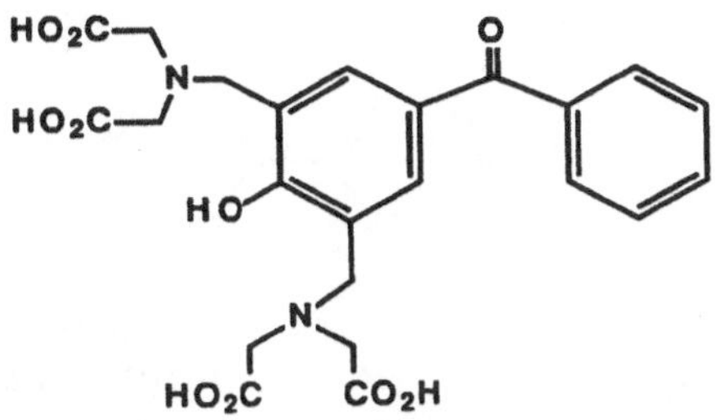

3

Coumarin model **4** successfully meets the above consideration. Normally, coumarins are highly fluorescent, which renders them inappropriate choices as lanthanide sensitizers. However, Farid and collaborators demonstrated that incorporation of a keto group in the 3-position of the coumarin nucleus converted the molecule to an excellent triplet sensitizer [9]. For example, ketocoumarin **5** has a triplet energy of 58.0 kcal/mol with $\Phi_{isc} = 0.94$. The quantum efficiency of the Eu^{+3} chelate of coumarin **4** in basic buffer is thirteen times that of the Eu^{+3} complex of benzophenone **3** (Table 1). In addition, the coumarins absorb further towards the visible region and have higher extinction coefficients than the benzophenones.

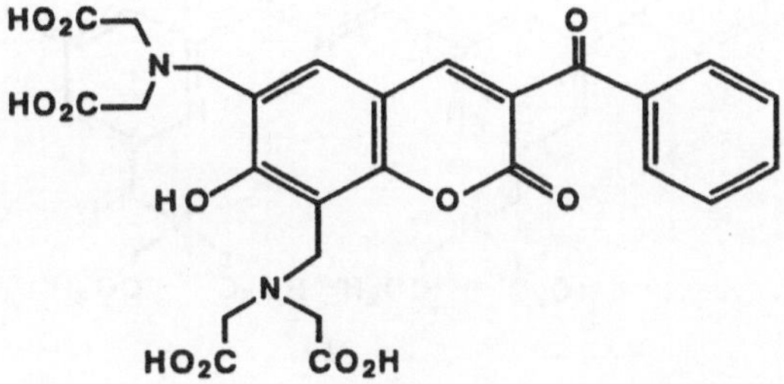

4

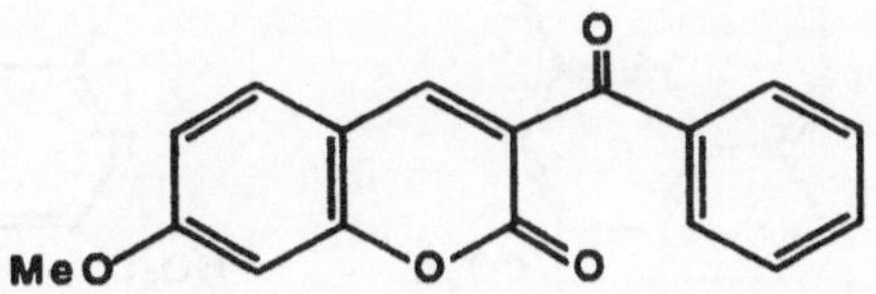

5

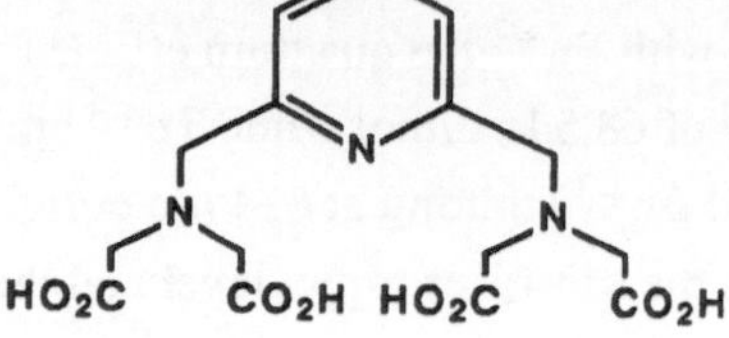

6

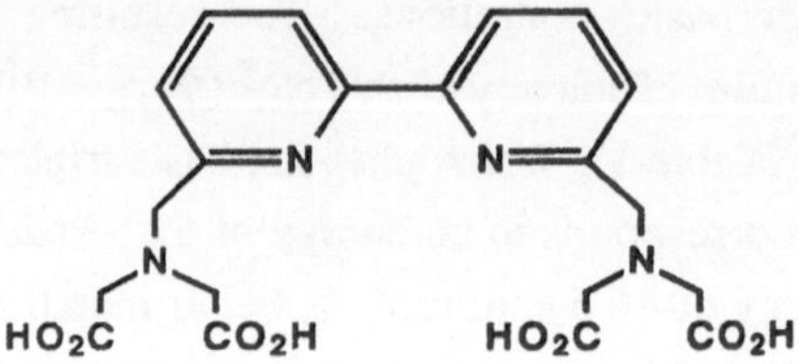

7

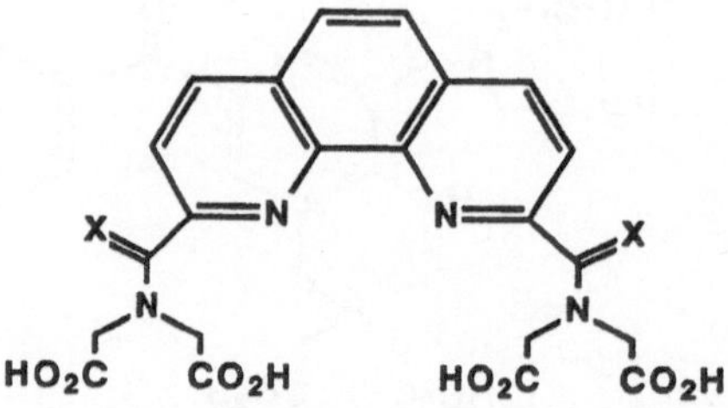

8: X = O
9: X = H$_2$

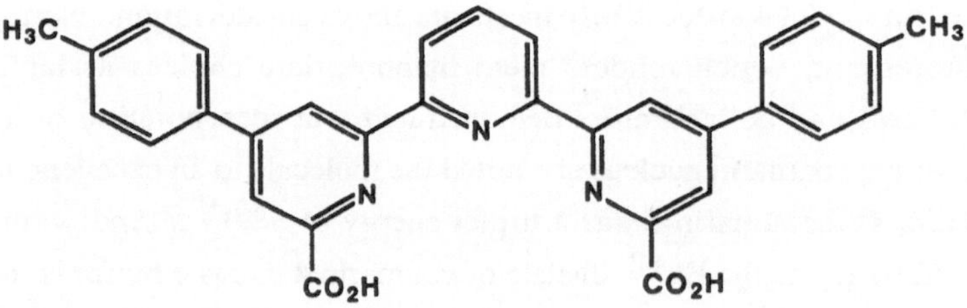

10

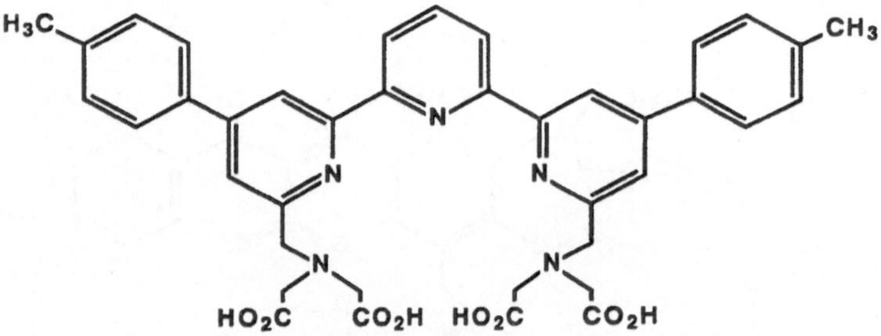

11

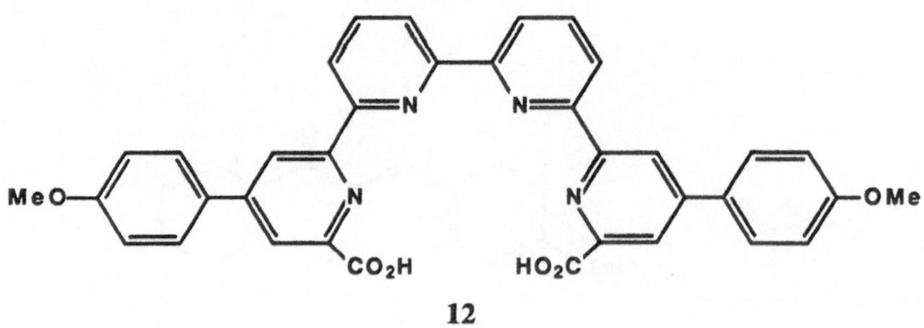

12

Table 1. Photophysical properties of Eu^{+3} chelates[a].

Host	λmax (nm)	ε ($M^{-1} cm^{-1}$)	Φ_{Eu}		τ_{Eu} (ms)
3	320	16000	0.003	b	
4	396	27000	0.04		b
6	265	7400	b		0.48
7	304	14000	b		0.71
10	340	23000	0.08		0.35
11	339	14000	0.40		1.5
12	361	33000	0.14		1.0

[a]All properties obtained with fully complexed material in aqueous buffer at [ligand] $\approx 10^{-6}$ M.
[b]Not determined.

Extension of the lanthanide sensitizer classes to substituted oligo-pyridines and phenanthrolines [10] has been our most fruitful approach. A large number of hosts have been synthesized, of which **6-12** provide representative samples. Relevant photophysical data are detailed in Table 1.

The binding ability for some of the ligands towards Tb^{+3} and Eu^{+3} has been determined. Association constants for hosts **3** and **4** were measured by the technique of Schwarzenbach and coworkers [11]. Hosts **3** and **4** had log K_a values for Tb^{+3} of 16.8 and 16.4, respectively, while corresponding log K_a values for Eu^{+3} were 16.7 and 16.3. Luminescence titration experiments permitted the determination of the K_a between Eu^{+3} and terpyridine diacid **10** (ca. 10^6 M^{-1}).

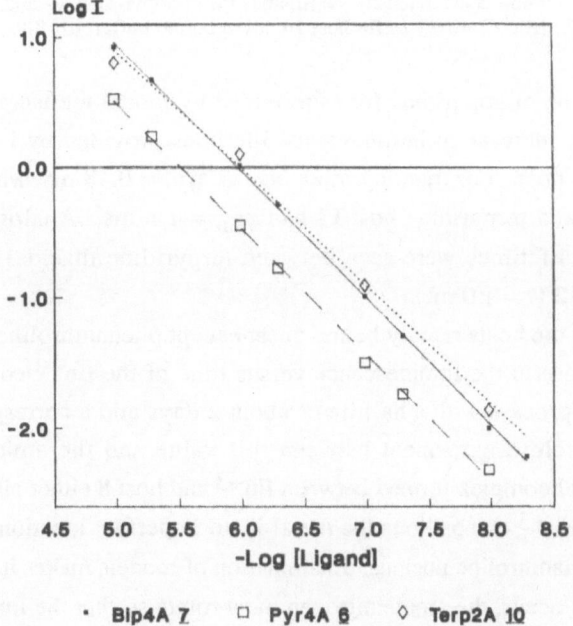

Figure 4. Relative Tb^{+3} luminescence intensity vs. [ligand] for oligopyridine di- and tetraacids. $[Tb^{+3}] = 10$ [ligand] for hosts **6** and **7**. $[Tb^{+3}] = 10^{-4}$ M for host **10**, all in borate buffer, pH 8.5.

Figure 4 shows the linear relationship between Tb^{+3} luminescence and ligand concentration for three of the sensitizer classes. Figure 5 is a luminescence versus concentration study for relevant model hosts from four of the sensitizer classes complexed with Eu^{+3}. Complete formation of the Tb^{+3} and Eu^{+3} complexes of terpyridine diacid 10 was sustained by maintaining the lanthanide concentration at 10^{-4} M.

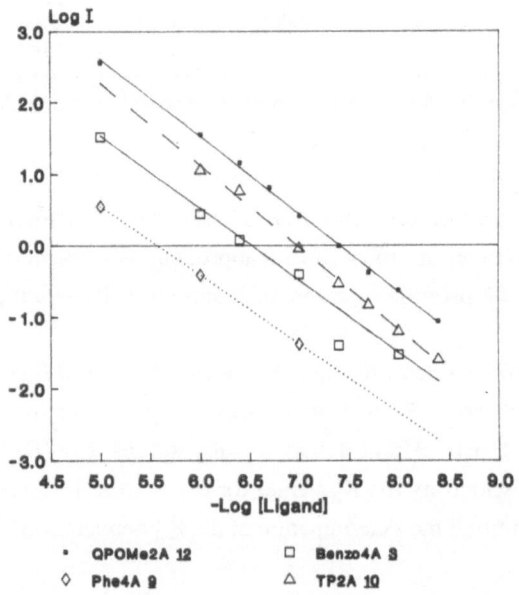

Figure 5. Relative Eu^{+3} luminescence intensity vs. [ligand] for oligopyridine di- and tetraacids. $[Eu^{+3}] = 10$ [ligand] for hosts 6 and 7. $[Eu^{+3}] = 10^{-4}$ M for host 10, all in borate buffer, pH 8.5.

Comparison of the oligopyridine series provided by model tetraacid hosts 6 [12], 7 [12], and 11 illustrates the increase in luminescence lifetimes provided by increasing numbers of chelation sites in the host. For monopyridine host 6, $\tau_{Eu} = 0.48$ ms, while bipyridine host 7 had $\tau_{Eu} = 0.71$ ms and terpyridine host 11 had $\tau_{Eu} = 1.5$ ms. Analogous increases in the Eu^{+3} luminescence lifetimes were seen between terpyridine diacid 10 ($\tau = 0.35$ ms) and tetrapyridine diacid 12 ($\tau = 1.0$ ms).

Interestingly, all the hosts rapidly bound guest except phenanthroline bisamide 8. Kinetic analysis of the increase in the luminescence versus time of the Eu^{+3} complex of bisamide 8 showed a first order process with a halflife of about 2 days and a corresponding E_a of ca. 21 kcal/mol [10]. The close agreement between this value and the amide resonance energy suggests that the initial complex formed between Eu^{+3} and host 8 either allows additional water to interact with the Eu^{+3} or positions the metal in an imperfect location for effective energy transfer from the phenanthroline nucleus. Examination of models makes it clear that in order for effective chelation to occur, the amide nitrogen must rotate so that the lone pair of electrons is orthogonal to the carbonyl double bond. There is an attendant 10-fold increase in luminescence intensity.

LUMINESCENCE ENHANCEMENT

Incorporation of 2,9-dicarboxy-1,10-phenanthroline **13** [14] as a co-ligand greatly enhanced the luminescence of the primary complex of Eu^{+3} with several of the chelating agents, for example, the benzophenone tetraacid **15**. Primary complexors with high binding constants toward Eu^{+3} were deliberately chosen to avoid the possibility of the phenanthroline co-ligand removing the metal in a transchelation process. When the primary complex **15**-Eu^{+3} was held constant at 2×10^{-7} M, a dramatic increase in luminescence as a function of the concentration of the added co-ligand was observed (Figure 6). Figure 7 illustrates the changes in the emission spectrum of 9×10^{-5} M benzophenone **14** at four different co-ligand concentrations.

13

14: R = Me, R' = H, R" = NHCOMe
15: R = Me, R' = H, R" = NHCSNHT4OMe
16: R = H, R' = CONHT4OMe, R" = H
(NHT4OMe is Thyroxin methyl ester)

Apparently, unlike the primary complex, the secondary complex is formed with a relatively low binding constant. Luminescence increased to a point with added co-ligand, then began to decrease because the concentration of the diacid was high enough to absorb most of the incident light at the interface of the spectrofluorimeter cell, rendering transfer to the photomultiplier problematic. High concentration self-quenching by the co-ligand is also a potential problem. We see the co-ligand effect with chelators such as the benzophenone tetraacids which have good binding constants for Eu^{+3} but still have sites available on the Eu^{+3} for additional chelation to occur. Presumably, the secondary binding involves loss of water from the solvation spheres of the metal which results in more efficient luminescence. Both EDTA and DTPA also show the co-ligand enhancement effect.

BIFUNCTIONAL CHELATORS

Several bifunctional chelators have been synthesized. Representative are hosts **14-19**. Host **14** is a model bifunctional chelator while chelators **15-18** have an attached thyroxin methyl ester as the hapten. Thyroxin is a hormone involved in determining metabolic rate with a normal physiological concentration range of 5.8×10^{-8} to 1.6×10^{-7} M. Terpyridine diacid **19** has two attached dilantin molecules as haptens. Dilantin is an anticonvulsant and antiepileptic drug, normally effective from $4.0-8.0 \times 10^{-5}$ M.

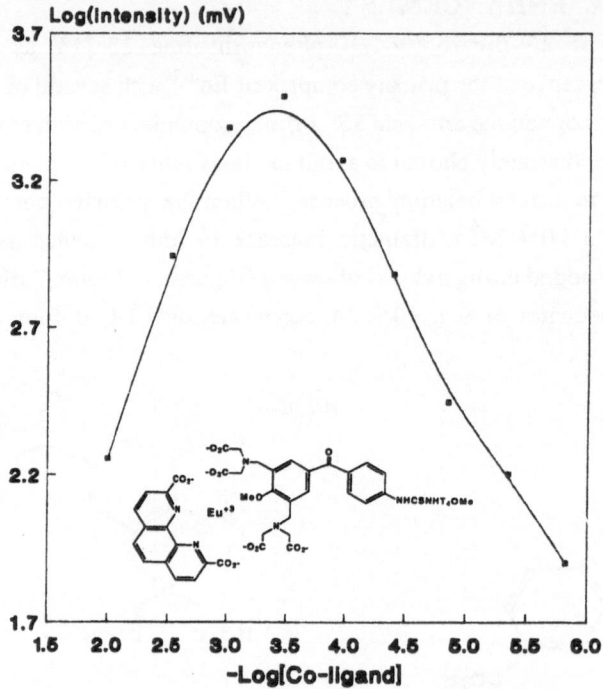

Figure 6. Co-ligand luminescence enhancement. [Benzophenone **15**-Eu^{+3}] = 2 x10^{-7} M in borate buffer, pH = 8.5. Time-delayed luminescence. Each point is the average of two determinations.

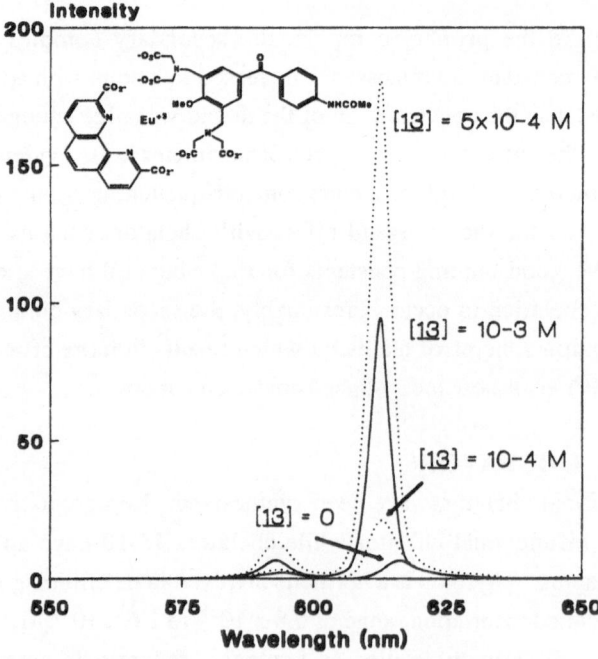

Figure 7. Emission spectra of benzophenone **14** as a function of [co-ligand **13**]. [Benzophenone **14**] = 9 x 10^{-5} M. Glycine acetate buffer, pH = 8.7.

17: R = CONHT4OMe, R' = H
18: R = H, R' = NHCSNT4OMe

19

Figure 8 shows the time-delayed luminescence signal for Eu^{+3} labelled benzo-phenone-thyroxin **16** as a function of added unlabelled thyroxin in competition for

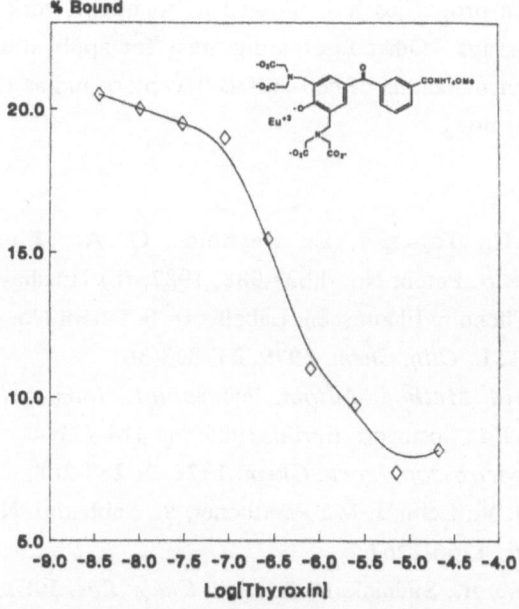

Figure 8. Dose-responsive curve for thyroxin. [Anti-thyroxin antibody] = 4.2 x 10^{-8} M on polystyrene beads. [Host **16**] = 2.0 x 10^{-7} M. Stripped human serum (50%) in glycine acetate buffer, pH = 8.7. Time-delayed luminescence signal determined with a hand-built phosphorimeter. Each point is the average of two determinations.

anti-thyroxin antibody. The anti-thyroxin antibody was present at 4.2×10^{-8} M on polystyrene beads. The benzophenone-thyroxin Eu^{+3} complex was maintained at the constant concentration of 2×10^{-7} M. For each determination, the labelled and unlabelled thyroxins in 50% stripped human serum and glycine acetate buffer were allowed to compete for antibody during a four minute incubation time, followed by washing with buffer to remove unbound material. Signal was obtained by reading the bound, labelled thyroxin-Eu^{+3} adduct **16** with a hand-built time-resolved flash phosphorimeter.

Although the results summarized in Figure 8 validate the concept of using these novel hosts for immunoluminescence assays, improvements are necessary. The anti-thyroxin antibody has a low affinity for the labelled thyroxin analog **16** ($K_d = 2.5 \times 10^{-7}$ M). The linker between the hapten and the sensitizer probably needs to be extended [15]. The low affinity of the anti-thyroxin antibody for the labelled thyroxin plus the short linker provide reasonable explanations for the dynamic range of the assay being shifted to relatively high concentrations of thyroxin and for the low percentage of bound material. In addition, the non-specific binding is reasonably high (8%). Finally, the benzophenone based Eu^{+3} chelators produce rather low intensity luminescence signals due to their poor quantum efficiencies. Efforts to overcome these problems are underway in studies utilizing improved antibodies with bifunctional chelators **16-19** as well as new chelators with other biologically important haptens.

CONCLUSIONS

The development of powerful lanthanide bifunctional chelators which remain stable in water with good luminescence properties has allowed us to begin work in developing novel immunoluminescence assays. Other fascinating areas for applications of these chelators include their use in the visualization of cell-surface receptors and as DNA probes. Work is progressing in these areas now.

References

1. (a) Hinshaw, J. C.; Toner, J. L.; Reynolds, G. A. Fluorescent Labels for Immunoassays, U. S. Patent No. 4,637,988, 1987; (b) Hinshaw, J. C.; Toner, J. L.; Reynolds, G. A. Phenolic Fluorescent Labels, U. S. Patent No. 4,637,988, 1987.

2. Soini, E.; Hemmilä, I. *Clin. Chem.* **1979**, *25*, 353-361.

3. Soini, E. In *Rapid Method Autom. Microbiol. Immunol., 4th Int. Symp.*; Habermehl, K.-O., Ed.; Springer: Berlin, 1985; pp 414-421.

4. Sinha, A. P. B. *Spectroscop. Inorg. Chem.* **1971**, *2*, 255-288.

5. Alpha, B.; Balzani, V.; Lehn, J.-M.; Perathoner, S.; Sabbatini, N. *Angew. Chem., Int. Ed. Engl.* **1987**, *26*, 1266-1267.

6. Horrocks, W. DeW., Jr.; Sudnick, D. R. *Acc. Chem. Res.* **1981**, *14*, 383-392.

7. (a) Soini, K.; Kojola, H. *Clin. Chem.* **1983**, *29*, 65-68; (b) Hemmilä, D. S.; Mukkala, V.-M.; Siitari, H.; Lövgren, T. *Anal. Biochem.* **1984**, *137*, 335-343; (c) Lövgren, T.; Hemmilä, I.; Pettersson, K.; Eskola, J. U.; Bertoft, E. *Talanta* **1984**, *31*, 909-919.

8.   (a) Blasse, G.; Dirksen, G. J.; Sabbatini, N.; Perathoner, S.; Lehn, J.-M.; Alpha,B. *J. Chem. Phys.* **1988**, *92*, 2419-2422; (b) Blasse, G.; Dirksen, G. J.; Van Der Voort, D.; Sabbatini, N.; Perathoner, S.; Lehn, J.-M.; Alpha, B. *Chem. Phys. Lett.* **1988**, *146*, 347-351; (c) Alpha, B.; Lehn, J.-M.; Mathis. G. *Angew. Chem., Int. Ed. Engl.* **1987**, *26*, 266-267; (d) Sabbatini, N.; Perathoner, S.; Balzani, V.; Alpha, B.; Lehn, J.-M. *NATO ASI Ser., Ser. C, (Supramol. Photochem.)* **1987**, *214*, 187-206; (e) Lehn, J.-M. *NATO ASI Ser., Ser. C, (Supramol. Photochem.)* **1987**, *241*, 29-43.

9.   Specht, D. P.; Martic, P. A.; Farid, S. *Tetrahedron* **1982**, *38*, 1203-1211.

10. Wheelwright, E. J.; Spedding, F. H.; Schwarzenbach, G. *J. Am. Chem. Soc.* **1953**, *75*, 4196-4201.

11. Vögtle, F.; Ohm, C. *Chem. Ber.* **1984**, *117*, 948-954.

12. Ohm, C.; Vögtle, F. *Chem. Ber.* **1985**, *118*, 22-27.

13. Chandler, C. J.; Deady, L. W.; Reiss, J. A. *J. Heterocyclic Chem.* **1981**, *18*, 599-601.

CATALYTIC REACTIONS OF MACROCYCLIC NICKEL(II) COMPLEXES

Cynthia J. Burrows

Department of Chemistry
State University of New York at Stony Brook
Stony Brook, NY 11794-3400 (USA)

SUMMARY

Certain square planar Ni(II) complexes are active as catalysts for hydrocarbon oxidation reactions including alkene epoxidation, oxidative C=C bond cleavage and hydroxylation. The reactions are highly dependent upon the structure of the ligand encapsulating Ni(II) and upon the terminal oxidant. Mechanistic studies of oxidations using nickel-cyclam **1** and iodosylbenzene provide interesting comparisons with cytochrome P_{450} model catalysts. Higher turnover rates are observed with nickel-salen **2** as catalyst through the use of hypochlorite under phase transfer conditions. A third series of catalysts is based upon dioxocyclam complexes of Ni(II) **3** which are derived from amino acids. These complexes are effective with OCl^- but not with PhIO.

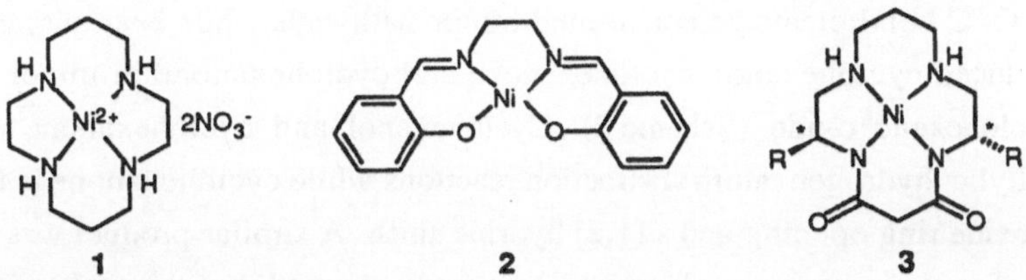

INTRODUCTION

Nature's hydroxylation catalyst cytochrome P_{450} is a heme-containing enzyme capable of transferring a single oxygen atom from dioxygen to a variety of hydrocarbon substrates with high regio- and stereoselectivity [1]. To organic chemists, this is an enviable achievement. While a number of metal porphyrin complexes serve as models for the enzymatic system, there remains a tremendous need to develop new oxidation catalysts of high turnover capability, low cost and tailored substrate specificity. Our goal was to investigate new metal ions and new non-porphyrinic ligands which might offer greater synthetic flexibility in the design of catalysts.

At the outset of our study, we were intrigued by the fact that Ni(II) was essentially absent from the area of oxidation catalysis. This was especially surprising in view of the large body of information available on stabilization of high oxidation states of nickel by polyaza macrocyclic ligands in particular [2]. For example, the saturated tetraamine cyclam forms a very stable complex with Ni(II) 1, and its square planar ligand field helps favor formation of Ni(III) at a modest oxidation potential (0.6 V) [2]. Since catalytic cycles involving transition metal-promoted oxidations generally involve an increase in the formal oxidation state of the metal ion, we reasoned that [Ni(II)-cyclam]$^{2+}$ would be a likely candidate for study.

ALKENE EPOXIDATION CATALYZED BY Ni(II)-CYCLAM

Using iodosylbenzene as terminal oxidant, we found that a variety of aromatic and aliphatic alkenes were oxidized [3,4]. Typical reaction conditions involved the use of 0.1 mmol Ni(II) catalyst, 0.5 mmol alkene and 2 mmol PhIO in 5 mL dry CH_3CN (Scheme 1). In most cases, the terminal oxidant was consumed within 1-2 hours, although a considerable amount of alkene starting material remained. In order to obtain high yields (based on alkene), a large excess (20 equiv.) of PhIO was required.

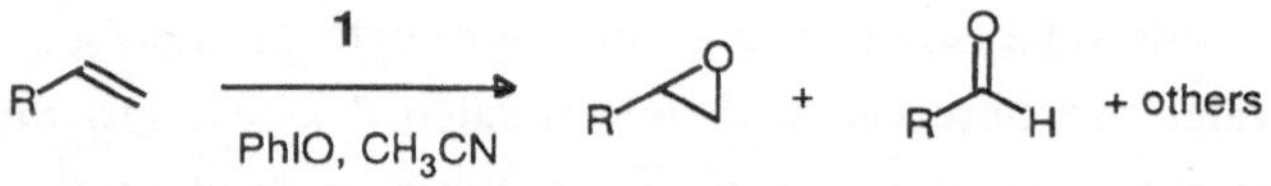

Scheme 1

In all cases, epoxides were the major products while allylic oxidation, rearrangement products and C=C bond cleavage constituted minor pathways. For example, cyclohexene oxidation produced cyclohexanol, cyclohexenone and cyclohexanone in minor amounts in addition to cyclohexene oxide (Scheme 2). Cyclohexanol and cyclohexanone presumably arose from allylic hydrogen atom abstraction reactions while cyclohexanone is formally the product of epoxide ring opening and a [1,2] hydride shift. A similar product was observed in the oxidation of various styrenes where a kinetic study showed that the carbonyl compound produced arose directly from an intermediate in the reaction as opposed to a ring opening of the epoxide product. The proposed mechanism (vide infra) must account for this.

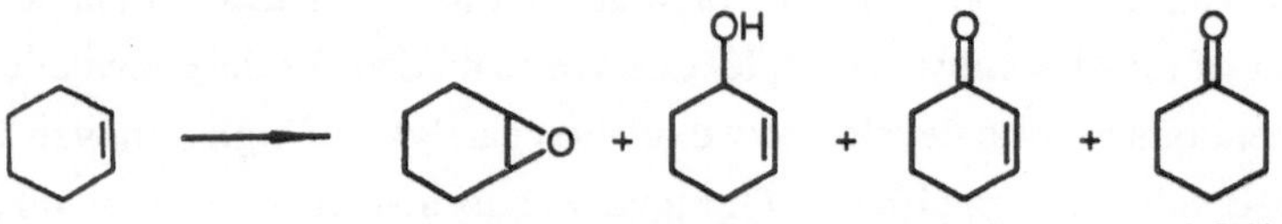

Scheme 2

Oxidation of *cis*- and *trans*-stilbenes in the Ni-cyclam/PhIO system yielded a myriad of products (Scheme 3). Although such a product mixture is undesirable for applications in synthesis, it is quite valuable in mechanistic studies because of the number of clues provided for elucidation of the reaction pathway. For example, the formation of about 50% *trans*-stilbene oxide from *cis*-stilbene suggests a pathway with about 50% retention of configuration. Complete loss of stereochemistry would have produced nearly 100% *trans*-epoxide, the thermodynamic product.

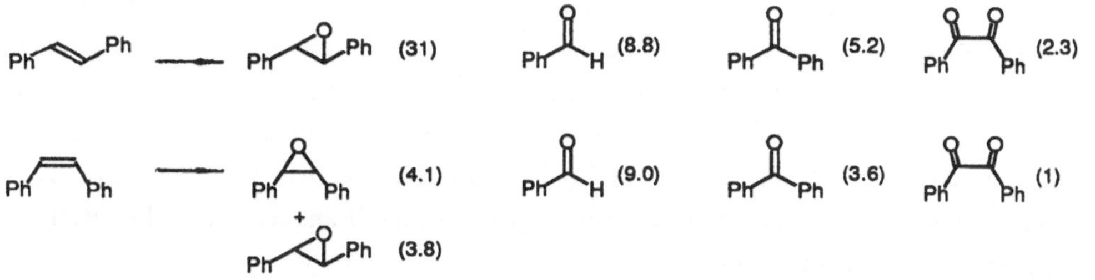

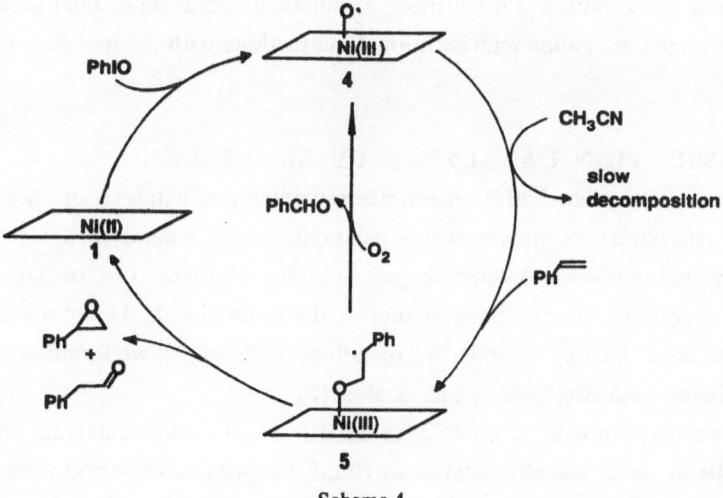

Scheme 3

The formation of benzaldehyde from phenyl-substituted alkenes was found to be dependent upon the presence of dioxygen. Rigorous exclusion of O_2 from the reaction mixture virtually eliminated this product. Benzophenone may arise from a [1,2] phenyl migration similar to the hydride shifts observed in other cases. The higher reactivity of *trans*- vs. *cis*-stilbene is also a clue since this behavior is opposite to that of metalloporphyrin catalysts. This pattern may be expected if the oxygen atom adding to the alkene has considerable radical character [4] and if there are no geometric restrictions imposed by the ligand.

These studies as well as the observation of a substituent effect for *para*-substituted styrenes of $\rho^+ = -0.83$ have formed the basis for speculation about reactive intermediates in the reaction pathway. It is proposed that reaction of PhIO with Ni(II)-cyclam leads to an oxidized

Scheme 4

nickel species (Scheme 4). By analogy to metalloporphyrin catalysts, it is attractive to propose a nickel-oxo intermediate **4** with radical character on oxygen. Such a species would be expected to be highly reactive, and indeed, similar studies by Kochi [5] suggest that the solvent CH_3CN is hydroxylated under these reaction conditions. At present, however, the structure of **4** as a reactive intermediate remains speculative.

Addition of **4** to an alkene in a stepwise fashion would generate the alkyl radical **5**. The usual fate of **5** is presumed to be reductive elimination to yield an epoxide and regenerate catalyst **1**. Studies with PhI ^{18}O and $^{18}O_2$ suggest a novel reaction pathway for **5** in the presence of dioxygen [2] (Scheme 5). Trapping of **5** by O_2 would lead to the peroxy radical **6** which could decompose via a pericyclic mechanism yielding two carbonyl compounds from the original alkene and regenerating the nickel-oxo species **4**. An interesting ramification of this postulate is that oxidative C=C bond cleavage, or effectively, ozonolysis, could be carried out with stoichiometric amounts of O_2 and only catalytic amounts of PhIO, an expensive and inconvenient oxidant. Further studies are in progress to exploit this feature of the reaction and to characterize reactive intermediates.

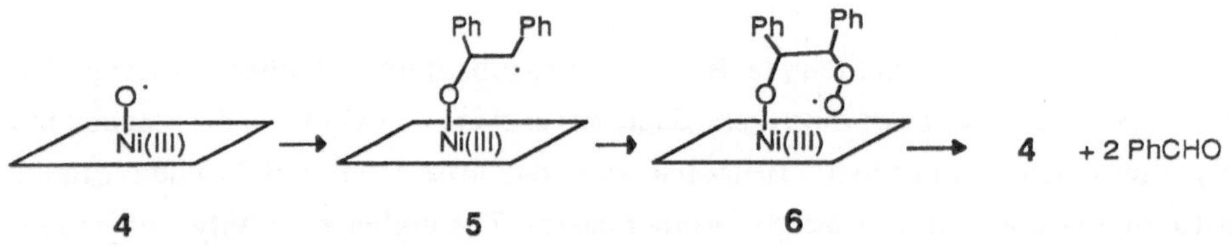

Scheme 5

Overall, the Ni(II)-cyclam/PhIO oxidation system differs from metalloporphyrin reactions in its stereoselectivity, stereospecificity, and sensitivity to O_2, although the electronic substituent effects and the products formed are similar. The porposed radical intermediates in this mechanism have led to new ideas for synthetic transformations beyond simple hydroxylation or epoxidation. Furthermore, synthetic elaboration of the cyclam ligand may generate functionalized ligands with additional interactions with the metal or alkene substrate [6].

ALKENE OXIDATION CATALYZED BY Ni(II)-SALEN

While the Ni(II)-cyclam/PhIO system demonstrated that catalysis of alkene epoxidation was possible with Ni(II) complexes, it was desirable to seek a catalytic system which used a less exotic terminal oxidant and which displayed higher turnover. One of the least expensive strong oxidants available is sodium hypochlorite, domestic bleach. Using reaction conditions similar to those developed by Meunier for metalloporphyrins [7], we found efficient catalysis of alkene oxidation with Ni(II)-salen **2** as catalyst [8].

In a typical experiment, 4 mmol alkene, 0.1 mmol nickel catalyst, and 0.15 mmol $PhCH_2NBu_3Br$ as phase transfer catalyst in 10 mL CH_2Cl_2 were stirred with 20 mL 0.77M aqueous NaOCl at pH 12.5 (Scheme 6). Using these reaction conditions, up to 40 turnovers were observed in the oxidations of a number of aliphatic and aromatic alkenes. Examples are shown in Scheme 7 with yields of epoxides based on starting alkene listed first followed by percent conversion of alkene to oxidised products. Phenyl-substituted alkenes such as styrene were converted to PhCHO in relatively large amounts in addition to the corresponding epoxide. In these cases, formation of C=C bond cleavage products was not dependent upon the presence of exogenous O_2. Aliphatic alkenes yielded chlorinated compounds as the major side product in addition to epoxides.

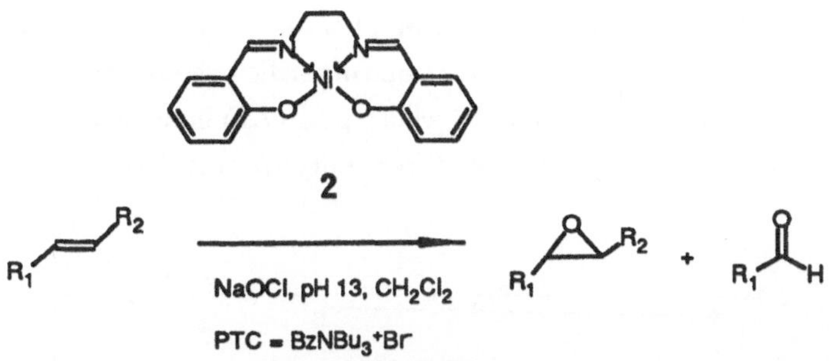

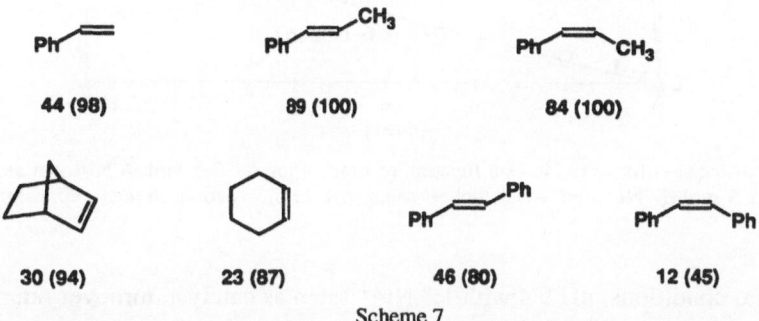

2

$R_1\diagup\!\!=\diagdown^{R_2}$ $\xrightarrow[\text{PTC = BzNBu}_3{}^+\text{Br}]{\text{NaOCl, pH 13, CH}_2\text{Cl}_2}$ $R_1\diagup\!\!\triangle\!\!\diagdown^{R_2}$ + $R_1\diagup\!\!\overset{O}{\underset{}{}}\!\!{}^H$

Scheme 6

Epoxide yield (% conversion)..........Ni salen / NaOCl

Ph━━	Ph━━CH$_3$	Ph━━CH$_3$
44 (98)	89 (100)	84 (100)

(norbornene)	(cyclohexene)	Ph━━Ph	Ph━━Ph
30 (94)	23 (87)	46 (80)	12 (45)

Scheme 7

The rate of oxidation of styrene using NaOCl at pH 12.5 showed an induction period of about 2-3 hours. At a lower pH (9-10), there was no induction period, and the reaction occured quite rapidly with an optimum pH of 9.4. Similar behavior was seen for oxidation of *trans*-stilbene; rather than displaying an induction period, however, the reaction rate was slow at pH 12.5 for the entire 18 hours period of observation. (See Figure 1, Plot A).

Also during the oxidation reactions, a black precipitate was observed to form before the onset of substantial oxidation product formation. When this precipitate was isolated and added to a CH$_2$Cl$_2$ solution of an alkene, no hydrocarbon oxidation occurred, indicating that this material is not responsible for epoxidation. Nickel peroxide, NiO_2, is known to be formed upon treatment of Ni(II) salts with basic hypochlorite and is a likely explanation for the black precipitate. It has been used as an oxidant in electron transfer-type oxidations but does not typically participate in oxygen atom transfer reactions [9].

Decomposition of Ni^{2+} from this salen complex and formation of NiO_2 depletes the concentration of catalyst. To overcome this, excess salen ligand (1-3 equiv. per equiv. catalyst) was added to the reaction mixture. Surprisingly, this had a very dramatic influence on the rate of the reaction (see Figure 1, Plot B). Approximately 40 turnovers of *trans*-stilbene to *trans*-stilbene oxide exclusively occurred in less than 20 minutes. However, the black precipitate of NiO_2 was still observed early in the course of the reaction. In order to further probe the effect of added salen, a number of phenols were tested as additives. Both *o*-salicylaldehyde and its *para* isomer led to the same dramatic rate enhancement. Other phenols had no effect. These included *p*-nitrophenol whose pK_a (7.2) is intermediate between *para*-(7.6) and *o*-salicylaldehyde (6.8), and 2,6-di-*tert*-butyl-*p*-cresol (BHT), which is readily oxidised.

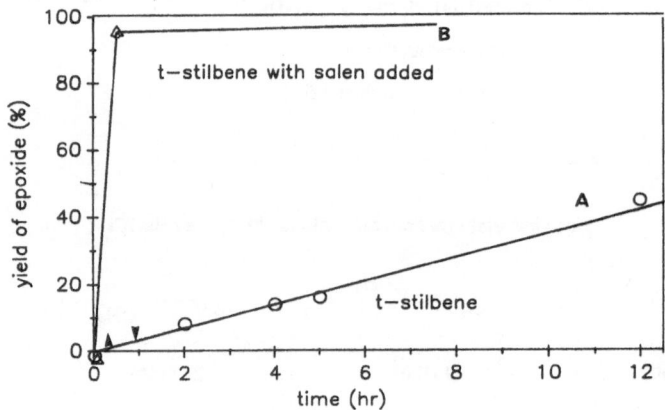

Figure 1. Yield of *trans*-stilbene oxide as a function of time. Plot A: 2.5 mol % Ni-salen as catalyst, pH 12.5. Plot B: 2.5 mol % Ni-salen + 5.0 mol % salen, pH 12.5. Arrows indicate appearance of black precipitate.

Under ideal conditions, pH 9.4 with 1:3 Ni^{2+}:salen as catalyst, turnover rates up to 150 per 15 minutes have been observed. Although a detailed explanation of the effect of pH and additives awaits further study, we postulate that a protonated species is involved in a rate-determining step with formation of either HOCl or a nickel hydroxide species. Epoxidation by HOCl alone does not explain these results since the reaction rate falls off below pH 9 where higher concentrations of HOCl are formed and because the background epoxidation of alkenes by HOCl is consistently low in all of our reaction conditions. It is plausible, though, that HOCl is the species that interacts with Ni(II)-salen or that an intermediate nickel-oxo species must be protonated before oxygen atoms transfer to an alkene. If the medium is too acidic the reaction rate decreases perhaps because protonation of salen would lead to decomplexation of Ni^{2+}. At any rate, the observation of high turnover rates paves the way for future development of practical oxidation systems.

ALKENE OXIDATION CATALYZED BY Ni(II)-DIOXOCYCLAMS

Our third generation nickel catalyst combined some of the features of both cyclam and salen. The dioxocyclam macrocycle, like cyclam, brings together four nitrogen ligands in a semi-rigid square planar array. Upon complexation of Ni^{2+}, the amide nitrogens of the ligand are deprotonated, yielding a Ni(II) complex 3, which is charge neutral, as is Ni(II)-salen. The redox behavior of the parent compound, 3 R = H, has been studied by Kimura [10]. We have devised a new synthetic route from amino acids which leads to disubstituted macrocycles in which the side chains, R, are the side chains of the starting amino acids [11]. Dioxocyclams derived from phenylalanine (R = CH_2Ph) and leucine (R = sec-butyl) show high solubility in CH_2Cl_2.

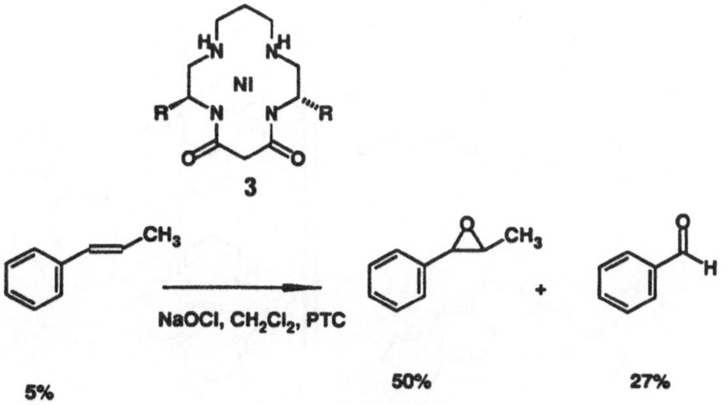

Scheme 8

Using NaOCl under conditions identical to those described above, we observed rapid epoxidation of trans-β-methylstyrene as well as PhCHO and $PhCO_2H$ as side products [12] (Scheme 8). The Ni(II)-dioxocyclam complexes displayed low solubility in CH_3CN but tests in CH_2Cl_2 showed that they were ineffective as oxidation catalysts when iodosylbenzene was used a terminal oxidant. For a series of dioxocyclams of varying CH_2Cl_2 solubility as a function of R, the efficacy of the Ni(II) complexes in alkene oxidation catalysis with hypochlorite was directly related to their methylene chloride/water partition coefficient. Although these complexes are optically active and display C_2 symmetry, no enantiomeric excess was detected in the epoxide product. This is perhaps not surprising in view of the small steric perturbation of one alkyl group per face of the macrocycle. Other more highly substituted macrocycles are currently under study.

CONCLUSIONS

No simple Ni^{2+} salts have been found to catalyze alkene oxidation [13]. In the course of our studies, a number of macrocyclic Ni(II) complexes have also failed. These include complexes of tetraphenylporphyrin (formally a 16-membered ring), [15]-ane-N_4 (the 15-membered ring analogy of cyclam), and tetra-N-methyl-cyclam (Scheme 9).

Other macrocyclic ligands that do give rise to catalytic Ni(II) complexes are the 13-membered ring analogs of **1** and **3**, [13]-ane-N_4 and dioxo-[13]-ane-N_4 (R = CH_2Ph). What then are the ligand requirements for formation of a Ni(II) coordination compound capable of oxidation catalysis? Our observation is that all Ni(II) complexes found to be active oxidation catalysts are diamagnetic square planar complexes. This is typically observed for 13- and 14-membered macrocyclic ligands with a small cavity size and a strong ligand field. So far, all high spin, octahedral Ni(II) complexes (NiTPP, Ni[15]-ane-N_4, and Ni(tmc)) have been ineffective. The empirical rule is this: yellow complexes (low spin, diamagnetic) might work, purple ones (high spin, octahedral) won't.

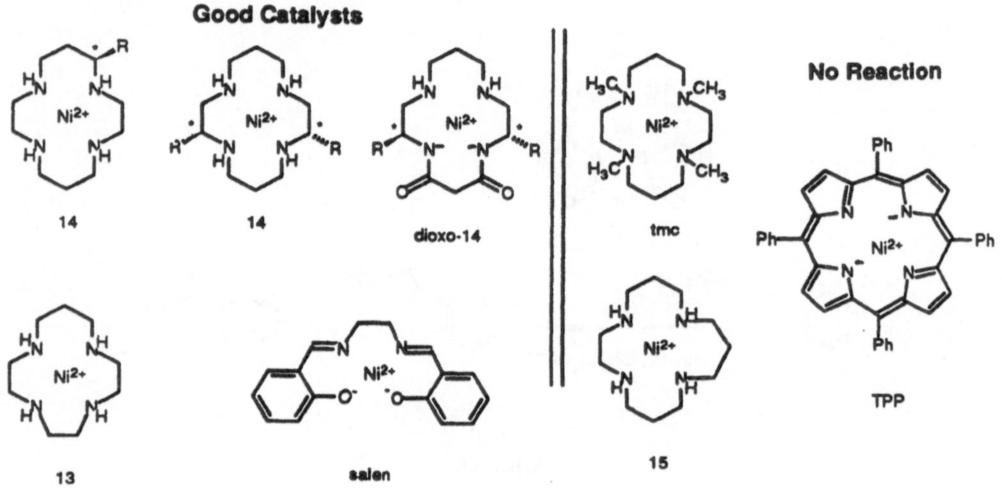

Scheme 9

A reasonable conclusion is that factors which give rise to a square planar coordination geometry for Ni(II), such as short nickel-nitrogen bonds and strong donor groups (N and O) may also be the factors which favor formation of a nickel-oxo species in which stabilization of a higher oxidation state of nickel is required. This concept fits nicely into the molecular recognition theme since a ligand can be specifically tailored to interact with a particular metal oxidation state or reactive intermediate in oxidation catalysis. This, in theory, is similar to an enzymatic approach to catalysis in which an organic receptor is designed to bind specifically to the transition state of a reaction pathway. In practice, nature has already made use of a macrocyclic nickel complex, factor F_{430}, to perform redox chemistry in the methanogen enzyme S-methyl-CoM reductase [14]. Further study will bring chemists closer to true mimickry of these enzymatic processes.

Acknowledgements. I am grateful to the co-workers whose names appear in the references cited for their contributions to this research. Finanical support by the National Science Foundation (CHE-8706616) and the National Institutes of Health (GM-34841) is gratefully acknowledged.

References

1. Ortiz de Montellano, P. R., Ed. *Cyctochrome P-450: Structure, Mechanism and Biochemistry*; Plenum: New York, 1976.

2. Busch, D. H. *Acc. Chem. Res.* **1978**, *11*, 392.

3. Kinneary, J. F.; Wagler, T. R.; Burrows, C. J. *Tetrahedron Lett.* **1988**, *29*, 877.

4. Kinneary, J. F.; Albert, J. S.; Burrows, C. J. *J. Am. Chem. Soc.* **1988**, *110*, 6124.

5. Koola, J. D.; Kochi, J. K. *Inorg. Chem.* **1987**, *26*, 908.

6. Wagler, T. R.; Burrows, C. J. *J. Chem. Soc., Chem. Commun.* **1987**, 277.

7. Meunier, B.; Guilmet, E.; De Carvalho, M. E.; Poilblanc, R. *J. Am. Chem. Soc.* **1984**, *106*, 6668.

8. Yoon, H.; Burrows, C. J. *J. Am. Chem. Soc.* **1988**, *110*, 4087.

9. George, M. V.; Balachandran, K. S. *Chem. Rev.* **1975**, *75*, 491.

10. Kimura, E. *J. Coord. Chem.* **1986**, *15*, 1.

11. Wagler, T. R.; Burrows, C. J. *Tetrahedron Lett.* **1988**, *29*, 5091.

12. Wagler, T. R.; Fang, Y.; Burrows, C. J. *J. Org. Chem.* **1989**, *54*, 1584.

13. Vanatta, R. B.; Franklin, C. C.; Valentine, J. S. *Inorg. Chem.* **1984**, *25*, 4121.

14. Pfaltz, von A.; Jaun, B.; Fässler, A.; Eschenmoser, A.; Jaenchen, R.; Gilles, H. H.; Diekert, G.; Thauer, R. K. *Helv. Chim. Acta* **1982**, *65*, 828.

SHAPE SELECTIVE OXIDATION AS A MECHANISTIC PROBE

Kenneth S. Suslick* and Bruce R. Cook

School of Chemical Sciences
University of Illinois at Urbana-Champaign
505 S. Mathews Avenue
Urbana, IL 61801 (USA)

SUMMARY

The manganese and iron complexes of the bis-pocket porphyrin (5,10,15,20-tetrakis-(2',4',6'-triphenyl)phenylporphyrin, H_2TTPPP) are shown to be shape selective catalysts for the hydroxylation of alkanes and the epoxidation of nonconjugated dienes with a wide variety of oxidants. The selectivities are independent of the choice of oxidant, demonstrating that the mechanism of epoxidation is very similar for all oxidant systems studied. The selectivities for terminal hydroxylation of n-alkanes are very similar to those for terminal epoxidation of 1,4-dienes and are dominated by the steric demands of the metalloporphyrin catalyst. There is, however, a dramatic diminution in selectivity for iron versus manganese. This means that both metals cannot be generating a common metalloxetane imtermediate. The selectivities observed for MnTTPPP(OAc) for both 4-vinyl-1-cyclohexene and limonene are very high and approach those of limonene-induced ctyochrome P450.

INTRODUCTION

In order to elucidate the reaction pathways for the oxidation of hydrocarbons by cytochrome P450, intense efforts have been made to generate similar catalytic behavior with synthetic metalloporphyrin complexes [1-5]. Recently, we found that the manganese and iron complexes of our extremely sterically hindered bis-pocket porphyrin (5,10,15,20-(2',4',6'-triphenyl)phenylporphyrin, H_2TTPPP) show dramatic shape selectivity for alkane hydroxylation [5]. We report here the extension of these studies to alkene epoxidations using a wide variety of oxidant systems for both iron and manganese porphyrins. For each metal, the selectivities are independent of the choice of oxidant, demonstrating that the mechanism of epoxidation is very similar for all oxidant systems examined. In contrast, there is a substantial decrease in shape selectivity shown by iron porphyrins compared to manganese. It is likely, therefore, that at least two pathways for metalloporphyrin epoxidation exist: one which is sterically demanding, and a second which is not. The selectivities observed for MnTTPPP(OAc) for 4-vinyl-1-cyclohexene and limonene are very high and approach those of cytochrome P450.

For alkene epoxidation catalyzed by metalloporphyrins, various reaction intermediates have been proposed. The two mechanisms most discussed involve either a metalloxetane intermediate [2h,3a-c] or an electron transfer step resulting in a metal-bound radical or carbocation intermediate [2i-l]. Some of the proposed reaction pathways are shown in Figure 1.

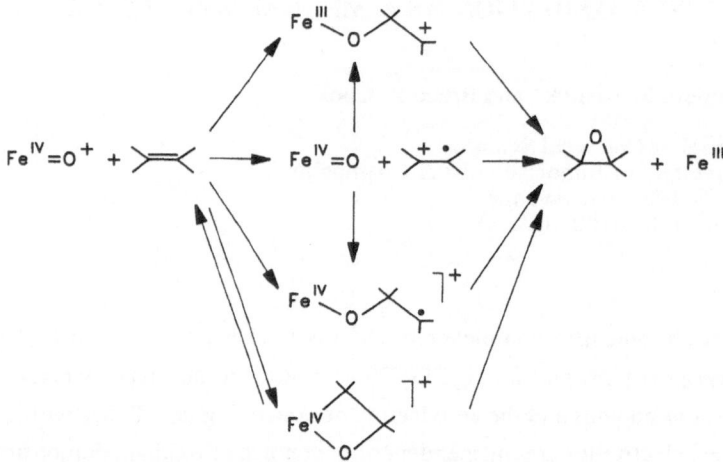

Figure 1. Some of the proposed mechanisms for metalloporphyrin-catalyzed epoxidation.

RESULTS AND DISCUSSION

We have used three manganese porphyrin complexes with increasing steric demands as hydroxylation and epoxidation catalysts (Figure 2) and have examined the shape selective epoxidation of 4-vinyl-1-cyclohexene as a mechanistic probe of the oxidation process. As shown in Figure 3, relative yield of the external epoxide increases as the metalloporphyrin catalyst becomes more encumbered. The substantial selectivity observed with MnTTPPP(OAc) is essentially independent of oxidant used and insensitive to the addition of nitrogenous bases. This is consistent with initial oxygen atom transfer from the oxidant to form a monomeric O=Mn(V) and demonstrates that epoxidation proceeds through very similar transition states for all of these epoxidation systems.

To our surprise, selectivity is not significantly altered by addition of ligands (such as imidazoles). Although these ligands have been used to increase reaction rates dramatically, they do not signficantly alter the geometry or properties of the transition state. Indeed, there is no evidence for their coordination to the catalytically active species. It is established, however, that metal coordination does occur to the catalyst precursor [M(Porph)(X)], even for complexes of the sterically hindered porphyrin TTPPP [6].

An interesting comparison is made in Figure 4 between the shape selectivities of diene epoxidation versus alkane hydroxylation. As expected, the unpocketed porphyrin [Mn(TPP(OAc)] strongly prefers the more substituted double bond (i.e. the more electron-rich)

Figure 2. Computer generated molecular models of an unhindered porphyrin (5,10,15,20-tetraphenylporphyrin, H$_2$TPP), a shallow-pocketed porphyrin (5,10,15,20-(2',4',6'trimethoxy)phenylporphyrin, H$_2$TTMPP), and an extremely sterically hindered bis-pocket porphyrin (5,10,15,20-(2',4',6'-triphenyl)phenylporphyrin, H$_2$TTPPP). For clarity, the atomic radii shown are only 0.8 of the van der Waals radii.

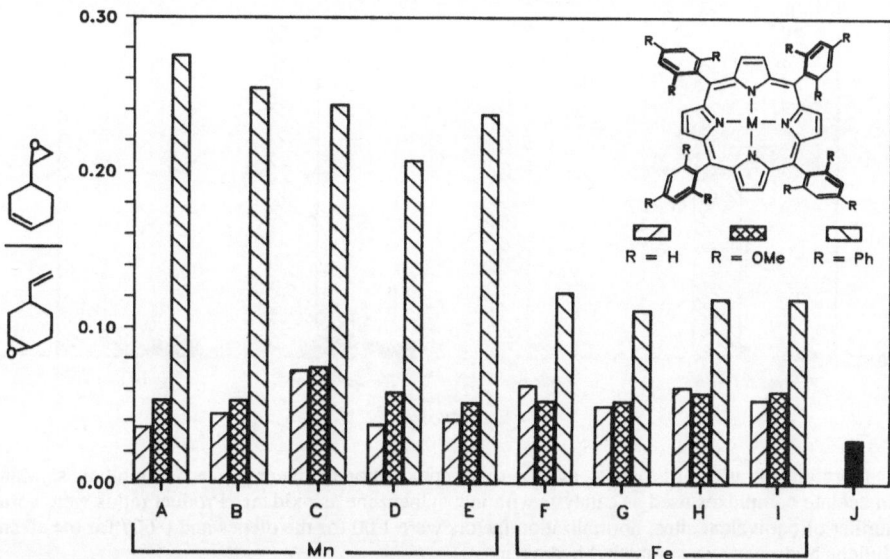

Figure 3. Ratio of external epoxide to ring epoxide for 4-vinyl-1-cyclohexene with various oxidant systems. All reaction solutions were 1 mM in metalloporphyrin and 2 M in 4-vinyl-1-cyclohexene in methylene chloride. All oxidants were prepared and used as cited. Reactions were run for 30 min (except A for 15 min. and E for 6 hrs). Products were analyzed by capillary G. C. and G.C./M.S. against known standards [8].

A.	Mn(porph)(OAc)	+	NaOCl(aq) + 20 mM 4'-(imidazol-1-yl)acetopheone [3c]
B.	Mn(porph)(OAc)	+	iodosylbenzene + 20 mM ImAcP
C.	Mn(porph)(OAc)	+	iodosylbenzene
D.	Mn(porph)(OAc)	+	1,1,1-trifluoroethanol solubilzed iodosylbenzene [2l]
E.	Mn(porph)(OAc)	+	O$_2$/H$_2$ + Pt colloid [3m]
F.	Fe(porph)(OAc)	+	pentafluoroiodosylbenzene [2l]
G	Fe(porph)(OAc)	+	iodosylbenzene
H	Fe(porph)(OAc)	+	1,1,1-trifluoroethanol solubilized iodosylbenzene
I.	Fe(porph)(OAc)	+	iodosylbenzene + 20 mM ImAcP
J.	Uncatalyzed epoxidation by *m*-chloroperoxybenzoic acid		

for diene epoxidation and the methylene positions over the terminal methyl group for alkane hydroxylation. The steric demands of the porphyrin, however, can reverse this natural predilection. MnTTPPP(OAc) shows significant preference for the ends of linear dienes or *n*-alkanes. Of special note is the quantitative similarity in the shape selectivity of the dienes and the alkanes. These selectivities are clearly influenced by the steric constraints of the porphyrin pocket and the shape of the substrate. The bis-pocket porphyrin shows similar selectivities for substrates that have similar shapes, whether for epoxidation or for hydroxylation.

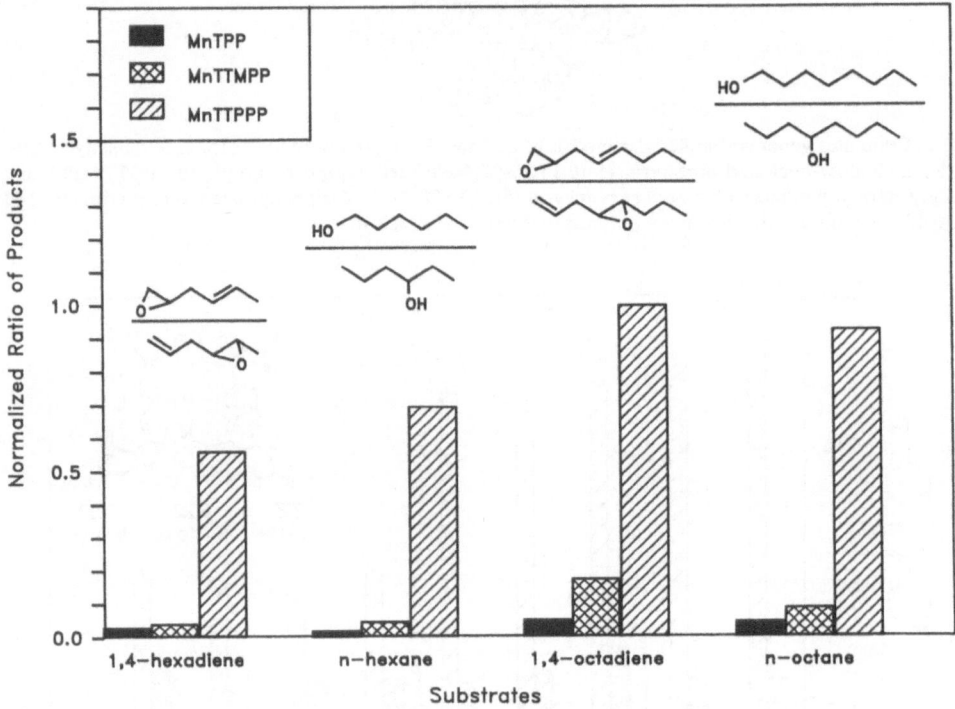

Figure 4. Normalized ratios of products, as shown, for epoxidation of linear dienes and n-alkanes. Manganese porphyrin acetate complexes used as catalysts with iodosylbenzene as oxidant. Product ratios were normalized by the number of equivalent sites; normalization factors were 1.00 for the dienes and 0.667 for the alkanes (i.e. four methylene hydrogens vs. six methyl hydrogens).

The iron(III) complexes of these prophyrins also show shape selectivity for epoxidation, albeit less so than manganese. Similar observations were made in the hydroxylation of alkanes [5]. This indicates a less sterically demanding intermediate or transition state for the iron catalyzed epoxidations relative to manganese. Possibilities include the ring-open radical or carbocation intermediate expected for an electron transfer mechanism [2i-l]. If epoxidation required a true metalloxetane *intermediate* (as compared to a transition state) then very similar selectivities would then be expected for both Fe and Mn complexes. Since the selectivities are very different, the reaction mechanism cannot rely exclusively on a putative metalloxetane intermediate.

Iron epoxidation selectivities are also independent of oxidant used, and insensitive to the addition of imidazole. This indicates that the iron-catalyzed epoxidations all go through the same set of mechanisms. If multiple pathways exist, then the partitioning among them is not significantly altered by the choice of oxidant or the addition of nitrogenous bases.

The epoxidation of limonene was also examined with representative oxidant systems of both iron and manganese. The selectivity differences between iron and manganese are quite dramatic for this sterically more encumbered diene. The ratio of external epoxide to ring epoxide is 0.22 for MnTPP(OAc), 0.49 for MnTTMPP(OAc), and 1.63 for MnTTPPP(OAc); but only 0.16 for FeTPP(OAc), 0.18 for FeTTMPP(OAc), and 0.35 for FeTTPPP(OAc). Indeed, the selectivity observed for MnTTPPP(OAc) catalyzed epoxidation is the largest by far ever reported for epoxidation with any synthetic oxidant system.

Certain isozymes of the heme protein cytochrome P450 are capable of dramatic regioselectivity for both hydroxylation [5b,7] and epoxidation [8]. Comparisons of epoxidation selectivity between our synthetic metalloporphyrins and the enzyme are shown in Figure 5. As the steric demands of the active site of the oxidizing system are increased from uncatalyzed m-chloroperoxybenzoic acid to the very bulky MnTTPPP(OAc), the selectivity for external epoxidation dramatically increases. MnTTPPP(OAc) is nearly as selective as the isozyme of P450 induced by limonene.

Although the details of the steric interactions undoubtedly differ between the enzyme [8] and our synthetic analogs, the total steric interaction with these two dienes must be quite similar in magnitude for both MnTTPPP(OAc) and cytochrome P450.

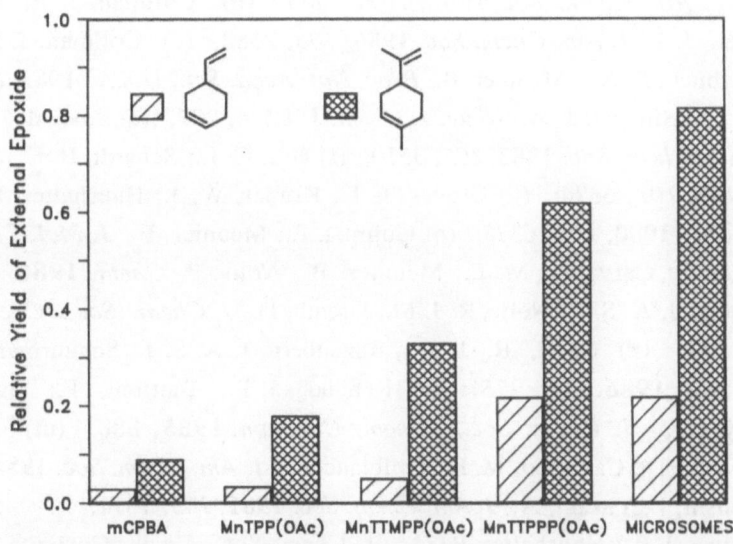

Figure 5. Comparisons of relative yields of external to ring epoxides for 4-vinyl-1-cyclohexene and limonene. The epoxidations were run as before with uncatalyzed *meta*-chlorobenzoic acid, Mn(porph)(OAc) + NaOCl(aq) + 20 mM ImAcP, or limonene-induced rat liver microsomes [7a] as oxidants. For the enzyme comparison, the corresponding epoxide and diol (from epoxide hydrolysis) products were summed.

Acknowledgements. This work was supported by grants from the National Institutes of Health and the American Heart Association. K.S.S. gratefully acknowledges receipt of an N.I.H Research Career Development Award and of a Sloan Foundation Research Fellowship.

References

1. For recent reviews see, (a) Suslick, K. S. "Shape Selective Hydrocarbon Oxidation" in *Activation and Functionalization of Enzymes*; Hill, C. L., ed.; Wiley Publishers, New York, 1989. (b) Meunier, B. *Bull. Soc. Chim. Fr.* **1986**, 578.

2. (a) Groves, J. T.; Nemo, T. E.; Myers, R. S. *J. Am. Chem. Soc.* **1979**, *101*, 1032; (b) Groves, J. T.; Kruper, W. J.; Nemo, T. E.; Myers, R. S. *J. Mol. Cat.* **1980**, *7*, 169; (c) Groves, J. T.; Watanabe, Y.; McMurray, T. J. *J. Am. Chem. Soc.* **1983**, *105*, 4489; (d) Groves, J. T.; Nemo, T. E. *J. Am. Chem. Soc.* **1983**, *105*, 5786; (e) Groves, J. T.; Myers, R. S. *J.* Am. Chem. Soc. **1983**, *105*, 5791; (f) Lindsay-Smith, J. R.; Steath, P. R. *J. Chem. Soc., Perkin Trans. 2* **1982**, 1009; (g) Nee, M. W.; Bruice, T. C. *J. Am. Chem. Soc.* **1982**, *104*, 6123; (h) Collman, J. P.; Kodadek, T.; Raybuck, S. A.; Brauman, J. I. *J. Am. Chem. Soc.* **1985**, *107*, 4343; (i) Fontecave, M.; Mansuy, D. *J. Chem. Soc., Chem. Commun.* **1984**, 879; (j) Traylor, T. G.; Nakano, T.; Dunlap, B. E.; Traylor, P. S.; Dolphin, D. *J. Am. Chem. Soc.* **1986**, *108*, 2782; (k) Traylor, T. G.; Iamamoto, Y.; Nakano, T. *J. Am. Chem. Soc.* **1986**, *108*, 3529; (l) Traylor, T. G.; Mursters, J. C., Jr.; Nakano, T.; Dunlap, B. E. *J. Am. Chem. Soc.* **1985**, *107*, 5537; (m) Nappa, M. J.; Tolman, C. A. *Inorg. Chem.* **1985**, *24*, 4711; (n) Groves, J. T.; Neumann, R. *J. Am. Chem. Soc.* **1987**, *109*, 5045.

3. (a) Collman, J. P.; Brauman, J. I.; Meunier, B.; Hayashi, T.; Kodadek, T.; Raybuck, S. A. *J. Am. Chem. Soc.* **1985**, *107*, 2000; (b) Collman, J. P.; Kodadek, T.; Brauman, J. I. *J. Am. Chem. Soc.* **1986**, *108*, 2588; (c) Collman, J. P.; Kodadek, T.; Raybuck, S. A.; Meunier, B. *Proc. Nat. Acad. Sci., U.S.A.* **1983**, *80*, 7039; (d) Hill, C. L.; Smegal, J. A. *Nouv. J. Chem.* **1982**, *6*, 287; (e) Smegal, J. A.; Hill, C. L. *J. Am. Chem. Soc.* **1983**, *105*, 3510; (f) Hill, C. L.; Schardt, B. C. *J. Am. Chem. Soc.* **1980**, *102*, 6374; (g) Groves, J. T.; Kruper, W. J.; Haushalter, R. C. *J. Am. Chem. Soc.* **1980**, *102*, 6375; (h) Guilmet, E.; Meunier, B. *J. Mol. Cat.* **1984**, *23*, 115; (i) De Carvalho, M. E.; Meunier, B. *Nouv. J. Chem.* **1986**, *10*, 223; (j) Razenberg, J. A. S. J.; Nolte, R. J. M.; Drenth, D. *J. Chem. Soc., Chem. Commun.* **1986**, 277; (k) Nolte, R. J. M.; Razenberg, J. A. S. J.; Schuurman, R. *J. Am. Chem. Soc.* **1986**, *108*, 2751; (l) Renaud, J. P.; Battioni, P.; Bartoli., J. F.; Mansuy, D. *J. Chem. Soc., Chem. Commun.* **1985**, 888; (m) Meunier, B.; Guilmet, E.; De Carvalho, M. E.; Poilblanc, R. *J. Am. Chem. Soc.* **1984**, *106*, 6668; (n) Tabushi, I.; Yazaki, A. *J. Am. Chem. Soc.* **1981**, *103*, 7371.

4. (a) Groves, J. T.; Haushalter, R. C. *J. Chem. Soc., Chem. Commun.* **1981**, 1165; (b) Groves, J. T.; Kruper, W. J., Jr.; Haushalter, R. C.; Butler, W. M. *Inorg. Chem.* **1982**, *21*, 1363; (c) Marchon, J.-C.; Ramasseul, R. *J. Chem. Soc., Chem. Commun.* **1988**, 298.

5. (a) Suslick, K. S.; Cook, B. R.; Fox, M. M. *J. Chem. Soc., Chem. Commun.* **1985**, 211; (b) Cook, B. R.; Reinhert, T. J.; Suslick, K. S. *J. Am. Chem. Soc.* **1986**, *108*, 7281; (c) Suslick, K. S.; Cook, B. R. *J. Chem. Soc., Chem. Commun.* **1987**, 200.

6. (a) Suslick, K. S.; Fox, M. M.; Reinhert, T. J. *J. Am. Chem. Soc.* **1984**, *106*, 4522; (b) Suslick, K. S.; Fox, M. M. *J. Am. Chem. Soc.* **1983**, *105*, 3507.

7. (a) McKenna, E. J.; Coon, M. J. *J. Biol. Chem.* **1970**, *245*, 3882; (b) Morohashi, K.; Sadano, H.; Okada, Y.; Omura, T. *J. Biochem.* **1983**, *93*, 413.

8. (a) Watabe, T.; Hiratsuka, A.; Ozawa, M.; Isobe, M. *Xenobiotica* **1980**, *11*, 333; (b) Carlson, R. G.; Behn, N. S.; Cowles, C. *J. Org. Chem.* **1971**, *36*, 3832.

9. Collins, J. R.; Loew, G. H. *J. Biol. Chem.* **1988**, *263*, 3164.

NOVEL CHROMOGENIC IONOPHORES FOR DETERMINATION OF SODIUM AND POTASSIUM IN BIOLOGICAL FLUIDS: SYNTHESIS AND APPLICATIONS

Donald J. Cram†, Eddy Chapoteau§, Bronislaw P. Czech§, Carl R. Gebauer§, Roger C. Helgeson†, and Anand Kumar§

†Department of Chemistry and Biochemistry
University of California
Los Angeles, California 90024 (USA)

§Technicon Instruments Corporation
Tarrytown, New York 10591 (USA)

Although the majority of clinical chemistry tests run on automated clinical analyzers are based on spectrophotometric determinations, sodium and potassium assays generally involve the use of ion-selective electrodes or flame photometry for direct analysis of both ions in blood and other body fluids. Therefore, it would be highly desirable to develop spectrophotometric methods for sodium and potassium which were readily adaptable to clinical analyzers.

The foundation for spectrophotometric assay of alkali metal cations was laid by the pioneering works in macrocyclic chemistry by Pedersen, Lehn and Cram. A great number of macrocyclic ionophores which are capable of selectively complexing alkali metal cations were designed and synthesized by these and other researchers during the last 20 years.

For more than a decade synthetic chromoionophores which respond to cation complexation with a change in color have attracted considerable attention. During this time, a large number of chromogenic compounds for determination of alkali and alkaline earth metal cations have been reported [1,2]. In 1977, Takagi and Ueno reported the first compound in which chromogenic properties were imparted to the ionophore by covalent linkage to a chromophoric group [3]. About the same time, Vögtle and Dix synthesized a series of non-ionizable chromoionophores based on aniline-derived crown ethers [4]. Either a shift in the acid-base equilibrium of the chromophore containing an ionizable proton or a direct influence on the electronic structure of the chromophoric group can produce a spectral shift in the chromophore, both of which result from cation complexation. The change in absorbance at a designated wavelength is proportional to cation concentration and can be measured.

Inclusion Phenomena and Molecular Recognition
Edited by J. Atwood
Plenum Press, New York, 1990

In an alternative approach, a non-chromogenic ionophore is used in a combination with an anionic dye (e.g. bromocresol green). A cation of interest is selectively extracted into an organic solution containing the ionophore together with the dye. The amount of the extracted dye which is proportional to the cation concentration is measured [5]. This approach has recently been used in commercial products by Abbott Vision™ [6] and Ames Seralyzer™ [7] for determination of potassium in blood.

The efforts to use chromogenic compounds for colorimetric assays of sodium and potassium in biological fluids have heretofore been unsuccessful for two following reasons. First, most of the macrocyclic ionophores require low dielectric organic solvents to bind ions efficiently. If used with clinical specimens such as blood serum, this generally causes protein precipitation. Secondly, they fail to meet the high selectivity and sensitivity requirements for clinical applications. In addition, the presence of organic solvent could be harmful to the typical analyzer's tubing system. Thus, the use of organic solvents in sodium and potassium assays is highly undesirable.

Chart 1 shows several chromogenic compounds which represent progress made in the design and synthesis of chromoionophores over the twelve year period. To obtain compounds which would show stronger binding, better selectivity, earlier flexible structures based on crown ethers (Takagi, Ueno [3], Vögtle [4] and Misumi [8]) were replaced by more rigid and preorganized ligands (Klink [9], Ogawa [10], Kasai [11] and Cram [12]).

Technicon became interested in clinical applications of chromoionophores in the early eighties. As a results of a very vigorous collaboration between Technicon and Professor Richard A. Bartsch at Texas Tech University over fifty novel chromogenic compounds were synthesized. Some representative chromoionophores synthesized at Texas Tech are shown in Chart 2. Those novel structures included a modified Ueno's chromogenic crown, a series of proton-ionizable crowns with different points of attachment to the chromophoric group, a chromogenic bis-crown, and chromogenic cryptands containing either ionizable or non-ionizable chromophores.

All of these novel chromogenic compounds were extensively studied at Technicon in single-phase, water-miscible organic solvent systems and were found to give insufficient selectivity and sensitivity for potassium and sodium determination in blood serum [13].

Recently, Cram discussed the importance of molecular preorganization in design of macrocyclic ionophores [14]. According to Cram, the shortcomings of crowns and cryptands can perhaps be best understood from the differences in the free energy of complexation. The free energy of complexation measured in chloroform for 18-crown-6 is only about 11 kcal mol^{-1}. This value increases by almost seven orders of magnitude in K for more "preorganized for complexation" cryptand 2.2.2 (18 kcal mol^{-1}). A fully organized spherand which incorporates six anisyl units exhibits a free energy of complexation higher than 23 kcal mol^{-1}. According to Cram, the free energy of binding can decrease by as much as 11 kcal mol^{-1} when changing from chloroform to water. The solvent effect is less pronounced in the case of ligands with structures that are more preorganized for binding.

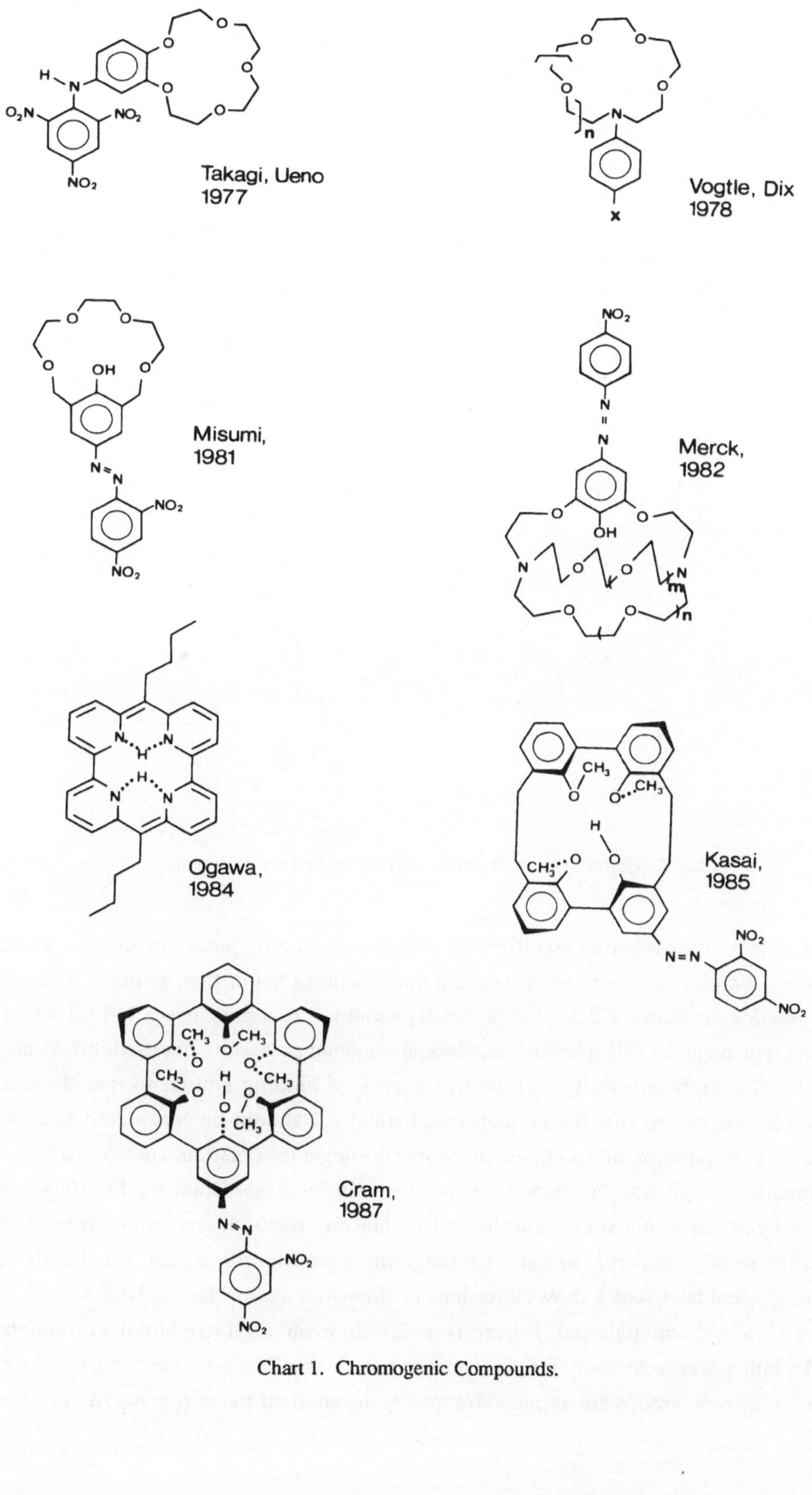

Takagi, Ueno
1977

Vogtle, Dix
1978

Misumi,
1981

Merck,
1982

Ogawa,
1984

Kasai,
1985

Cram,
1987

Chart 1. Chromogenic Compounds.

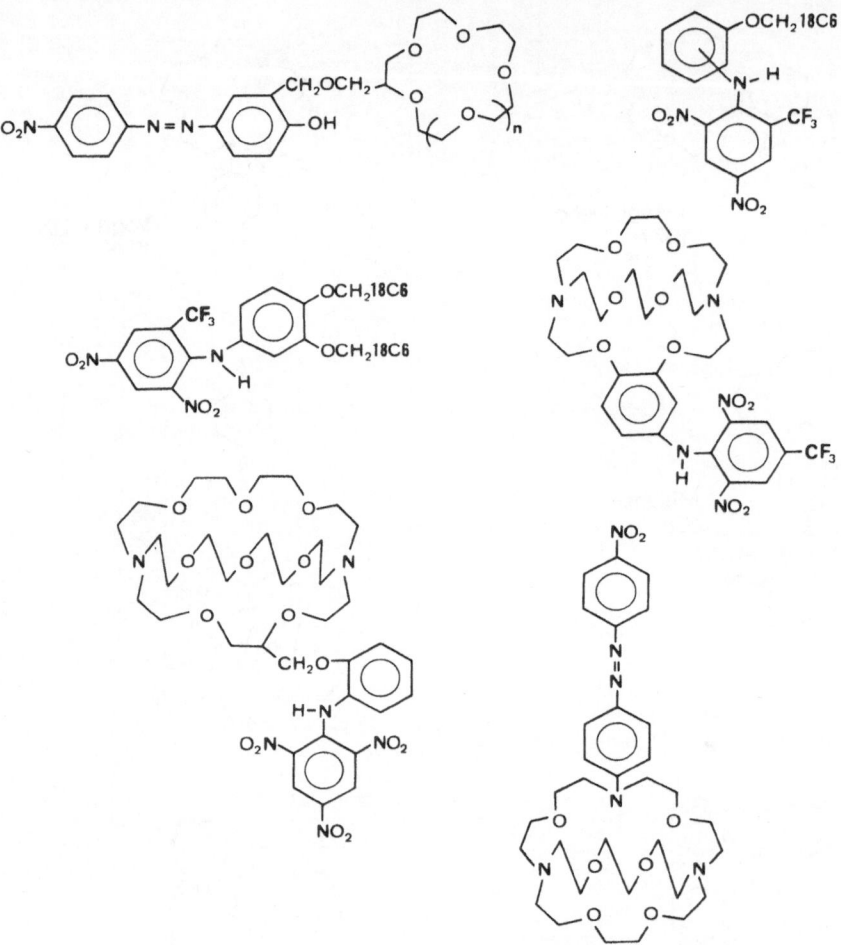

Chart 2. Chromogenic compounds made at Texas Tech University.

Also, selectivity increases significantly with the degree of preorganization. Thus, in chloroform, 18-crown-6 complexes potassium only 70 times better than sodium. This factor reaches 440 for cryptand 2.2.2. Cryptahemispherand 2.2, which has a hybrid structure between a cryptand and a full spherand, exhibits an excellent potassium over sodium selectivity of 11,000. This high selectivity and the free energy of binding similar to that showed by cryptand 2.2.2 suggested that the cryptahemispherand 2.2 should be considered as a prime candidate for incorporation into a chromoionophore designed for potassium determination.

An intensive collaboration which developed between Professor Cram and Dr. Helgeson of the University of California at Los Angeles and Technicon's research group quickly resulted in the synthesis of several novel chromogenic compounds based on hemi- and cryptahemispherands. Chart 3 shows three novel compounds synthesized at UCLA.

Surprisingly, hemispherand 1 bearing a 2,4-dinitrophenylazophenol chromophore exhibited a rather poor selectivity for potassium over sodium. One possible explanation could be that four alkoxy groups are required for preorganization of the cavity and for selectivity

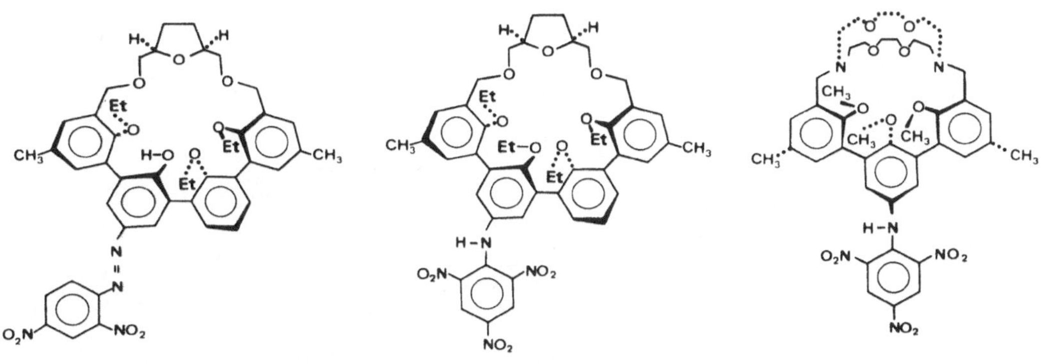

Chart 3.

of K$^+$ over Na$^+$. It is also possible that an interaction between a cation and the phenolic hydroxyl, observed before with crown ethers and cryptands, destroys the selectivity. In fact, hemispherand 2 containing four ethoxyl groups and a picrylamine chromophore was found to be a good binder and highly selective for potassium. Unfortunately, both compounds 1 and 2 showed a significant decrease in binding strength and sensitivity going from organic to aqueous solutions and hence were unsuitable for our purposes. Chromogenic cryptahemispherand 3 designed to complex potassium appeared to be especially interesting. It was first synthesized in a multistep process in about 4% total yield (Scheme 1). In this classical approach, a great deal of synthetic work was done to prepare a key-intermediate diacid which was then converted into the diacid chloride. This was followed by high dilution cyclization with diaza-18-crown-6 to give the diamide cryptaspherand. The nitro group of the diamide was reduced with hydrogen and the two amide groups were reduced with a borane-dimethyl sulfide complex. The complexed borane was displaced from the triamine with sodium chloride and coupling with picryl chloride produced 3·NaCl.

The elegant but lengthy procedure was then replaced with a less conventional shorter approach which allows the production of 3 in kilogram quantities in a 25% total yield. This shorter method is proprietary and will not be discussed in this paper.

Figure 1 shows absorption maxima of the acid (HL) and base form (L$^-$) for 3. Compound 3 which is prepared and used as a sodium complex readily distinguishes colorimetrically between Na$^+$ and K$^+$ at the proper pH. Complexation of potassium acidifies the system driving the chromophore pK_a from 7.75 (3·Na$^+$) to 7.05 (3·K$^+$) (adjusted to zero ionic strength). Selectivity of potassium over sodium is about 1500:1. Response to the change in potassium ion concentration at pH 7.3 is shown in Figure 2.

An analogous sodium reagent was prepared in a similar manner. This incorporates a diaza-12-crown-4 moiety and is isolated as a lithium complex. The pK_a values for the lithium and sodium complexes are 7.85 and 6.95, respectively. Selectivity of Na$^+$ over K$^+$ is exceptionally high and cannot be measured since the potassium response is at the level of instrumental noise.

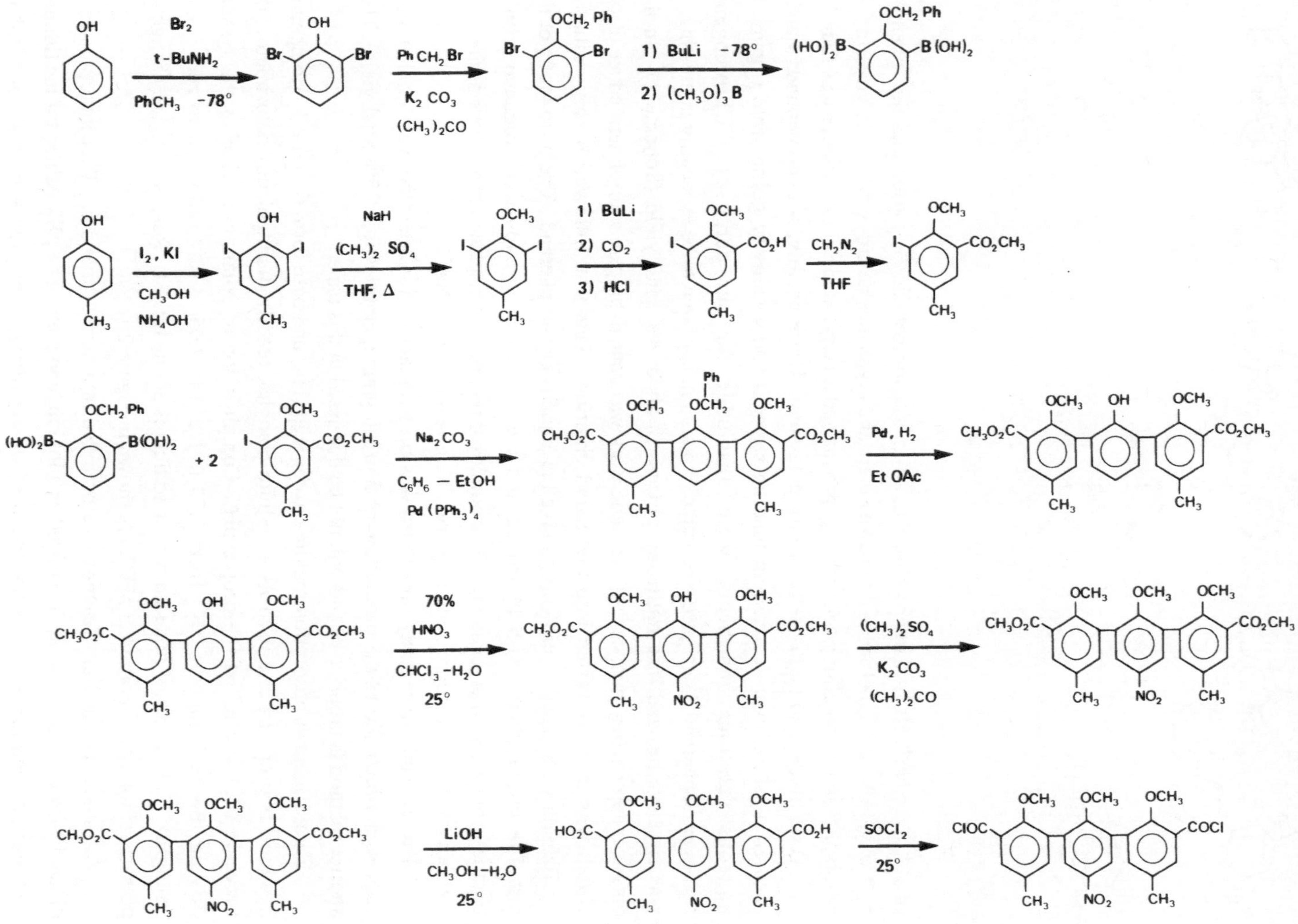
Br2
t-BuNH2
PhCH3 -78°
PhCH2Br
K2CO3
(CH3)2CO
1) BuLi -78°
2) (CH3O)3B
I2, KI
CH3OH
NH4OH
NaH
(CH3)2SO4
THF, Δ
1) BuLi
2) CO2
3) HCl
CH2N2
THF
+ 2
Na2CO3
C6H6 — EtOH
Pd(PPh3)4
Pd, H2
EtOAc
70%
HNO3
CHCl3-H2O
25°
(CH3)2SO4
K2CO3
(CH3)2CO
LiOH
CH3OH-H2O
25°
SOCl2
25°

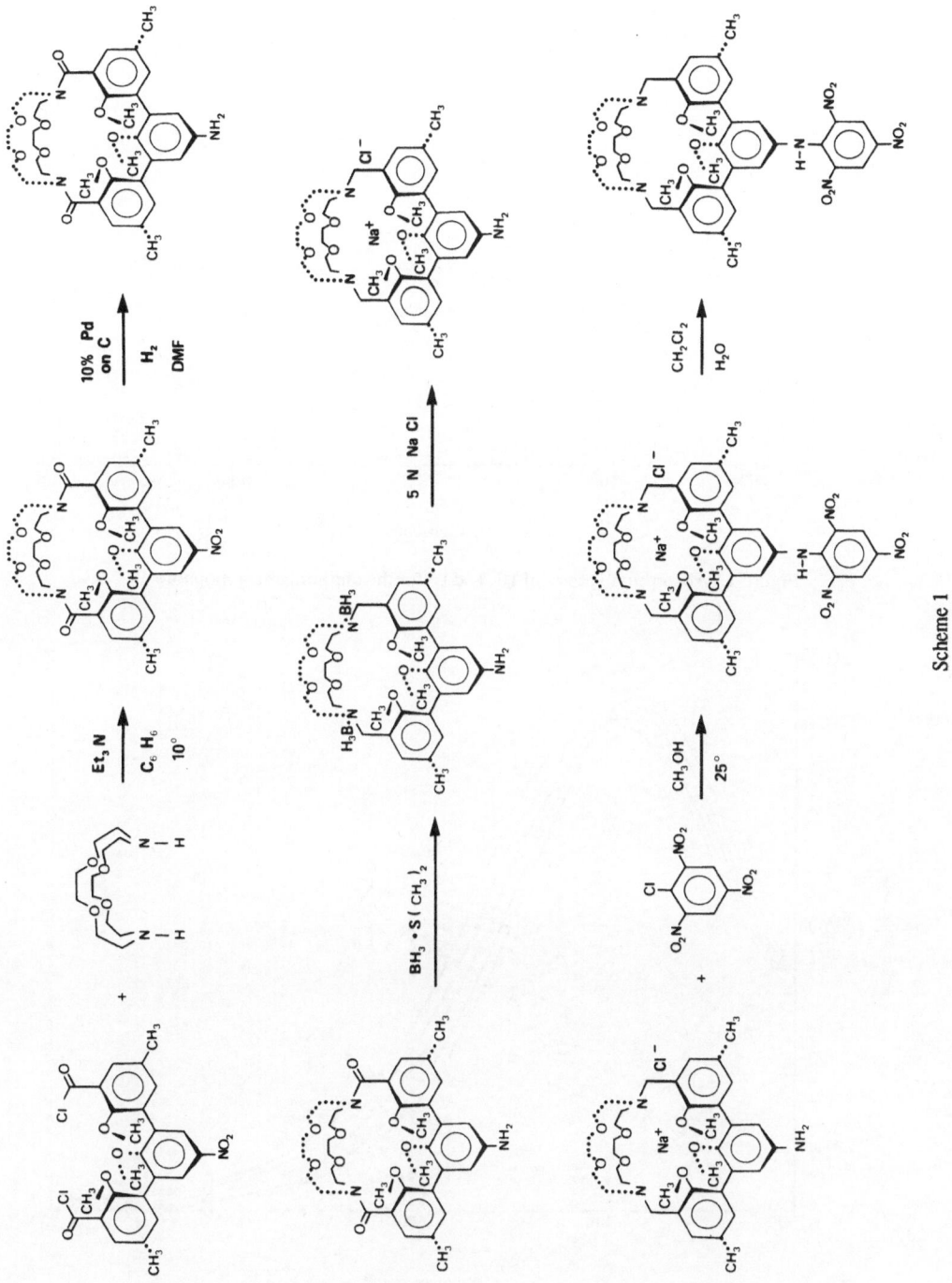

Scheme 1

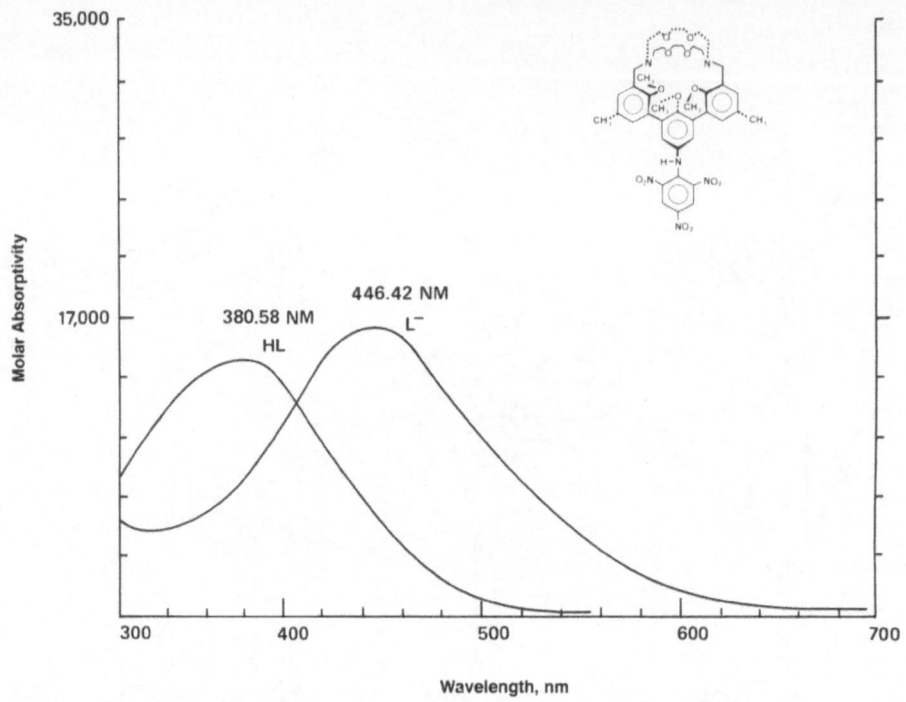

Figure 1. Absorption spectra of HL and L$^-$ for the chromogenic ionophore.

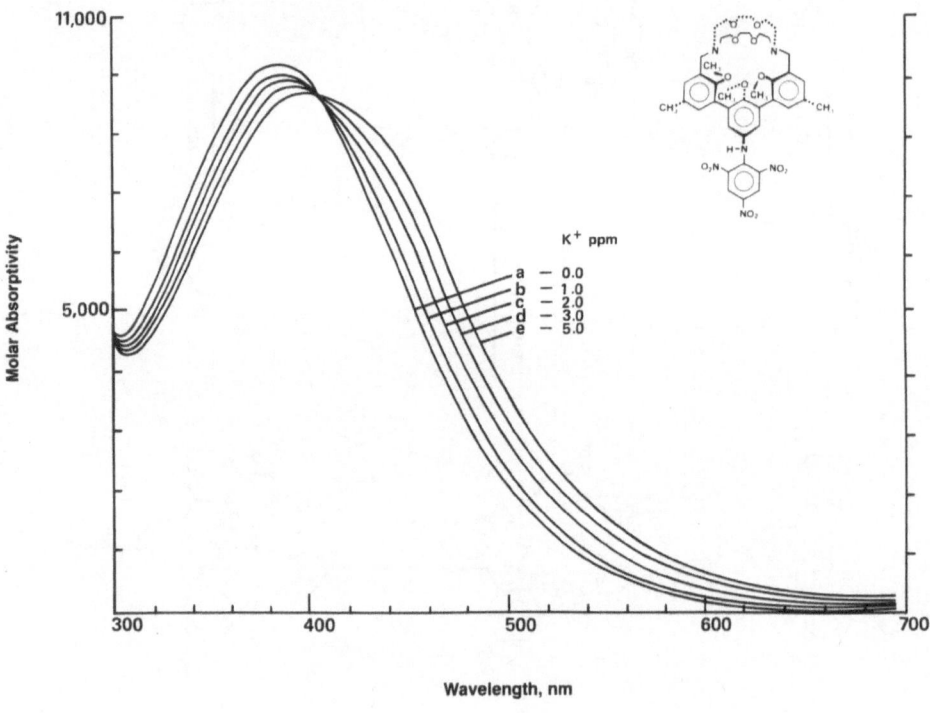

Figure 2. Potassium response in a pH 7.3 aqueous buffer.

Both spectroscopic methods for sodium and potassium determination show good correlation with the corresponding ISE methods. The potassium assay is an equilibrium method and the sodium assay is a two-point kinetic method. More detailed results of the analytical evaluation of new colorimetric reagents were published elsewhere [15]. The reagents based on these novel chromoionophores will be marketed by Technicon Instruments Corporation under the trade name "ChromoLyte".

References

1. Takagi, M.; Ueno, K. *Top. Curr. Chem.* **1984**, *121*, 39.
2. Lohr, H.-G.; Vögtle, F. *Acc. Chem. Res.* **1985**, *18*, 65.
3. Takagi, M.; Nakamura, H.; Ueno, K. *Anal. Lett.* **1977**, *10*, 1115.
4. Dix, J.; Vögtle, F. *Angew. Chem., Int. Ed. Engl.* **1978**, *17*, 857.
5. Nakamura, H.; Nishida, H.; Takagi, M.; Ueno, K. *Bunseki Kagaku* **1982**, *31*, E131.
6. Wong, S. T.; Chong, J.; Spoo, J. *Clin. Chem.* **1987**, *33*, 1005.
7. Wong, D.; Charlton, S.; Lau, A.; Fleming, R.; Atkinson, J.; Makowski, E.; Hemmes, P.; Frank, G. *Clin. Chem.* **1984**, *30*, 962.
8. Nakashima, K.; Nakatsuji, S.; Akiyama, S.; Kaneda, T.; Misumi, S. *Chem. Lett.* **1982**, 1781.
9. Klink, R.; Bodart, D.; Lehn, J.-M.; Helfert, B.; Bitsch, R. (Merck Patent GmbH) Eur. Pat. 85,320.
10. Ogawa, S.; Narushima, R.; Arai, Y. *J. Am. Chem. Soc.* **1984**, *106*, 5760.
11. Misumi, S.; Kai, Y.; Morii, H.; Miki, K.; Kasai, N.; Kaneda, T.; Umeda, S.; Tanigawa, H. *J. Am. Chem. Soc.* **1985**, *107*, 4802.
12. Cram, D. J.; Carmack, R. A.; Helgeson, R. C. *J. Am. Chem. Soc.* **1988**, *110*, 571.
13. Czech, B. P.; Gebauer, C. R.; Kumar, A.; Sy-Icaza, S.; Barczak, C.; Chapoteau, E.; Babb, D. A.; Czech, A.; Bartsch, R. A. *Clin. Chem.* **1986**, *32*, 1173.
14. Cram, D. J. *Angew. Chem., Int. Ed. Engl.* **1986**, *25*, 1039.
15. Kumar, A.; Chapoteau, E.; Czech, B. P.; Gebauer, C. R.; Chimenti, M. Z.; Raimondo, O. *Clin. Chem.* **1988**, *34*, 1709.

FROM REAL CHYMOTRYPSIN TO ARTIFICIAL CHYMOTRYPSIN
MYRON L. BENDER'S LEGACY

Valerian T. D'Souza

Department of Chemsitry
Universiy of Missouri-St. Louis
St. Louis
MO 63121 (USA)

On July 29th, 1988 one of the greatest scientific minds of our time, Myron L. Bender, passed away. On the occasion of this conference, one needs to look at his achievements which have led us to the understanding of enzymes and syntheses of molecules which can mimic real enzymes. He was recently able to produce a complete artificial enzyme from a thorough understanding of the structure and mechanism of action of a naturally occurring enzyme. An account of this achievement follows.

Myron L. Bender started his scientific endeavors at a time when enzymes were still thought to possess vital powers which came from life to enable them to bring about chemical transformations with 10^{12} acceleration and very high specificity. He, an organic chemist, decided to accept the challenge of uncovering their magical powers and show that these molecules followed simple rules of organic chemistry. The enzyme he decided to work on was chymotrypsin since this enzyme had been crystallized, fully characterized and was available in pure form (Figure 1). He subjected this enzyme to the rigors of physical organic chemistry of which he was a master. He published extensively and an account in the *Annual Review of Biochemistry* in 1965 [1] explained the powers of this enzyme in terms of rules of chemistry. This article was so authoritative that it was published as a book in Russian language under the authorship of "Bendeb". Having achieved this he decided to show that these powerful catalysts could be mimicked in the laboratory and do not have to come from life forces. Chymotrypsin, a large and complicated molecule with a molecular weight of 24,800 and 245 amino acids, is an archetype of 24 serine proteases. However, Bender attributed the powers of this enzyme, like other enzymes, to two aspects, binding and catalysis [2]. He proceeded to investigate systems to mimic these two characteristics independently and then combine them to produce the synergistic effect of the enzyme.

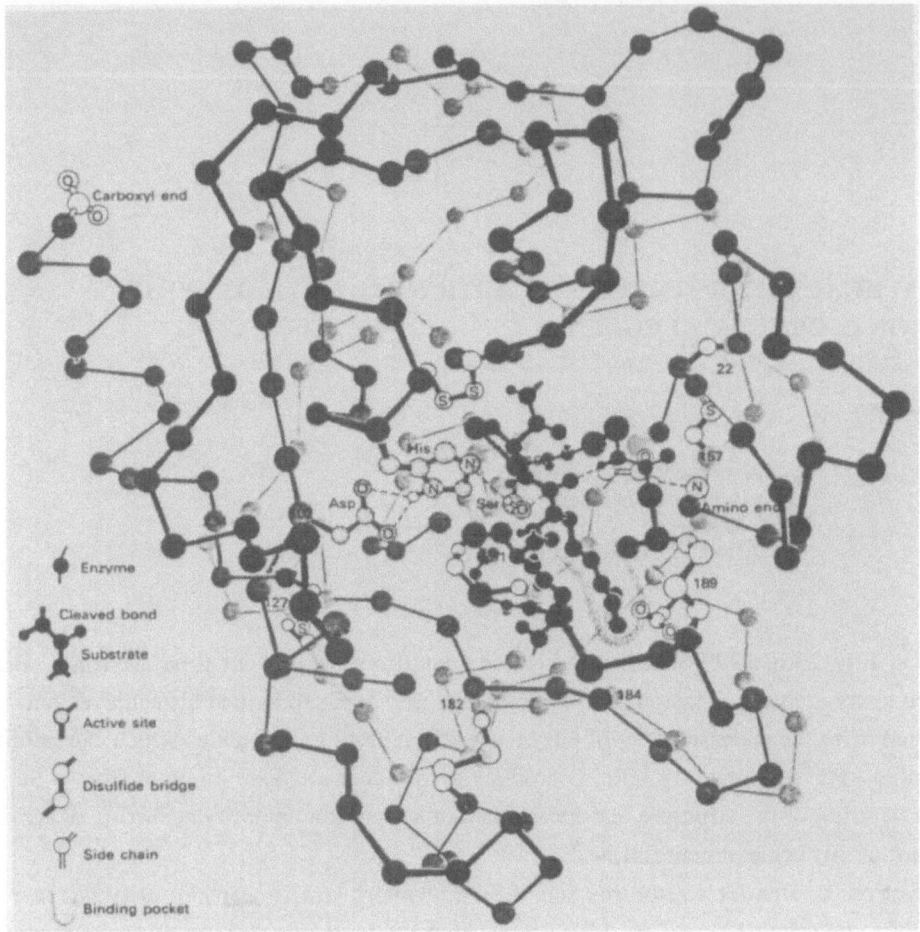

Figure 1. Structure of chymotrypsin.

The binding site in chymotrypsin is a hydrophobic cavity approximately 4 by 6 Å in cross section and 12 Å deep and can snugly fit an aromatic ring. To mimic the binding site of chymotrypsin Myron Bender chose to investigate cyclodextrins – oligosaccarides consisting of 6,7 or 8 glucose units bound in a cyclic fashion in α-1,4-linkages. They form doughnut shaped molecules in which all the secondary hydroxyl groups lie on one side and all the primary hydroxyls lie on the other side of the doughnut. The inside of the molecule is lined with a ring of -CH- units, a ring of glycosidic oxygens and another ring of -CH- units making it a hydrophobic cavity. This cavity is 4.5, 6.0 and 7.5 Å in α-, β- and γ-cyclodextrins respectively. These molecules are capable of binding other small organic molecules [3].

A typical kinetic investigation he carried out which showed binding is the effect of cyclodextrin concentration on the hydrolysis of *p*-nitrophenyl acetate (Figure 2). It followed Michaelis-Menten kinetics similar to that of chymotrypsin. The same data could be treated by a variant of a Lineweaver-Burk plot. This indicated that the accelerations in the presence of cyclodextrin were due to the formation of Michaelis-Menten type complexes prior to reaction.

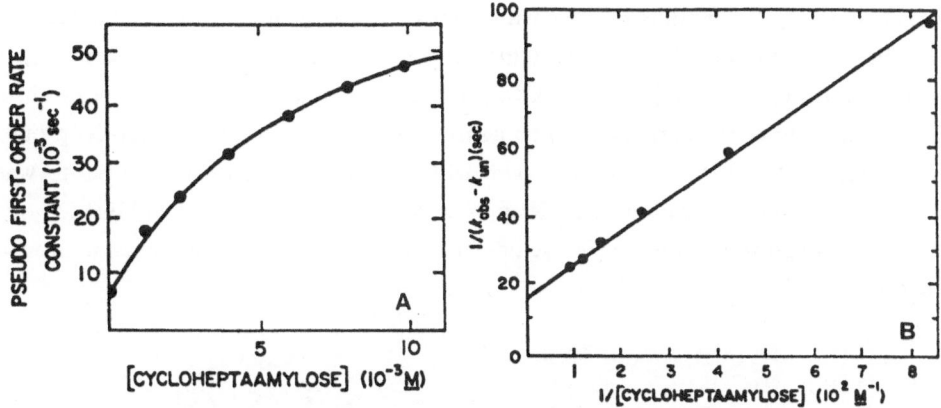

Figure 2. (A) Rate constants of hydrolysis of *p*-nitrophenylacetate versus β-cyclodextrin concentration, (B) double reciprocal plot of the same data.

The effect of cyclodextrins on a series of substituted phenyl acetates produced chaos, leading to what we consider to be the worlds worst Hammett plot as opposed to a normal Hammett plot, produced by an acyclic analog of cyclodextrin, methyl glucoside (FIgure 3). From this he made two important conclusions about cyclodextrin catalyzed hydrolyses of phenyl esters: (1) electronic effects are not important, and (2) steric effects control these reactions since it can be seen that the hydrolytic acceleration by cyclodextrin on any *m*-substituted phenyl ester is larger than on the corresponding *p*-substituted phenyl ester. As seen in the

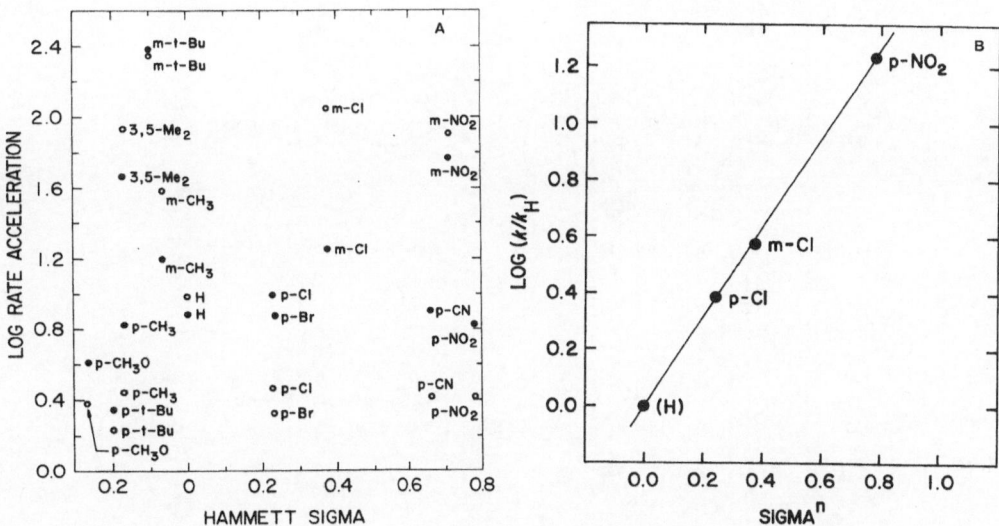

Figure 3. Hammett plots of acceleration of hydrolyses of substituted phenyl esters in the presence of (A) β-cyclodextrin and (B) methyl glucoside.

models (Figure 4), the binding of *m*-substituted acetates to cyclodextrins orients the carbonyl carbon atom of the ester substrate toward the oxygen atom of the secondary hydroxyl group of the cyclodextrin for a nucleophilic attack which is fast, whereas the complexes of *p*-substituted phenyl acetates have the carbonyl carbon atom far from the secondary hydroxyl groups and thus reaction is slow [4]. This is consistent with the 'Lock and Key' theory put forward by Emil Fischer in 1892 to explain specificity of enzymatic catalysis. The publication of these results in 1967 spurred an interest in cyclodextrin chemistry that has continued to grow to the present level.

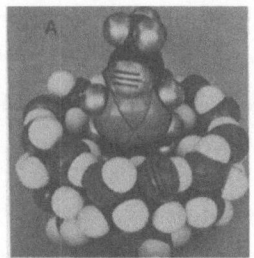 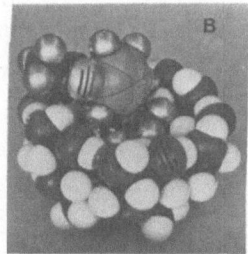

Figure 4. CPK models of β-cyclodextrin complexes with (A) *m-t*-butylphenyl acetate and (B) *p-t*-butylphenyl acetate.

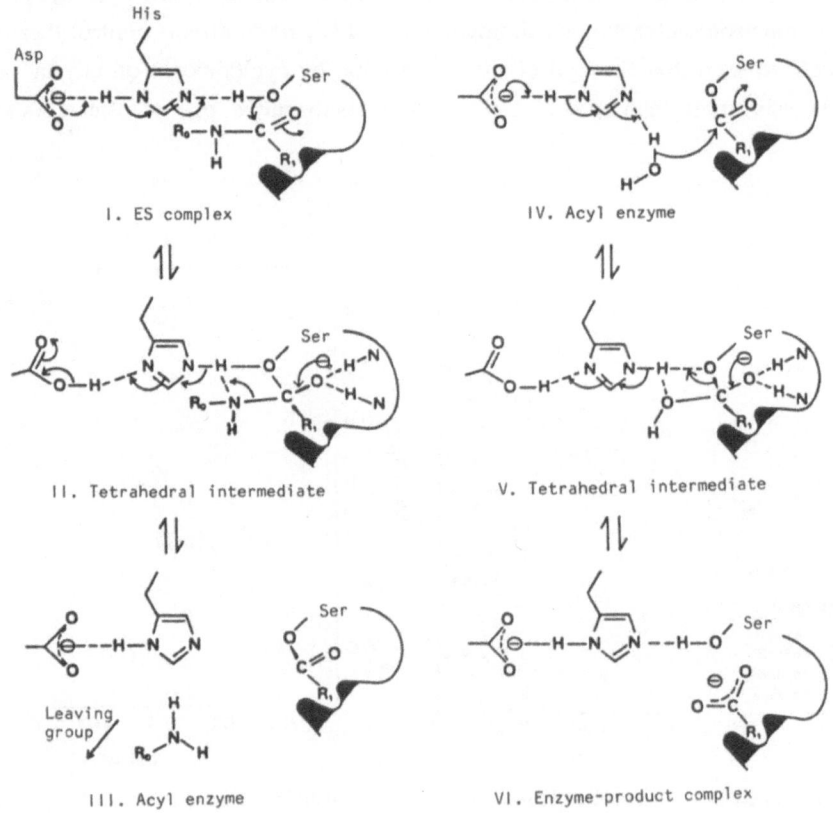

Figure 5. Mechanism of action of chymotrypsin.

230

Myron Bender then turned to the mechanism of action of chymotrypsin. Out of the 245 amino acids of chymotrypsin only three amino acids are involved in the chemistry of catalysis. They are: (1) serine 195, (2) histidine 57, and (3) aspartate 102. However, it is the functional groups that these amino acids carry that are responsible for the chemistry of catalysis rather than the amino acids themselves. These functional groups are respectively (1) the hydroxyl group, (2) the imidazolyl group and (3) the carboxylate ion.

The mechanism by which chymotrypsin hydrolyzes the bound amide or the ester substrate, now well known as the 'proton transfer relay' mechanism, involves two steps, acylation and deacylation (Figure 5). In acylation, two proton transfers, one by carboxylate ion and another by imidazole, increase the nucleophilicity of the hydroxyl oxygen atom of serine and thus enhance its **nucleophilic attack on the carbonyl carbon atom of the bound** substrate. The products of the first step are an amine and an acyl-enzyme. Deacylation occurs via the same two proton transfers, increasing the nucleophilicity of the hydroxyl group of the water molecule, which attacks the carbonyl carbon atom of the acyl enzyme ester. Thus, the products of the second step are a carboxylic acid and the original enzyme which can turnover and catalyze another molecule of the substrate [5].

In order to mimic the mechanism of action of chymotrypsin Myron Bender designed a model of acyl-chymotrypsin so that the contributions from binding could be bypassed. It consists of *endo*-[4'(5')-imidazolyl-bicyclo[2.2.1]hept-*endo*-2-yl *trans*-cinnamate, a rigid norborane backbone to hold the functional groups [6] (Figure 6A). The imidazole and the ester functionality were attached at the *endo, endo*-2,4 positions so that they would be at the right distance and in the correct orientation for a general base catalysis as is the case for real acyl-chymotrypsin. A computer aided picture of the molecule indicated that the distance between the carbonyl carbon and the nitrogen of the imidazole group was about 2.6 Å which was rather large for a nucleophilic attack by the imidazole group. On the other hand, insertion

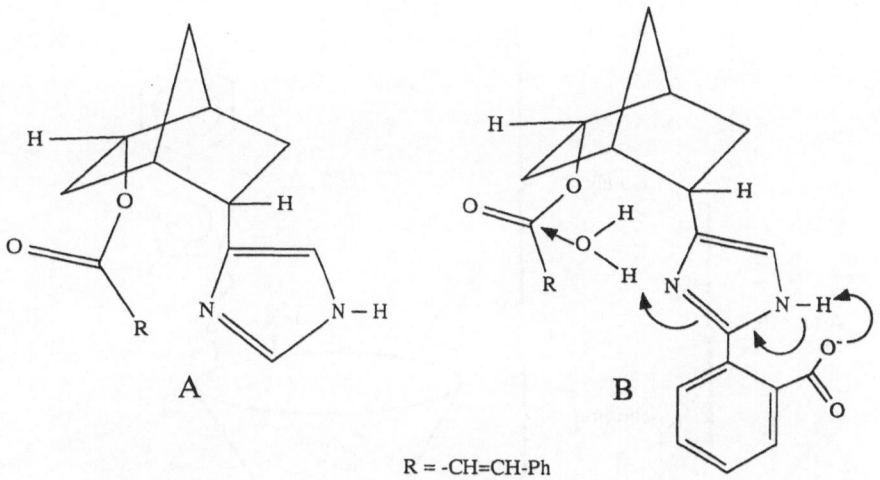

R = -CH=CH-Ph

Figure 6. Acyl-chymotrypsin models consisting of (A) cinnamoyl ester and imidazole group and (B) A plus an intramolecular carboxylate ion.

of a molecule of water between the two functionalities results in distances of 1.37 Å between the carbonyl carbon atom and the oxygen atom of the water molecule, and 1.6 Å between the hydrogen atom of the water molecule and the nitrogen atom of the imidazole group, making it ideal for a general base catalysis by imidazole. A solvent isotope effect ($kH_2O/kD_2O = 3$) indicated that the imidazole is indeed acting as a general base. Interestingly, the deacylation of cinnamoyl-chymotrypsin has a value of ($kH_2O/kD_2O = 2.5$). Although, mechanistically this was satisfying, the acceleration in the rate of hydrolysis of the ester was modest. Upon realizing that the third component of the active site, a carboxylate ion, was missing from the model, the deacylation was investigated in the presence of benzoate ion. This apparently simple addition brought about a large acceleration in the rate of deacylation. Dioxane, added to simulate the apolar character of the active site, further increased the rate of deacylation. It was possible to accelerate the reaction 2,500 times by using a concentration of 0.5 M benzoate ion in 0.42 mole fraction dioxane-water mixture. This acceleration was about 1,500 times slower than the real enzyme [7], a fact which was attributed to the intermolecularity of the carboxylate ion used. Thus, a new model using an intramolecular benzoate ion was designed and synthesized (*endo,endo*-5-[2-(2-carboxyphenyl)-4-(5)-imidazolyl]bicyclo[2.2.1]hept-2-yl*trans*-cinnamate, Figure 6B). The model accelerated the rate of deacylation by 154,000 and was equal to 1/18th the rate of hydrolysis of the cinnamoyl chymotrypsin [3]. Thus, a system capable of mimicking the mechanism of action of chymotrypsin was designed and synthesized.

Having thus achieved the synthesis of systems to mimic binding and catalysis of chymotrypsin, Myron Bender decided to build a complete artificial enzyme by covalently attaching the catalytic site that had been so successfully designed to the secondary side of a cyclodextrin since it was established that the bound substrate would have its carbonyl function at the secondary side of cyclodextrin (Figure 7).

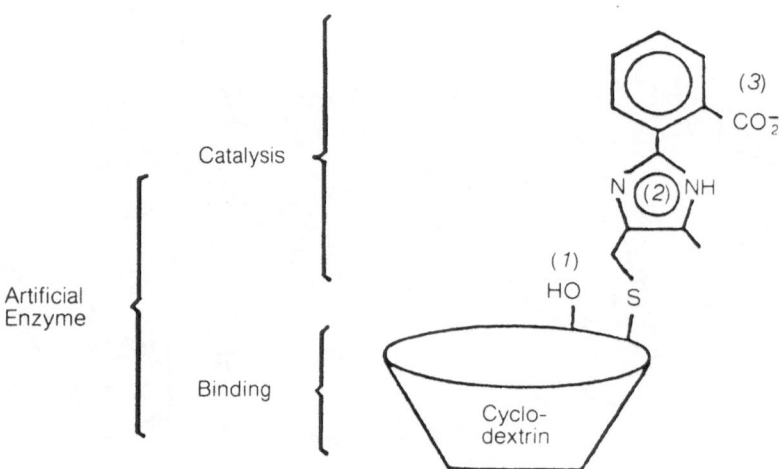

Figure 7. Complete artificial enzyme – 1, 2 and 3 indicate the functional groups.

The first synthesis was achieved by the reaction of β-cyclodextrin-2,3-epoxide (obtained from the reaction of β-cyclodextrin-2-tosylate by the reaction of ammonium bicarbonate) with o-[4(5)mercatomethyl-4(5)-methylimidazol-2-yl] benzoic acid [9]. The artificial enzyme of chymotrypsin based cyclodextrin is named "benzyme" by Dr. X. L. Lu. We have now achieved synthesis of α-, β- and γ-cyclodextrins, respectively [10].

The β-benzyme model was examined for its catalytic activity in m-t-butylphenyl acetate. The hydrolysis of more than 10 equivalents of substrate indicates there is indeed a turnover. Lineweaver-Burk plots indicate that the artificial enzyme is comparable to the real enzyme in its activity. The solvent isotope effect ($kH_2O/kD_2O = 3$) shows that it is a general base catalyzed hydrolysis. Hydrolysis of m-t-butylphenyl trimethylacetate by β-cyclodextrin indicates that there is a presteady state (acylation) and a steady state (deacylation and turnover). These results indicate that the artificial enzyme is similar to the real enzyme in its mechanism of action. The mechanism, assuming 1:1 complex formation between the artificial enzyme and the substrate, is shown in Figure 8.

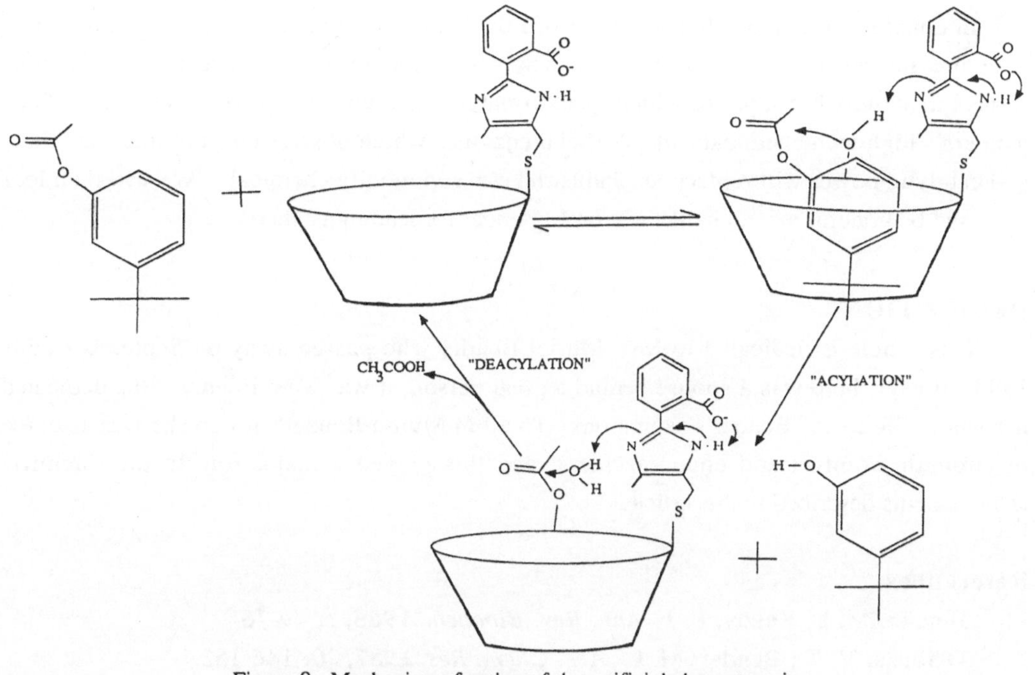

Figure 8. Mechanism of action of the artificial chymotrypsin.

The main difference between the artificial enzyme and the real enzyme is in their stabilities. The artificial enzyme is stable, and is in fact more active, at higher temperatures whereas the real enzyme denatures around 50°C (Figure 9A). The artificial enzyme is also stable to pH changes. It begins to lose activity only at very low pH (<2) when glycosidic bonds are hydrolyzed. The real chymotrypsin on the other hand is sensitive to pH changes and denatures at neutral as well as high pH ranges (Figure 9B). Thus, this artificial enzyme is useful under harsh conditions of temperature and pH [11].

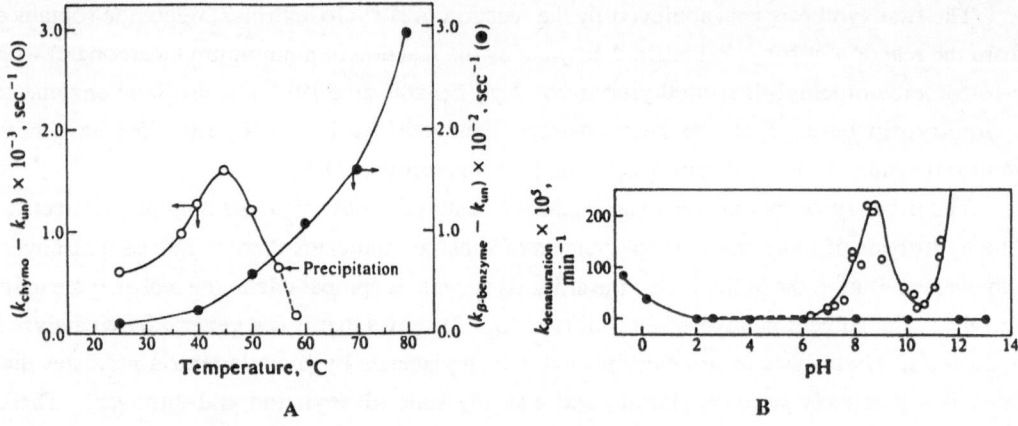

Figure 9. (A) Thermal stability of the artificial (•) and natural (o) enzymes and (B) pH stability of the artifical (•) and natural (o) enzymes.

In conclusion, Myron Bender has shown us that by studying an enzyme in terms of its structure and detailed mechanism, one can synthesize catalysts similar to the enzyme with respect to activity but superior with respect to stability and utility. As we move into the 'high tech era', 'high tech chemicals' like artificial enzymes which possess enzyme-like specificity and catalytic power will replace 'the industrial era' commodity chemicals. We will then look back and be grateful to people like Myron L. Bender for leading us there.

DEDICATION

This article is dedicated to Mrs. Muriel Bender who passed away on September 26th, 1988. If ever there was a spouse behind a great person, it was Mrs. Bender. She dedicated her whole life to Dr. Bender's aspirations. To all in Myron Bender's group she was a source of strength, comfort and encouragement, and this played a major role in the scientific achievements described in this article.

References

1.   Bender, M. L; Kezdy, F. J. *Ann. Rev. Biochem.* **1965**, *1*, 49-76.

2.   D'Souza, V. T.; Bender, M. L. *Acc. Chem. Res.* **1987**, *20*, 146-152.

3.   Bender, M. L.; Komiyama, M. *Cyclodextrin Chemistry*; Springer-Verlag: New York, 1978.

4.   VanEtten, R. L.; Sebastian, J. F.; Clowes, G. A.; Bender, M. L. *J. Am. Chem. Soc.* **1967**, *89*, 3242-3253.

5.   Walsh, C. *Enzymatic Reaction Mechanisms*; W. H. Freeman Company: New York, 1979, pp. 53-108.

6.   Komiyama, M.; Roesel, T. R.; Bender, M. L.; Utaka, M.; Takeda, A. *Proc. Natl. Acad. Sci. USA* **1977**, *74*, 23-25.

7. Komiyama, M.; Bender, M. L.; Utaka, M.; Takeda, A. *Proc. Natl. Acad. Sci. USA* **1977**, *74*, 2634-2638.

8. Mallick, I. M.; D'Souza, V. T.; Yamaguchi, M.; Lee, J.; Chalabi, P.; Gadwood, R. C.; Bender, M. L. *J. Am. Chem. Soc.* **1984**, *106*, 7252-7254.

9. D'Souza, V. T.; Hanabusa, K.; O'Leary, T.; Gadwood, R. C.; Bender, M. L. *Biochem. Biophys. Res. Commun.* **1985**, *129*, 727-732.

10. Bender, M. L.; D'Souza, V. T.; Lu, X. *Trends in Biotech.* **1986**, *4*, 132-135.

11. D'Souza, V. T.; Lu, X. L.; Ginger, R. D.; Bender, M. L. *Proc. Natl. Acad. Sci. USA* **1987**, *84*, 673-674.

MOLECULAR RECOGNITION BY CYCLODEXTRIN HOSTS: APPROACH TO NEW FUNCTIONS

Iwao Tabushi[a], Kazuo Yamamura*, and Yasuhisa Kuroda

Department of Synthetic Chemistry
Kyoto University
Sakyo-ku, Kyoto 606 (Japan)

[a]Deceased March 22, 1987

INTRODUCTION

The design of inclusion hosts simulating the behavior of enzymes or receptors is one of the major goals of the current research in biomimetic or bioorganic chemistry. Greatly accelerated rates for catalytic reactions are observed for cyclodextrins which form inclusion complexes with hydrophobic guest molecules. Modified cyclodextrins with reactive functional groups adjacent to the binding site are useful for the specific molecular recognition.

Enzymes or receptors usually have (at least) two important functional groups. Consequently it is important and necessary to have two different functional groups in certain fixed spatial orientation, for preparing artificial enzymes or receptors. "Regiospecific capping" of β-cyclodextrin (Table 1) for the purpose of regiospecific introduction of two (different) functional groups has been developed [1]. The process of "disulfonate capping" is kinetically fast and geometrically favored for the transannular difunctionalization, and AB, AC, and AD-regiomer caps can be converted into regiospecifically difunctionalized cyclodextrins with two functional groups at **Z** (zusammen), **R** (rechtwinkling), and **E** (entgegen) orientations, respectively. Several new enzyme/receptor characteristics have been attempted by using molecular recognition of new functionalized cyclodextrins.

EFFICIENT ELECTRON TRANSFER CATALYSIS VIA CYCLODEXTRIN INCLUSION

6-(8α-S-riboflavo)-α-cyclodextrin **1** was prepared from 6-(SH)-α-cyclodextrin and 8α-bromoriboflavin, as a flavocyclodextrin model [2]. Very rapid N-alkyldihydronicotin-amide (RANH) reduction of **1** was observed under anaerobic aqueous conditions. At 2.5×10^{-4} to 1.0×10^{-3} M concentrations of n-hexyl-NAH, saturation kinetics were observed (Equation 1), giving $k_2 = 0.5$ sec^{-1} and K_a (association constant) = 2500 M^{-1}. Similarly, for i-propyl-NAH, values of $k_2 = 0.36$ sec^{-1} and $K_a = 260$ M^{-1} were obtained. The overall

Inclusion Phenomena and Molecular Recognition
Edited by J. Atwood
Plenum Press, New York, 1990

Table 1. Three possible regioisomers of capped cyclodextrin and regioisomer distribution in the crude cap mixture of β-cyclodextrin.

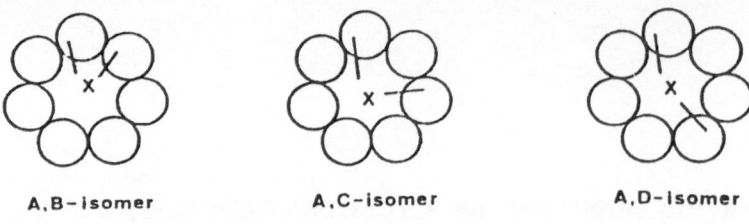

A,B-Isomer   A,C-Isomer   A,D-Isomer

capping reagent	[β-CD] (10^{-3}M)	T°(C)	Time(h)	Ratio(Anal.Yield%) AC	:	AD
	3.2	60	2	92		8
	5.9			93		7
(benzophenone)	18			94		6
	35			92		8
	18	27	4	89(40)		11(5)
	2.9	60	2	11(3)		89(23)
(stilbene)	18			23(4)		77(14)
	44			26(6)		74(16)
	3.3	27	20	8(4)		92(41)
—◯—CH₂—◯—	18	60	2	66		34
—◯—O—◯—	18	60	2	52		48
—◯◯—	18	60	2	7		93
				AB	:	AC
—◯—	44	25		97		3

1

efficiency of dihydroflavin (FlH$_2$) production, $k_2 \cdot K_a$ is 1200 M^{-1} sec^{-1} for **1**, which is very large (Table 2). Catalytic electron transfer from RNAH to Mn(III)TPPS (tetraphenyl-porphyrin tetrasulfonate) for the rate determining electron transfer at pH 7.4 in an aqueous solution was observed for **1**. The observed catalytic constant with n-hexyl-NAH in the presence of **1** was 200 M^{-1}sec^{-1}, being 5.5 times faster than MnTPPS-FMN-MeNAH, the $k_2 \cdot K_a$ value of which was 36 M^{-1}sec^{-1}.

$$Fl + RNAH \underset{}{\overset{K_a}{\rightleftharpoons}} Fl\,RNAH \xrightarrow{k_2} RNA + FlH_2 \tag{1}$$

Table 2. Flavocyclodextrin and reduction rate of flavin by NAH.

flavin	reductant	K_a(M^{-1})	k_2 (s^{-1})	$k_2 \cdot K_a$ (M^{-1} s^{-1})
NADH FMN oxidoreductase	NADH	21000	15.5	326000
8a-S-flavopapain	benzyl NAH	10300	0.0054	56
8a-S-flavoCD (1)	*n*-hexyl NAH	2500	0.5	1200
	iso-propyl NAH	260	0.36	94
	benzyl NAH	1050	0.06	63

AB-substituted-α-cyclodextrin was converted to flavometalloporphyrinato-α-cyclo-dextrin **2**. The manganese complex **2** was reduced with *n*-hexyl-NAH, and a large association constant (K_a = 1.4 x 10^{-3} M^{-1}) and very rapid electron transfer (k_2 = 0.8 sec^{-1}) were observed. The rapid electron transfer is due mainly to the strong association of *n*-hexyl-NAH with **1** and good orientation. The $k_2 \cdot K_a$ value is 1.1 x 10^{-3} M^{-1}sec^{-1} which is about 50-fold larger than the second order rate constant k_2 of the Mn(III)TPPS-FMN-RNAH system (Table 3).

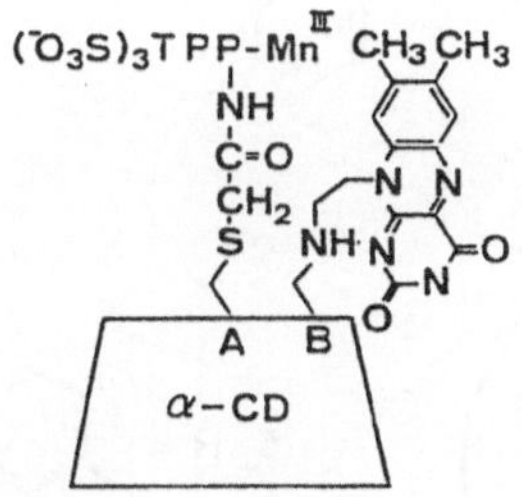

2

Table 3. Reduction rate of Mn(III) porphyrin at 25°C, pH 7.4, 0.1 M phosphate buffer.

Mn(III)porphyrin	Electron transfer catalyst	Electron source	k_2 (M^{-1}s^{-1})
Mn(III)TPPS (2 x 10^{-6} M)	FMN (2 x 10^{-6} M)	Me-NAH (1-10 x 10^{-4} M)	2.5 x 10 [a]
Mn(III)TPPS (2 x 10^{-6} M)	Fl-α-CD (2 x 10^{-6} M)	*n*-hexyl-NAH (1-10 x 10^{-4} M)	2.5 x 10^2 [b]
	Fl-α-CD-P.Mn (2) (2 x 10-6 M)	*n*-hexyl-NAH (2-250 x 10^{-4} M)	1.1 x 10^3 [b]
Mn(III)TPPS (2 x 10^{-6} M)	FMNH (2 x 10^{-5} M)		10^5 [c]

[a]2nd order rate constant of Mn(III)TPPS and *n*-hexyl-NAH.
[b]$k_{cat} \cdot K_a$, k_{cat} = 0.8 s^{-1}, K_a = 1.4 x 10^3 M^{-1}.
[c]2nd order rate constant of Mn(III)TPPS and FMNH.

CYCLODEXTRIN SANDWICHED Fe₄S₄ CLUSTER

A new type of cyclodextrin, cyclodextrin sandwiched Fe_4S_4 cluster **3** has been prepared from 6A,6D-biphenyl-4,4'-disulfonyl capped β-cyclodextrin via conversion to AD-dimercaptocyclodextrin followed by ligand exchange reaction with $[Fe_4S_4(S\text{-}t\text{-}Bu)_4]$-$[n\text{-}Bu_4N]_2$ [3]. This synthetic cluster is stable in an aqueous solution without addition of any stabilizing reagents and the physical properties including redox properties are shown in Table 4. This type of synthetic cluster may have a great advantage in a sense that the hydrophobic pockets of cyclodextrins may be available as recognition sites for substrates such as donors and/or acceptors.

$[Fe_4S_4(SR)_4]\ (n\text{-}Bu_4N)_2$

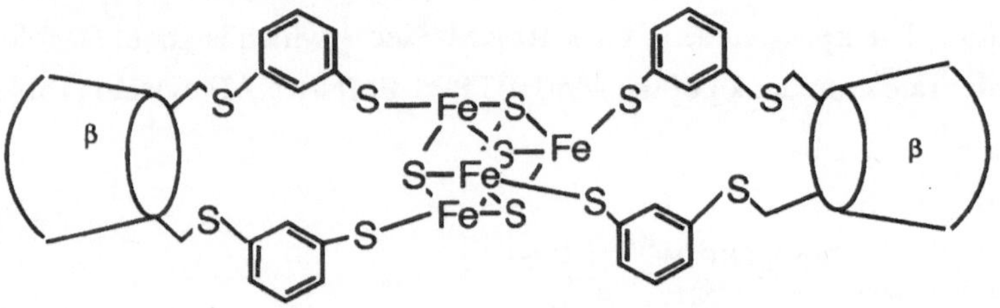

3

4: R =

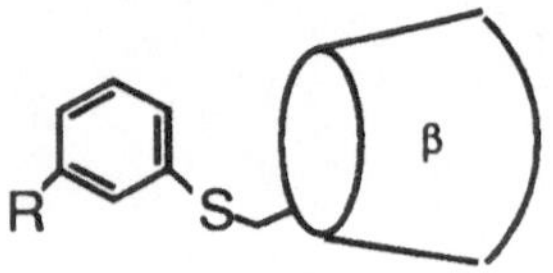

5: R =

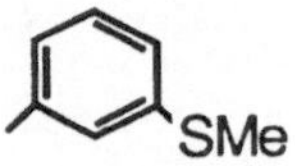

Table 4. Absorption maximum, half-life[a] and $E_{1/2}$ of synthetic iron-sulfur cluster.

	3	4	5
In H₂O (10mM phosphate buffer, pH 7.0)	428 nm (17000) 120 h -0.58 V	468 nm (17000) 70 h -0.55 V	468 nm[b] 5.6 h[b] -0.54 V[b]
in DMF	441 nm (17500) c -0.94 V[d]	458 nm (17000) c -0.93 V[d]	463 nm (17300) c -0.92 V[d]

[a]By monitoring the decreases of the characteristic absoprtions of clusters at 420-470 nm.
[b]5% DMF-95% 10 mM phosphate buffer, pH 7.0.
[c]No appreciable spectral changes were observed for several days.
[d]100 mM $n\text{-}Bu_4NBr$.

ENHANCEMENT OF PROTON DISSOCIATION OF AMIDE CAP VIA INCLUSION OF HYDROPHOBIC CATIONIC GUESTS

2,9-Phenanthridinone capped β-cyclodextrin is a new regioselective amide cap which has been prepared via capping technique (Figure 1). Its amide moiety is ionizable in aqueous solution, and its pK_a value was determined to be 11.2 by UV titration. The apparent association constant K_a for the binding of p-bromobenzyltrimethylammonium by the proton

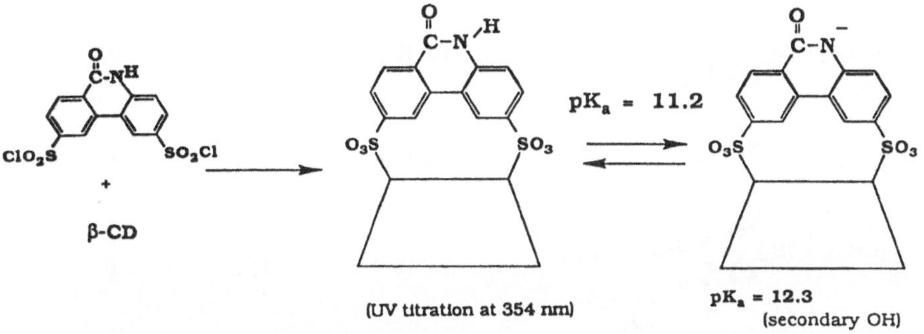

AH-Cap   A⁻-Cap

$pK_a = 11.2$

(UV titration at 354 nm) $pK_a = 12.3$
                                  (secondary OH)

Figure 1. 2,9-Phenanthridinone cap (amide cap) with ionizable amide moiety.

dissociated form of the amide cap (anion cap) was $K_{assoc} = 3300$ M^{-1} at pH 12.2 (above pK_a), which is larger than the $K_{assoc} = 1900$ M^{-1} at pH 6.86 (below pK_a). This indicates that the host-guest electrostatic interaction is significant (Figure 2). As a consequence, the proton dissociation of the amide cap would be enhanced, if the host-guest electrostatic interaction operates significantly in the hydrophobic cavity via inclusion. This was the case for the amide cap, and the pK_a was decreased to 10.0 with $\Delta pK_a = -0.2$ (Table 5) in the

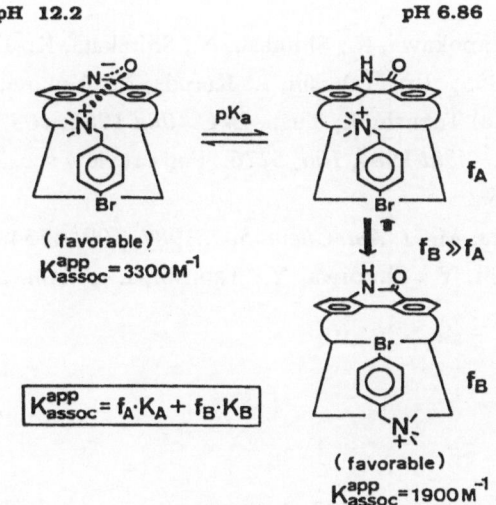

pH 12.2   pH 6.86

(favorable)
$K_{assoc}^{app} = 3300$ M^{-1}

$f_B \gg f_A$

(favorable)
$K_{assoc}^{app} = 1900$ M^{-1}

$$K_{assoc}^{app} = f_A \cdot K_A + f_B \cdot K_B$$

Figure 2. Apparent association constant for binding of p-bromobenzyltrimethylammonium by amide cap.

Table 5. pKa of amide cap and its enhancement via inclusion of hydrophobic cationic guest molecules.

Guest	Conc (M)	pK_a	ΔpK_a
none		11.2	
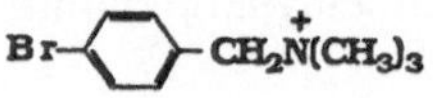 Br—⟨⟩—CH$_2$N$^+$(CH$_3$)$_3$	9×10^{-3}	11.0	-0.2
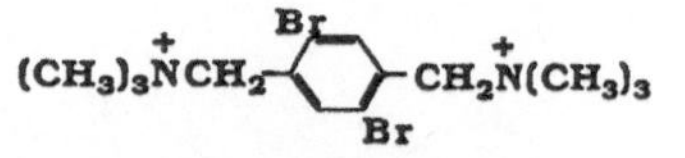 (CH$_3$)$_3$N$^+$CH$_2$—⟨Br, Br⟩—CH$_2$N$^+$(CH$_3$)$_3$	7×10^{-3}	10.7	-0.5
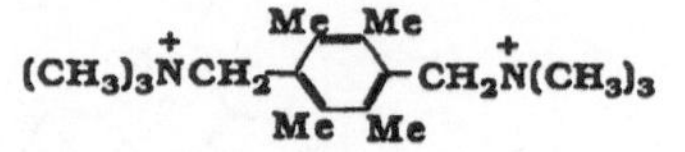 (CH$_3$)$_3$N$^+$CH$_2$—⟨Me Me, Me Me⟩—CH$_2$N$^+$(CH$_3$)$_3$	1×10^{-3}	10.7	-0.5
 N⟨⟩N	2×10^{-2}	11.2	0
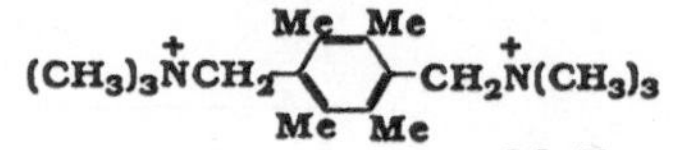 (CH$_3$)$_3$N$^+$CH$_2$—⟨Me Me, Me Me⟩—CH$_2$N$^+$(CH$_3$)$_3$	1×10^{-3}	11.2	0
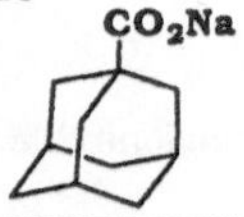 (adamantane)CO$_2$Na	2.8×10^{-3}	11.2	0

coexistence of p-bromobenzyltrimethylammonium. Much larger enhancement of proton dissociation with a ΔpK_a = -0.5 has also been achieved, when diammonium compound was employed as the guest molecule (Table 5). A change of physical property such as the ionic state of an artificial receptor provides a new interesting aspect of function triggered by molecular recognition.

References

1. (a) Tabushi, I.; Shimokawa, K.; Shimazu, N.; Shirakata, K.; Fujita, K. *J. Am. Chem. Soc.* **1978**, *98*, 7855; (b) Tabushi, I.; Kuroda, Y.; Yokota, K.; Yuan, L. C. *ibid* **1981**, *103*, 711; (c) Tabushi, I.; Yuan, L. C. *ibid* **1981**, *103*, 3574; (d) Yamamura, K.; Nabeshima, T. *ibid* **1984**, *106*, 5276. For a review see, Tabushi, I. *Acc. Chem. Res.* **1982**, *15*, 66.
2. Tabushi, I.; Kodera, M. *J. Am. Chem. Soc.* **1987**, *109*, 4734.
3. Kuroda, Y.; Sasaki, Y.; Shiroiwa, Y.; Tabushi, I. *J. Am. Chem. Soc.* **1988**, *110*, 4049.

FLUORESCENCE AND CIRCULAR DICHROISM STUDIES ON
MOLECULAR AND CHIRAL RECOGNITION BY CYCLODEXTRINS

Koji Kano

Faculty of Engineering
Doshisha University
Kamikyo-ku, Kyoto 602 (Japan)

SUMMARY

Chiral recognition of several aromatics by cyclodextrins has been studied by fluorescence and circular dichroism spectroscopy. Pyrene forms a chiral excimer in the gamma-cyclodextrin cavity where the primary hydroxyl group side seems to be the recognition site. (S)-Helixed bilirubin is bound preferentially to beta- and gamma-cyclodextrins through hydrogen bonding. No essentially important role of the cyclodextrin cavities is found for the enantioselective complexation between cyclodextrins and bilirubin. Noncyclic oligosaccharides also show their abilities to recognize the chirality of bilirubin. Permethylated beta-cyclodextrin is the most effective host molecule to recognize the chiralities of the binaphthyl derivatives.

INTRODUCTION

One of the current research subjects of cyclodextrin (CDx) chemistry is molecular recognition by those oligosaccharides. Since CDxs are cyclic oligosaccharides composed of alpha-glucopyranoses, the molecular recognition should be realized mainly by relative size and hydrophobicity and hydrogen bonding ability between host and guest molecules. In some cases, CDxs can be used for chiral recognition because the cavities of CDxs provide chiral environments. In order to understand the mechanism for molecular recognition including chiral recognition, it is very important to know the structure of the resulting inclusion complex. Although the structure of a crystalline inclusion complex can be determined easily by X-ray analysis, it is considerably more difficult to elucidate the structure of an inclusion complex in aqueous solution.

We would like to discuss the structures of the inclusion complexes of pyrene and its related compounds at the initial part of this article and then to report the chiral recognition of pyrene excimer, bilirubin, and binaphthyl derivatives by CDxs. As the host molecules, we used usual beta- and gamma-CDxs, heptakis-(2,6-di-O-methyl)-beta-CDx (DM-beta-CDx), heptakis-(2,3,6-tri-O-methyl)-beta-CDx (TM-beta-CDx), 6-O-alpha-glucosyl-beta-CDx

Inclusion Phenomena and Molecular Recognition
Edited by J. Atwood
Plenum Press, New York, 1990

(G1-beta), 6-O-alpha-maltosyl-beta-CDx (G2-beta), and 6-O-alpha-maltosyl-gamma-CDx (G2-gamma). The branched CDxs were kindly provided by Dr. Y. Yoshimura at Tokuyama Soda Company, Japan.

BINDING SITE OF PYRENE AND ITS RELATED COMPOUNDS [1]

We used circular dichroism (CD) and fluorescence spectroscopy to elucidate the structures of the inclusion complexes in solutions.

The CD spectrum of pyrene included in the beta-CDx cavity was measured. The signs of the Cotton effect at the ^1L_a and ^1B_b transition bands of pyrene were positive and negative, respectively. The theoretical considerations [2,3] indicate the formation of an axial inclusion complex of beta-CDx and pyrene. On the other hand, the signs were completely opposite for the pyrene-G2-beta system. It may be assumed that an equatorial inclusion complex is formed for this system.

In order to check the effect of the maltosyl group on the induced CD, we measured the CD spectra of acridine which can be included completely into the beta-CDx cavity. No difference was observed in CD spectra between beta-CDx and G2-beta. The signs of the induced CD (ICD) were negative at the ^1L_a transition band and positive at the ^1L_b band. The CD spectra, therefore, indicate the formation of an axial inclusion complex of beta-CDx and acridine. The maltosyl group does not affect the CD spectrum for such complexes.

From the CD spectral data, it can be concluded that the pyrene-beta-CDx complex is the axial complex and the pyrene-G2-beta complex is the equatorial complex. The molecular dimensions of the beta-CDx and pyrene molecules indicate that a pyrene molecule cannot be included completely into the beta-CDx cavity. Pyrene, therefore, should be bound shallowly to the rim of the beta-CDx cavity.

There are two sides of the cavity, i.e., the narrower primary hydroxyl group side and the wider secondary hydroxyl group side. Which side is the binding site of pyrene?

G1-beta and G2-beta played very important roles to answer this question. The CPK molecular model suggests that the maltosyl group of G2-beta can cap the CDx cavity via hydrogen bonds. Of course, another possible structure of G2-beta is an open-form where no interaction exists between the CDx cavity and the maltosyl group. If G2-beta takes the capped form, the secondary hydroxyl group side is the only binding site of pyrene for this CDx. In the case of G1-beta, the glucosyl group cannot cap the cavity.

Remarkable differences between beta-CDx and the branched CDxs were observed in the fluorescence quenching of pyrene by trimethylamine in these CDx solutions. Previously, we reported that the pyrene fluorescence in the aqueous beta-CDx solution is quenched statistically by forming the pyrene-trimethyl-amine-beta-CDx three-component complex [4,5].

The results of the fluorescence quenching of pyrene by trimethylamine in water containing three kinds of CDxs were treated by the Stern-Volmer plots. In the case of beta-CDx, a saturation-type Stern-Volmer plot was obtained. In the case of G2-beta, however, the fluorescence was not quenched at all at lower quencher concentrations and the abrupt quenching started at higher quencher concentrations. The most effective fluorescence quenching took place in the presence of G1-beta. In order to understand such interesting behavior, we examined the fluorescence quenching of naphthalene derivatives. Molecular models predict that naphthalene is included completely into the beta-CDx cavity while 1-methylnaphthalene and acenaphthene having groups which cause steric hindrances form the shallow inclusion complexes similar to that of pyrene. The results for acenaphthene were essentially the same as those for pyrene. Namely, the fluorescence quenching does not occur at lower quencher concentrations in the aqueous G2-beta solution. In contrast, the fluorescence quenching of naphthalene was completely different, no difference being observed between beta- and branched beta-CDxs. It can be concluded, therefore, that the novel difference in the fluorescence quenching between beta- and branched beta-CDxs is observed in the case where the guest molecule is bound shallowly to CDx.

The fluorescence quenching of pyrene and acenaphthene can be explained as follows. The pyrene molecule is bound preferentially to the more hydrophobic primary hydroxyl group side of beta-CDx to form a shallow inclusion complex. Trimethylamine can penetrate into the remaining space of the cavity of the pyrene capped beta-CDx to form a three-component complex, which is nonfluorescent. In this case, the pyrene long axis is parallel to the C_7 symmetry axis of beta-CDx. On the other hand, pyrene cannot be bound to the primary hydroxyl group side of G2-beta because this side is capped by the maltosyl group through hydrogen bonding. Unwillingly, pyrene may bind to the more hydrophilic secondary hydroxyl group side to form the equatorial complex. Since both cavity sides of this equatorial complex are capped, trimethylamine cannot penetrate into the cavity. This may be the reason why the emissions from pyrene and acenaphthene are not quenched at lower trimethylamine concentrations in the aqueous G2-beta solutions. At higher trimethylamine concentrations, the trimethylamine molecules are bound to the maltosyl group via hydrogen bonds leading to the release of capping. The release of capping upon addition of trimethylamine was supported by the ^{1}H NMR data. Consequently, we concluded that hydrophobic pyrene is bound preferentially to the narrow but hydrophobic primary hydroxyl group side of beta-CDx.

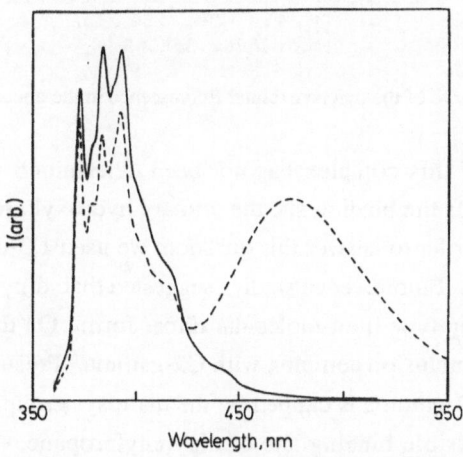

Figure 1. Fluorescence spectra of pyrene in water containing beta- (–) and gamma-CDxs (---).

CHIRAL PYRENE DIMER FORMED IN gamma-CDx CAVITY[1,6]

Figure 1 shows the fluorescence spectra of pyrene in the aqueous beta- and gamma-CDx solutions. In the case of beta-CDx, only monomer emission of pyrene was observed. Pyrene, however, shows both monomer and excimer-like emissions in the gamma-CDx solution. The fluorescence decay curve of the pyrene excimer formed in the gamma-CDx cavity is shown in Figure 2. Interestingly, a very fast rise of the pyrene excimer was observed, suggesting that the pyrene molecules can move in the gamma-CDx cavity to form an excimer state. In other words, the pyrene dimer in the gamma-CDx cavity has a very loose structure.

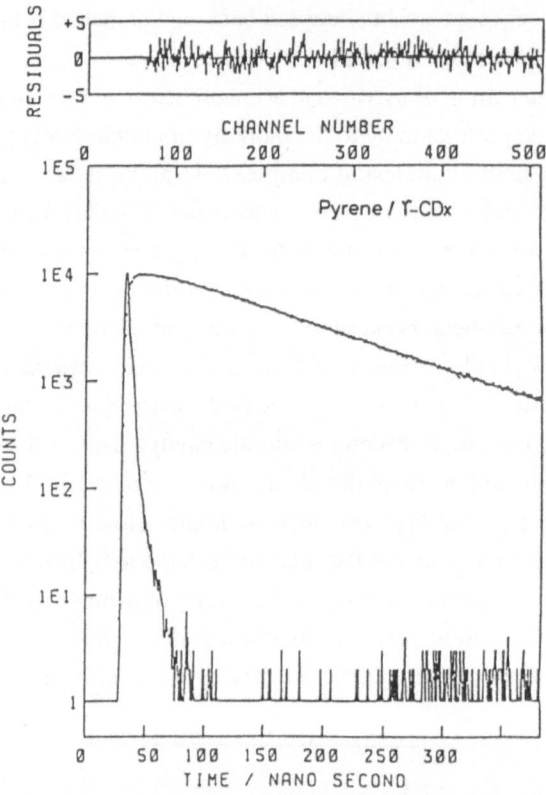

Figure. 2. Rise and decay curve of the pyrene excimer fluorescence in the aqueous gamma-CDx solution.

The stoichiometry of this complex has not been determined yet. The problem is the binding site of the dimer. Is the binding site the primary hydroxyl group side or the secondary hydroxyl group side? In order to answer this question, we used 1,3-di(1-pyrenyl)propane as a model of the pyrene dimer. Fluorescence studies suggested that dipyrenylpropane is bound to gamma-CDx in its stacking type intra-molecular dimer form. On the other hand, dipyrenyl-propane did not form the inclusion complex with G2-gamma. Presumably, since the primary hydroxyl group side of G2-gamma is capped by the maltosyl group, the secondary hydroxyl group side is the only possible binding site of dipyrenylpropane. The secondary hydroxyl group side of G2-gamma, however, seems to be too hydrophilic to interact with the very hydrophobic dipyrenylpropane. It may be assumed, therefore, that the binding site of dipyrenylpropane as well as the pyrene dimer is the primary hydroxyl group side of gamma-CDx.

Of course, the face to face dimer of pyrene is achiral. The twisted pyrene dimer, however, has chirality. Circularly polarized fluorescence (CPF) spectroscopy is a very good method to prove the formation of asymmetric bichromophoric fluorescent dimer [7,8]. The sample is excited by natural light and the difference in the intensity between left-handed polarized fluorescence and right-handed polarized fluorescence is detected. We measured the CPF spectra of the aqueous pyrene solutions containing gamma-CDx and G2-gamma. Very intense CPF signals were observed at the pyrene excimer fluorescence region while no signal was detected at the monomer fluorescence region, indicating that the pyrene dimers in the CDx cavities are asymmetric. Interestingly, the sign of the CPF signal observed for gamma-CDx was positive while that for G2-gamma was negative. The pyrene dimer formed in the gamma-CDx cavity seems to have an opposite chirality to that in the G2-gamma. Unfortunately, the absolute configuration of the pyrene dimer cannot be determined from CPF. Then we measured CD spectra. The results were somewhat complex, but the oppositely signed bisignated Cotton effects were observed for gamma-CDx and G2-gamma. From the analysis of the CD spectra, it can be concluded that the pyrene dimer having a left-handed chirality is a preferable guest for gamma-CDx while the pyrene dimer having a right-handed chirality is preferred for G2-gamma. Presumably, such a difference in the chiral recognition between gamma-CDx and G2-gamma is ascribed to the difference in binding site. 1-Ethylpyrene also forms the chiral dimers in the gamma-CDx and G2-gamma cavities. 1,1'-Dinaphthylpropane forms a chiral intramolecular dimer in the gamma-CDx cavity.

CHIRAL BILIRUBIN COMPLEXES OF CDxs AND NONCYCLIC OLIGOSACCHARIDES [9]

The guest molecules so far described bind with CDxs via hydrophobic interaction. Hydrogen bonding can also be used for the binding force. For example, hydroquinone forms a stable inclusion complex with beta-CDx and it is very hard to extract hydroquinone from the aqueous CDx solution with nonpolar solvents such as benzene and ether. The signs of the ICD spectra of hydroquinone in aqueous beta-, DM-beta-, and TM-beta-CDx solutions indicate that hydroquinone forms the axial inclusion complexes with beta- and DM-beta-CDxs and the equatorial inclusion complex with TM-beta-CDx.

Bilirubin is also one of the guest molecules which is bound with CDx through hydrogen bonding. Bilirubin is a bichromophoric tetrapyrrole and cytotoxic pigment of jaundice. Bilirubin does not have any asymmetric center but the intramolecular hydrogen bonds provide the chirality of this molecule. Of course, bilirubin is not optically active in homogeneous solution.

As reported by Lightner et al. [10], the beautiful bisignated CD spectra of bilirubin were measured in water containing alpha-, beta-, and gamma-CDxs. The exciton coupling theory indicates that (S)-helixed bilirubin is bound preferentially to CDxs. Lightner et al. have not discussed the binding force and the role of the cavity of CDxs for chiral recognition. We studied in more detail the interaction of bilirubin with cyclic and noncyclic oligosaccharides.

The CD spectra of bilirubin in water (pH 10.8) were measured in the presence of beta-, DM-beta-, and TM-beta-CDxs. The intense - to + bisignated CD signal was observed only in the case of beta-CDx. The CD signal was diminished in the presence of methylated beta-CDxs. These results suggest that the hydrogen bond is the main binding force for enantioselective complexation of bilirubin with CDxs.

The next problem is the role of the cavity of CDx for enantioselective binding of bilirubin. Then the CD spectral study was carried out by using noncyclic di- and oligosaccharides. Fairly intense - to + bisignated CD signal of bilirubin was observed in water containing maltoheptoase, which is an open form of beta-CDx. The CD spectrum was essentially the same as that in the aqueous beta-CDx solution. This means that the cyclic cavity of CDx does not play an essential role for enantioselective binding.

No CD signal was observed in the glucose solution. Bilirubin in the disaccharide solution, however, showed the bisignated Cotton effect. The CD spectrum indicates that (S)-helixed bilirubin is bound preferentially to maltose.

The saccharides so far described are the alpha[1→4]-linked di- and oligosaccharides. Cellobiose and cellotetraose are the beta-[1→4]-linked saccharides. In these saccharide solutions, bilirubin showed the + to - bisignated Cotton effect. This means that bilirubin having (R)-helixed conformation is bound preferentially to the beta[1->4]-linked saccharides. In the case of beta[1->4]galactoside linked disaccharide, lactose, bilirubin shows the - to + bisignated Cotton effect.

As described above, the enantioselective binding of bilirubin and saccharides is not so simple and we cannot explain the mechanism clearly. The system described here, however, is one of the simplest systems of the chiral recognition by the saccharides.

CHIRAL RECOGNITION OF BINAPHTHYL DERIVATIVES BY CDxs

Armstrong and his co-workers have developed a CDx-bearing HPLC column which can separate several optical isomers [11]. They reported that the CDx-column does not work for separation of the optical isomers of a simple binaphthyl derivative, 1,1'-binaphthyl-2,2'-diyl hydrogen phosphate. In the present work, we found that TM-beta-CDx is the effective host molecule to recognize the enantiomers of the binaphthyl derivatives. The following guests were employed:

The racemate of 2,2'-dimethoxy-1,1'-binaphthyl is completely insoluble in water. After dimethoxybinaphthyl was added into the aqueous TM-beta-CDx solution, the solution was filtered. The residue was extracted by acetonitrile. The CD spectra of the resulting two solutions were measured. The aqueous filtrate showed the + to - bisignated CD signals, indicating that the (S)-isomer of this guest molecule is bound preferentially to TM-beta-CDx. Meanwhile, the acetonitrile solution showed the - to + bisignated CD signals, suggesting the (R)-enantiomer hardly forms an inclusion complex with TM-beta-CDx. The similar result was obtained for the dimethoxybinaphthyl-beta-CDx system but the ability of beta-CDx to recognize the chirality was much lower than TM-beta-CDx.

TM-beta-CDxs can also recognize the chirality of 1,1'-bi-2-naphthol. The fluorescence measurements indicate that the (S)-enantiomer was a preferable guest molecule for TM-beta-CDx.

The most remarkable enantioselective complexation was observed for the 1,1'-binaphthyl-2,2'-diyl hydrogen phosphate-TM-beta-CDx system. The binding constant for the 1:1 complex of the (S)-enantiomer and TM-beta-CDx was 8-times larger than that of the (R)-isomer. In the case of beta-CDx, the 2:1 complex of the host and binaphthyl phosphate was formed. The K_1K_2 value for the (S)-isomer was 2-times larger than that for the (R)-isomer.

References

1. Kano, K.; Matsumoto, H.; Yoshimura, Y.; Hashimoto, S. *J. Am. Chem. Soc.* **1988**, *110*, 204 .

2. Harata, K.; Uedaira, H. *Bull. Chem. Soc. Jpn.* **1975**, *48*, 375.

3. Schipper, P. E.; Rodger, A. *J. Am. Chem. Soc.* **1983**, *105*, 4541.

4. Kano, K.; Takenoshita, I.; Ogawa, T. *Chem. Lett.* **1980**, 1035.

5. Kano, K.; Takenoshita, I.; Ogawa, T. *J. Phys. Chem.* **1982**, *86*, 1833.

6. Kano, K.; Matsumoto, H.; Hashimoto, S.; Sisido, M.; Imanishi, Y. *J. Am. Chem. Soc.* **1985**, *107*, 6117.

7. Brittain, H. G. *Molecular Luminescence Spectroscopy. Methods and Applications*; ed. Shulman, S. G., ed.; Wiley: New York, 1985; Vol. 1, Chapter 8.

8. Riehl, J. P.; Richardson, F. S. *Chem. Rev.* **1986**, *86*, 1.

9. Kano, K.; Yoshiyasu, K.; Hashimoto, S. *J. Chem. Soc., Chem. Commun.* **1988**, 801.

10. Lightner, D. A.; Grawronski, J. K.; Grawronska, K. *J. Am. Chem. Soc.* **1985**, *107*, 2456.

11. Armstrong, D. W.; Ward, T. J. ; Czech, A.; Czech, B. P.; Bartsch, R. A. *J. Org. Chem.* **1985**, *50*, 5556.

THE INTERACTIONS OF VESICLE-FORMING SURFACTANTS WITH CYCLODEXTRINS

Orlando Garcia, Pablo A. Quintela, Jodi M. Schuette, Rafael Vargas, Hae R. Yoon, and Angel E. Kaifer*

Department of Chemistry
University of Miami
Coral Gables, FL 33124 (USA)

INTRODUCTION

Cyclodextrins (CD) are cyclic glucopyranose oligomers having a characteristic toroidal shape [1-3]. These compounds form strong host-guest complexes with many organic molecules in aqueous media because of the hydrophobic nature of the inner cavity defined by the torus. As usual in host-guest complexation, the stability of the resulting complex depends, among many other factors, on the fit of the guest molecule within the cavity provided by the host [1-3]. The three unmodified cyclodextrins are the so-called α-, β-, and γ-CD which are composed of 6, 7, and 8 glucopyranose rings respectively. α-CD, the smallest of the three hosts, has a cavity diameter of about 4 Å which is ideal for binding straight alkyl chains and certain uncrowded phenyl moieties. We are particularly interested in the first type of binding because of the possibility of using cyclodextrins for the structural modification of surfactant aggregates. The interactions of surfactants with cyclodextrins have received scant attention in the literature. A few studies on the complexation of simple surfactants, like alkylsulfates or alkyltrimethylammonium salts, with α-CD and β-CD are available [4,5]. Therefore, we have recently started a research program aimed at the characterization of CD-surfactant complexes and their properties [6].

We have concentrated our efforts in the study of surfactants containing a redox-active group. These surfactants are commonly synthesized in our group because of our interest in redox reactions taking place in environments resembling biological membranes. The presence of the redox-active group in the surfactant structure confers electroactivity to the free surfactant and its CD complexes so that electrochemical techniques can be utilized in assessing their properties. So far we have synthesized surfactant derivatives containing either the 4,4'-bipyridinium (viologen) or the bis(cyclopentadienyl)iron(II) (ferrocene) units as electroactive groups. In this work we will describe briefly the complexation of both types of surfactants by CD's and utilize this information to discuss preliminary results on the complexation of double chain, vesicle-forming surfactants by CD's.

Inclusion Phenomena and Molecular Recognition
Edited by J. Atwood
Plenum Press, New York, 1990

COMPLEXATION OF SURFACTANT VIOLOGENS WITH CYCLODEXTRINS

The following two viologens were synthesized by the stepwise alkylation of 4,4'-bipyridine [6].

$$1 = C_{16}V^{2+}$$

$$2 = C_{18}V^{2+}$$

The electrochemisrty of viologens (V^{2+}) is characterized by two consecutive monoelectronic reductions yielding first a cation radical (V^+) and then a neutral molecule (V). In aqueous media, the voltammetric behavior of compounds **1** and **2** is complicated by the increase in hydrophobic character that accompanies both reduction steps. Thus, both **1** and **2** precipitate at the electrode surface upon reduction to their cation radicals. This precipitation can be eliminated by addition of a tenfold excess of α-CD to the solution. Under these conditions, the redox couple $C_{16}V^{2+/+}$ shows reversible, diffusion controlled voltammetric behavior. The fact that no kinetic effects are observed on the electrochemical reduction-oxidation processess of the viologen subunit suggests that the main interaction site of the surfactant viologen with the cyclodextrin host is the alkyl chain. Otherwise, if the complexation took place at the aromatic viologen moiety, one would expect to see the current controlled by the dissociation rate of the CD complex as reported by Evans et al. for the β-CD complex of ferrocenecarboxylic acid [7]. Similar results were also obtained for the longer chain viologen **2**. Thus, it is suggested that the association of these viologens with α-CD proceeds by the formation of a complex in which the alkyl chain of the surfactant viologen is included inside the hydrophobic cavity of the CD host, as represented in the scheme below.

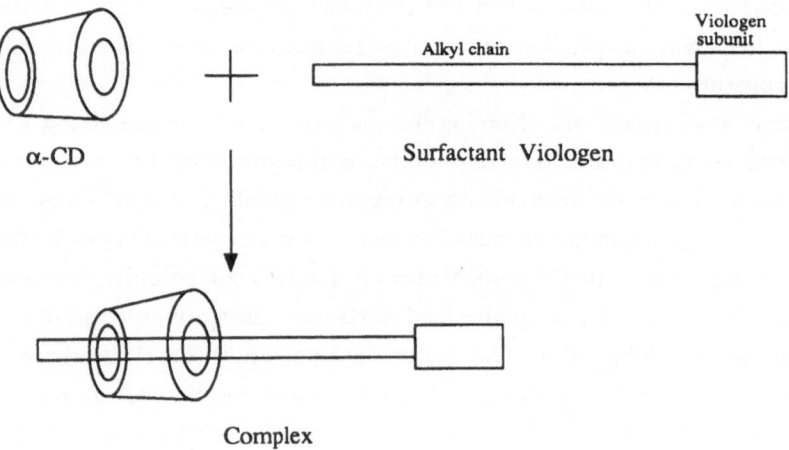

252

The formation of these complexes is thus driven by hydrophobic interactions. The α-CD-$C_{16}V^{2+}$ or the α-CD-$C_{18}V^{2+}$ complexes exhibit uncomplicated electrochemistry because the complexation of the alkyl tail by the cyclodextrin solubilizes the one-electron reduced forms of the viologens by decreasing their hydrophobic character. The reduction potential corresponding to the $V^{2+/+}$ couple in these complexes was found to be -0.66 V vs. SSCE (sodium chloride saturated calomel electrode) which is close to the reduction potential observed for methylviologen (-0.69 V vs. SSCE) in the same medium [8]. This provides further support for the proposed complexation site.

If the negative potential scan in the voltammetric experiments is extended to -1.0 V vs. SSCE, a second reduction wave can be observed. However, the distorted shape of this reduction peak (and its corresponding oxidation peak on the reverse scan) clearly suggests that diffusion is no longer the only process controlling the current. Therefore, it appears that the reduction of the α-CD-$C_{16}V^{+}$ species produces an insoluble compound that immediately precipitates on the electrode surface. The presence of β-CD also changes the voltammetric behavior of either $C_{16}V^{2+}$ or $C_{18}V^{2+}$. However, the first reduction couple of the β-CD complex does not exhibit reversible behavior at the scan rates surveyed (up to 1000 mV/s). In both cases, the reduction of the complex appears to be coupled to other processes such as precipitation or dissociation of the complex. This may indicate that β-CD, in contrast to α-CD, interacts with the electroactive viologen subunit.

An important aspect of the chemistry of viologens is the tendency of their cation radicals (V^{+}) to dimerize in aqueous media [9]. The extent of the dimerization reaction can be easily estimated by inspecting the visible spectrum of solutions of the cation radicals [10]. We have utilized SnO_2 optically transparent electrode (OTE's) to electrogenerate the cation radicals and record their visible spectra using a rapid-scan spectrophotometer. The results of these experiments are given in the Table 1.

Table 1. The wavelengths of maximum absorption (in nm) measured for electrogenerated $C_{16}V^{+}$ and $C_{18}V^{+}$ in 50 mM NaCl.

Conditions	$C_{16}V^{+}$	$C_{18}V^{+}$
No CD	365,560	365,560
Excess α-CD	395,602	395,602
Excess β-CD	365,560	365,560

As it is clearly seen from the Table the maxima observed in the presence of α-CD are different from those observed in all the other experimental conditions. According to previously published spectral data [10], in the presence of α-CD, the cation radicals do not undergo dimerization; however, either in the absence of CD or in the presence of excess β-CD, the spectral results indicate extensive dimerization for both cation radicals. Furthermore, in the latter cases the violet color characteristic of the cation radical was not observed to extend to the solution, instead it developed in a film at the electrode surface upon electrogeneration. By way of contrast, in the presence of α-CD, filming was not observed, and the blue color extended homogeneously into the solution.

These results on the dimerization of the cation radicals agree very well with the voltammetric results. These indicate the solubilization of the V^+ forms of the surfactant viologens only in the presence of α-CD and, indeed, under these conditions the cation radicals are observed to exist predominantly in the monomeric state. Addition of β-CD does not cause similar effects. For instance, β-CD does not prevent the filming of cation radical salts nor their subsequent dimerization/aggregation. Two lines of argumentation can be advanced to explain this behavior. First, β-CD has a bigger cavity that may at least partially include the viologen and, thus, its solubilizing effects may not be so important as in the case of the chain-including α-CD. Second, β-CD is less soluble in water than α-CD, so that the solubilizing effects of the former are expected to be inherently less important. In conclusion, all of our data in this area suggest that the single most important factor determining the effect of cyclodextrin complexation on the electrochemical and dimerization behavior of surfactant viologens **1** and **2** is the solubility of the viologen-CD complexes. More details on the interactions of cyclodextrins with these viologens have been published elsewhere [6,11].

THE COMPLEXATION OF SURFACTANT DERIVATIVES OF FERROCENE AND CYCLODEXTRINS

We have also assessed the complexation of surfactant derivatives containing the ferrocene subunit with α-CD. In particular we have synthesized and surveyed the following two compounds.

3, n = 7

4, n = 16

The electrochemistry of the ferrocene moiety results from its reversible one-electron oxidation. Thus the cyclic voltammogram of compound **3**, shown in Figure 1a, presents an oxidation peak and a corresponding reduction peak on the reverse scan. The voltammetric behavior is diffusion-controlled as evidenced by the linearity of a plot of anodic peak current vs. square root of scan rate. The potential difference between the two peak potentials is ca. 60 mV, as expected for a reversible couple, and the oxidation potential observed for the ferrocene subunit is consistent with other data published for ferrocene derivatives.

In the presence of 5 mM α-CD, the voltammetric response is essentially the same (Figure 1B). However, the peak currents are substantially decreased indicating a smaller diffusion coefficient (D_0) for the electroactive species. In fact the diffusion coefficient changes from 4.2 x 10^{-6} cm^2/s (no α-CD) down to 2.6 x 10^{-6} cm^2/s (5 mM α-CD). Since the associated changes in viscosity of the solution are too small to explain this decrease in D_0-values, it is suggested that α-CD binds the heptyl group in **3**, essentially in the same fashion as it does with the alkyl tails of the surfactant viologens.

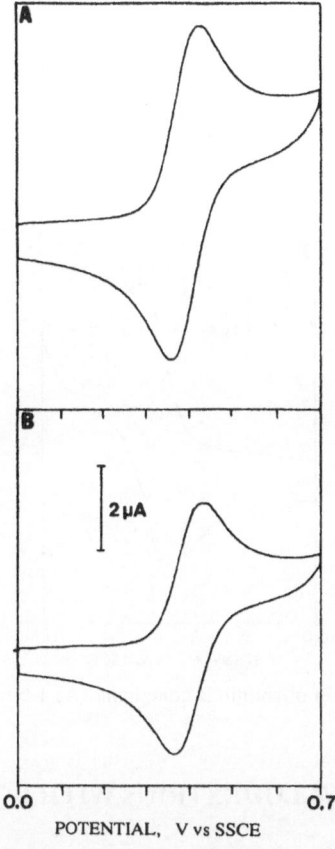

Figure 1. Cyclic voltammograms on Pt of solutions containing (A) 1.0 mM **3** + 50 mM NaCl; (B) same + 5 mM α-CD. Scan rate = 50 mV/s.

Similar experiments with the hexadecyl derivative **4**, indicate also complexation of the alkyl tail by the cyclodextrin. In this case, the voltammetric behavior of **4** is not diffusion controlled (see Figure 2A) probably because of the substantial hydrophobic character of this hexadecyl derivative which appears to favor the adsorption and/or micellization of this compound. However, the addition of a five-fold excess of α-CD to the solution eliminates these complications and the resulting voltammogram (Figure 2B) has all the characteristics corresponding to a reversible, diffusion controlled redox couple.

We are currently conducting experiments to elucidate the possibility of compound **4** forming complexes with more than one cyclodextrin molecule. CPK models indicate that there are no steric impediments to such a structure. In principle the analysis of apparent diffusion coefficient data as a function of added α-CD concentration for compounds **3** and **4** should shed some light into this problem. However, such an analysis is hindered by the complicated voltammetric behavior of compound **4** in the presence of α-CD concentrations below the 5 mM level.

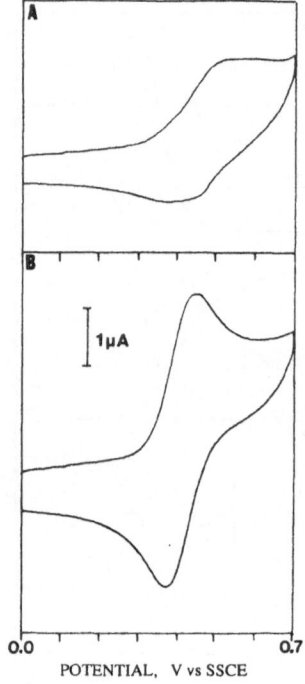

Figure 2. Cyclic voltammograms on Pt of solutions containing (A) 1.0 mM **4** + 50 mM NaCl; (B) same + 5 mM α-CD. Scan rate = 50 mV/s.

THE INTERACTIONS OF CYCLODEXTRINS WITH DOUBLE-CHAIN SURFACTANTS

After exploring the complexation of single-chain surfactants by cyclodextrins it was only natural to carry out similar studies with double-chain surfactants of the type utilized to build vesicles. This is particularly interesting to us because of the structural similarity of the vesicle membranes to the lipidic backbone of biological membranes [12].

For reasons of synthetic accessibility, we decided to start the study with the commercially available surfactants dicetyl phosphate (DCP) and dioctadecyldimethylammonium bromide (DODAB) whose structures are represented below.

The addition of excess α-CD to a well sonicated dispersion of DODAB vesicles in water produces a drastic increase in the turbidity of the dispersion (see Figure 3). This turbidity increase, which is clearly related to the formation of a white precipitate, is not instantaneous; it rather exhibits an induction period. The precipitate was isolated, washed with water, dried under vacuum, dissolved in DMSO-D$_6$, and submitted to NMR analysis.

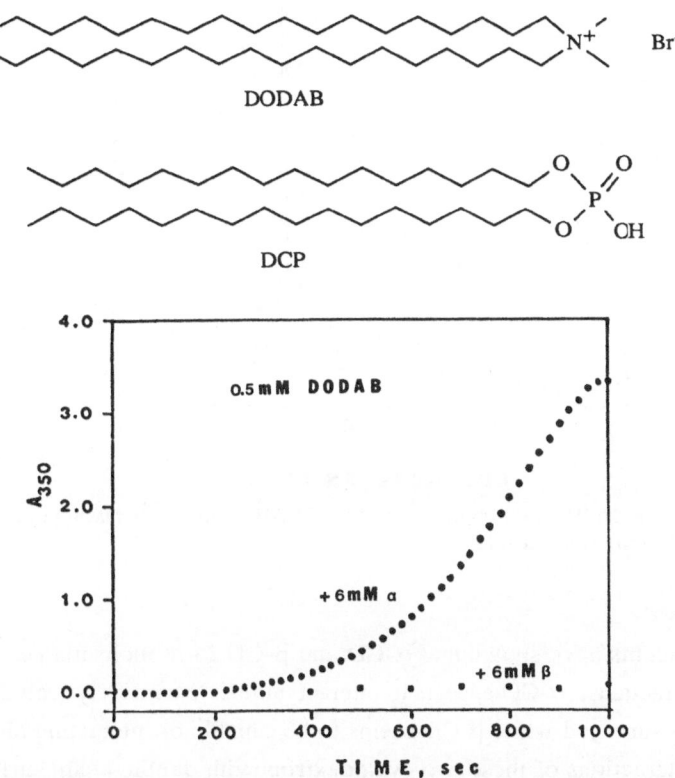

DODAB

DCP

Figure 3. Apparent absorbance (at 350 nm) as a function of time of a sonicated DODAB vesicle dispersion upon addition of cyclodextrins. (•) [DODAB]$_0$ = 0.5 mM, [α-CD]$_0$ = 6.6 mM; ( ) [DODAB]$_0$ = 0.5 mM, [β-CD]$_0$ = 6.6 mM. These concentrations are those prevalent after mixing the vesicle dispersion and the cyclodextrin studies. Optical pathway = 1.0 cm.

The ^{1}H NMR spectrum (Figure 4) shows all the peaks expected for a cyclodextrin-DODAB complex. Integration of the spectrum reveals a stoichiometry of 5 molecules of α-CD per molecule of DODAB. We have also prepared this complex using a procedure which does not involve sonication. In this procedure, a chloroform solution of DODAB is mixed with an aqueous solution of α-CD. The two-phase mixture is heated to evaporate the CHCl$_3$; after the evaporation is complete, the dispersion is allowed to cool down to yield a white precipitate that was found to be identical with that obtained in the vesicle experiments.

Control experiments with tetramethylammonium bromide failed to produce any precipitates upon addition of similar excesses of α-CD. Addition of β-CD to dispersions containing sonicated DODAB vesicles do not cause any turbidity changes in the same time scale (see Figure 4). However, after prolonged reaction times (several hours to several days) small amounts of white precipitates were isolated and found to have a 4:1 β-CD-DODAB stoichiometry. A similar reaction pattern was also exhibited by the other double-chain surfactant, DCP.

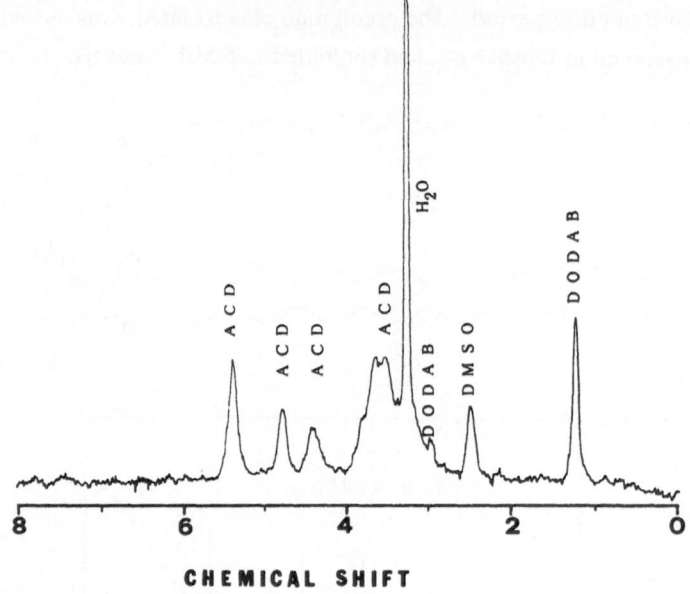

CHEMICAL SHIFT

Figure 4. 60 MHz ^{1}H NMR spectrum of the α-CD-DODAB complex in DMSO-d6. Cyclodextrin peak assignment were taken from reference 13.

CONCLUSION

Our experiments have shown that α-CD and β-CD form inclusion complexes with the alkyl tails of surfactants. α-CD appears to interact almost exclusively with the alkyl chain of the amphiphiles surveyed while β-CD seems to be capable of interacting also with the head groups. The interactions of these two cyclodextrins with double-chain surfactants are very complex because of the complicated aggregation behavior of these bilayer-forming amphiphiles [12]. However, our preliminary results demonstrate that several cyclodextrin hosts bind to the chains of one surfactant molecule. We are currently working on further defining the structure and properties of these complexes and exploring the potential application of these phenomena to drug release systems.

References

1. Szetli, J. *Cyclodextrins and their Inclusion Complexes*; Akademiai Kiado: Budapest, 1982.

2. Bender, M. L.; Koniyama, M. *Cyclodextrin Chemistry*; Springer-Verlag: Berlin, 1978.

3. Saenger, W. *Angew. Chem., Int. Ed. Engl.* **1980**, *19*, 344.

4. Okubo, T.; Kitano, H.; Ise, N. *J. Phys. Chem.* **1976**, *80*, 2661.

5. Satake, I.; Ikenoue, T.; Takeshita, T.; Hayakawa, K.; Maeda, T. *Bull. Chem. Soc. Jap.* **1985**, *58*, 2746.

6. Diaz, A.; Quintela, P. A.; Schuette, J. M.; Kaifer, A. E. *J. Phys. Chem.* **1988**, *92*, 3537.

7. Matsue, T.; Evans, D. H.; Osa, T.; Kobayashi, N. *J. Am. Chem. Soc.* **1985**, *107*, 3411.

8. Kaifer, A. E.; Bard, A. J. *J. Phys. Chem.* **1985**, *89*, 4876.

9. Bird, C. L.; Kuhn, A. T. *Chem. Soc. Rev.* **1981**, *10*, 49.

10. Kosower, E. M.; Cotter, J. L. *J. Am. Chem. Soc.* **1964**, *86*, 5524.

11. Kaifer, A. E.; Quintla, P. A.; Schuette, J. M. *J. Incl. Phenom.* **1989**, 7, 107.

12. Fendler, J. H. *Membrane Mimetic Chemistry*; Wiley: New York, 1982.

13. Casu, B.; Reggiani, M.; Gallo, G. G.; Vigevani, A. *Tetrahedron* **1966**, *22*, 3061.

BIOMEDICAL USES OF AMORPHOUS WATER-SOLUBLE DERIVATIVES OF CYCLODEXTRINS

Josef Pitha

National Institutes of Health, NIA/GRC
Baltimore, Maryland 21224 (USA)

Inclusion complexes based on cyclodextrins have recently become newsworthy and fashionable. The phenomenon itself is not exactly of recent origin: complexes of this type are only a slight modification of inclusion complexes formed by starch, a natural mode of storage in which plants store "guest" fatty acids in "host" carbohydrate, i.e. amylose, which is the main component of starch.

From a chemical point of view amylose is chain polymer glucose linked by $\alpha(1 \rightarrow 4)$ bonds and in an ordered state these chains are folded into helixes with about half a dozen glucose units per turn. Helix amylose has a peculiar arrangement in which all hydroxyls of glucoses are pointed outward and thus the interior of the helix is non-polar. Furthermore, the interior of the helix is roomy enough to accommodate hydrophobic guests of which free fatty acids are a natural example. Cyclodextrins are made from amylose by the enzymes cyclodextrin glycosyltransferases (EC 2.4.1.19) which detach one of the turns from the amylose helix (six, seven, or eight glucose units) and connect these into a circle forming α-, β-, or γ-cyclodextrin, respectively (Figure 1). These have hydroxyls arranged similarly to amylose and thus continue to be able to serve as host compounds for lipophiles.

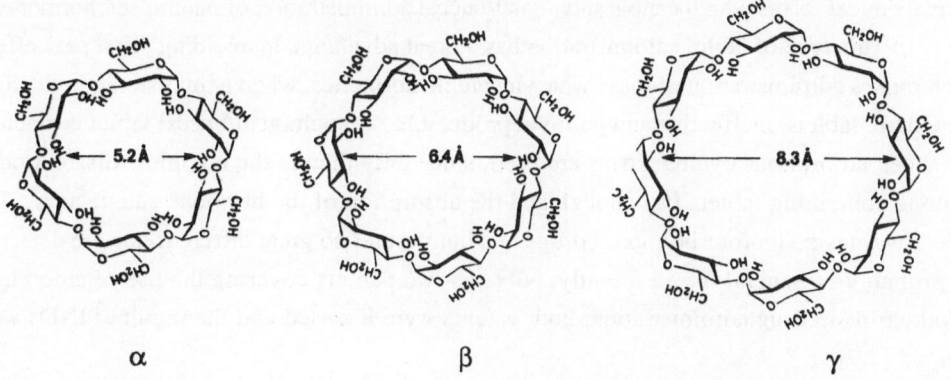

Figure 1. Structures of α-, β-, and γ-cyclodextrins. In hydroxypropyl-cyclodextrins some of the -OH groups of the parent compounds are converted to -O-CH$_2$-CHOH-CH$_3$ groups.

Inclusion Phenomena and Molecular Recognition
Edited by J. Atwood
Plenum Press, New York, 1990

Non-polar compounds may find a refuge from the hostility of water in the hydrophobic cavity of cyclodextrins, which compounds are again hydrophilic enough to get along well with the surrounding water. Since that refuge is without doors, entry or exit is a fast process. In addition to the obvious condition of the proper size (Figure 1) there is a requirement for the guest to have low polarity but otherwise there are no restrictions.

For a great number of applications the formation of solid complexes (cyclodextrin-guest) is all that is needed. Thus, a majority of the presently produced β-cyclodextrin (about 1000 tons per year) ends up as food flavor stabilizers. For example, when a vanillin:β-cyclodextrin complex is used in place of vanillin in instant pudding mixes, the loss of flavor through evaporation on the shelves of supermarkets is much slower. Pharmaceutical and pharmacological uses of cyclodextrins impose other requirements: they center mainly on dissolution of non-polar drugs and lipids, and the most critical requirement is the lack of toxic effects. Cyclodextrins are crystalline and thus have limited water solubility: α, 11.4%; β, 1.84%; γ, 20.4% (all data W/W at 25°C) [1] and complexes of cyclodextrins with lipophiles or drugs have invariably much lower solubility than the respective cyclodextrins themselves. It is obvious that to use as a solubilizer a compound which itself has quite limited water solubility is rather a futile endeavor. Furthermore, there is a problem with toxicity. While in oral administration no untoward effects were observed for any of the cyclodextrins, the most common, β-cyclodextrin, when administered intravenously, has an LD_{50} of about 250 mg/kg and harms the kidneys, probably by crystallizing there during the re-absorption process.

Fortunately, the problem with solubility and crystallinity can be remedied easily: non-selective substitution of cyclodextrins yields a mixture of derivatives and crystallinity is gone forever. Of course, the matter is not simple: the chemical modification must give a mixture of cyclodextrin derivatives which (a) still have the ability to form host-guest complexes, (b) are non-toxic, and (c) can be characterized to the satisfaction of the FDA. Eventually, in our work we settled on the condensation products of cyclodextrins with propylene oxide, the so-called 2-hydroxypropylcyclodextrins [2-6]. These compounds are easy to prepare and can be fully characterized by mass spectrometry [4-6].

Hydroxypropylcyclodextrins were used in two projects. In the first, we developed pharmaceutical forms which enable sublingual/buccal administration of natural sex hormones to man [7]. This type of tablet administration has a great advantage in avoiding "first pass effect" which makes administration of these water-insoluble hormones, when administered in the form of standard tablets, ineffective and poorly reproducible. A sublingual/buccal tablet is practical only when amorphous cyclodextrins are used, since only then is the complete dissolution of hormone containing tablets fast enough and the absorption of the hormone satisfactory. The process was tested on four hormones/drugs in a human and no great differences were detected; it is probably of general use. Presently, NIH has two patents covering the use of amorphous cyclodextrins in drug administration; both patents were licensed and the required INDs were obtained [8,9].

The second project involved intravenous administration of hydroxypropylcyclodextrin in order to help lipophilic deposits in an organism re-enter the circulation [10]. This re-entry eventually results in re-distribution, increased metabolism, and excretion of the lipophile from the body. The process was eventually used in a critically ill patient with familial hypervitaminosis A and levels of vitamin A and cholesterol were followed [11,12]. All required approvals for the treatment were obtained (IND approval by FDA, spring 1985). Altogether 30 g of hydroxypropyl-β-cyclodextrin was administered to the 20 kg patient over three days; treatment was given in May 1985 and the patient is alive. This is a dose of about 0.5 g/kg/day and that makes the importance of chemical modification obvious (intravenous LD_{50} of the natural β-cyclodextrin is 0.2 g/kg in male mice).

Andy Warhol declared that in the art world fame is a fifteen minute phenomenon. Biomedical research cannot be expected to have a slower pace than the arts and soon hydroxypropylcyclodextrins will be replaced by better compounds. Attempts to predict which compounds these will be must be left to dreamers: an experimentalist can only compare hydroxypropylcyclodextrins with compounds which were phased out and try to determine the differences between these; such a comparison follows.

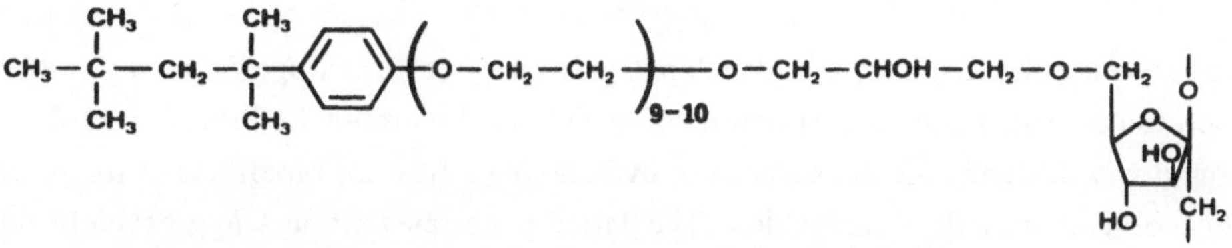

Figure 2. Structure of Triton X-100: inulin conjugate.

The first new solubilizer synthesized and evaluated in the author's laboratory was a chemical conjugate of the detergent Triton X-100 with a polysaccharide inulin (Figure 2) [13]. Typical detergents, like Triton X-100, interact with biomembranes by first entering the lipid bilayer; this process cannot be effectively regulated and toxic effects often occur. This is due primarily to the presence of micelles. Detergents, when used in meaningful concentrations, contain a micellar phase and thus Triton X-100 enters the cell as micelles which contain close to hundreds of molecules of detergent rather than entering as individual molecules. The above Triton X-100:inulin conjugate was designed and synthesized to fulfill the following criteria: (a) to prevent tendency of the detergent to form micelles and thus to enable the delivery of solubilizer gradually into cells and prevent their unrestricted distribution there; (b) to obtain a colorless preparation free of impurity. The conjugate indeed did not form micellar aggregates and acted upon biomembranes with the expected mildness-affecting only a partial solubilization of membrane bound proteins. Possibly this design was abandoned prematurely, but its disadvantage became quite obvious. The compound carries the stigma of a polydisperse macromolecule and is thus unacceptable to many researchers. Biochemists use detergents to solubilize and purify membrane proteins and thus introduction of another macromolecule into the system may create complications. There is some prejudice against macromolecules even in pharmaceutical applications; in that field it is necessary to characterize the preparation fully and this is invariably difficult and costly with any polydisperse macromolecular system.

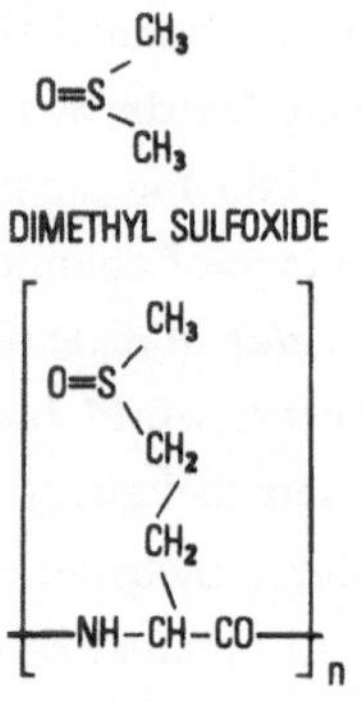

Figure 3. Structure of polymethionine sulfoxide.

The second solubilizing system developed was a polymeric analog of dimethyl sulfoxide, polymethionine sulfoxide, the structure of which is illustrated in Figure 3 [14]. This compound was designed for the purpose of overcoming untoward bioeffects of the powerful organic solvent, dimethyl sulfoxide. The latter penetrates tissues aggressively and is metabolized there with rapidity. Furthermore, dimethyl sulfoxide has somewhat mysterious effects on mammalian cells: some start to differentiate (e.g. erythroleukemic cells grown *in vitro* start to synthesize hemoglobin), whereas other cells fail in their expected differentiation in the presence of this solvent. The design addressed the above defects correctly: polymethionine sulfoxide does not have the above mentioned mysterious effects on cells, whereas it is a quite potent solubilizer of some drugs (Table 1). Unfortunately, again polymethionine sulfoxide is a polydisperse macromolecule and that lowers its general acceptance.

Table 1. Solubilization of lipophiles (μg/mL) by solubilizing agents (5%, W/W).

	no solubilizer	poly-L-methionine sulfoxide	2-hydroxypropyl-digitonin	2,6-di-O-methyl β-cyclodextrin
β-ionone	7.3	1400	2050	
retinol	<4	12	1600	330
chole-calciferol	<0.2	140	190	420

The third solubilizing system was based on digitonin (Figure 4). This saponin is a by-product of the isolation of cardiac glycosides from foxglove. Presently, digitonin is widely used by biochemists for the solubilization of functional hormonal receptors from membrane preparations, but its use is complicated by low solubility in water. We modified a commercial digitonin by reaction with propylene oxide, converting it into a mixture of hydroxypropyldigitonins. These derivatives dissolve easily and clearly in water and have reasonable solubilizing activity (Table 1). 2-Hydroxypropyldigitonin is indeed very mild in its effects on tissues: that is documented by its use to eliminate mycoplasma infection in cell culture. The mildness of this detergent enabled the finding of a concentration which is fatal for the small mycoplasmas while the large mammalian cells survive [15].

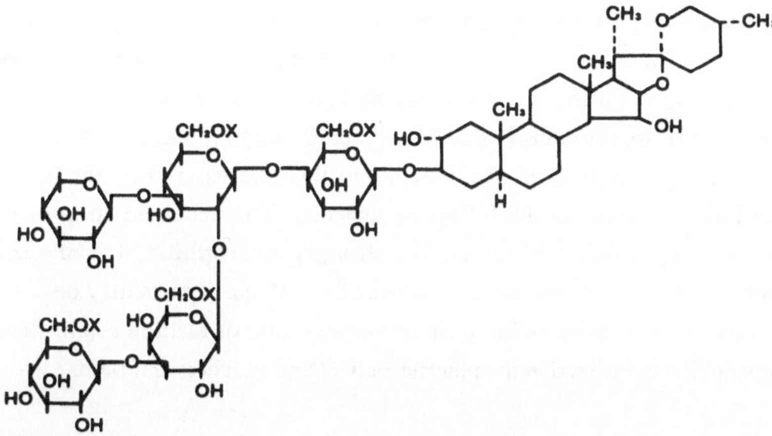

Figure 4. Structure of digitonin. The reaction with propylene oxide converts some of the -OX groups into -O-CH$_2$-CHOH-CH$_3$ groups.

A solubilizing system was also developed for retinoids which uses underivatized cyclodextrin as the host. In that system crystallization is prevented by attachment of a guest molecule, retinal, to an amorphous polysaccharide [16] (Figure 5). Upon bioevaluation, which in this case was the inhibition of growth of cancerous cells *in vitro*, the complexed drug was indeed different from the uncomplexed drug. The complexation of the drug slightly decreased

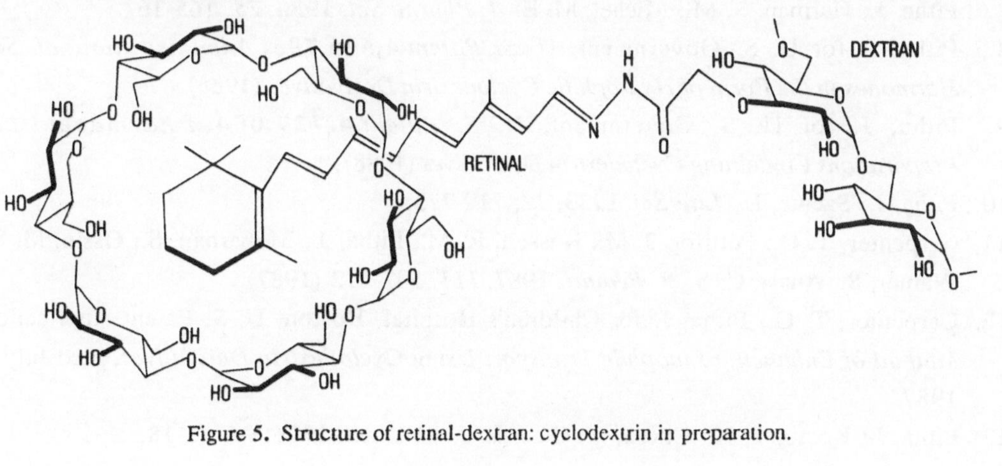

Figure 5. Structure of retinal-dextran: cyclodextrin in preparation.

its activity, but fortunately its toxicity was decreased considerably. Thus, the compound could be used at much higher concentrations and higher inhibition of growth of malignant cells could be obtained without lysis of cells and general toxicity. This system seems to have potential; nevertheless, it also carries the stigma of a macromolecular character.

Cyclodextrins may be derivatized in enumerable ways and the bioeffects may vary considerably. Thus, 2,6-dimethyl-β-cyclodextrin was used both in an attempt to improve oral absorption of testosterone [7] and, after intraperitoneal administration, for rescue from poisoning by retinoic acid [10], but in both of these applications much better effects could be obtained with the more polar 2-hydroxypropyl-β-cyclodextrin.

Each and every one of the systems described above had enough merit to deserve publication or patent protection. Nevertheless, not one of these systems raised enough interest to help it get the extended use which amorphous derivatives of cyclodextrins enjoy. Research itself and acceptance of its products is indeed subject to fashion; taking into account only the final results (i.e., factors of fashion are included in this analysis) amorphous cyclodextrins differ from the failed systems in the following aspects. The accepted solubilizers are: (a) amorphous solids very soluble in water, (b) strongly hydrophilic, (c) of relatively low molecular weight and have no tendency to form micelles. These aspects may be defects in other uses; nevertheless, a combination of the technical aspects and of fashion established the above preferences for solubilzers utilized in the pharmaceutical and biochemical fields.

References

1. Jozwiakowski, M. J.; Connors, K. A. *Carbohydr. Res.* **1985**, *143*, 51-59.
2. Pitha, J.; Pitha, J. *J. Pharm. Sci.* **1985**,*74*, 987-990.
3. Pitha, J. *J. Incl. Phenom.* **1984** 2, 477-485.
4. Pitha, J.; Milecki, J.; Fales, H.; Pannell, L.; Uekama, K. *Int. J. Pharmaceut.* **1986**, *29*, 73-82.
5. Pitha, J.; Szabo, L.; Fales, H. M. *Carbohydr. Res.* **1987**, *168*, 191-198.
6. Irie, T.; Fukunaga, K.; Yoshida, A.; Uekama, K.; Fales, H. M.; Pitha, J. *Pharm. Res.*, in press.
7. Pitha, J.; Harman, S. M.; Michel, M. E. *J. Pharm. Sci.* **1986**, *75*, 165-167.
8. Pitha, J. for U. S. Government, *U. S. Patent* **4,596,795**, *Administration of Sex Hormones in the Form of Hydrophilic Cyclodextrin Derivatives* (1986).
9. Pitha, J. for U. S. Government, *U. S. Patent* **4,727,064**, *Pharmaceutical Preparations Containing Cyclodextrin Derivatives* (1988).
10. Pitha, J.; Szente, L. *Life Sci.* **1983**, *32*, 719-723.
11. Carpenter, T. O.; Pettifor, J. M.; Russell, R. M.; Pitha, J.; Mobarhan, S.; Ossip, M. S.; Wainer, S.; Anast, C. S. *J. Pediatr.* **1987**, *111*, 507-512 (1987).
12. Carpenter, T. O.; Pitha, J. for Children's Hospital, Boston, U. S. Patent Application: *Method of Enhancing Lipophile Transport Using Cyclodextrin Derivatives*, filed July 1, 1987.
13. Pitha, J.; Kociolek, K.; Caron, M. G. *Eur. J. Biochem.* **1979**, *94*, 11-18.

14. Pitha, J.; Szente, L.; Greenberg, J. *J. Pharm. Sci.* **1983**, *72*, 665-668.

15. Proust, J. J.; Buchholz, M. A.; Nordin, A. A. *J. Immunol.* **1985**, *134*, 390-396.

16. Pitha, J.; Zawadzki, S.; Chytil, F.; Lotan, D.; lotan, R. *J. Natl. Cancer Inst.* **1980**, *65*, 1011-1015.

INCLUSION AND REACTION CHEMISTRY OF TWO
ANTHRACENE-APPENDED γ-CYCLODEXTRINS

Akihiko Ueno*, Fumio Moriwaki, Akiko Azuma, and Tetsuo Osa

Pharmaceutical Institute
Tohoku University
Aobayama, Sendai 980 (Japan)

SUMMARY

The anthracene moieties of regioisomers of 6A,6X-bis(anthracene-9-carbonyl)-γ-cyclodextrins undergo photodimerization, affording a *trans* photodimer for AC, AD, and AE isomers and a *cis* photodimer for the AB isomer. The photodimers of the AD and AE isomers were stable, but those of AB and AC were unstable, returning to the original anthracene monomers. The dissociation of the photodimers was suggested to be due to the inherent instability of the *cis* photodimer of the AB isomer and the strain-rich nature of the *trans* photodimer of the AC isomer. The photodimerization followed by hydrolysis of 6A,6X-bis-(anthracene-1-carbonyl)-γ-cyclodextrins afforded four photodimers of 1-anthracenecarboxylic acid, the relative yields of the photodimers being controlled by the locations of the anthracene moieties in the regioisomers. These results demonstrate that the γ-cyclodextrin template method is useful to make chemical reactions proceed stereoselectively as well as to produce new photochromic systems.

INTRODUCTION

α-, β-, and γ-cyclodextrins are composed of six, seven, and eight glucose units, respectively. They have a hydrophobic cavity, in which guest molecules can be included. The smaller cavities of α- and β-cyclodextrins are suitable for the accommodation of one molecule of benzene and naphthalene, respectively. However, the cavity of γ-cyclodextrin is too large to accommodate one molecule of these guests. In 1980, we found that γ-cyclodextrin can include 2 molecules of naphthalene derivatives [1]. This finding suggests that γ-cyclodextrin can be used as a molecular flask or container, in which two identical (A) or different (B) species

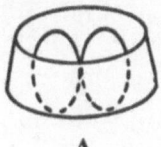

A

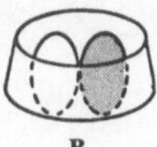

B

gather together to achieve specific interactions or reactions. On this basis, γ-cyclodextrin has been used to promote excimer formation, charge-transfer complex formation, and photodimerization of anthracene [2].

As an extension to this work, we prepared γ-cyclodextrin derivatives with one and two naphthyl moieties at the primary hydroxyl group side of the γ-cyclodextrin. When one naphthalene is attached to γ-cyclodextrin, the naphthyl moiety has been shown to enter the cavity together with a guest molecule and acts as a spacer (Equation 1), which narrows the large cyclodextrin cavity [3]. Generally, cyclodextrins have been considered to be rigid hosts with pre-determined cavity sizes. However, this induced-fit type of complex formation indicates that γ-cyclodextrin is converted into a flexible host, in which cavity size can be regulated by changing the bulkiness of the spacer moiety. On the other hand, two naphthalene moieties attached to γ-cyclodextrin have been shown to change their positions from inside to outside the cavity upon guest binding (Equation 2) [4,5]. We now wish to report the reaction and inclusion chemistry of γ-cyclodextrin derivatives, which have two anthracene moieties in place of the naphthalene.

(1)

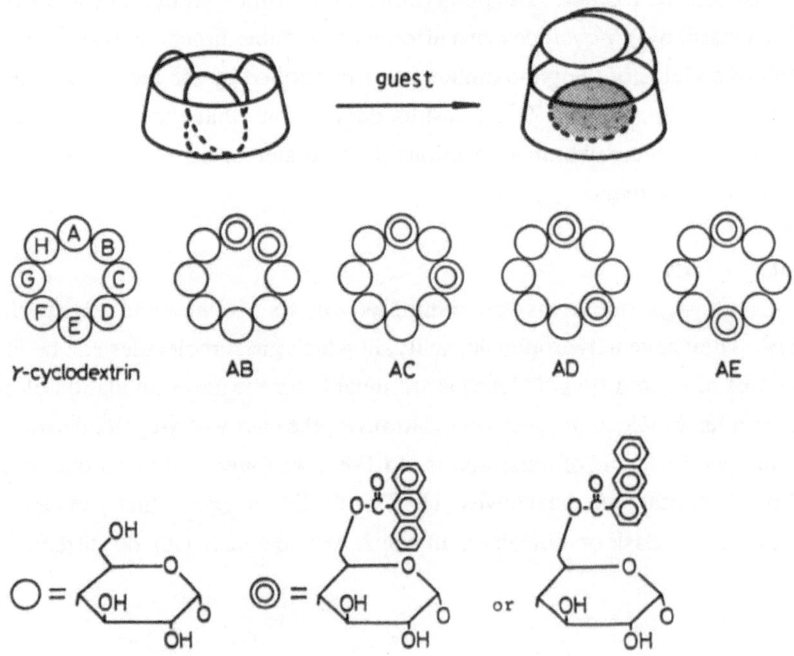

(2)

270

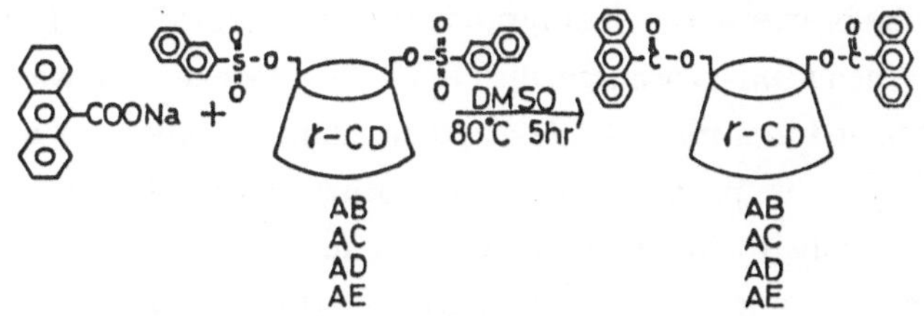

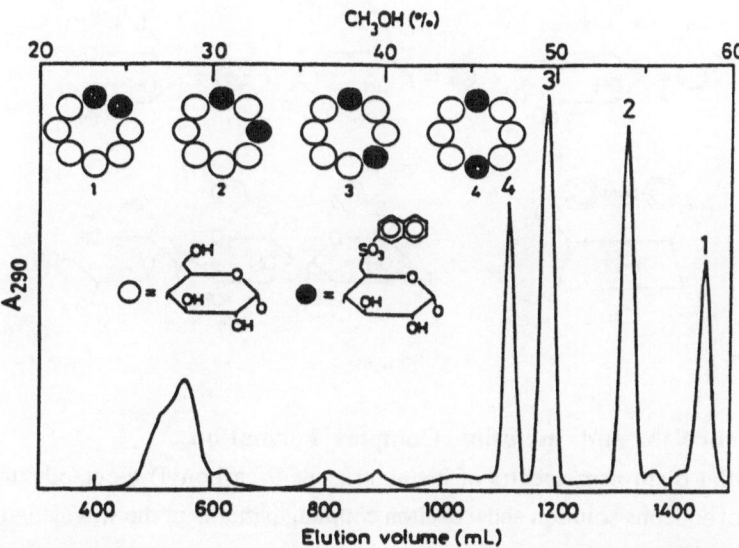

Scheme 1

INCLUSION PHENOMENA AND ANTHRACENE PHOTODIMERIZATION
OF 6A,6X-BIS-(ANTHRACENE-9-CARBONYL)-γ-CYCLODEXTRINS
Synthesis and Characterization

Designating the glucose units of γ-cyclodextrin as A, B, C, D, E, F, G, and H, we attached two 9-anthracenecarboxylate units to AB, AC, AD and AE glucose units. The compounds were prepared by the reactions of 6A,6X-bis(2-naphthalenesulfonyl)-γ-cyclodextrins (**1-4**) with sodium 9-anthracenecarboxylate in DMSO (Scheme 1). The regioisomers **1-4** were previously separated and isolated using HPLC (Figure 1) and they were characterized by reaction of each isomer with two equivalents of sodium thiophenolate in two-step reactions (Scheme 2). As can be seen from the Scheme, two mono-substituted intermediates should be formed by isomers AB, AC and AD, but only the one intermediate should arise from the AE isomer.

Figure 1. Separation of **1-4** by reversed phase HPLC. A linear gradient elution of aqueous 10% MeOH - aqueous 60% MeOH was applied.

Consequently, the isomer **4**, which only produced one intermediate on HPLC, was assigned as the AE isomer. When *trans*-azobenzenedisulfonyl-γ-cyclodextrin **16** was used in place of **4**, **14** and **15** were obtained in a ratio of 6:94. Examination of molecular models showed that only AE and AD isomers are possible from this azobenzene derivative, so **3** could be assigned as the AD isomer. Finally, hydrolysis of **12** by Taka-amylase produced the fragment **17**, so **1** was assigned as the AB isomer. As a result, **2** could be assigned as the AC isomer.

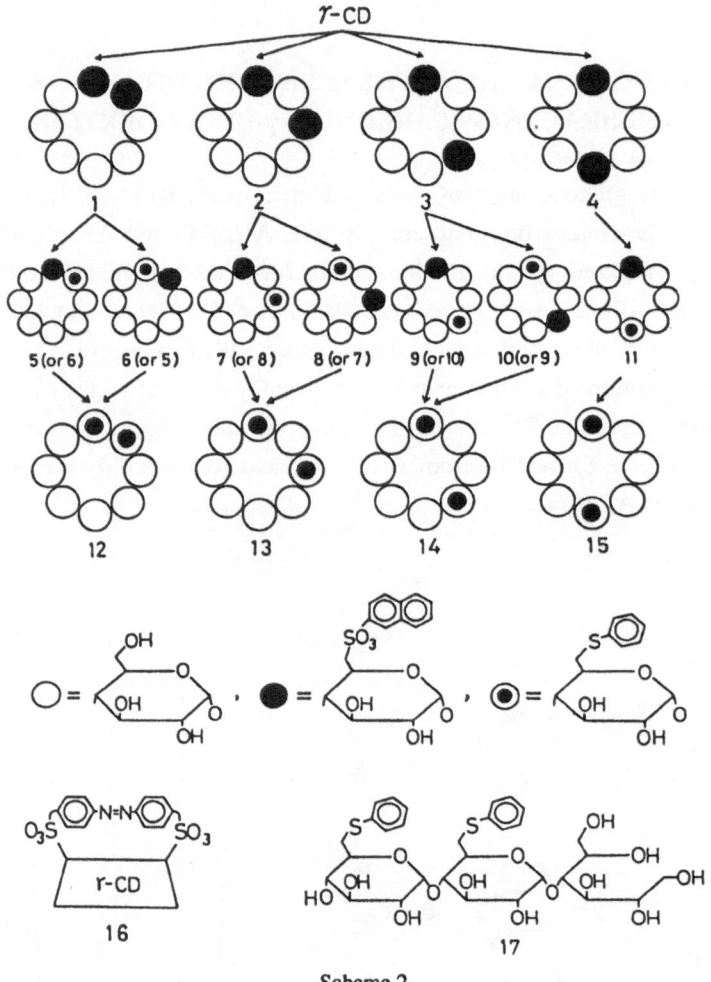

Scheme 2

Circular Dichroism and Inclusion Complex Formation

The circular dichroism spectra of bis(anthracene-9-carbonyl)-γ-cyclodextrins in a 10% ethylene glycol aqueous solution show exciton coupling patterns in the wavelength region from 220 to 300 nm. The positive and negative signs from longer wavelength regions for the AB, AC, and AE isomers indicate that the two anthracenes are oriented clockwise in the γ-cyclodextrin cavities of these substances. The pattern of the AD isomer is opposite in sign, indicating that the two anthracenes are oriented counterclockwise. The circular dichroism intensities of all of these isomers changed upon guest addition. This result indicates that the location of the anthracene moieties changes associated with guest binding in a similar fashion to that observed for two naphthalene-appended γ-cyclodextrins (Equation 2). The analysis of the circular dichroism variations upon complexation with 1-borneol gave host-guest association constants of 3400, 12000, 2500, and 1400 M^{-1} for the AB, AC, AD, and AE isomers, respectively. The value for the AC isomer is much larger than the values for the other isomers. This suggests that the intramolecular complexes of the AB, AD, and AE isomers are too stable to be converted into the intermolecular complexes, while the intramolecular complex of the AC isomer is not stable because of the difficulty in making two anthracenes arrange themselves in a parallel fashion.

Photodimerization of Anthracene in the Regioisomers

Photodimerization of anthracene proceeded in the regioisomers. Absorption spectra before and after irradiation are shown in Figure 2 together with the spectra taken during the course of recovery, which was either thermal (for the AB and AC isomers) or due to 260 nm irradiation (for the AD and AE isomers). It is interesting that photodimers of the AB and AC isomers are unstable, making the systems photochromic, while the photodimers of the AD and AE isomers are thermally stable. Table 1 shows the half-lives of the photodimers. The half-life of the photodimer of the AB isomer in a 10% ethylene glycol aqueous solution is 12.5 min and in methanol is much longer, with a value of 70.5 min. In comparison with the AB isomer, the photodimer of the AC isomer had longer half-lives, which are hardly affected by solvent. On the other hand, the photodimers of the AD and AE isomers are stable. Examination of molecular models suggests that the photodimer of the AB isomer is *cis*, whereas the photo-dimers of the other isomers are *trans*. Generally, anthracene derivatives with a substituent at

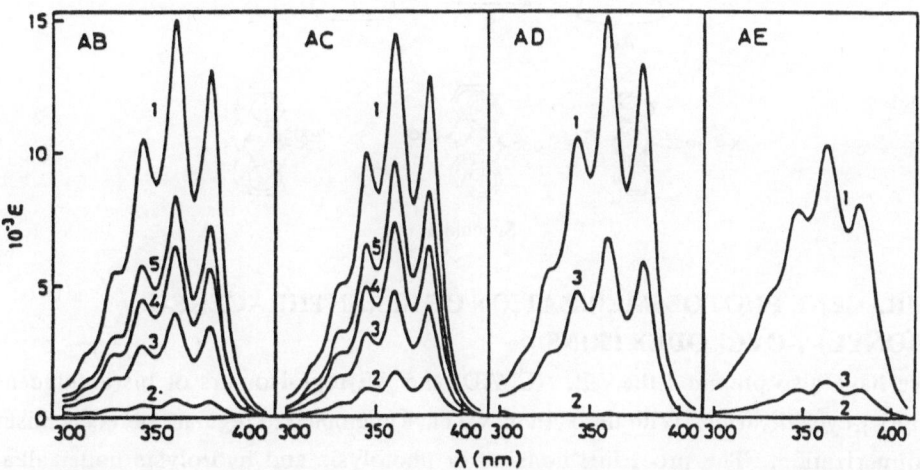

Figure 2. Absorption spectra of AB, AC, AD, and AE isomers in MeOH before (1) and after (2) irradiation (λ>300 nm). The spectra obtained by dissociation of the photodimers, which occurs thermally (AB: 3, 30 min; 4, 60 min; 5, 90 min.; AC: 3, 80 min; 4, 170 min; 5, 290 min) or photochemically (260 nm irradiation, 3 for AD and AE isomers) are also shown.

Table 1. Stereochemistry and half-lives of anthracene photodimers of bis(anthracene-9-carbonyl)-γ- cyclodextrins at 25°C[a].

Regioisomer	Solvent	Photodimer	Half-life (min)
AB	10% EG[b]	cis	12.5
AB	MeOH	cis	70.5
AC	10% EG	trans	275
AC	MeOH	trans	206
AD	10% EG,MeOH	trans	stable
AE	MeOH	trans	stable

[a]The concentration of isomers was 2.5 x 10^{-5} M.
[b]10% EG: 10% ethylene glycol aqueous solution.

carbon-9 only give *trans*-photodimers, and the *cis*-photodimer is not formed, probably due to steric repulsion between substituents. Therefore, the results obtained here are unique in that the *cis*-photodimer is formed for the AB isomer while the *trans*-photodimer of the AC isomer is unstable. The stereochemistry and reaction behavior of the AE and AB isomers are shown in Scheme 3. The AB isomer shows photochromic behavior, that is, both photodimerization and dissociation occur reversibly. The AC isomer also shows similar photochromic behavior in spite of its *trans* form probably because of the strain existing in the structure of its photodimer.

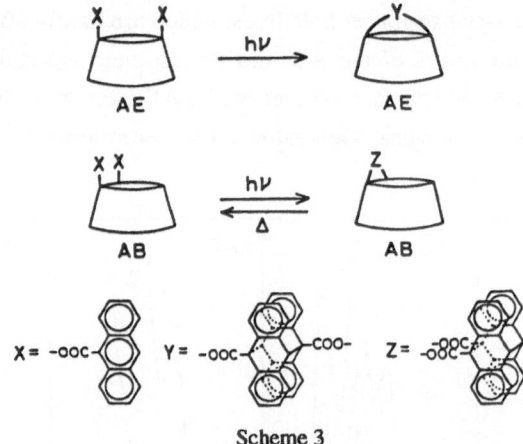

Scheme 3

ANTHRACENE PHOTODIMERIZATION OF BIS(ANTHRACENE-1-CARBONYL)-γ-CYCLODEXTRINS

We have also prepared the AB, AC, AD, and AE regioisomers of bis(anthracene-1-carbonyl)-γ-cyclodextrins. With this system, we have attempted to regulate stereochemistry in photodimerization. The procedure consists of photolysis and hydrolysis under alkaline conditions to cut out the photodimers (Equation 3).

Methanol was used as the solvent for photodimerization. The structure of the possible photodimers of 1-anthracenecarboxylic acid are the *anti*-head-tail, *anti*-head-head, *syn*-head-

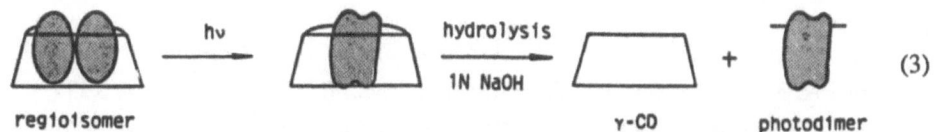

regioisomer   hυ   hydrolysis 1N NaOH   γ-CD + photodimer   (3)

head, and *syn*-head-tail. As the reference experiment, photodimerization of 1-anthracene-carboxylic acid was undertaken. The HPLC analysis gave four peaks **18-21**, which are numbered in the order of elution, with an area ratio of 48, 28, 4, 20% for **18**, **19**, **20**, and **21**, respectively. There was also a peak corresponding to a small amount of unreacted 1-anthracenecarboxylic acid. Although the characterization of the photodimers has not yet been completed, we have tentatively assigned these substances as *anti*-head-tail (**18**), *anti*-head-head (**19**), *syn*-head-head (**20**), and *syn*-head-tail (**21**). Table 2 shows the distribution of the photodimers. The product ratio depends markedly on the positions of the anthracene moieties in the γ-cyclodextrin framework. As expected, the AB and AC isomers gave preferentially *syn* photodimers while the AD and AE isomers gave mainly *anti* photodimers. These results suggest that the γ-cyclodextrin template method can be used to regulate the stereochemistry of organic reactions. Since in water, anthracene moieties are considered to be more tightly involved in the γ-cyclodextrin cavity, we hope that better stereoselectivity will be obtained.

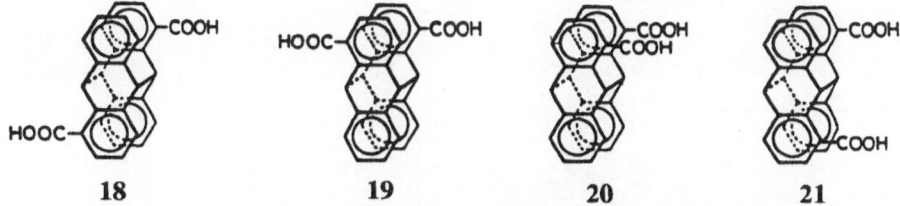

18   **19**   **20**   **21**

Table 2. Relative yields of photodimers of 1-anthracenecarboxylic acid.

	relative yield of photodimer (%)			
	18	**19**	**20**	**21**
1-ACA[a]	48	28	4	20
AB	0	2	66	32
AC	1	4	45	50
AD	10	71	4	15
AE	42	55	1	2

[a]1-anthracenecarboxylic acid.

CONCLUSION

Stereospecific photodimerization is well known to occur in crystals of some photoreactive monomers [6]. The present study has shown that similar stereochemical control in photo-dimerization can be attained in solution when aided by molecular templates such as γ-cyclodextrin. This new synthetic method might be widely applicable for the regulation of product stereochemistry in reactions where two species form new covalent bonds.

References

1. Ueno, A.; Takahashi, K.; Osa, T. *J. Chem. Soc., Chem. Commun.* **1980**, 921.
2. Ueno, A; Moriwaki, F.; Osa, T.; Hamada, F.; Murai, K. *Tetrahedron* **1987**, *43*, 1571.
3. Ueno, A.; Tomita, Y.; Osa, T. *J. Chem. Soc., Chem. Commun.* **1983**, 976.
4. Ueno, A; Moriwaki, F.; Osa, T.; Hamada, F.; Murai, K. *Bull. Chem. Soc. Jpn.* **1986**, *59*, 465.
5. Ueno, A; Moriwaki, F.; Osa, T.; Hamada, F.; Murai, K. *J. Am. Chem. Soc.* **1988**, *110*, 4323.
6. Ramamurthy, V.; Venkatesan, K. *Chem. Rev.* **1987**, *87*, 433.

COMPUTER SIMULATIONS OF HOST-GUEST COMPLEXES

A. K. Cheetham* and B. K. Peterson

University of Oxford
Chemical Crystallography Laboratory
9 Parks Road
Oxford, OX1 3PD (UK)

SUMMARY

The use of computer simulation procedures to model the behaviour of hydrocarbons and similar guest species in zeolites and other microporous materials is discussed. The prediction of the location of solute molecules, and the estimation of heats of adsorption and activation energies for diffusion by molecular mechanics (MM) procedures, are described. In addition, the use of Monte Carlo (MC) and molecular dynamics (MD) techniques to study sorbates at high loadings is reviewed, together with the calculation of diffusion coefficients by MD.

INTRODUCTION

Host-guest complexes are found widely in both inorganic and organic materials. Zeolites, for example, and other microporous silicates such as the clays and clathrasils, can play host to an extraordinary range of guest molecules, from monatomic gases to complex coordination compounds, and there are striking parallels between these systems and their organic counterparts. This paper describes the use of computer simulations to predict the behaviour of such complexes, focusing primarily on zeolite chemistry [1]. Previous reviews in this general area include a survey of the uses of molecular graphics [2], and a recent compilation of theoretical work on zeolites [3].

Simple force-field calculations based upon the use of atom-atom potentials [4,5,6] can be used to predict the location of a guest molecule in, say, a zeolite or clathrate. How reliable are these predictions? We shall examine the agreement between the computer simulations and the results of careful X-ray and neutron diffraction studies, and suggest ways in which the computations can be improved. Examples will include aromatic hydrocarbons adsorbed in zeolite cavities and alkyl ammonium ions in clays.

The behavior of sorbate molecules as a function of temperature and loading is also of interest. To what extent can we predict the distribution of adsorbed species at elevated temperatures, and, for hosts that permit molecular diffusion, their activation energies and diffusion coefficients? Monte Carlo (MC) methods offer a powerful means of probing the

Inclusion Phenomena and Molecular Recognition
Edited by J. Atwood
Plenum Press, New York, 1990

behaviour of molecules as a function of temperature and can be used to estimate heats of adsorption both at infinite dilution and at high sorbate concentrations, but in order to add the *time* dimension that is necessary for the prediction of diffusion coefficients, we must use molecular dynamics (MD). The scope and limitations of these approaches will be discussed with reference to some recent work on simple hydrocarbons in sodium zeolite Y.

Our long-term objective is to refine the calculations to the point where they can be used to simulate *reactivity* in inclusion compounds. This will require a combination of MC or MD methods with molecular orbital calculations, in order that we can assess the effect of the host lattice upon the energetics of a reaction. The quantum dynamics procedure of Car and Parrinello [7] may offer a means of tackling this daunting problem. For the moment, however, we shall simply describe some of the quantum mechanical work that is currently being carried out on host-guest complexes.

LOCATION OF ADSORBED MOLECULES

Perhaps the simplest objective that we can address is the prediction of the siting of an adsorbed molecule in a zeolite cavity at *low* temperatures. Under these circumstances, entropy effects can be ignored and we need only locate the global minimum for the interaction energy between the host and the guest. Several approximations are made:

(i)   the zeolite is assumed to be rigid and unperturbed by the presence of the solute molecule.

(ii) the hydrocarbon is assumed to be rigid and present in very low concentrations.

(iii) the interaction is described by a simple atom-atom potential, using, for example, the parameterisations (for different Si/Al ratios of the zeolite) of Kiselev [6].

(iv) The parameterisations are assumed to be transferable from one zeolite to another.

We have used a potential of the form:

$$\phi(\text{tot}) = \Sigma_{ij} \left[\frac{B}{r^{12}} - \frac{A}{r^6} + \frac{cq_iq_j}{r} \right] \qquad (1)$$

Values of A and B were determined semi-empirically [6] by fitting experimental data for the heat of adsorption of methane in zeolite Y (one of the faujasite family) and the charges for the electrostatic term were obtained from molecular orbital calculations on both the hydrocarbons and fragments of the zeolites.

Whereas, at ambient temperatures, the molecules are likely to be distributed over several sites, their observed positions at low temperatures should provide a useful test of the computer simulations since they should correspond to the positions of $\phi(\text{min})$. There have been virtually no single crystal diffraction studies of zeolite-guest complexes, but several recent studies by high resolution powder neutron diffraction have been reported. These include the location of Xe in zeolite rho [8], CO in zeolite A [9], benzene in zeolite Y [10], and pyridine in zeolite L [11]. In the latter instance, the molecule occupies a single site in the main channel, at

4K, coordinated to a potassium ion through the nitrogen of the pyridine ring. The molecule is thus able to form an acid-base complex with the cation whilst benefiting from a non-bonding interaction with the cavity wall.

Evaluation of the global minimum predicts the location and orientation of the molecule within 0.2 Å of the observed position, thus lending credence to the validity of both the experimental result and the simulation. The principal discrepancy is to be found in the N-K distance, which is estimated to be slightly longer than the observed value. This may stem from the neglect of any covalent contribution to the acid-base interaction.

In another recent comparison between experiment and theory, the location of anilinium ions in the interlamellar region of the clay, vermiculite, was explored. This is a more challenging problem, since the guest molecule is charged and the reliability of the simulation therefore rests heavily on the proper treatment of the long-range electrostatic interactions. Hydrogen bonding is also important. The calculation [12], which is based upon the methodology of Catlow and co-workers [13], involved energy-minimisation of the whole structure, permitting not only the optimisation of the guest location but also the relaxation of the host structure in response to the presence of the guest. Three different models were considered (Figure 1), but model 1 gives the lowest energy; in almost all respects, this simulated structure (Figure 2) is strikingly close to the three-dimensional crystal structure reported by Slade and Stone [14] (Table 1). These results augur well for the future use of simulations to probe host-guest complexes that may not readily be amenable to experimental study.

CALCULATION OF HEATS OF ADSORPTION AT INFINITE DILUTION

The evaluation of the heat of adsorption at, say, room temperature is a more complex problem since it requires us to evaluate Equation (1) for all orientations of the guest molecule, and at all positions in the cage. The following integrations, which describe the Boltzmann distribution of the molecule over the available energy levels, can then be performed:

$$I_1 = \int_V \exp[-\phi(tot)/RT]dv \tag{2}$$

$$I_2 = \int_V \phi(tot)\exp[-\phi(tot)/RT]dv \tag{3}$$

The internal energy of adsorption is then given by:

$$\Delta U(ads) = \phi(tot) = I_1/I_2 \tag{4}$$

This treatment accounts for entropy effects that distribute the molecule over an increasing number of higher enthalpy sites as the temperature is raised. For example, though the minimum energy, $\phi(min)$, of methane in zeolite Y is -23.0 kJ mol^{-1}, the molecule would only be expected to occupy the position with this energy at very low temperatures. At room temperature, $\Delta U(ads)$ is only -13.3 kJ mol^{-1}.

Table 1. Results of computer simulations of three structural models for the anilinium-vermiculite complex; the experimental data of Slade and Stone [14] are shown for comparison. ΔE represents the difference in energy between the preferred model and the alternatives shown in Figure 1.

Model	Lattice energy (eV)	ΔE (kJ mol^{-1})	a (Å)	b (Å)	c (Å)	β (^{o})
Expt.	--	--	5.330	9.268	14.892	97.02
1	-1184.621	0.0	5.412	9.373	14.971	96.86
2	-1184.181	42.4	5.414	9.385	14.923	97.40
3	-1184.360	25.2	5.414	9.384	18.358	111.24

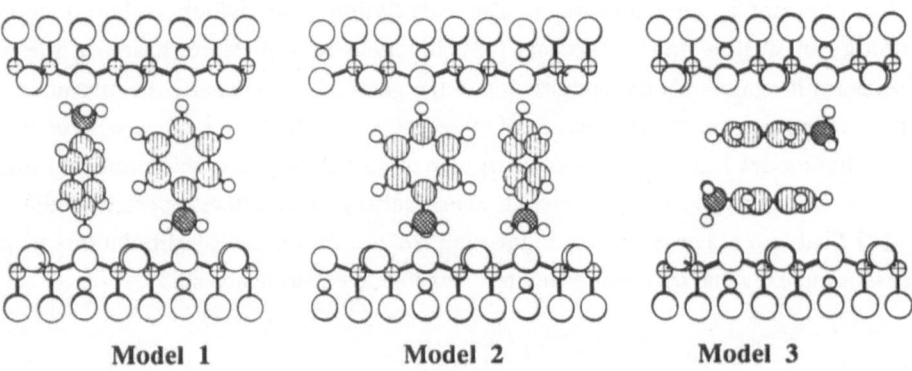

Model 1   **Model 2**   **Model 3**

Figure 1. Three possible configurations of the anilinium ion in vermiculite at the high density limit.

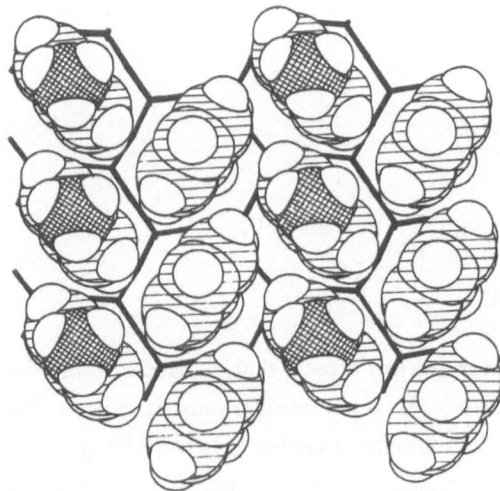

Figure 2. The calculated minimum energy packing arrangement, viewed perpendicular to the silicate layers.

In Table 2 we compare the calculated and experimental adsorption energies for a series of hydrocarbons in zeolite Y. As the molecules become larger, the energies increase. We emphasise that these values correspond to the energies of isolated molecules and are only valid at infinite dilution when intermolecular interactions can be ignored. In a later section we discuss the extension of this treatment to cavities with multiple occupancy.

Table 2. Internal energies of adsorption of hydrocarbons in zeolite Y (kJ mol^{-1}) [15].

	exptl.	calc.
CH_4	13.3	15.2
C_2H_6	21.5	23.3
C_3H_8	30.1	32.3
C_4H_{10}	35.2	37.4

ACTIVATION ENERGIES FOR DIFFUSION

Activation energies are also of interest since they provide a means of rationalising some of the factors that control hydrocarbon mobility in zeolites. Bezus et al. [5] have calculated E_a by taking the difference between the minimum energy position, $\phi(min)$, and the lowest potential barrier to diffusion. Given that the molecules have only a low probability of occupying the global minimum position at ambient temperatures, the approach appears to be over-simplistic.

A slightly more rigorous approach is to take the difference between $\phi(tot)$ and the weighted-average energy of the molecule as it passes through the ring into the next cage, $\phi(ring)$. In the case of propane in zeolite Y, the predicted value of the activation energy is then 13 kJ mol^{-1}, compared with the experimental value of 9.8 ± 1.5 kJ mol^{-1} [16]. One of the interesting conclusions that emerge from such calculations is that high activation energies often stem from large values for the heat of adsorption, rather than high ring barriers. A comparison of diffusion rates for 1,3,5-trimethylbenzene and its trimethylaniline analogue, in zeolite Y, provides an interesting example. The sizes and shapes of the two molecules are almost identical, but their experimental activation energies are 38 and 71 kJ mol^{-1}, respectively. Calculations show that much of this difference stems from the stronger interaction of the more polar amine molecule with the zeolite framework, leading to a substantially more negative value for $\phi(tot)$ [15].

COMPUTER SIMULATION AT HIGH SORBATE CONCENTRATIONS

Many of the interesting properties of zeolites are dependent upon cooperative effects between sorbate molecules. The internal energy of adsorption, for instance, will have components due both to the adsorbate/zeolite interactions and to the adsorbate/adsorbate interactions. In such a case, the simple formulae (2-4) must be evaluated over all configuration space and one must turn to more sophisticated approaches to determine the properties of interest. The techniques of Monte Carlo and Molecular Dynamics as used in liquid state theory are also applicable here. MD also offers the opportunity to study dynamic processes such as diffusion, a topic that will be addressed in the next section.

From statistical mechanics we know that in a system where N simple particles are enclosed in a volume V, at a temperature T (the canonical ensemble), the average of any time independent property is given by:

$$< F(r^N) >> = \frac{\int F(r^N) \exp(-U(r^N)/kT) \, dr^N}{\int \exp(-U(r^N)/kT) dr^N} \tag{5}$$

where U is the total potential energy, including interactions between all particles in the system. The integral in (5) extends over the possible positions of all N particles. In all of the MC simulations performed so far, the static lattice approximation has been used and so the atoms comprising the zeolite crystal remain fixed and provide an external field to the energy in (5). It is clear that the future will bring a coupling of simulations *of the zeolite lattice* [17] and simulations of guest molecules within the pore spaces. Equations (2-4) are the special case of (5) where only one particle is involved in a system that provides an external field, ϕ(tot), and the property being averaged is the interaction of the particle with that field. From the form of (5) we can see that

$$e^{-U(r^N)/kT} \tag{6}$$

is proportional to the probability of a configuration with energy U occurring and the denominator (the configurational part of the partition function) is a normalisation constant. So (5) represents an average over possible states which occur with the given (Boltzmann) probability distribution. If one generates a sequence of configurations with that probability distribution, the average of a given property becomes just [18]:

$$< F> = \frac{\sum_{i=1}^{M} F_i}{M} \tag{7}$$

which is the usual average. This is precisely the procedure in the Monte Carlo simulation of molecular systems: one generates a sequence of configurations of the system with a given probability distribution and averages the quantities of interest over those configurations. The method of generating the configurations is due to Metropolis et al. [19] and hence one speaks of the Metropolis Monte Carlo scheme. In the canonical ensemble, a configuration is generated from a previous one by attempting to move a particle a random distance and then accepting the attempted move with probability exp(-U/kT). This is done on a computer by always accepting the new configuration if its energy is less than the previous one. If the new configuration's energy is greater than the old one, we compare the value of exp($-\Delta U/kT$), where ΔU is the difference in energy between the new and old configurations, with a random number between 0 and 1, and accept the new configuration if the Boltzmann factor of the energy is greater than the random number. In a typical simulation of a zeolite adsorption system, this procedure would be carried out several thousands of times for every particle in the system. If molecules with shape are being studied, rotations must also be considered in the attempted configurations.

The first use of Monte Carlo Simulations in the study of adsorption in zeolites appears to be that by Stroud et al. [20]. They studied methane in zeolite 5A and calculated thermodynamic properties such as the isosteric heat of adsorption and the heat capacity. They also calculated the adsorption isotherm, though by a tedious method using coupling parameters and simulations for systems with potentials other than the one of interest. Kretschmer and Fiedler [21] also performed some early Monte Carlo work, simulating alkanes in zeolites. However, their method was restricted to one molecule per cavity and therefore corresponds to the ideal gas limit. They were particularly interested in the configurations of the sorbate molecules within the zeolite cavities.

More recently, Yashonath et al. [22] found good agreement between MC simulations and experimental results for the heat of adsorption of methane at zero coverage in sodium zeolite Y, and Smit and den Ouden [23] have done zero-coverage MC studies of methane in the zeolites faujasite, mordenite, and ZSM-5. Both studies used a potential of the type given in (1). A very interesting feature of Smit and den Ouden's work was that they varied the Si/Al ratio in mordenite and found a sharp change in the heat of adsorption at Si/Al $\approx$ 6.7. They were able to rationalise this result in terms of the blocking of high energy adsorption sites by sodium cations. This prediction has yet to be verified by experiment.

It is sometimes advantageous to hold fixed in a simulation a set of properties other than N, V, and T. A useful example is to hold the chemical potential, μ, fixed rather than N. Since two phases are in equilibrium if their chemical potentials are equal, the properties of a bulk gas in equilibrium with the adsorption system of interest can be found through the equation of state of that gas. A simulation with applied parameters (μ,V,T) is done in the grand canonical ensemble and is termed Grand Canonical Monte Carlo (GCMC) [18]. In such a simulation, not only are new configurations chosen with different particle positions, but configurations with different *numbers* of particles are considered. This results from the fact that a distribution function that is not equal to (6) must be used in the grand canonical ensemble. GCMC simulations of a fluid in a zeolite correspond very well to the experimental situation where one has the adsorbed fluid in coexistence with the bulk fluid outside the zeolite crystals. A typical experiment involves choosing the bulk gas pressure (which determines μ) and determining (gravimetrically perhaps) the amount of material adsorbed in the solid sample (N).

Woods and Rowlinson [24] have recently performed just such a simulation for xenon and methane adsorbed in zeolites X and Y. They used crystallographically determined coordinates for the zeolites and compared adsorption isotherms and heats of adsorption with experimental data. These workers used fluid-fluid potential parameters available in the literature and fitted the fluid-zeolite potential parameters to zero-coverage data. Although they fitted the potentials at one temperature and used them at a higher temperature, they were able to reproduce all of the qualitative features of the heat of adsorption and adsorption isotherm throughout the range of zeolite cavity occupancies. The potential they used was a particularly simple one, including only the Lennard-Jones terms and a constant "background" correction term. Even with this simple potential, including no explicit polarization or electrostatic terms, many of the features of the experiments were reproduced semi-quantitatively.

Soto and Myers [25] had previously used GCMC to study hard sphere and Lennard-Jones fluids in zeolite 13X. They were the first to use this method for zeolite adsorption and were able to demonstrate its usefulness. They did not achieve a level of agreement with experiment comparable with that obtained by Woods and Rowlinson, but instead chose to use simple potentials to study the qualitative effects of including various interactions. Among their conclusions was that the hard sphere model (used in many earlier theories of adsorption) worked well for the heat of adsorption (because the energy is mostly determined by the fluid-zeolite contributions), but not for the adsorption isotherm (because the fluid-fluid interactions determine the chemical potential).

DYMAMICS OF SORBATE MATERIALS

The MC methods mentioned above are very useful for probing the configuration space of a molecular system, but to obtain information about the time dependence of such systems, MD must be used. MD also has a long history in the study of bulk liquids and gases [18] and the methods developed there should easily transfer to the study of adsorption in zeolites. Molecular dynamics is most simply the solution of Newton's equations of motion for N particles in a specifed volume and with a specified total energy. It is very similar to MC except that the system evolves new configurations naturally in time. The average of a quantity F is then represented as

$$< F> = \frac{\sum_{i=1}^{M} F(i\delta t)}{M} \tag{8}$$

where the index i denotes states of the system at consecutive multiples of the time step used in the discrete representation of the dynamic equations. The averages are usually carried out for many thousands of time steps resulting in an 'experiment' lasting a few hundred picoseconds. Quantities such as the diffusivity, which depend on the state of the system at more than one time, can also be calculated. The diffusivity is usually found from the Einstein diffusion equation,

$$<\Delta x^2> = 6Dt \tag{9}$$

where Δx is the displacement of a particle from its initial position and the brackets denote averages over numbers of particles or over separate experiments. It will be especially enlightening to compare the values of D calculated from simulations with results from pulsed field gradient NMR experiments.

The only MD study to appear so far is that of Yashonath et al. [26]. They studied methane in NaY and investigated the effect of temperature on the mobility of the sorbate. They used the complicated RMK [27] potential to model the methane/methane interactions and Lennard-Jones plus electrostatic terms to model the methane/zeolite interactions. Only one loading was studied (6 molecules per cage), but they studied several temperatures from 50K to 300K.

Yashonath et al. calculated cage- and site-residence times for the methane molecules at the various temperatures. As expected, the methane becomes much more mobile at higher temperatures, with a large drop in the residence times being obtained in the range 50K to 150K. At all of the temperatures studied, the molecules remained close to the walls of the cages and hence the mode of transport is surface diffusion. The trajectory at 300K was analyzed with (9) and $D = 2.0 \times 10^{-8}$ m^2/s was found, which compares well with an experimental (NMR) value of 1.5×10^{-8} m^2/s.

One feature of the Yashonath et al. [26] work reminds us that simulation methods have their limits. At the lowest temperature studied (50K), the cage residence time was of the same magnitude as the length of a simulation run (≈ 25ps). When this is the case, the simulation run is not long enough to give a reliable estimate of the diffusivity or the residence time. This will be true any time that the time scales that govern the phenomena of interest are longer than the amount of time than can be afforded for a simulation. This is further emphasized by the fact that they found different results at 50K, depending on whether the sample was heated from the minimum energy (0K) configuration or cooled from a higher temperature configuration. If there are metastable states in the vicinity and the lifetime of these states is of the same order as the simulation time, the system can be trapped in them. This problem is particularly acute at low temperatures when mobilities are low.

Some of the flexibility offered by the various MC methods can also be obtained by MD. Constant temperature, rather than constant energy, simulations can be performed via a variety of methods [18]. Also, the chemical potential can be determined (at least for systems with not too high a density) via the potential distribution theory of Widom [28]. These conveniences, along with its dynamic nature, endow MD with the properties of a very useful tool. Future developments in this area will involve the extension of the treatment to flexible sorbates and non-rigid hosts.

CONCLUSION

The MC and MD studies of zeolites presented above are good examples of the kinds of information obtainable by simulation. One can use the same model potentials that are used in theories to test the ideas and/or approximations in those theories. Alternatively, one can fit the potentials to easily obtainable experimental data and then use the simulations to predict quantities that are not readily accessible by experiment. Similarly, experimental results can be rationalised in terms of spatial and structural configurations on the molecular level. If suitable potentials can be derived, one could even perform computations on zeolites that have *not yet been synthesized*. The good agreement between the simulations performed so far and experimental measurements for a wide range of properties ensures that computer simulations will be of increasing importance in understanding and predicting the behaviour of host-guest complexes. However, a great deal remains to be done. Future work must explore some of the approximations that are often made, in particular the rigidity of both the framework and the guest molecules. The results on the vermiculite-anilinium complex demonstrate that this is feasible. Another area of concern is the choice of parameters for the atom-atom potential. Recent studies of the location of *para*-xylene in silicalite draw very different conclusions, simply because the parameterisations were different [29,30]. In some instances, the choice can clearly have a profound effect.

To end on a positive note, however, we observe that an area that has yet to be addressed in detail is the simulation of *catalytic reactions* involving host-guest complexes in zeolites and other systems. This, of course, requires the use of quantum mechanical methods. The reaction profile of xylene isomerisation has been studied without taking account of the influence of the aluminosilicate cage [31,32], and the energetics of zeolite fragments have been investigated by *ab initio* methods [33]. In particular, the siting of aluminium [34] and the O-H bond energy [35] have been examined. Some useful progress has therefore been made and we look forward to exciting advances in this area during the next five years.

References

1.   *Zeolite Chemistry and Catalysis*; Rabo J. A., ed.; Am. Chem. Soc. Monograph 171, (1976).

2.   Ramdas, S.; Thomas, J. M.; Betteridge, P. W.; Cheetham, A. K.; Davies, E. K. *Angew. Chem., Int. Ed. Engl.* **1984**, *23*, 671.

3.   Ramdas, S. *J. Computer-Aided Mol. Design* **1988**, *2*, 137.

4.   Kiselev, A. V.; Pham Quang, Du *J. Chem. Soc., Faraday Trans. 2* **1981**, *77*, 1.

5.   Bezus, A. G.; Kocirik, M.; Lopatkin, A. A. *Zeolites* **1984**, *4*, 346.

6.   Kiselev, A. V.; Bezus, A. G.; Lopatkin, A. A.; Pham Quang, Du *J. Chem. Soc., Faraday Trans. 2* **1978**, *74*, 367.

7.   Car, R.; Parrinello, M. *Phys. Rev. Lett.* **1985**, *55*, 2471.

8.   Wright, P. A.; Thomas, J. M.; Ramdas, S.; Cheetham, A. K. *J. Chem. Soc., Chem. Commun.* **1984**, 1338.

9.   Adams, J. M.; Haselden, D. A. *J. Solid State Chem.* **1984**, *55*, 209.

10.   Fitch, A. N.; Jobic, H.; Renouprez, A. *J. Chem. Soc., Chem. Commun.* **1985**, 284.

11.   Wright, P. A.; Thomas, J. M.; Cheetham, A. K.; Nowak, A. *Nature* **1985**, *318*, 611.

12.   Gale, J.; Cheetham, A. K.; Thomas, J. M.; Jackson, R. A.; Catlow, C. R. A., to be published.

13.   Stone, P. A.; Slade, P. G. *Clays & Clay Miner.* **1985**, *33*, 200.

14.   Jackson, R. A.; Catlow, C. R. A. *Molecular Simulations* **1988**, 1, 207.

15. Nowak, A.; Cheetham, A. K. In *New Developments in Zeolite Science and Technology, Proc 7th Int. Zeolite Conf;* Murakami, M.; Iijima, A.; Ward, J. W., Eds.; Kodansha and Elsevier, 1986; 475.

16. Pfeifer, H. *Sitzungberichte der Akademie der Wissenschaften der DDR;* Akademie Verlag, Berlin; 1981, IN, 16.

17. Demontis, P.; Suffritti, G. B.; Quartieri, S.; Fois, E. S; Gamba, A. *J. Phys. Chem.* **1988**, *92*, 867.

18. Allen, M. P.; Tildesley, D. J. *Computer Simulation of Liquids*; Clarendon: Oxford, 1987.

19. Metropolis, N.; Rosenbluth, A. W.; Rosenbluth, M. N.; Teller, A. H.; Teller, E. *J. Chem. Phys.* **1953**, *21*, 1087.

20. Stroud, H. J. F.; Richard, E.; Limcharoen, P.; Parsonage, N. G. *J. Chem. Soc., Faraday Trans.* **1976**, *72*, 942.

21. Kretschmer, R. G.; Fiedler, K. *Z. Phys. Chem.* **1977**, *258*, 1045.

22. Yashonath, S.; Thomas, J. M.; Nowak, A.; Cheetham, A. K. *Nature* **1988**, *331*, 601.

23. Smit, B.; den Ouden, J. J. *J. Phys. Chem.* **1988**, *92*, 7169..

24. Woods, G. B.; Rowlison, J. S. *J. Chem. Soc., Faraday Trans. 2* **1989**, *85*, 765.

25. Soto, J. L.; Myers, A. L. *Mol. Phys.* **1981**, *42*, 971.

26. Yashonath, S.; Demontis, P.; Klein, M. L. *Chem. Phys. Lett.* **1988**, *153*, 551.

27. Penner, A. R.; Meinender, N.; Tabisz, G. C. *Mol. Phys.* **1985**, *54*, 479.

28. Widom, B. *J. Phys. Chem.* **1982**, *86*, 869.

29. Pickett, S. D.; Nowak, A.; Cheetham, A. K.; Thomas, J. M. *Molecular Simulation* **1989**, *2*, 353.

30. Reischman, P. T.; Schmitt, K. D.; Olson, D. H. *J. Phys. Chem.* **1988**, *92*, 5165.

31. Corma, A.; Cortés, A.; Nebot, I.; Tomás, F. *J. Catal.* **1979**, *57*, 444.

32. Nebot, I.; Tomás, F.; Zabala, I. *J. Catal.* **1981**, *71*, 41.

33. van Beest, B. W. H.; Verbeek, J.; van Santen, R. A. *Catalysis Lett.* **1988**, *1*, 147.

34. Derouane, E. G.; Fripiat, J. G. *Zeolites* **1985**, *5*, 165.

35. Vetrivel, R.; Catlow C. R. A.;. Colbourn, E. A. *Proc Roy. Soc.* **1988**, *417A*, 81.

PHOTOCHEMISTRY OF ORGANIC MOLECULES ADSORBED ON FAUJASITE ZEOLITES: STERIC EFFECTS ON PRODUCT DISTRIBUTIONS

Nicholas J. Turro

Chemistry Department
Columbia University
New York, NY 10027 (USA)

SUMMARY

Zeolites are robust, crystalline, porous aluminosilicates possessing an enormous internal surface area that is capable of adsorbing large quantities of guest molecules, the size and shape of whose structures allow them to pass from the external to the internal zeolitic surface and to diffuse onto the internal surface. The framework composition, the presence of cations associated with the framework, and the topology of the void space internal to the zeolite all contrive to imbue these materials with special properties that contribute to their widespread use as catalysts, ion exchange materials and molecular sieves. Photochemical probes have been developed to explore the structure of zeolites near the site of adsorption and to examine the dynamics of reactions of molecules adsorbed on the internal zeolitic surface. The basic composition of the internal framework of alkali ion exchanged zeolites is extremely chemically inert. As a result, very reactive photochemically generated intermediates such as radicals do not react with the framework, but react with each other via radical-radical reactions. However, in contrast to the results of radical-radical reactions in solution, the products of radical-radical reactions in zeolites may be controlled by the geometry and occurrence of the zeolite surface. In this report, we review the structure of zeolites in general and then survey the structure of an important class of zeolites, the faujasites. We then show how a photochemical probe, the photochemistry of dibenzyl ketone, can yield information on how intracrystalline dynamics can be influenced by cation type, cation number density and coadsorbed guests and, in turn, how intracrystalline dynamics can determine the products of photoreactions.

THE ZEOLITE PARADIGM

Zeolites are fascinating materials whose unusual chemical and physical properties derive from their porous yet crystalline structure [1-6].

Inclusion Phenomena and Molecular Recognition
Edited by J. Atwood
Plenum Press, New York, 1990

The critical zeolite properties that are explored in this investigation are: (1) the geometric properties of the pore system and the overall three-dimensional network of the void space that allows rotational and diffusional motion of guest molecules; (2) steric properties induced by exchangeable cations associated with the framework; (3) electronic properties of the exchangeable cations; (4) the number density of the exchangeable cations. Each of these attributes can contribute to modify the course of photochemical reactions of organic molecules adsorbed on zeolitic surfaces by controlling the rotational and diffusional characteristics of reactive intermediates produced by light absorption.

ZEOLITE COMPOSITION

The typical composition (for dehydrated material) of the "classical" aluminosilicate zeolites may be represented by the empirical formula:

$$M^+(AlO_2)^-(SiO_2)_nA_m$$

where M^+ denotes an exchangeable singly charged cation (which can also be replaced by 1/2 the number of M^{2+} or 1/3 the number of M^{3+} cations) and A is a physioadsorbed guest molecule (such as water or an organic molecule). The value of n may be almost any whole number greater than one. The number density of cations is determined by the number of aluminum atoms and by the charge of the cations (each aluminum atom contributes a single negative charge that must be compensated for by a cation to maintain electroneutrality). The constitutional or "framework" structure of zeolites is based on an infinitely extending three-dimensional network of AlO_4 and SiO_4 tetrahedra that are linked to each other by shared oxygen atoms.

GEOMETRIC AND TOPOLOGICAL ATTRIBUTES OF THE VOID SPACE OF THE FAUJASITE (X AND Y) ZEOLITES

The void space topologies of zeolites result from the surface generated by the zeolite framework structure. With modern computers the internal void space is easy to visualize [7], i.e., computer graphics representations of the void space of a faujasite type zeolite. One network of pores consists of relatively large and roughly spherical cavities, termed supercages, that possess a diameter of about 13 Å [1]. The supercages are linked by four tetrahedrally disposed, roughly cyclindrical pores which act as windows to the supercage. The free diameter of one of these windows is about 8 Å. The faujasite zeolite's internal topology is one of the most open of all known zeolite structures and, therefore, presents many opportuntites to investigate diffusional and rotational processes within the internal surface.

DIFFUSION IN ZEOLITES; THE BASIS OF SIEVING AND CATALYTIC ACTION

The enormous internal free surface (usually more than 1000 times the external surface of an individual particle) of zeolites may be conveniently classified in terms of a local porous structure consisting of channels (cyclindrical shapes) and cages (spherical shapes) which are connected to one another through intersections containing "windows" or "pores" that determine which molecules can diffuse through the internal surface. Similar windows presumably occur at the external surface of the zeolite crystal and determine which molecules can access the internal surface. Molecular motions such as diffusion and rotation which occur on the internal surface are at the heart of the sieving and catalytic action of zeolites. In catalytic action the ability of a reactant to diffuse to an active site is a critical step in the reaction sequence. In zeolitic structures, the geometry associated with the size and shape of the porous structure, in addition to the chemical and steric effects associated with the framework cations and adsorbed molecules, can control the diffusional and rotational motions of reactants within the zeolite. The intimate interactions between the size and shape of the reactant species and the dimension, geometry and the chemical species occupying the channels and cages will play a dominant role in determining the catalytic effectiveness of a zeolite. The very same features will determine the molecular sieving characteristics.

PHOTOCHEMICAL PROBES OF THE MOBILITY OF MOLECULES ADSORBED ON ZEOLITES

Photochemistry provides a powerful and versatile means of probing the mobility of species adsorbed on surfaces [9,10]. The basic reason for this power is that the absorption of light can produce, instantaneously on the time scale of diffusion, reactive intermediates whose chemistry is totally determined by their mobility on the surface of the porous solid. With proper selection of the reactant species, **information concerning the mobility of the precursor reactive intermediates can be locked into the structure of the stable, isolable products.** In such cases [11-15] product analysis provides a simple, yet elegant method of obtaining information on the dynamics of motion of molecules adsorbed on the zeolites. The formation and the structures of isomers from a geminate radical pair generated by photochemical excitation can be employed to examine the rotational and diffusional motion of radical pairs generated on a zeolite surface. It is important to note that the zeolite material is a strong light scatterer, but does not significantly absorb the light employed to excite the adsorbed organic molecules investigated, i.e., the non-scattered light penetrates the zeolite particles and is absorbed by the molecules adsorbed on the surface.

FURTHER DETAILS ON THE STRUCTURE OF THE X AND Y ZEOLITES

The extent of diffusional and rotational motion that occurs for probe molecules can be varied and examined for the X and Y zeolites as a function of the following parameters:

i)   *The nature of the exchangeable cations* [2,16-17]. The charge of the cation may be kept constant, and the atomic number of the cation may be varied. For example, the X or Y zeolites containing alkali ions Li, Na, K, Rb, and Cs may be compared. The ionic diameters in the alkali ions (Li = 1.4 Å, Na = 1.9 Å, K = 2.7 Å, Rb = 3.0 Å, and Cs = 3.4 Å) increase by over a factor of 2, implying an increase in volume of a factor of about an order of magnitude in going from Li to Cs. The biggest change, however, is expected in going from Li to K, with a smaller change in going from K to Cs.

ii)   *The number density of exchangeable cations in a cage.* The number of cations in a cage may be varied in one of two ways: either variation of the charge of the ion, or variation of the Si/Al ratio. A di-cation such as Mg^{2+} can neutralize the negative charge of two Al atoms, whereas a mono-cation such as Na^+ can neutralize only one negative charge. Thus a cage contains one half the number of di-cations as mono-cations. Similarly, if the number of Al atoms is cut in half (as is roughly the case in going from the X zeolite to the Y zeolite) the number of cations required to neutralize the framework negative charge is decreased by a factor of two.

iii) *Spectator-guest molecules in a cage.* Molecules whose kinetic diameter is about 8 Å or smaller are able to pass through the windows of the X and Y zeolites and be adsorbed within the internal surface. Water [18] and benzene [19-21], for example, can be added as "spectator" guest molecules which do not participate in a reaction sequence in a direct chemical manner, but may strongly influence the course of reaction of another adsorbed reactant by controlling factors such as the site of substrate adsorption or by influencing the diffusional or rotational motion of the reactants. The number and position of these guest molecules within a supercage may also be varied.

iv)   *The location of exchangeable cations and spectator guest molecules in a cage.* There is no guarantee that an exchanged cation will position itself in the same location as its predecessor. Indeed, in some cases it is highly unlikely that this will be the case. An ion or spectator guest that positions itself at or near a window may exert a significant influence on diffusional processes in and out of the supercage. An ion or spectator guest that positions itself inside the supercage may exert a significant influence on diffusional and rotational motions within the supercage.

ION LOCATIONS IN THE MX AND MY FAUJASITE ZEOLITES

The sites available for occupancy of exchangeable ions in the MX and MY families may be classified in terms of their locations within the supercage framework. According to convention [1], three typical sites have been recognized: (a) Site I consists of the "inside" hexagonal prisms of the wall of the supercage. These pores, which make up the sodalite building blocks of the faujasite structures, have ca. 2.4 Å openings so that the cations at these sites will not be accessible to a probe whose size compels it to be located in the supercage. (b) Site II consists of the "external" six membered rings of the sodalite which form the walls of the supercage containing dibenzyl ketone (DBK) molecules. From information in the literature [19-21], approximately four benzene molecules may be adsorbed into a supercage and the adsorbed benzenes are associated with the Site II cations [19-21]. (c) Site III consists of four or five cations which cannot interact significantly with specific rings of oxygen atoms or negative aluminum atoms and, as a result, the Site III cations are not localized and possess a strong binding affinity.

Given the occurrence of three sites, we now recognize three general and important features of variation of the Si/Al ratio of the faujasite composition: (a) As the ratio increases, the number density of the cations decreases, because there are fewer (negatively charged) Al atoms in the framework. (b) As the ratio increases, the steric inhibitions to rotational and diffusional motion within the supercage will decrease, because the number density of cations will decrease. (c) As the ratio increases, the polarity of the supercage will increase. Furthermore, we shall assume that the Site III cations, being unable to find specific site binding, will be mobile and serve as "marbles" that take up space within the supercage. Starting from this hypothesis we shall design experiments to test it and to determine when and if factors other than steric effects are significant in determining the products of photochemical reactions of molecules adsorbed in the supercage.

PHOTOCHEMISTRY OF DIBENZYL KETONE ADSORBED ON MX AND MY ZEOLITES

We have developed a well-established paradigm which expresses the "static structure" of faujasite zeolite. Below we shall develop one which expresses the "dynamic structure" of a photochemically generated radical pair. We shall merge the two paradigms to generate a working hybrid paradigm which can drive experiments to obtain information on the dynamics of photochemically generated radical pairs adsorbed on the internal surface of faujasite zeolites. This working paradign will pre-suppose no novel feature of either the photochemical system or the zeolite system, unless compelled to by experimental information. The simplicity of the working paradigm, of course, prevents quantitative experimental comparison with theory, but the simplicity is blessed with qualitative precision of prediction which must precede any quantitative modelling.

THE PHOTOCHEMICAL PARADIGM

The probe structure, dibenzyl ketone, was selected from consideration of the quantitative aspects of the size/shape constraints of the faujasite zeolites, from the global and local void space topologies and from the ability to manipulate framework, cation, and guest properties, experimentally. The mechanism of photolysis of dibenzyl ketone (hereafter referred to as DBK) has been firmly established [22-24] and proceeds in two important stages.

First we consider the global overall mechanism and then the details of the structures of the products which will lead to reporters of zeolite structure and of rotational and diffusional motion of small molecules on the zeolite internal surface. The absorption of a photon cleaves the ketone into two fragments (ACO/B), an acyl (carbonyl containing), and a benzyl radical, termed a **primary geminate radical pair**. The acyl radical is known to persist for about 100 ns and then decarbonylates to produce a second benzyl radical [26]. *The decarbonylation reaction serves as a clock which can monitor reaction rates and molecular motion.* This primary geminate pair can undergo geminate coupling processes in competition with decarbonylation. The geminate combination processes will regenerate the starting structure (ACOB) or an isomer (ACOB') of the starting structure. If decarbonylation occurs, a secondary geminate pair of benzyl radicals (A/B) is produced. This pair may undergo a competition between geminate coupling and diffusional separation to form free radicals (A+B). In the systems under investigation the structure of the products and their absolute and relative yields are the data needed to infer the manner in which adsorption on the zeolite surface controls the chemistry of the various radicals. Let us now see how this statement is justified mechanistically.

Consider the primary geminate pair (ACO/B) for the specific case of dibenzyl ketone. If the primary geminate radical pair can rotate within the 100 ns time window allowed by the rate of decarbonylation, the carbonyl fragment can attach itself to the *ortho* or *para* position of the benzyl radical to yield oDBK and pDBK, respectively. If the primary radical pair can diffuse apart and remain apart for the 100 ns, decarbonylation occurs and a secondary geminate radical pair (A/B) is produced. If the mechanistic ideas are correct, the ratio of oDBK to pDBK provides a simple probe of the rotational degrees of freedom available to the primary radical pair.

Let us now consider the possible fates of the secondary geminate radical pair. Since the product of reaction, DPE, does not contain the carbonyl group, we can easily exploit the structures of the products of the photolysis of DBK to provide information concerning the difusional motion of benzene-like molecules adsorbed on the zeolite surface and on the ability of the zeolitic cavities and channels to serve as "cages" which promote geminate combination. If the mechanistic ideas proposed are correct, the ratio of DPE to the (sum of) isomeric DBK products provides a simple probe of the diffusional and rotational motion of the primary radical pair.

The zeolite systems selected for detailed investigation by the DBK probe were the X and Y faujasites. These systems possess a common internal surface topology but differ in the Si/Al composition of the framework, with the X zeolite possessing a lower ratio than the Y zeolite (Si/Al = 1.2 and 2.4, respectively). As a result of the differing compositions, the X zeolite possesses about 4 to 5 Site III cations per supercage, whereas the Y zeolite possesses less than one cation per supercage [27]. Thus, it is expected that **the steric factors experienced by DBK adsorbed in the supercage of an X zeolite will be much more severe than those experienced by a DBK molecule adsorbed in a Y zeolite.** Experimentally, this expectation can be tested by simply examining the ratio of products from the photolysis of DBK adsorbed in a X or Y zeolite, both of which have been fully exchanged with the same cation. Based on the steric effect hypothesis, the products which require more rotational and/or diffusional degrees of freedom will be strongly favored when DBK is photolyzed in the Y zeolite. The influence of electrostatic and related electronic (dispersion and quadrupolar) effects on the course of reaction can be tested by examining the products of photolysis of DBK adsorbed on a MX (or MY) zeolite as a function of the exchangeable cations and coadsorbed "inert" gases. In this case, it is expected that if electronic factors are important, steric factors alone will not be able to explain the results. For example, the steric factors are qualitatively clear as one proceeds from Li to Na to K exchanged zeolites, whereas the electronic factors cannot be qualitatively predicted.

PHOTOLYSIS OF KETONES ADSORBED IN ALKALI ION EXCHANGED FAUJASITES

Only a brief outline of the salient results will be given here and the reader is referred to the original literature for details [9,11,13,14]. The photochemistry of DBK adsorbed on faujasite zeolites was examined under various conditions. Under all conditions the amount of DBK adsorbed on the zeolites was sufficiently low that only one out of every several supercages would contain a DBK molecule. Controls of various sorts including the results described below provide convincing evidence that essentially all of the observed photochemistry occurred from species adsorbed on the internal surface of the zeolite.

Table1. Product distributions from the photolysis of dibenzyl ketone adsorbed on ion exchanged X and Y zeolites under vacuum.

Zeolite	DPE	pDBK	oDBK
LiX	80%	16%	3%
NaX	55%	26%	17%
KX	40%	16%	40%
LiY	100%	0%	0%
NaY	95%	5%	0%
KY	94%	4%	2%

For an overview of the results, consider the Table which shows how the products of photolysis of DBK may be varied in a "catalytic" manner by varying zeolite characteristics. First consider the samples in the absence of any added guest molecules. The major product of photolysis of DBK adsorbed on MY is DPE (>90%) and is completely independent of M for the samples in the absence of added guests. In contrast, the major product of photolysis of DBK adsorbed on MX zeolites is strongly dependent on M for the samples in the absence of added guests (Table), with DPE the major product for LiX, pDBK a significant product for NaX and oDBK the major product for KX.

The effect of added guest molecules or of variation of the exchangeable cations on the product distributions is remarkable. Addition of benzene vapor causes the photochemistry to change from decarbonylation to formation of coupling products of the **primary radical pair**, i.e., coupling now occurs faster than decarbonylation. Thus, we can conclude that the diffusional and rotational motion of the primary radical pairs has been tremendously inhibited by the benzene molecules that fill the supercage. A similar result occurs upon changing the cation from Na to K (Table). In this case the yield of coupling products generated from the primary pair increased substantially relative to that found for NaX. With LiX, the yield of primary pair coupling decreases substantially relative to NaX. These results reflect the fact that **the steric constraints of the cations influence the rotational and diffusional motions of the radicals in the supercage.** Li being smaller than Na allows more freedom of motion, and K being larger than Na provides more constraints on motion of the radicals. Furthermore, a lack of any significant difference in the product distribution (Table) in the MY samples (M = Li, Na, or K) provides support for the proposal that steric effects dominate the rotational and diffusional dynamics of the primary and secondary radical pairs and thereby determine the product distributions.

CONCLUSION

The products produced by the photoexcitation of dibenzyl ketone adsorbed on faujasite (X and Y) zeolites have been shown to be sensitive to variations in zeolite structure such as the Si/Al ratio of the framework composition, the number density of the exchangeable cations and the nature (size) of the exchangeable cation. In addition, additives such as permanent gases and organic molecules may exert a profound influence on the product distributions. All of the results are consistent with the hypothesis that steric factors play a dominant role in determining the product distribution. In its simplest form, the paradigm which rationalizes the results is as follows. The photochemistry of DBK adsorbed on the internal surface of faujasite zeolites is the same as that of DBK in homogeneous solvents. The competition between rotation, diffusion and decarbonylation determines the observed products. The dynamics of these three processes are modulated by the zeolite topology, the zeolite composition, the cation type and cation number density, in addition to the number density and type of co-adsorbed guest molecules. In particular, the ability of the zeolite attributes to allow or to inhibit the three competing processes are directly related to steric factors. It is unlikely that steric factors will always dominate product ratios. Indeed, a recent system has provided evidence for the role of electrostatic effects when the guest molecule possesses significant nucleophilic groups such as ether linkages [15]. A second caveat should be made concerning the effect of coverage and sample preparation on the observed results. There appears to be a small percent of "hyperactive" or defect sites associated with almost any porous solid. These sites bind more strongly to guest molecules than the ordinary sites which make up the bulk of the surface. At higher coverage (often only a few percent) the defect sites are completely filled and major sites for the observed chemistry are the ordinary sites. In the case of the photolysis of DBK adsorbed on faujasite zeolites, we believe that it is photochemistry associated with the ordinary sites that we are observing.

Acknowledgements. The author thanks the AFOSR, the DOE, and the NSF for their generous support of this research.

References

1. Breck, D. W. *Zeolite Molecular Sieves*; Wiley: New York, 1974.
2. Avgul, N. N.; Bezus, A. G.; Dzhigit, O. M. *Adv. Chem. Ser.* **1971**, *102*, 185.
3. Thomas, J. M. In *Chemistry and Physics of Solid Surfaces*; Vanselow, R.; Howe, R., Eds.; Springer-Verlag: Berlin, 1986; p. 107.
4. Barrer, R. M. *Pure Appl. Chem.* **1986**, *58*, 1317.
5. Derouane, E. G. In *Intercalation Chemistry*; Whittingham, M. S.; Jacobson, A. J., Eds.; Academic Press: New York, 1982; p. 101.
6. Flanigen, E. M. In *Proceedings of the Fifth InternationalConference on Zeolites*; Rees, L. V. C., Ed.; Heyden: London, 1980, p.760.
7. Ramdas, S.; Thomas, J. M.; Betteridge, P. W.; Cheetham, A. K.; Davies, E. K. *Angew. Chem., Int. Ed. Engl.* **1984**, *23*, 671.
8. Suzuki, I.; Oki, S.; Namba, S. *J. Catal.* **1986**, *100*, 219.
9. Turro, N. J. *Pure Appl. Chem.* **1986**, *58*, 1219.
10. Thomas, J. K. *J. Phys. Chem.* **1987**, *91*, 267.
11. Turro, N. J.; Cheng, C.-C.; Abrams, L.; Corbin, D. R. *J. Am. Chem. Soc.* **1987**, *109*, 2449.
12. Frederick, B.; Johnston, L. J.; de Mayo, P.; Wong, S.-K. *Can. J. Chem.* **1984**, *62*, 403.
13. Turro, N. J.; Cheng, C.-C.; Lei, X.-G.; Flanigen, E. M. *J. Am. Chem. Soc.* **1985**, *107*, 3739.
14. Turro, N. J.; Zhang, Z. *Tetrahedron Lett.* **1987**, 5637.
15. Corbin, D. R.; Eaton, D. F.; Ramamurthy, V. *J. Am. Chem. Soc.* **1988**, *110*, 4848.
16. Sherry, H. S. *J. Phys. Chem.* **1966**, *70*, 1158.
17. Onaka, M.; Ishikawa, K.; Izumi, Y. *J. Incl. Phenom.* **1984**, *2*, 359.
18. Dubinin, M. M.; Isirikyan, A. A.; Rakhmatkariev, G. U.; Serpinskii, V. V. *Bull. Acad. USSR Div. Chem.* **1973**, 900.
19. Lechert, H. L.; Wittern, K. -P. Ber. *Bunsenges. Phys. Chem.* **1978**, *82*, 1054.
20. Jobic, H.; Renouprez, A.; Fitch, A. N.; Lauter, H. J. *J. Chem. Soc., Faraday Trans. 1* **1987**, *83*, 3199.
21. Unland, M. L.; Freeman, J. J. *J. Phys. Chem.* **1978**, *82*, 1036.
22. (a) Engel, P. S. *J. Am. Chem. Soc.* **1970**, *92*, 6074; (b) Robbins, W. K.; Eastman, R. H. *ibid* **1970**, *92*, 6076, 6077.

23. Turro, N. J.; Weed, C. G. *J. Am. Chem. Soc.* **1983**, *105*, 1861.

24. Turro, N. J.; Kraeutler, B. *Acc. Chem. Res.* **1980**, *13*, 369.

25. Turro, N. J.; Zhang, Z., unpublished results.

26. Gould, I. R.; Baretz, B. H.; Turro, N. J. *J. Phys. Chem.* **1987**, *91*, 925.

27. Resing, H. A.; Wade, C. G. In *Magnetic Resonance in Colloid and Interface Science*; ACS Symp. Series 34: Washington D.C. 1976; p. 36.

HIGH RESOLUTION SOLID STATE NMR STUDIES OF THE EFFECT OF TEMPERATURE AND ADSORBED ORGANIC MOLECULES ON ZEOLITE LATTICE STRUCTURES

C. A. Fyfe*, G. T. Kokotailo, H. Gies, and H. Strobl

Department of Chemistry
University of British Columbia
Vancouver, British Columbia (Canada)

INTRODUCTION

Zeolite Catalysts and Molecular Sieves

Zeolites are an important class of inorganic materials which are widely used as sorbents, ion exchangers, catalysts and catalyst supports [1-6]. The unique feature of zeolites which makes them so useful in catalytic reactions and sorption processes is their selective accessibility to organic molecules due to host-guest interactions. Their lattice structures are very open three dimensional frameworks with extremely uniform pore systems which control the size and shape of the organic molecules which can enter and diffuse through the lattice. When catalytic reactions take place within the structure, control is exerted on both the reactant and the product molecules. The general formula for an aluminosilicate zeolite system is as given in Equation (1) where the stoichiometry of the elements present is usually presented in terms of the ratios of the oxides. Because Si and Al differ in atomic charge, an extra-lattice cationic charge M^x must be present for each aluminum in the lattice to preserve electrical neutrality (these cations may be easily ion exchanged). In addition, water of hydration will usually be present but is not part of the aluminosilicate lattice framework.

$$(M^x)_{y/x}[(AlO_2)_y(SiO_2)_{1-y}].nH_2O \qquad (1)$$

The catalytic activity of zeolites is generated by converting them to an "acid form" by heating the "ammonium" form of the zeolite (where the cations are NH_4^+) to 450°C which causes decomposition of the ammonium ions as in Equation (2) to yield a material where the

$$NH_4^+ \rightarrow NH_3\uparrow + H^+ \qquad (2)$$

Inclusion Phenomena and Molecular Recognition
Edited by J. Atwood
Plenum Press, New York, 1990

extra-lattice cations are (at least formally) H^+. The converted system will now act as a very powerful catalyst in acid-catalyzed reactions of hydrocarbons such as isomerizations, alkylations and hydrogen transfers.

The lattice frameworks themselves are built up from AlO_4^{5-} and SiO_4^{4-} tetrahedra linked through common oxygen atoms (the Al and Si atoms are often referred to as T-atoms). Figure 1 shows how the lattice frameworks of several common zeolites can be assembled from a common building unit. Thus, Figure 1A shows the sodalite cage subunit which has faces of 4- and 6-membered rings where the vertices are Si or Al T-atoms joined by linking oxygens (not shown). When two of these are joined with bridging oxygens via 4-membered rings with bridging oxygens, the structure shown in Figure 1B is formed and linking of two of these units together in a similar manner gives the structure in Figure 1C. This is the basic lattice structure of zeolite-A, commonly used as a "molecular sieve" in laboratory applications. There is a large central cavity accessible via three orthogonal straight channels. It is this pore and channel structure which gives the lattice its unique size and shape selectivity. Depending on the size of the cations in the cavities, molecules of different sizes will be adsorbed and accommodated in the lattice: Na^+ 4 Å, K^+ 3 Å, Ca^{2+} 8 Å. Figures 1D and 1E show two other zeolites made up from the same basic building block. 1E shows the lattice structure of faujasite (also known as zeolite-X or zeolite-Y if obtained by synthesis) widely used in petroleum refining. There is again a large central cavity, but in this case the channels are no longer straight. When the units are joined via four-membered rings but without bridging oxygens (Figure 1D), the sodalite

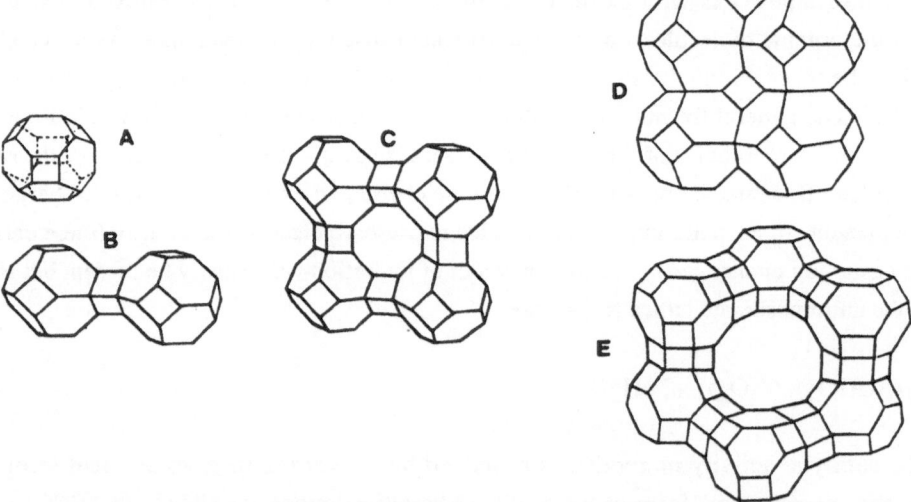

Figure 1. (A) The truncated octahedron building block (also termed "sodalite cage" or "β-cage"). Tetrahedral atoms (usually Si or Al) are located at the corners of the polygons with oxygen atoms halfway between them, (B) illustration of the linkage, through double four-membered rings of two truncated octahedra, (C) the structure of zeolite-A, (D) the sodalite structure formed by linking the truncated octahedra through six membered rings with no bridging oxygens, (E) the faujasite structure (also known as zeolites X and Y depending upon Si/Al ratio) formed by linking cages via six-membered rings with bridging oxygens.

lattice is formed. In this case the structure is completely space filling (the central 'cavity' being identical to the building unit) and this system shows no catalytic or sieving properties. It is a very stable lattice and relatively common in nature.

The traditional technique by which structural investigations of zeolites have been carried out is X-ray diffraction [7]. However, although zeolites are usually very highly crystalline materials, they exhibit several characteristics which greatly limit the structural information which may be obtained from them by diffraction measurements. Firstly, they are usually microcrystalline, yielding crystals too small (of the order of microns in most cases) for single crystal diffraction studies and most structure determinations must be attempted from much more limited powder diffraction data [7]. Secondly, since Si and Al have relatively high atomic masses and differ by only one mass unit (28,27) respectively, their X-ray scattering factors are almost identical and it is often not possible to clearly distinguish them. An additional complication is that the distribution of Si and Al atoms in zeolites is usually disordered and only an average structure results from diffraction experiments. In general, diffraction measurements will at best define the overall lattice, but not the placement of the individual Si and Al T-atoms within this structure. Although the data from powder refinements can be greatly improved by the use of the synchrotron X-ray sources and Rietveld analysis techniques [8] and indeed single crystal investigations of microcrystallites may also be possible using synchrotron X-ray sources in the future [9], at present combined information from a variety of techniques is necessary to deduce the lattice structures of these materials. In this regard, high-resolution solid state NMR spectroscopy has emerged in recent years as an important complementary technique to diffraction investigations [10], reflecting as it does local environments and short range orderings compared with the long range orderings and periodicities to which diffraction techniques are most sensitive.

Solid State NMR Spectroscopy

The NMR spectra of solids with abundant nuclei (e.g. ^{1}H and ^{19}F) are usually broad and featureless (Figure 2) due to the dominance of the direct dipole-dipole interactions between

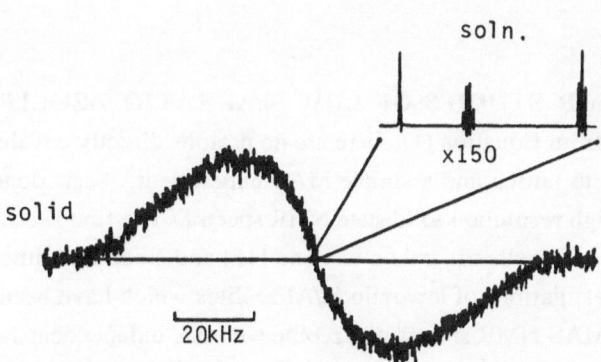

Figure 2. Solid state proton NMR spectrum of solid ethanol with the corresponding high-resolution solution NMR spectrum shown inset.

the nuclei whereas in solution these dipolar interactions are averaged to zero due to rapid thermal motion, resulting in the familiar high resolution spectra (as shown inset in the figure). The situation is simplified considerably by "diluting" the nuclei in the solid matrix either by working with nuclei of low natural isotopic abundance (e.g. ^{13}C, 1%; ^{29}Si, 4.6%) or by physical dilution. With dilution, the homonuclear dipolar interactions between the "dilute" nuclei, which drop off very rapidly with internuclear distance, become negligible. Heteronuclear dipolar interactions between protons and dilute nuclei in the system can be removed by application of a high power proton RF decoupling field. The remaining interaction in the solid state is due to the chemical shielding of the nucleus by the surrounding electrons and since this shielding is a three-dimensional quantity, the chemical shift will be anisotropic (i.e. dependent on the orientation of the nucleus with respect to the magnetic field). The chemical shift anisotropy, H_{CSA}, is observed as a broad signal due to the random orientations of the crystallites. This pattern can be averaged by "magic angle" spinning [11] of the sample at 54° 44' to the magnetic field vector yielding the isotropic chemical shift similar to that produced by random motion in solution. In addition, cross polarization techniques can be used to enhance the dilute nucleus magnetization from that of the protons in the system [12]. The combined techniques of dipolar decoupling, cross polarization (CP) and magic angle spinning (MAS) thus yield high resolution (CPMAS) spectra for dilute nuclei in the solid state [13]. An important point is that in many inorganic systems, a very considerable simplification exists in that often there are no protons present which are directly incorporated into the lattice (e.g. zeolites, glasses and chalcogenides) and the only mechanism of line-broadening is the chemical shift anisotropy, which as discussed above can be removed by MAS. Further, cross-polarization is not possible and the removal of the need for high-powered decoupling means that it is possible to use a conventional high-resolution spectrometer and the experiment is thus easily within reach of most NMR spectroscopists. In many cases, there are substantial advantages of working at high fields including the superior homogeneity and stability of superconducting magnet systems and a probe for such experiments has been described [14]. Examples of spectra of different nuclei obtained in this way for a variety of inorganic materials have been given.

SOLID STATE NMR STUDIES OF LOW Si/Al RATIO ZEOLITES

As can be seen from Equation (1), there are no protons directly covalently bonded to the T-atoms of the zeolite lattice and a simple MAS experiment, best done at high field, is adequate to obtain high resolution solid state NMR spectra. The first spectra of this type were obtained by Lippmaa, Engelhardt and co-workers [15] and since that time there have been a large number of investigations of low ratio Si/Al zeolites which have been reviewed [16,17]. In general, for ^{29}Si MAS NMR spectra of a zeolite with one independent T-atom and Si/Al>1, Lippmaa showed that the spectra were sensitive to the local silicon environment, that is to the distribution of Si and Al in the first coordination sphere: Si(4Al), Si(3Al, 1Si), Si(2Al, 2Si), Si(1Al, 3Si), Si(4Si). This and sensitivity to local environment is shown in Figure 3 for the

preserved as shown by the XRD patterns in Figure 7. The upfield shift of the Si(0Al) line is consistent with the reduction in chemical shift distribution arising from the disordering of Si and Al in the local environment [22].

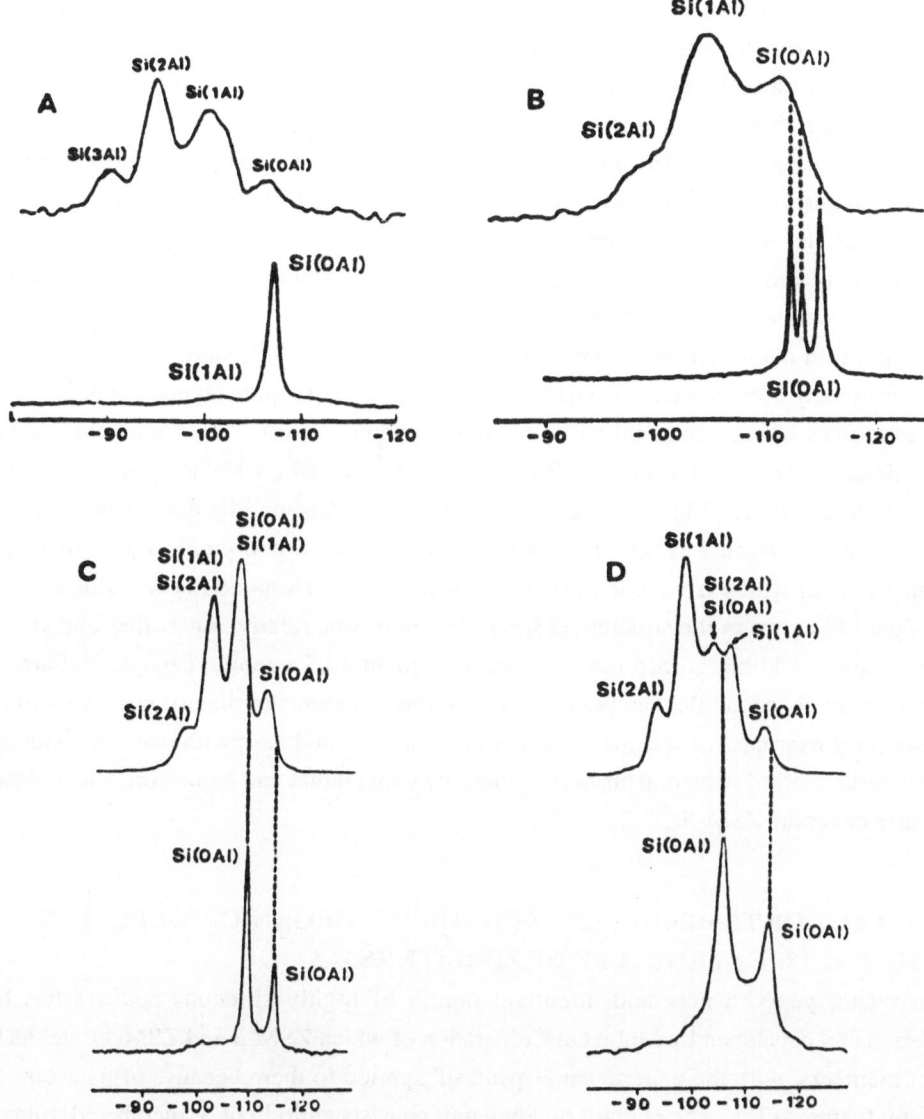

Figure 8. ^{29}Si MAS NMR spectra of the highly siliceous forms and corresponding low Si/Al ratio forms of (A) zeolite Y, (B) mordenite, (C) offretite and (D) omega [23,24].

The effects discussed above for zeolite A are quite general. Examples of the spectra for a variety of typical zeolites are shown in Figure 8 [23,24]. In the cases of faujasite and zeolite A, there is a single unique lattice site and the completely siliceous analogs both show single sharp signals. The multiple-resonances in the ^{29}Si MAS NMR spectra of dealuminated offretite, omega and mordenite can be directly related to the crystallographically inequivalent T-sites in the unit cells. Thus, the mordenite structure has $16T_1$, $16T_2$, $8T_3$, and $8T_4$ sites.

The three resonances in the NMR spectrum with relative intensities of 2:1:3 may be assigned using the average secant of the TOT angles to $16T_1$, $8T_2$, $24(T_2 + T_3)$. The spectrum of zeolite omega which has the mazzite structure with $24T_1$ and $12T_2$ sites shows only two resonances with relative intensities 2:1 which may thus be completely unambiguously assigned. Similarly, the two resonances in the offretite spectrum may be assigned to the $12T_1$ and $6T_2$ sites in the unit cell. Similar assignments may be made for other highly siliceous zeolites, giving a direct correlation between NMR and XRD experiments and data of this type may be useful in the future for the determination of unknown zeolite structures when used together with quantitative correlations of the shifts with the local silicon geometries.

However, the spectra in Figure 8 also indicate that the interpretation of the ^{29}Si MAS NMR spectra of zeolites with more than one independent T-site in the unit cell may be more complex than we have indicated to this point as the effects of site inequivalence as well as that of Al in the first coordination sphere must now be considered. (It should be noted that the systems originally investigated in the deduction of Figure 4 were quite atypical in this regard as they had lattices which contained only one single unique T-site.) The implications of these observations to the calculation of Si/Al ratios from ^{29}Si MAS NMR spectra have been discussed elsewhere [25]. In the context of the present symposium, the most important aspect of the spectra of Figure 8 is that the removal of lattice aluminum yields very narrow ^{29}Si resonances, all of which are due to Si(0Al, 4Si) and which correspond to crystallographically independent Si atoms in the structure [23,24]. The numbers, relative intensities and shifts of the resonances yield direct information as to the structure of a zeolite catalyst and are very sensitive to small and subtle changes in the lattice due to defects or distortions. We will now examine some examples of the application of these spectra in detail, with particular emphasis on the investigation of structural changes induced by temperature and sorbed organic molecules in the case of zeolite ZSM-5.

THE EFFECT OF TEMPERATURE AND SORBED ORGANIC MOLECULES ON THE LATTICE STRUCTURE OF ZEOLITE ZSM-5

In recent years, a new and important family of highly siliceous zeolites has been discovered and developed by Mobil Oil Corporation of which ZSM-5 and ZSM-11 are the best known members, with the generic name 'pentasil' applied to them because of their common structural features [26]. The pentasil building unit consists entirely of 5 membered rings and is shown in Figure 9A. Joining these units in columns (Figure 9B) and then joining layers of columns such that the neighbouring layers are related by inversion symmetry produces the zeolite ZSM-5 shown in Figure 9D. Joining the columns so that the layers are related by a reflection plane yields the closely related ZSM-11 lattice structure (Figure 9C). Because the two structures differ in only one projection and so many of the repeat distances are similar, the powder X-ray diffraction patterns are very similar. The ^{29}Si MAS NMR spectra of highly siliceous samples of these two materials are, however, quite different since they reflect the local Si environments and provide very sensitive 'fingerprints' of the two unit cells [27]. The ^{29}Si

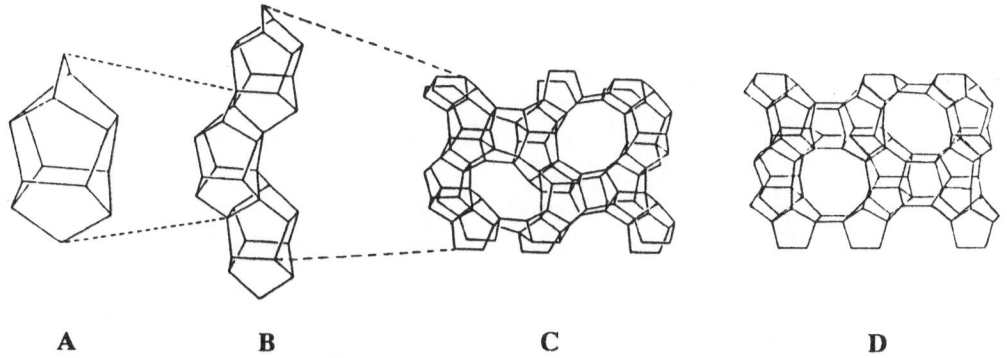

A   **B**   **C**   **D**

Figure 9. Schematic representations of (A) the pentasil unit, (B) a chain formed by interlocking pentasil units, and the structures of (C) zeolite ZSM-11 and (D) zeolite ZSM-5.

MAS NMR spectrum of ZSM-5 is shown in Figure 10. The relative intensities of the completely resolved low and high field resonances in a highly siliceous sample to the total spectral intensity reveal a total of 24 independent T-atoms in the unit cell of the (calcined) ZSM-5 lattice. The sensitivity of the ^{29}Si MAS NMR spectra of highly siliceous ZSM-5 to small changes in the local environments of the silicon nuclei makes it possible to monitor small and subtle changes in the structure induced by sorbed organic molecules and by temperature changes. Thus, the ^{29}Si MAS NMR spectrum of dealuminated ZSM-5 is changed by the sorption of small amounts of p-xylene and other organic molecules (Figure 11) with small

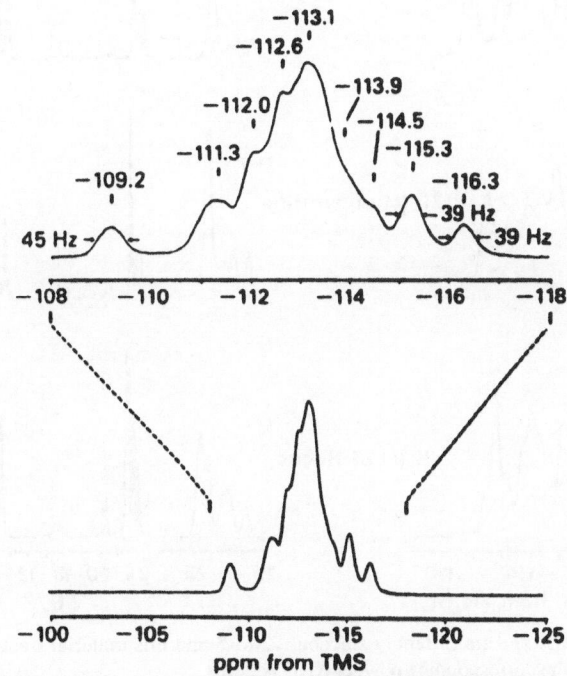

Figure 10. ^{29}Si MAS NMR spectrum (79.6 MHz) of highly siliceous zeolite ZSM-5 [25].

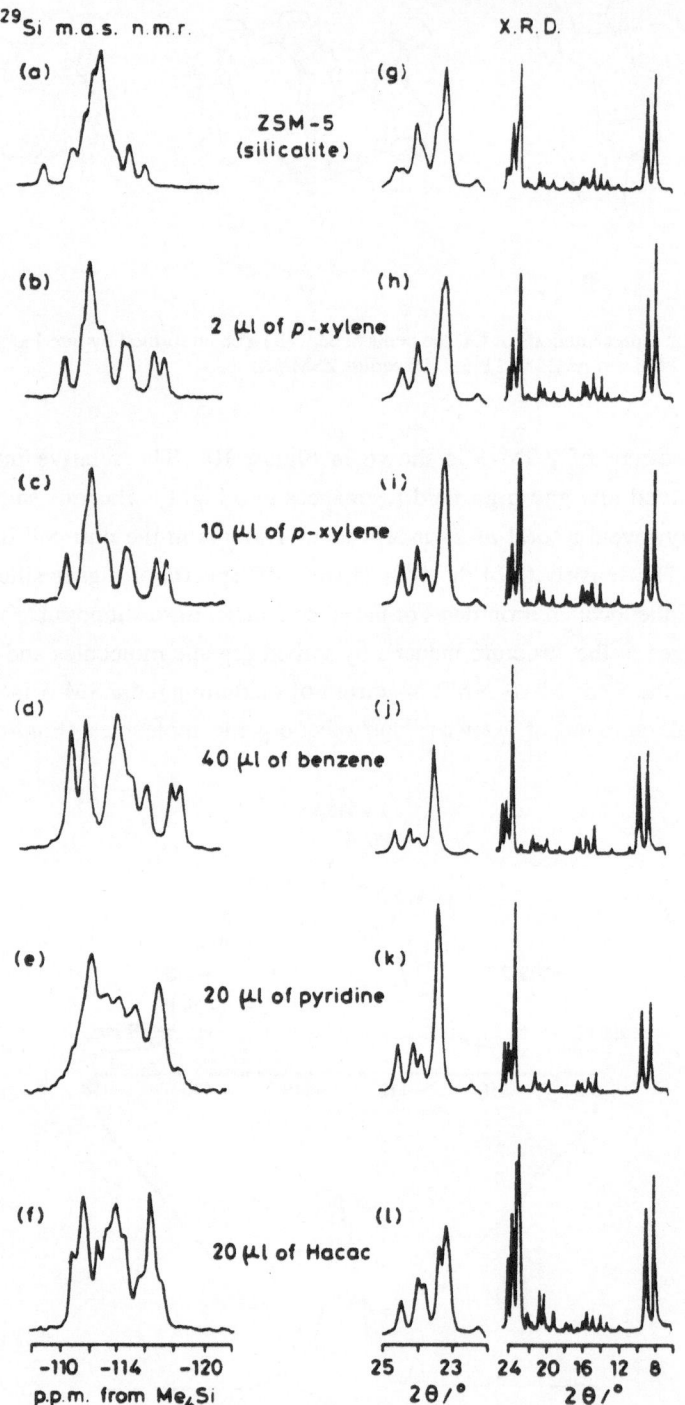

Figure 11. ^{29}Si MAS NMR spectra of highly siliceous ZSM-5 and this material treated with various organic molecules (a-f) as well as the corresponding powder XRD patterns.

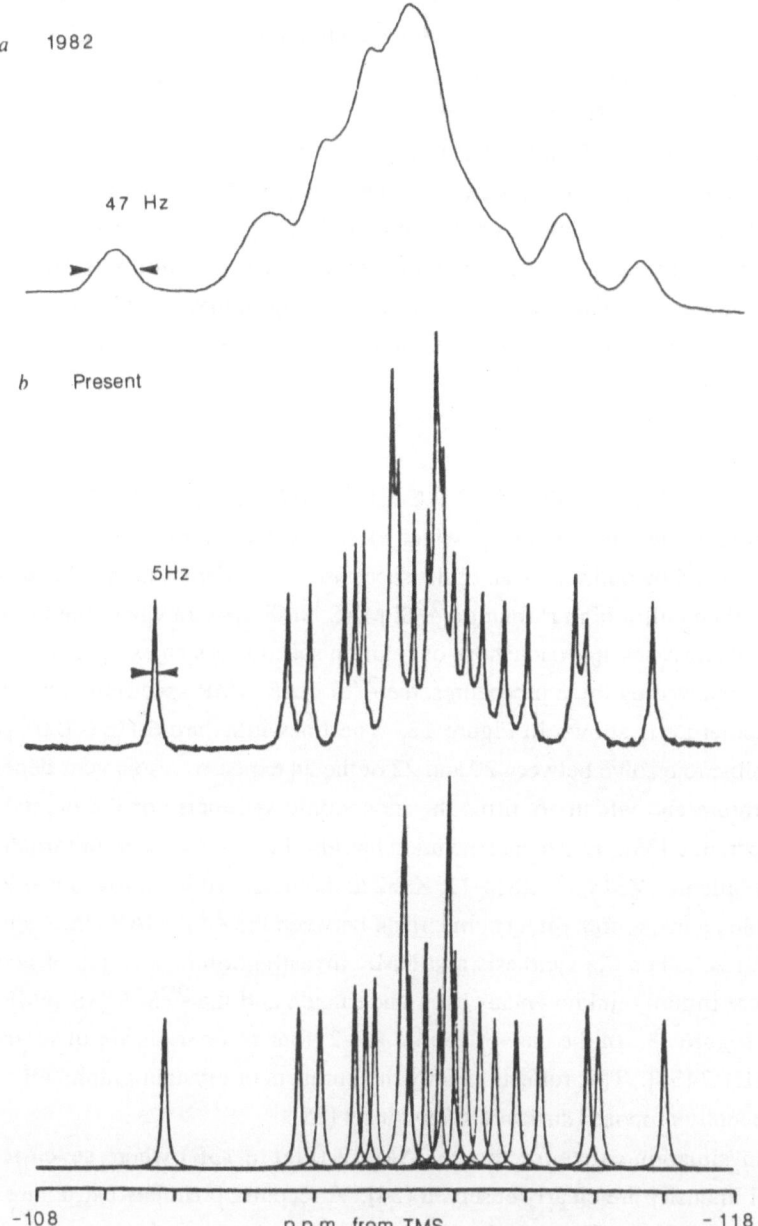

Figure 12. ^{29}Si MAS NMR spectra of highly crystalline ZSM-5, 6.0 Hz line broadening.

related changes in the corresponding XRD patterns and different organic molecules give characteristic limiting spectra. Molecules which are very small or too large to be accommodated in the lattice (e.g. *o*-xylene) cause no major changes in the spectra [29,30]. The XRD patterns show small changes indicating transformation from monoclinic to orthorhombic symmetry with the loss of the characteristic 'doublet' at $2\theta = 23.3°$ and $2\theta = 23.8°$ which clearly indicates, in agreement with the NMR data, that structural changes in the lattice are occurring which are characteristic of the sorbate, while the sample crystallinity is maintained intact. All of the above changes are completely reversible on removal of the organic guests. The lattice of siliceous ZSM-5 is also affected by temperature [31,32]. ^{29}Si MAS NMR spectra recorded between 300 and 377K show gradual changes with small movements of some resonances which may be due to a general expansion of the lattice and a discrete change between 355K and 370K which, together with the limiting high temperature ^{29}Si spectrum, indicates that the spectral changes are due to a phase transition from monoclinic to orthorhombic symmetry. The exact details of these changes, like those induced by the organic sorbates, are somewhat obscured by the limited resolution obtainable, even with these highly siliceous materials.

FURTHER STUDIES WITH 'ULTRA-HIGH RESOLUTION' MAS NMR

By careful hydrothermal treatment which effects both aluminum removal and the healing of lattice defects and by optimizing all of the spectroscopic variables in the NMR experiment, it is possible to obtain ultra-high resolution ^{29}Si MAS NMR spectra where the linewidths of the resonances in some cases approach those of solution spectra. As an example of improvements which can be achieved by these procedures, the ^{29}Si MAS NMR spectrum of zeolite ZSM-5 at ambient temperature is shown in Figure 12. The linewidths are 6 Hz (≈ 0.07 ppm), which makes it possible to resolve between 20 and 22 of the 24 expected resonances depending on the exact temperature and which confirms the monoclinic symmetry of the crystal structure at ambient temperature [33]. Excellent resolution has also been achieved with samples of zeolites Y, A, beta, mordenite, ZSM-11, ZSM-12, KZ-2 and others. This improved resolution clearly makes possible an even more direct comparison between the ^{29}Si MAS NMR spectra and the lattice structures. Thus the synthesis and NMR investigation of a series of zeolites whose structures were initially unknown has been undertaken and the ^{29}Si MAS NMR spectra are presented in Figure 13. In the case of zeolite KZ-2, four resonances are observed of relative intensities 2:1:1:2 [34]. This reflects exactly the numbers of crystallographically inequivalent sites in the recently proposed structure shown inset [35].

A similar situation exists for zeolite ZSM-12 (Figure 13B) where seven resonances of exactly equal intensity are clearly observed [34]. A recently postulated structure shown inset has exactly this number of inequivalent sites as indicated, although recent synchrotron data indicates that there must be a doubling of one of the unit cell dimensions [36,37]. We do know, however, that there must be seven equally-populated crystallographically inequivalent sites. In the case of zeolite beta (Figure 13C) the ^{29}Si MAS NMR spectrum showed clearly a minimum of nine resonances and a very recently proposed structure has used exactly this number in the asymmetric unit [38].

The greatly improved spectral resolution obtained for these ^{29}Si MAS NMR spectra now makes it possible to investigate the effects of temperature and sorbed organics and representative spectra are shown in Figure 14. It is clear that the presence of organics induces changes in the lattice which are quite characteristic of the organic sorbate present [38].

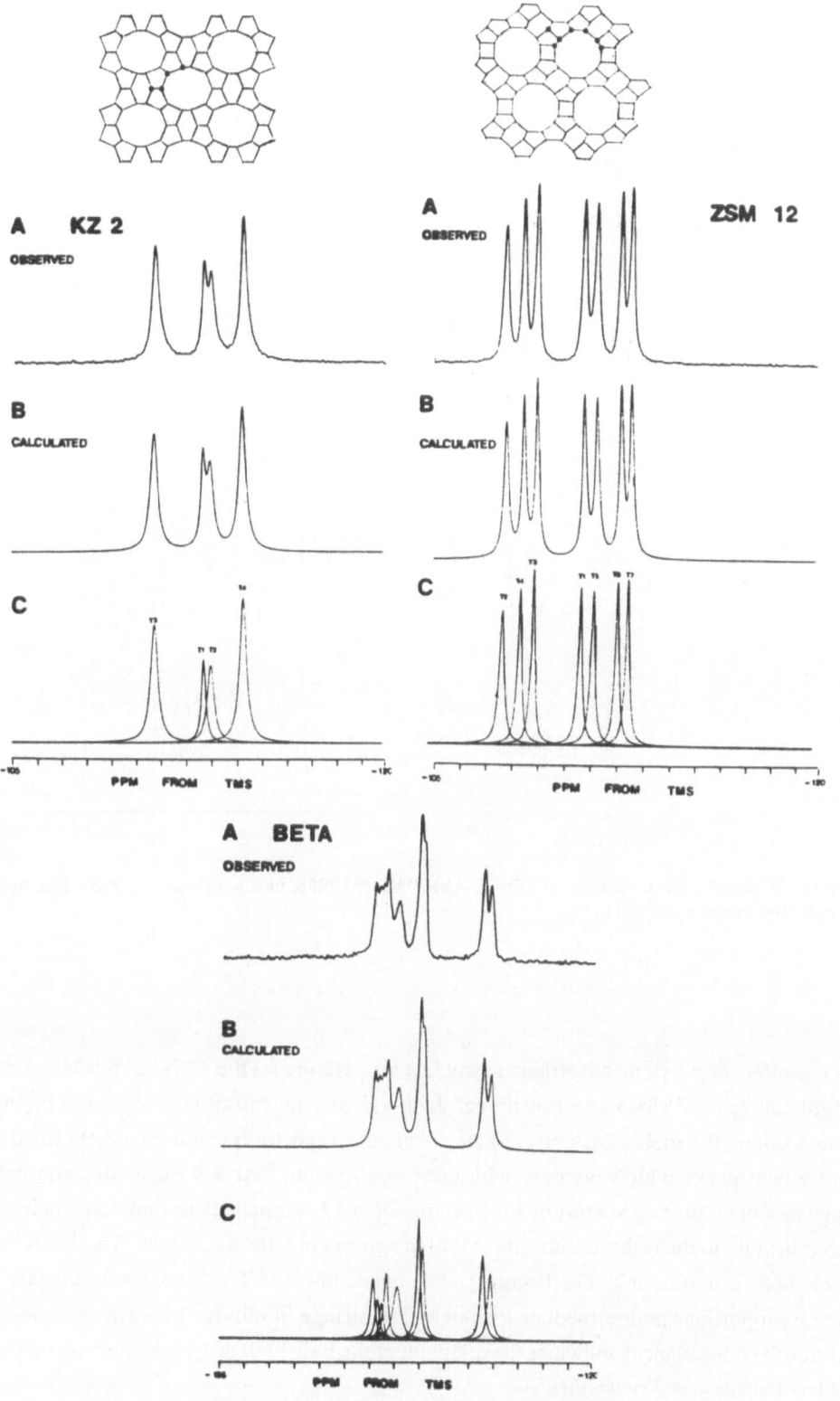

Figure 13. ^{29}Si MAS NMR spectra of zeolites KZ-2, ZSM-12 and Beta [34].

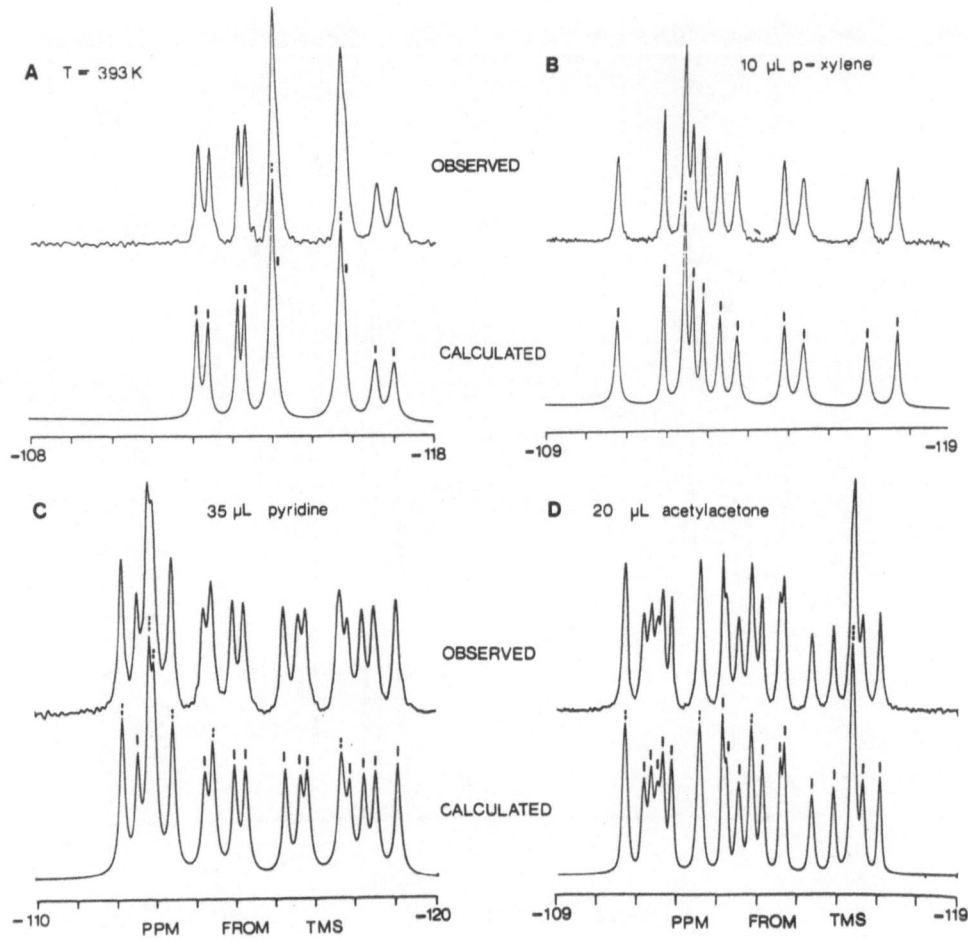

Figure 14. ^{29}Si MAS NMR spectra of ZSM-5. (A) 393K, (B) 298K plus *p*-xylene, (C) 298K plus pyridine, (D) 298K plus acetylacetone [34].

The effect of *p*-xylene adsorption at ambient temperature on the ^{29}Si MAS NMR spectrum of highly siliceous ZSM-5 as a function of the *p*-xylene concentration is shown in Figure 15. At low loading (0.4 molecules/unit cell) the effect on the spectrum is minimal, with small shifts in individual peaks which increase with increased loading. At 1.6 molecules/unit cell the change to a new limiting spectrum which shows only 12 resonances is complete, indicating a phase transition to the orthorhombic form [38] in agreement with XRD data. The NMR spectra indicate both orthorhombic (12 T-atoms) and monoclinic (24 T-atoms) forms are present in different proportions at intermediate loadings. The change in relative intensity of some of the better resolved resonances indicates the midpoint of the transition at 1 molecule/unit cell with a complete transition at 2 molecules/unit cell.

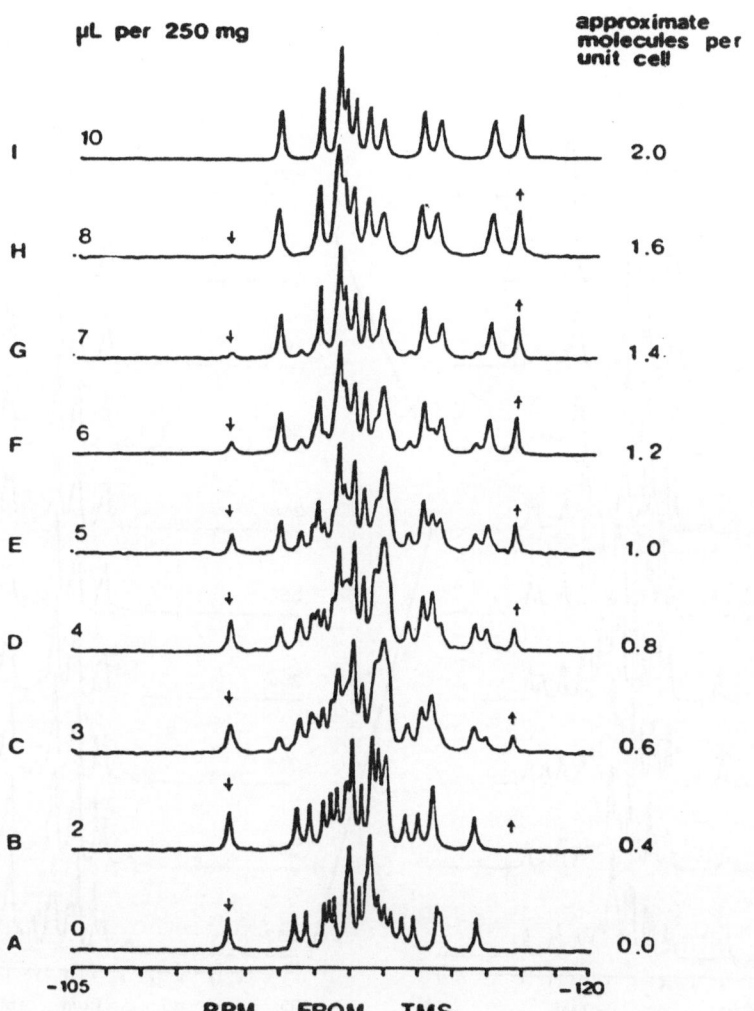

μL per 250 mg

approximate molecules per unit cell

Figure 15. Effect of *p*-xylene on ZSM-5 at ambient temperature [38].

315

As in the case of the sorbed species, the effect of raising the temperature is to induce a phase transformation from monoclinic to orthorhombic. Detailed spectra at 10° intervals shown in Figure 16 show gradual shifts of individual resonances up to 353K with a rapid change between 353 and 363K. This is confirmed by synchrotron X-ray diffraction analysis of the phase transformation as a function of temperature [40], which also shows that above the transition temperature only a single phase (the orthorhombic) persists.

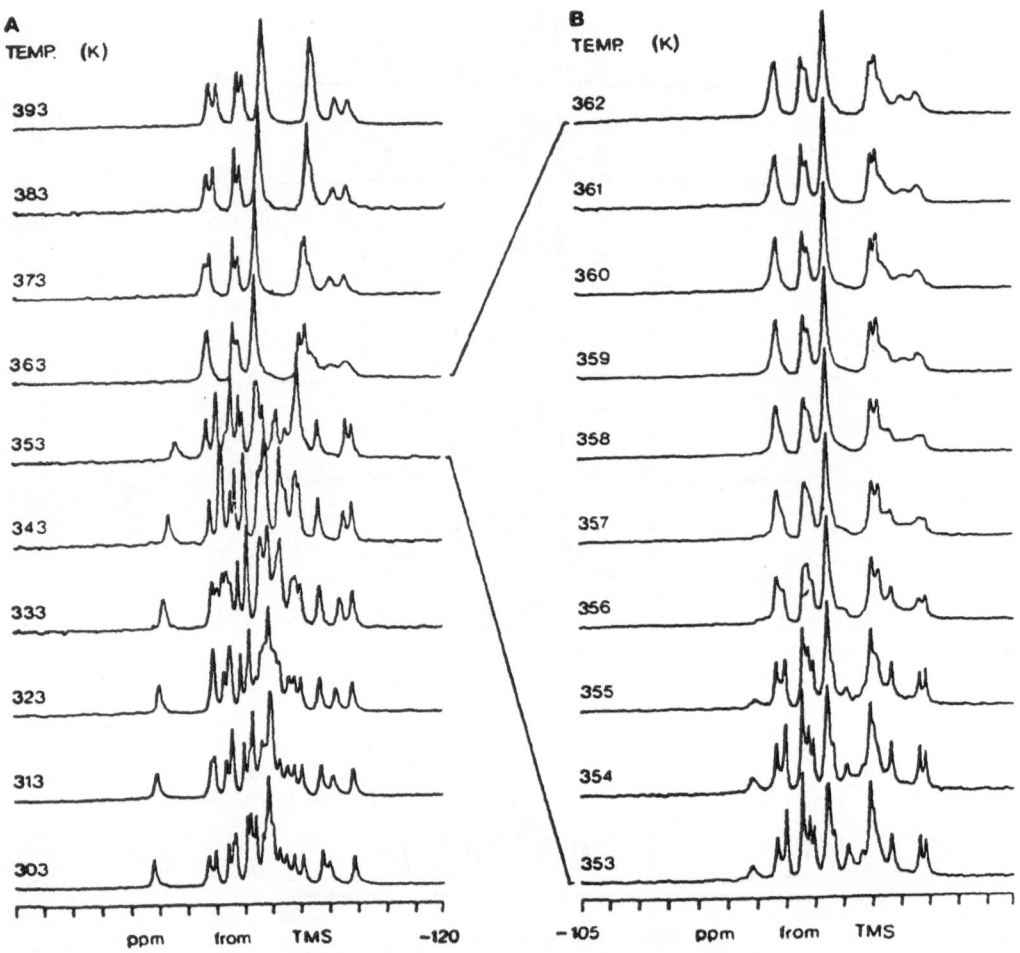

Figure 16. Effect of temperature on ZSM-5 [38].

The combined effect of temperature and sorbate as shown in Figure 17 is to lower the temperature of the phase transition and increase its width. In the phase transition temperature range, both phases now coexist with all the material being crystalline. Above the transition temperature, there are only minor changes reflecting lattice expansion as a function of temperature within the orthorhombic (12 T-atom) phase [38]. The NMR data may now be used in a quite unique way to construct a three-dimensional phase diagram, that is, the

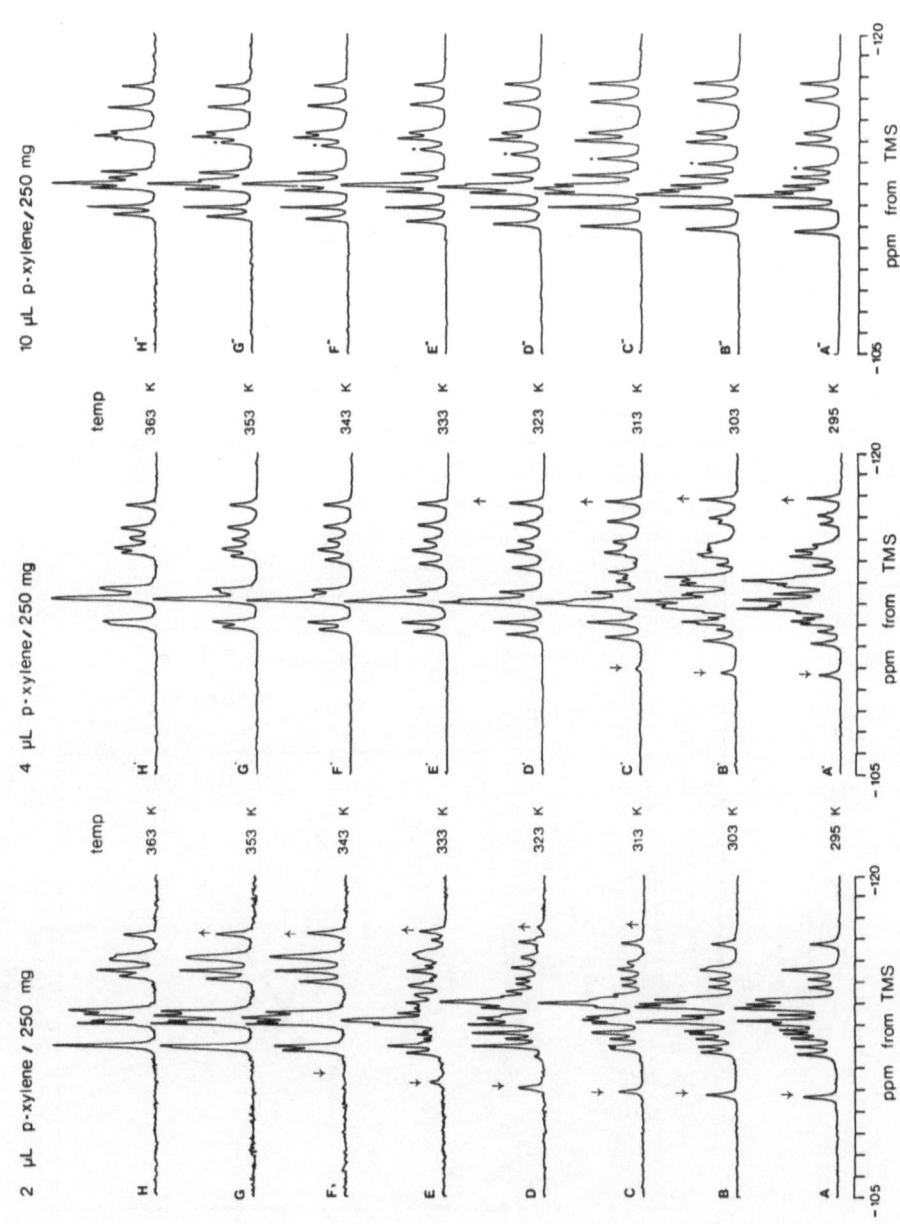

Figure 17. Combined effects of *p*-xylene and temperature on the ^{29}Si MAS NMR spectrum of ZSM-5 under the conditions listed [38].

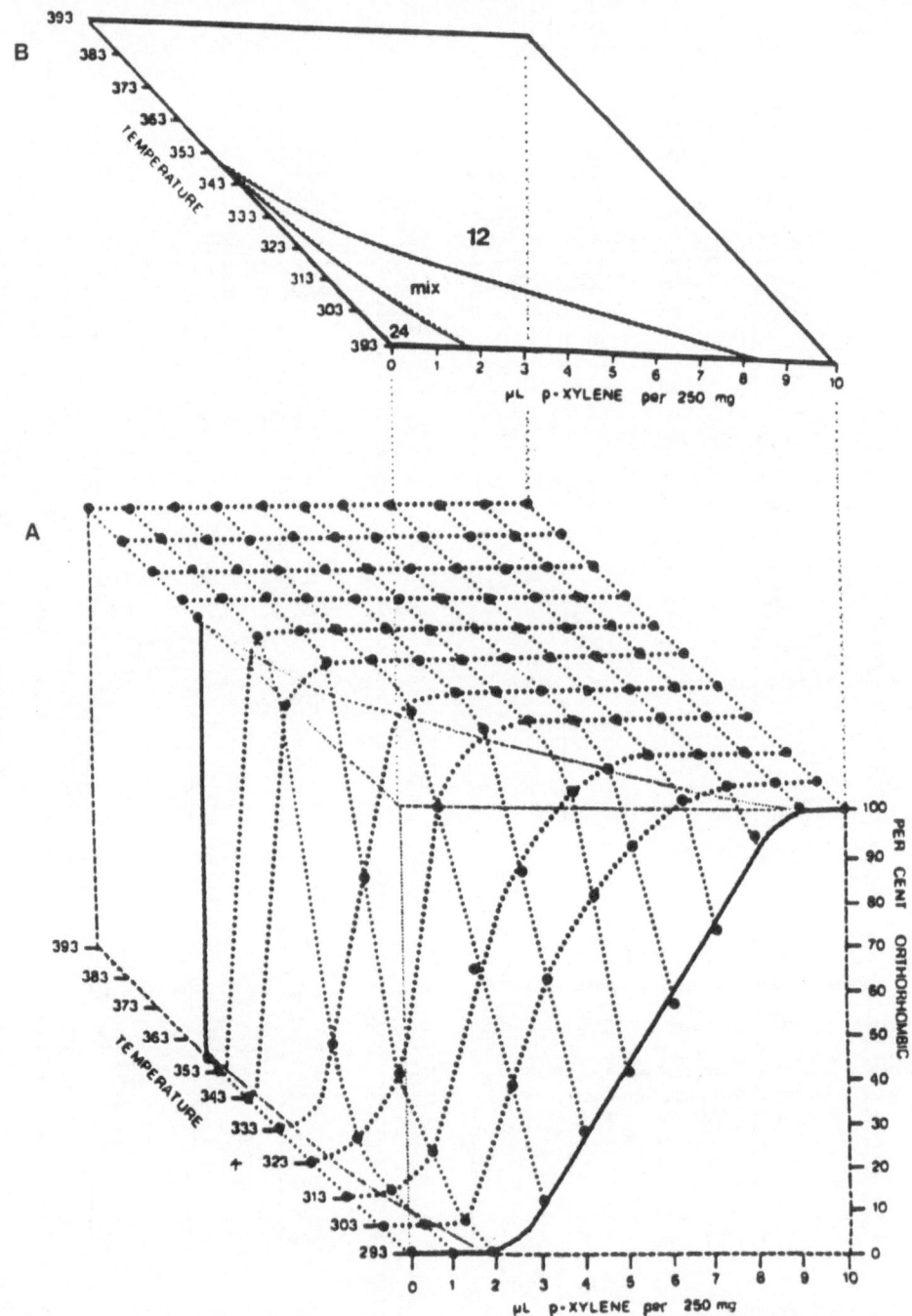

Figure 18. Effects of *p*-xylene and temperature on the crystal symmetry of ZSM-5 [38].

quantitative determination of the percentage of orthorhombic form present as a function of both sorbate concentration and temperature as shown in Figure 18 [39]. This can now be used to determine the limiting conditions under which the different structures exist in phase-pure forms and synchrotron X-ray data may be collected on exactly these phases as previously characterized by NMR. These data have been obtained and are currently being refined. However, NMR can provide further information which is very helpful in obtaining the most useful and easily interpretable X-ray data. Thus, the structure of most chemical interest is the one where sorbed p-xylene has induced the change to the orthorhombic form as this structure will provide direct and unambiguous information regarding the nature of the interactions between the guest molecule and the host lattice. However, the p-xylene molecule is composed of light atoms and their contribution to the diffraction data will be mainly by affecting the intensities of the reflections. This problem may be alleviated by NMR studies while providing additional information regarding the nature of the host-guest interaction. Thus, Figure 14 above shows that the spectra are, in general, very sensitive to the nature of the organic molecule present, presumably reflecting directly the host-guest interactions present. From studies in physical organic chemistry, it is well established that chlorine atoms and methyl

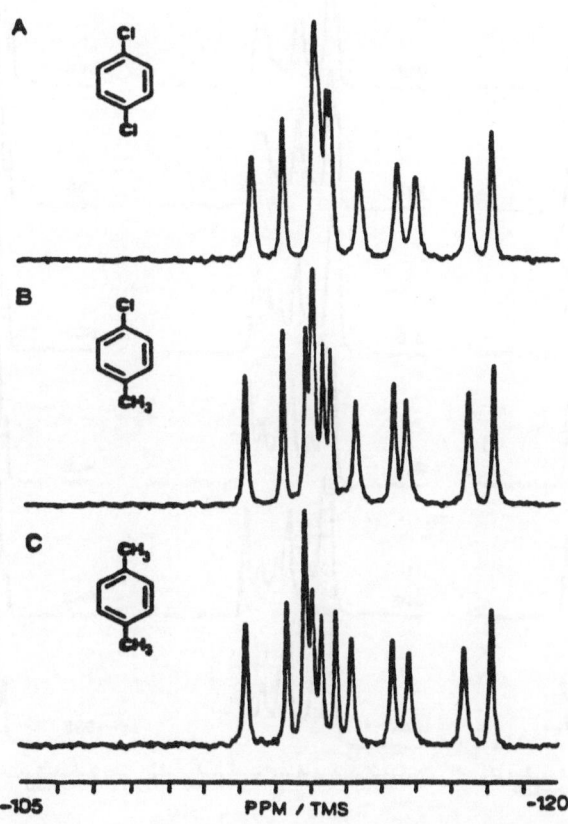

Figure 19. ^{29}Si MAS NMR spectra of ZSM-5 in the presence of, (A) p-dichlorobenzene, (B) p-chlorotoluene, (C) p-xylene [39].

groups exhibit identical steric effects and this may be used to probe the interactions. Thus, the spectra (Figure 19) from *p*-xylene, *p*-chlorotoluene and *p*-dichlorobenzene in ZSM-5 are identical, indicating that, at least for hydrocarbon molecules, the interaction depends only on the size and shape of the organic [39]. Further, the phase diagrams of the form of Figure 18 are absolutely identical and X-ray studies may now be carried out on the *p*-dichlorobenzene sorbate where the organic molecule, because of the large number of electrons in the chlorine atoms, will make a much larger contribution to the diffraction pattern. Refinement of these data are in progress [40].

OTHER SYSTEMS

Other structural investigations may be complicated by temperature induced phase transitions. Thus, Figure 20 shows the variable temperature ^{29}Si MAS NMR spectra of a pure

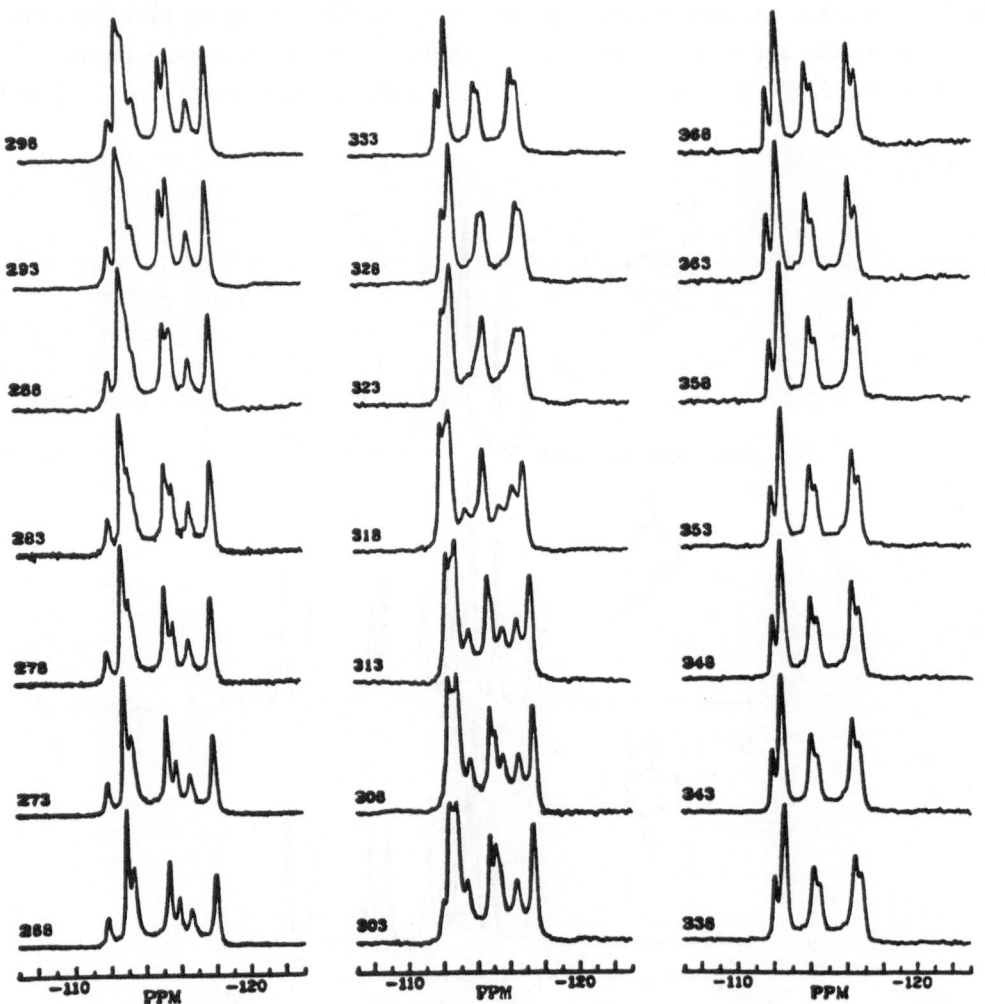

Figure 20. ^{29}Si MAS NMR spectra (79.6 MHz) of zeolite ZSM-11 recorded at the temperatures indicated. Each spectrum is the result of 720 scans with a delay time of 10 seconds [42].

sample of zeolite ZSM-11 synthesized free of any intergrowths of ZSM-5 in our laboratory [41]. Narrow resonances are observed in all of the spectra indicating a high degree of crystallinity and local order. The postulated lattice structure space group has seven inequivalent sites of relative proportions 1:1:2:2:2:2:2 [42]. Inspection of Figure 6 shows that there are too many resonances in the room temperature (293K) spectrum and in fact the system is in an intermediate state between two limiting structures at ambient temperature. Using the data of Figure 6 as our guide, synchrotron powder X-ray data have been obtained in the high and low temperature limits. X-ray data obtained at 373K refined smoothly (R = 4.4%) in the tetragonal space group I4m2 yielding the postulated structure with reasonable values for all bond lengths and angles [43]. Refinement of the low temperature limiting structure is currently in progress [40]. The relative intensities of the limiting high temperature spectrum are in exact agreement with the postulated structure while those of the low temperature form indicate a lowering of the symmetry of the lattice with a basic structural unit which still contains a total of twelve T-sites. Preliminary investigations of the effects of organics on ZSM-11 have been carried out and indicate that changes in this lattice also occur which reflect the nature of the organic sorbate present [44].

CONCLUSION

In this lecture we have examined the application of solid state NMR techniques to the study of zeolite lattice structures with particular emphasis on the investigation of the effects of temperature and sorbed organic molecules. The results obtained are complementary to those from diffraction techniques being sensitive to local orderings and geometric factors in contrast to the long range orderings and periodicities to which diffraction techniques are sensitive. From a careful optimization of both techniques and the samples themselves, detailed information and sorbate-lattice interactions can be obtained which will lead to a more fundamental understanding of the fundamental nature of these interactions.

Acknowledgements. The authors acknowledge the financial assistance of the Natural Sciences and Engineering Research Council of Canada in the form of Operating and Equipment Grants (C.A.F.) and a Graduate Scholarship (H.S.). H.G. and G.T.K. acknowledge the financial support of the Alexander von Humboldt Foundation. C.A.F. acknowledges the award of a Killam Research Fellowship by the Canada Council.

References

1. Barrer, R. M. *Zeolites and Clay Minerals as Sorbents and Molecular Sieves*; Academic Press: New York, 1978.
2. Ward, J. W. *J. Catal.* **1969**, *13*, 321.
3. Rabo, J. A. *Catalysis by Zeolites*; Elsevier: Amsterdam, 1980.
4. Jacobs, P. A. *Carboniogenic Activity of Zeolites*; Elsevier: Amsterdam, 1977.
5. Meier, W. M. In *Molecular Sieves*; Soc. Chem. Ind. London, 1968.
6. Breck, D. W. *Zeolite Molecular Sieves*; Wiley Interscience: New York, 1974.

7. Meier, W. M.; Olson, D. H. *Atlas of Zeolite Structure Types*; Structure Commission of the International Zeolite Association, 1978.

8. (a) McCusker, L. B.; Baerlocher, C. *Proc. Sixth Int. Conf. Zeolites*, Reno, 1983; (b) Rietveld, H. M. *J. Appl. Cryst.* **1969**, *2*, 65.

9. Eisenberg, P.; Newman, J. B.; Leonowicz, M. E.; Vaughan, D. E. W. *Nature (London)* **1984**, *309*, 45.

10. Englehardt, G.; Michel, D. *High Resolution Solid State NMR of Zeolites and Related Systems*; John Wiley & Sons: London, 1987.

11. Andrew, E. R.; Bradbury, A.; Eades. R. G. *Nature (London)* **1958**, *182*, 1659.

12. Pines, A.; Gibby, M. G.; Waugh, J. S. *Chem. Phys. Lett.* **1972**, *15*, 373.

13. Schaefer, J.; Stejskal, E. O. *J. Am. Chem. Soc.* **1976**, *98*, 1031.

14. Fyfe, C. A.; Gobbi, G. C.; Hartmann, J. S.; Lenkinski, R. E.; O'Brien, J. H.; Beange, E. R.; Smith, M. A. R. *J. Mag. Reson.* **1982**, *47*, 168.

15. (a) Lippmaa, E.; Magi, M.; Samoson, A.; Grimmer, A. R.; Engelhardt, G. *J. Am. Chem. Soc.* **1980**, *102*, 4889; (b) Lippmaa, E.; Magi, M.; Samoson, A.; Tarmak, M.; Engelhardt, G. *J. Am. Chem. Soc.* **1981**, *103*, 4992.

16. Fyfe, C. A.; Thomas, J. M.; Klinowski, J.; Gobbi, C. G. *Angew. Chem.* **1983**, *95*, 257; and *Angew. Chem., Int. Ed. Engl.* **1983**, *22*, 259.

17. A recent and very complete complilation of the results of NMR studies of zeolites is given in reference 10.

18. Fyfe, C. A.; Gobbi, G. C.; Kennedy, G. J.; Graham, J. D.; Ozubko, R. S.; Murphy, W. A.; Bothnerby, D. J.; Chesnick, A. S. *Zeolites* **1985**, *5*, 179.

19. Klinowski, J.; Rmdas, S.; Thomas, J. M.; Fyfe, C. A.; Hartmann, J. S. *J. Chem. Soc., Faraday Trans. 2* **1983**, *78*, 1025.

20. Loewenstein, W. *Amer. Mineralog.* **1954**, *39*, 92.

21. Kerr, G. T. *J. Phys. Chem.* **1967**, *71*, 4165.

22. Fyfe, C. A.; Kennedy, G. J.; Kokotailo, G. T.; DeSchutter, C. T. *J. Chem. Soc., Chem. Commun.* **1984**, 1093.

23. Fyfe, C. A.; Gobbi, G. C.; Murphy, W. J.; Ozubko, R. S.; Slack, D. A. *Chem. Lett.* **1983**, 1547.

24. Fyfe, C. A.; Gobbi, G. C.; Murphy, W. J.; Ozubko, R. S.; Slack, D. A. *J. Am. Chem. Soc.* **1984**, *106*, 4435.

25. Fyfe, C. A.; Gobbi, G. C.; Kennedy, G. J.; Graham, J. D.; Ozubko, R. Z.; Murphy, W. J.; Bothner-By, A.; Dadok, J.; Chesnick, A. S. *Zeolites* **1985**, *5*, 179.

26. (a) U. S. Patent 3,702,886; (b) Kokotailo, G. T.; Lawton, S. L.; Olson, D. H.; Meier, W. *Nature (London)* **1978**, *272*, 437; (c) Kokotailo, G. T.; Meier, W. M. *Spec. Publ. Soc. Chem.* **1980**, *33*, 133.

27. Fyfe, C. A.; Kokotailo, G. T.; Kennedy, G. J.; DeSchutter, C. T. *J. Chem. Soc., Chem. Commun.* **1985**, 306.

28. Fyfe, C. A.; Gobbi, G. C.; Klinowski, J.; Thomas, J. M.; Ramdas, S. *Nature (London)* **1982**, *296*, 530.

29. Fyfe, C. A.; Kennedy, G. J.; DeSchutter, C. T.; Kokotailo, G. T. *J. Chem. Soc., Chem. Commun.* **1984**, 541.

30. West, G. W. *Aust. J. Chem.* **1984**, *37*, 455.

31. Fyfe, C. A.; Kennedy, G. J.; Kokotailo, G. T.; Lyerla, J. R.; Fleming, W. W. *J. Chem. Soc., Chem. Commun.* **1985**, 740.

32. Hay, D. G.; Jaeger, H.; West, G. W. *J. Phys. Chem.* **1985**, *89*, 1070.

33. Fyfe, C. A.; O'Brien, J. H.; Strobl, H. *Nature (London)* **1987**, *363*, 6110.

34. Fyfe, C. A.; Strobl, H.; Kokotailo, G. T.; Pasztor, C. T.; Barlow, G. E.; Bradley, S. *Zeolites* **1988**, *8*, 132.

35. (a) Barri, S. A. J.; Smith, G. W.; White, D.; Young, D. *Nature* **1984**, *312*, 533; (b) Highcock, R. M.; Smith, G. W.; Wood, D. *Acta Crystallogr.* **1985**, *C41*, 1391; (c) Kokotailo, G. T.; Schlenker, J. L.; Dwyer, F. G.; Valyocsik, E. W. *Zeolites* **1985**, *5*, 349; (d) Marler, B. *Zeolites* **1987**, *7*, 393.

36. Lapierre, R. B.; Rohrman Jr., A. C.; Schlenker, J. L.; Wood, J. D.; Rubin, M. K.; Rohrbough, W. J. *Zeolites* **1985**, *5*, 346.

37. Gies, H.; Fyfe, C. A.; Kokotailo, G. T.; Marler, B.; Cox, J. E. *J. Phys. Chem.*, in press.

38. Fyfe, C. A.; Strobl, H.; Kokotailo, G. T.; Kennedy, G. J.; Barlow, G. E. *J. Am. Chem. Soc.* **1988**, 110, 3373.

39. Fyfe, C. A.; Gies, H.; Kokotailo, G. T.; Pasztor, C.; Strobl, H.; Cox, D. E. *Can. J. Chem.* **1988**, *66*, 1942.

40. Gies, H.; Fyfe, C. A.; Strobl, H.; Kokotailo, G. T.; Feng, Y.; Cox, D. E. Unpublished results.

41. Fyfe, C. A.; Kokotailo, G. T.; Kennedy, G. J..; DeSchutter, C. T. *J. Chem. Soc., Chem. Commun.* **1985**, 306.

42. Kokotailo, G. T.; Chu, P.; Lawton, S. L.; Meier, W. M. *Nature (London)* **1978**, *275*, 119.

43. Fyfe, C. A.; Gies, H.; Kokotailo, G. T.; Pasztor, C.; Strobl, H.; Cox, D. E. *J. Am. Chem. Soc.* **1989**, *111*, 2470.

44. Kennedy, G. J. Ph.D. Thesis, University of Guelph, 1985.

MOTION OF ORGANIC SPECIES OCCLUDED OR SORBED WITHIN ZEOLITES [1]

J. M. Newsam*†, B. G. Silbernagel†, M. T. Melchior†,
T. O. Brun§, and F. Trouw§

†Exxon Research and Engineering Company
Route 22 East
Annandale, NJ 08801 (USA)

§Intense Pulsed Neutron Source
Argonne National Laboratory, Argonne, IL 60439 (USA)

SUMMARY

Containment of organic species within microporous materials such as zeolites influences their modes of molecular motion. This manifests itself in a range of ways, including a reduction in the volume that is readily accessible to the molecules by translation, constraints upon the molecular conformational space, restricted modes of molecular reorientation, or altered vibrational frequencies. The perspectives on certain of the motional properties of sorbates and of tetramethylammonium cations within zeolites provided by simple molecular modelling, neutron powder diffraction, ^{13}C and $^2H(D)$ NMR, and inelastic and quasielastic neutron scattering are outlined and illustrated by selected recent results.

INTRODUCTION

Interest in the behavior of organic species within zeolites [2-7] arises from several different perspectives. Zeolites as catalysts and catalyst supports enjoy a dominant role in heterogeneous catalysis, particularly in petroleum and petrochemical processing [8]. Zeolites as sorbents and molecular sieves offer unique selectivities [3,9]. In addition, from a more fundamental standpoint, zeolites provide a medium in which the properties and phase behavior of organic or inorganic species can be studied in constrained environments of defined shapes, sizes and dimensionalities. Containment of sorbates or organic cations within microporous materials influences their modes of molecular motion. This manifests itself in a range of ways, including a reduction in the volume that is readily accessible to the molecules by translation, constraints upon the molecular conformational space, restricted modes of molecular reorientation that reflect the nature of interactions with internal surfaces, or alteration of the vibrational frequencies. Each of these aspects can be studied and exploited experimentally.

Inclusion Phenomena and Molecular Recognition
Edited by J. Atwood
Plenum Press, New York, 1990

We consider here a small number of the probes that are appropriate for studying the molecular motion of organic species within zeolites, focussing on the relationship between the influence of the zeolite on the motion of the sorbed or occluded phase, and the parameters that are measured experimentally. We illustrate these interrelationships by recent results of interactive molecular modelling, neutron powder diffraction, ^{13}C and $^2H(D)$ nuclear magnetic resonance (NMR), and inelastic and quasielastic neutron scattering.

A ROLE FOR MOLECULAR MODELLING USING ATOM-ATOM POTENTIALS

Computer modelling will ultimately provide the framework within which the results of experimental measures of molecular motion are interpreted. Currently, molecular graphics and molecular modelling methods provide a means of visualizing various types of molecular motion within zeolites and, in favorable circumstances, of estimating certain interaction energies [10,11]. The usual starting point for such computer simulations is a set of atomic coordinates determined from diffraction experiments for (at least) the framework components. Although developments in full energy minimization techniques promise to improve the reliability with which framework atomic coordinates can be computed based on a knowledge of composition and topology [12,13], the present degree of agreement between simulation and experiment is limited. The simpler, distance least squares approach [14,15] presently continues to yield better estimates of framework geometries, although subject also to limited agreement with experiment (due, in part, to neglect of the influence of non-framework species).

Several interactive molecular graphics packages are suitable for constructing models of zeolite structures from coordinates for atoms comprising the asymmetric unit deduced from modelling or crystallographic data [11,16]. An organic component(s) can then be introduced into this model, and various physical parameters computed as a function of its position as it is manipulated within the zeolite framework. For convenience, it is usually assumed that the framework geometry is unaltered by the presence and/or motion of the sorbed or occluded component (although depending on framework flexibility, considerable changes in structure can accompany sorption - desorption, see e.g. [17]). Parameters readily calculated include separation distances (and hence rigid-sphere steric constraints), angles and atom-atom potential energy sums. The atom-atom potential approach to modelling zeolite-sorbate systems has been pursued for several years, most notably by Kiselev and coworkers [11,18,19]. The interaction energy between a pair of atoms is described in terms of Lennard-Jones 6-12 potential (or some modification thereof), with parameters that are computed from atomic properties, or optimized so as to best reproduce experimental heats of sorption. The total energy of a given zeolite-sorbate configuration is taken as the sum of the individual atom-atom contributions, cross-summed over all atoms in the sorbate, and all framework (and non-framework) atoms out to a reasonable cut-off distance (frequently, the framework oxygen atoms alone are considered, for simplicity). Coulombic terms can be included, but in direct space summation converge only slowly with distance and are often omitted. It is encouraging that, despite these various simplifications, these methods have reproduced approximately observed isosteric sorption heats for some simple molecules in zeolites, such as, for example, the *n*-alkanes in sodium zeolites X and Y [19]. Considerable scope remains, however, for improving both the modelling methods and the parameterizations of the potential functions.

These modelling methods compute relative energies for different sorbate locations and conformations. The state of minimum computed energy predicts the configuration that will be predominantly populated at low temperature. Comparison with low-temperature structure results (see below) then provides a gauge of the validity of the methods, and, potentially, one means of improving the parameterization of the energy functions. The number of full structure determinations of zeolite-sorbate systems at low temperature, however, remains small. Powder neutron diffraction and molecular modelling were used jointly in studying pyridine in gallosilicate zeolite L [20]. In sodium zeolite Y at 4.2K, even at low loading levels, two distinct sites for benzene were observed, one in a capping position above a site II Na$^+$ cation, and the second lying in the 12-ring window [21]. In the former case the dominant interaction is between the positive cation and the π electron density of the benzene ring. In the latter case, the site stability arises from van der Waals interactions between the benzene protons and the oxygen atoms defining the 12-ring window, supplemented by those with adjacent benzene molecules in the site II capping positions on either side of the window. Although the 12-ring window site corresponds to a computed local energy minimum (Figure 1), calculations have not yet reproduced the relative energies of these two types of benzene site. An earlier atom-atom potential modelling study of benzene in sodium zeolite X [22] (isostructural with sodium zeolite Y, but of lower framework Si:Al ratio and, hence, with a larger total sodium cation population) had predicted a minimum energy site different from either of the observed positions, with the benzene apparently interacting with a site III Na$^+$ cation. The location predicted by recent modelling work for benzene in potassium zeolite L [23], likewise, is very different from that actually observed [24].

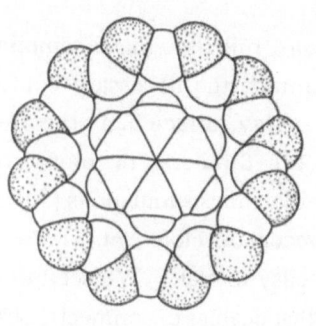
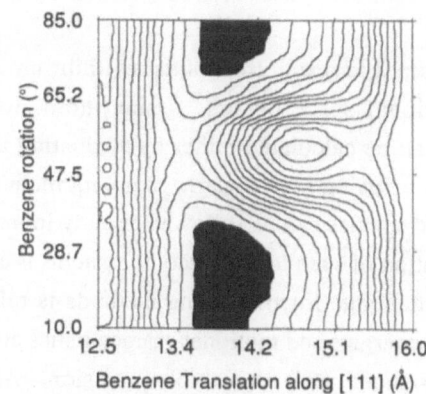

Figure 1. Siting of benzene in the 12-ring window site of the FAU-framework [21] (left), and a portion of the energy surface for benzene in the vicinity of such a site calculated using a simple atom-atom potential approach [33] (right - contour spacing 0.1 kcal mol^{-1}; lowest energy contour indicated by the shading).

Part of the difficulty in estimating energy minimum configurations reflects the problem of placing the different categories of interactions, sorbate-framework, sorbate-non-framework cation, and sorbate-sorbate, on the same absolute scale. In more favorable cases, however, where, for example, the full series of accessible sites involves merely differing degrees of interaction between the sorbate and the framework oxygen atoms, more sensible estimates of configurations can apparently be derived. Topical examples include studies of the behavior of aromatic molecules such as benzene and toluene in silicalite [25]. Studies of local sorbate reorientations or conformational changes can also be achieved with semiquantitative reliability. In the benzene-zeolite L system, for example, the benzene molecule is observed to occupy a capping position above the channel wall potassium cation [24] (Figure 2). The dominant interaction is the π electron density-K^+ cation interaction, which is insensitive to rotation of the benzene molecule about the unique axis. The strength of the non-bonded interactions between the benzene protons and the framework oxygen atoms, however, varies slightly with orientation. This gives rise to what is computed and observed to be a single preferred orientation at 78K, but with a relatively small activation barrier to 60° reorientations of the molecule [26].

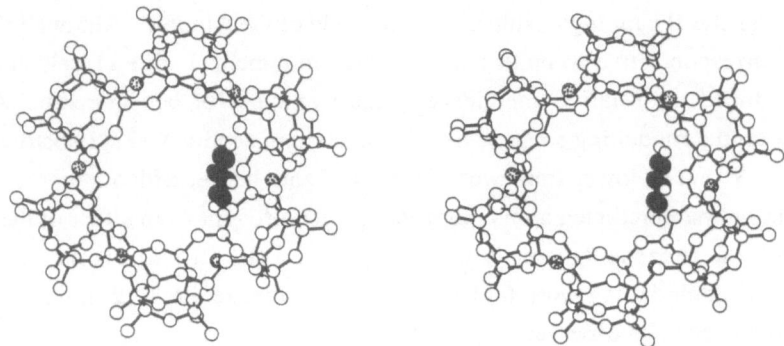

Figure 2. Stereoview illustrating the capping site for benzene above the channel wall potassium cation in zeolite L as determined by neutron powder diffraction [24] (K^+-speckled; C-solid; Si/Al, O, H-open).

At normal temperatures, under equilibrium conditions, full Boltzmann sampling of the configurational space occurs. Using atom-atom potentials, the necessary configuration integrals can be calculated either by evaluating the full energy surface in a gridwise fashion [18,22], or by using Monte Carlo sampling methods [27-29]. Similarly, the motion of sorbed or occluded species can be followed directly in molecular dynamics simulations [30], although the application of such tools to zeolite systems is a rather recent development [31,32].

The challenge for modelling methods is relatively easy to state. We seek to compute interaction energies and motional characteristics at a level that enables experimental observables to be reproduced with reasonable precision. Although our predictive capabilities remain primitive, present molecular modelling methods can already afford considerable insight when parameters are optimized for a specific system, or when relatively specific questions about the zeolite-sorbate system can be phrased.

OBSERVATION OF STATIC ZEOLITE-SORBATE STRUCTURES –
NEUTRON POWDER DIFFRACTION

Diffraction experiments do not directly reveal molecular motion, but rather provide structural data that is averaged over the full volume of the sample and over the full timescale of the data acquisition (typically several hours). Nevertheless, the general structural characteristics determine the scope for molecular mobility, and, as above, diffraction studies provide essential input for computer modelling work. Most synthetic zeolites are microcrystalline, with typical particle sizes of <5μm, necessitating a general reliance on powder diffraction techniques. The requirements both of sensitivity to scattering by the sorbed phase and of data accumulation at low temperatures determine that neutron diffraction is preferred. The size and complexity of zeolite structures, and materials related difficulties, make the application of neutron powder diffraction methods [33,34] far from routine. Nevertheless, successful studies of a small number of zeolite-hydrocarbon sorbate systems have been reported, including benzene in zeolites sodium Y [21], potassium L [24] (Figure 2) and ZSM-5 [35], and pyridine in gallosilicate zeolite L [18]. Information about the precise locations and conformations of organic species occluded during synthesis is scarce, although, for example, work on high silica sodalite containing trapped ethylene glycol [36], a single crystal X-ray study of ZSM-5 containing the tetrapropylammonium cation [37], and a powder X-ray diffraction study of tetramethylammonium (TMA) sodalite [38] have been described.

Good definition of the conformation and location of a sorbate or templating species is possible only when its degree of mobility is restricted and, hence, the sorbate appears well localized in the diffraction analyses. Frequently the organic species are subject to extensive static and/or dynamic disorder, even at relatively low temperatures, preventing the derivation of definitive structural data. In such cases, the perspectives provided by direct probes of molecular motion can be helpful in developing a full structural picture.

RESTRICTED TRANSLATIONAL FREEDOM

For an atom, under observation, that is moving within the confines of an enclosing cage, the observation will be an average over the points accessible by the atom, if the time scale of motion is short compared to the characteristic time for the measurement. For an observable which depends on some function of the separation from the perimeter that decays rapidly with distance, such as an induced dipole-induced dipole interaction, the observable will be a direct measure of the proportion of the total volume that is within some short distance of the surface. The observable will scale directly with some (mean) pore dimension. This dynamic picture provides one way of interpreting the variation observed in the magnitude of the ^{13}C chemical shift in tetramethylammonium (TMA) cations housed within zeolite cages of varying sizes (Figure 3). The ^{13}C chemical shifts observed for TMA^+ in TMA-sodalite and in ZK-4 at low Si:Al ratios are approximately equal, and are found to be characteristic of TMA^+ occluded within a sodalite cage [39]. For ZK-4 zeolites at higher Si:Al ratios, in which TMA^+ occupancy of the supercage also occurs, an additional peak appears in the ^{13}C magic angle spinning NMR (MAS NMR) spectrum. Slightly different chemical shifts are, indeed, observed depending on whether the supercage is singly or doubly occupied. This correlation between ^{13}C chemical shift and cage size [39] has been extended to a range of other materials, in which the size of the cage is intermediate [40,41] (Figure 3). The dynamic picture is, it should be emphasized, not the only suggested interpretation of this correlation, and a model based on the effective induced dipole-induced dipole interaction between a polarizable atom and a curved cage surface also reproduces the observed linear correlation between ^{13}C chemical shift and mean cage dimension [42].

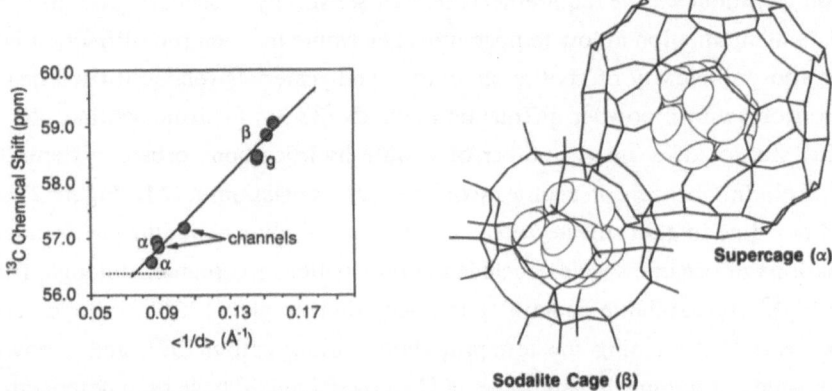

Figure 3. Representations of TMA[+] cations in the sodalite (β) and supercage (α) of zeolite ZK-4 (right), and plot of ^{13}C chemical shift against the mean of the reciprocal cage dimension for TMA[+] cations in various zeolites (left) [40]. The data points correspond to TMA[+] cations in the sodalite cages of sodium and lithium ZK-4 (similar chemical shifts) and TMA-sodalite (β); the gmelinite cages of zeolites omega and offretite (g); the supercages of zeolites ZK-4 and Y (α); and the continuous channels of zeolites offretite and omega. The dashed horizontal line indicates the chemical shift observed for free TMA[+] cations in solution [40].

The number of zeolites synthesized containing occluded TMA[+] cations is limited and, because of its size, TMA[+] can be introduced by ion-exchange into only the largest pore systems. The scope for using the TMA[+] [13]C chemical shift as a probe of new pore systems is therefore limited. The technique has, however, been successfully applied to characterizing intergrowths and associated pore blockage in synthetic offretites [40]. The OFF-framework of offretite (with linear 12-ring channels and gmelinite cages) can be described in terms of parallel stackings of single 6-ring units in an AABAAB sequence [43]. The ERI-framework of erionite (with elongated erionite cages) is constructed similarly from 6-ring units, but in an AABAAC sequence [43]. Presumably reflecting, at least in part, their similar stacking sequences, the two topologies frequently intergrow, synthetic zeolite T being a well-known example. The OFF-ERI intergrowths or stacking disorder give, depending on the inter-fault spacing, cages of varying dimensions. The ^{13}C MAS NMR spectra of TMA[+] cations occluded within faulted synthetic offretites thus comprise broad, overlapped peaks that reflect the numbers of the various cages ranging in size from the gmelinite cage through the erionite cage to the continuous OFF-framework channel. Interpretation of such spectra in terms of discrete decomposed intensities for the different components then provides quantitative information on the characteristics of the stacking disorder [40].

The chemical shift of ^{129}Xe is also found to correlate linearly with the mean dimension of the cage within which it resides [42,44], consistent with the determining factor being induced dipole-induced dipole interactions. In some senses the use of ^{129}Xe NMR has advantages for xenon can be introduced into the cage structures of many zeolites and related microporous crystalline solids either during crystallization [45] or by sorption [44], typically under applied external pressure.

CONSTRAINED MOLECULAR REORIENTATION – ^{2}H(D) NMR

Deuterium, ^{2}H(D), is a quadrupolar, I = 1 nucleus and, in a hydrocarbon, its resonance properties are dominated by the interaction between the ^{2}H(D) quadrupole moment and the electric field gradient at the nucleus (arising from the asymmetry of the C-D bond). The energy differences between the -1 and 0, and 0 and 1 states depend on the angle, θ, of the C-D bond with respect to the external magnetic field. For a polycrystalline sample in which the hydrocarbon molecules are static, a horned powder spectrum is observed, reflecting the population distribution of C-D bond orientations. Molecular motion that is faster than the inverse of the quadrupolar coupling constant, τ, (roughly 10^{-6} s), leads to a reduction in the effective quadrupolar coupling constant, resulting, most noticeably, in a reduction in the separation between the two horns in the spectrum (which correspond respectively to the θ = 90º orientations for the -1 → 0 and 0 → 1 states). For fast motion, the magnitude of the reduction is related directly to the angle between the C-D bond direction and the rotation axis, β, as $[3\cos^2\beta - 1]/2$. Thus, rotation about the magic angle (β = 54.74º) completely averages out the relative orientation with respect to the field, resulting in zero effective coupling and hence a narrow signal. Rotation about an axis perpendicular to the C-D bond, β = 90º, leads to a reduction factor of exactly one half, etc. Clearly, therefore, the appearance of the ^{2}H(D) spectrum can reveal the character of the molecular motion, at least for time scales $<<10^{-6}$ s.

Following initial experiments on *para*-xylene in zeolite ZSM-5 [46,47], measurements on a series of systems have been described, including benzene in zeolites X [48,49] and ZSM-5 [47,49]. The constraints on the character of the molecular motions imposed by containment within the zeolite micropores prove to be a rich area of study. A series of studies on the three C-6 hydrocarbons benzene, *n*-hexane, and cyclohexane sorbed at various loading levels in potassium zeolite L [26] provides a good illustration of the types of information that are relatively readily accessible by the technique.

Over a range of loading levels and temperatures, -100° ≤ T ≤ 23°C, benzene (C_6D_6) molecular motion is dominated by spinning in the plane of the molecule. The ^{2}H(D) spectrum shows a reduction in the horn splitting by a factor of 2.0 compared with the static molecular value, consistent with reorientation about an axis at β = 90° that is faster than ≈10^{-6}s. This mode of motion is consistent with the benzene location observed by neutron powder diffraction [24]. Sorbed *n*-hexane (C_6D_{14}) also shows no evidence for isotropic motion but gives rise to two distinct spectral components, associated respectively with the methyl and methylene deuterons [26] (Figure 4). Describing the *n*-hexane molecular motion solely in terms of rot-ations about the tetrahedral carbon-carbon bonds reproduces the observed ^{2}H(D) quadrupolar coupling strengths. This mode of reorientation implies the presence of a significant, but non-specific zeolite-sorbate interaction, that maintains the n-hexane in conformations that support the zeolite-sorbate interaction, while permitting C-C bond reorientations. Sorbed cyclohexane apparently provides an example where competition between the internal degrees of freedom and the zeolite-sorbate interaction gives rise to interesting dynamical effects [26]. Close to 0°C a narrow ^{2}H(D) signal is observed, but above and below 0°C there is a residual quadrupole interaction. This particular system highlights that although ^{2}H(D) NMR spectra of zeolite-sorbate systems are relatively readily obtainable, interpretations are not always completely straightforward. Detailed simulations of the line-shapes anticipated in the presence of various modes and time-scales of motion may be necessary to enable definitive conclusions to be drawn. The ^{2}H(D) NMR experiment itself is also considerably richer than suggested by this discussion of the line-shape alone. The temperature dependence of the line-shape and/or spin-lattice relaxation time, for example, can help in enabling a self-consistent picture of the sorbate dynamics to be derived [26].

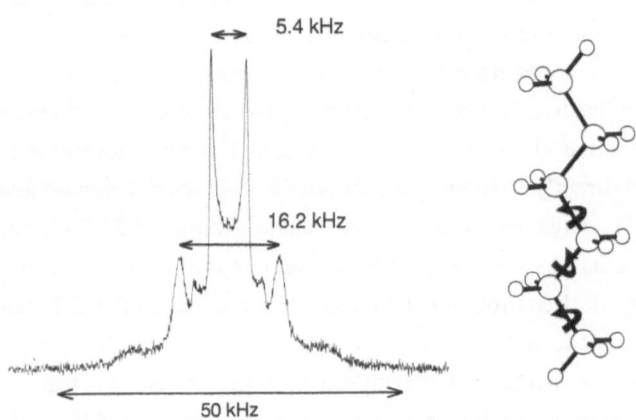

5.4 kHz

16.2 kHz

50 kHz

Figure 4. The ^{2}(H)D NMR spectrum for C_6D_{14} in potassium zeolite L [26] (left) has distinct resonances for the methyl and methylene deuterons. The magnitudes of the effective quadrupolar coupling constants can be interpreted in terms of rotations about the individual C-C bonds [26] (right).

ALTERED VIBRATIONAL FREQUENCIES – INELASTIC NEUTRON SCATTERING

Containment within the zeolite has various mechanisms whereby molecular vibrational frequencies can be altered. For example, steric constraints associated with close collision with framework or non-framework species can modify the shape of the potential wells, particularly at shorter separations, or close interactions with framework or non-framework species (such as the π electron density - K^+ cation interaction between benzene and zeolite L) can influence the molecular force constants. Studies of the vibrational properties of sorbed or occluded organic species within zeolites have predominantly relied on infrared and Raman spectroscopies. Incoherent inelastic neutron scattering (IINS), as a technique, has been less well applied, but provides certain advantages for studying these types of system [see, e.g. 50,51].

The incoherent neutron scattering cross-section for hydrogen is extremely large compared to that of other elements. The low proton mass implies that the extent of hydrogen atom motion associated with vibrational modes is relatively large. For these reasons, modes involving hydrogen atom motion dominate the IINS spectra. In addition, the neutron scattering process is not subject to the selection rules that determine the observability of modes in optical spectroscopies, and full quantitative treatment of the IINS spectra is possible. Facilities for performing IINS measurements are accessible at most of the major neutron scattering centers.

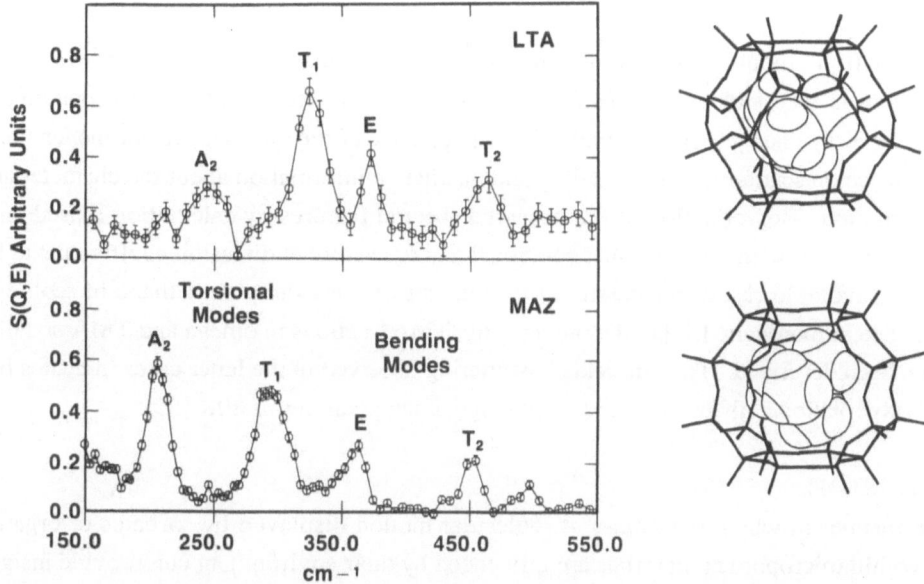

Figure 5. The inelastic neutron scattering spectra for TMA^+ cations in zeolites ZK-4 (labelled LTA) (at Si:Al = 1:3; only sodalite cages contain TMA^+) and omega (MAZ) [53], and representations of the TMA^+ cation occluded in a sodalite (upper right) and a gmelinite cage (lower right).

The vibrational spectrum of the tetramethylammonium cation in the region 150-550 cm^{-1} contains both torsional and vibrational modes. The v_8 and v_{19} vibrational modes of E and T_2 symmetry involve C-N-C bond angle bending. These modes are Raman active and have been studied for TMA^+ in several zeolite environments, although little change in frequency is observed [52]. The v_4 and v_{12} torsional modes involve partial rotation about C-N bonds and form respectively a singlet (A_2) and a triplet (T_1) which are both Raman inactive. These torsional modes are directly observed in the IINS spectra and prove to be quite sensitive to the character of the TMA^+ cation environment.

In the halide salts, TMAX, X = Cl, Br, I, the torsional mode frequencies evolve smoothly to lower force constants in the series Cl-Br-I [53]. The crystal structures are similar, with each TMA^+ cation tetrahedrally coordinated by the halide ions. The change in torsional frequencies correlates with the anion polarizabilities. The TMA^+ torsional frequencies observed in zeolites

[54,55] (Figure 5) indicate a weaker interaction between the TMA^+ cation and its environment. The torsional frequencies also vary significantly depending on the size of the cage within which the TMA^+ cation is occluded. Thus, the IINS spectra, $150 \leq v \leq 550 \text{ cm}^{-1}$, for TMA-sodalite [55], SAPO-20 [55] and zeolite ZK-4 (at a framework Si:Al = 1:3) [54] are similar, and characteristic of TMA^+ in a sodalite cage. Slightly lower torsional frequencies are observed for TMA^+ cations occluded in the gmelinite cages of zeolite omega [54], with still lower values observed for the torsional modes ascribed to TMA^+ cations in the supercage of zeolite ZK-4 [55]. The correlation suggested by these observations is further strengthened by estimates of torsional frequencies computed for the free TMA^+ ion based on a Hartree-Fock ab initio treatment [54].

Information about translational and rotational diffusion of sorbed or occluded hydrocarbons is accessible in measurements in the quasielastic regime, that corresponding to very small energy changes, $\Delta E < 1 \text{ meV}$. The appearance of the quasielastic component, and its dependence on scattering angle readily yield qualitative information about the character and extent of motion. Reproduction of the observed spectral features by calculation (based on a suitable model for the motion) enables rotational and translational diffusion coefficients to be deduced in favorable cases. Successful measurements of, for example, methane in zeolite A [56], benzene in mordenite [57] and more recently, TMA^+ cations in omega and TMA-sodalite [55] have been described. The quasielastic scattering observed in the latter cases indicates full TMA^+ body rotational diffusion over a wide range of temperatures $\geq 80K$ [55].

CONCLUSION

The manner in which the modes of molecular motion displayed by sorbates or organic cations within microporous materials are influenced by their environment can provide insight into the character of the interaction between the organic component and the host. Several techniques can be used to study the motions of sorbed or occluded organic components, with each appropriate for a particular subset of motional constraints. We have illustrated the use of interactive molecular modelling, neutron powder diffraction, ^{13}C and $^{2}(H)D$ nuclear magnetic resonance (NMR), and inelastic and quasielastic neutron scattering. The studies cited, however, underline that while individual techniques can afford definitive conclusions about molecular motion in certain cases, the most detailed pictures are produced only when a range of different techniques is in concert brought to bear on a particular problem.

Acknowledgments. We thank the many individuals who have contributed to our own studies that are mentioned here, most particularly A. R. Garcia, R. H. Jarman, D. E. W. Vaughan, A. J. Jacobson, R. Hulme, L. E. Iton, L. A. Curtiss and the staff of the Intense Pulsed Neutron Source at Argonne National Laboratory.

References

1. Work supported in part by the U. S. Department of Energy, BES-Materials Sciences, under contract W-31-109-ENG-38.

2.	Breck, D. W. *Zeolite Molecular Sieves: Structure, Chemistry and Use*; Wiley and Sons: London, 1973. Reprinted, R. E. Krieger: Malabar Fl, 1984.

3.	Barrer, R. M. *Zeolites and Clay Minerals as Sorbents and Molecular Sieves*; Academic Press: London, 1978.

4.	Barrer, R. M. *Hydrothermal Chemistry of Zeolites*; Academic Press: London, 1982.

5.	Olson, D.; Bisio, A. Eds. *Proc. Sixth Int. Zeolite Conf.*; Butterworths: Surrey, UK, 1984.

6.	Murakami, Y.; Iijima, A.; Ward, J. W. Eds. *New Developments in Zeolite Science and Technology*; Kodansha: Tokyo - Elsevier: Amsterdam, 1986.

7.	Newsam, J. M. *Science* **1986**, *231*, 1093-1099.

8.	Vaughan, D. E. W. In *Properties and Applications of Zeolites*; Chem. Soc.: London, 1980; Special Pub. No. 33, pp. 294-328.

9.	Ruthven, D. M. *Principles of Adsorption and Adsorption Processes*; Wiley-Interscience: New York, 1984.

10.	Suffritti, G.; Gamba, A. *Int. Rev. Phys. Chem.* **1987**, *6*, 299-314.

11.	Newsam, J. M. *Aspects of Zeolite Structure Modeling and Graphics*, in preparation.

12.	Jackson, R. A.; Catlow, C. R. A. *Molec. Simulation*, in press.

13.	Ooms, G.; van Santen, R. A.; den Ouden, C. J. J.; Jackson, R. A.; Catlow, C. R. A. *J. Phys. Chem.* **1988**, *92*, 4462-4465.

14.	Meier, W. M.; Villiger, H. *Z. Kristallogr.* **1969**, *129*, 411-423.

15.	Baerlocher, Ch.; Hepp, A.; Meier, W. M. DLS-76 - *A Program for Simulation of Crystal Structures by Geometric Refinement* (ETH, Zurich Report, 1977).

16.	Ramdas, S.; Thomas, J. M.; Betteridge, P. W.; Cheetham, A. K.; Davies, E. K. *Angew. Chem., Int. Ed. Engl.* **1984**, *23*, 671-679.

17.	Fyfe, C. A.; Kennedy, G. J.; De Schutter, C. T.; Kokotailo, G. T. *J. Chem. Soc., Chem. Commun.* **1984**, 541-542.

18.	Kiselev, A. V. *J. Chem. Tech. Biotechnol.* **1979**, *29*, 673-685.

19.	Kiselev. A. V.; Du, P. Q. *J. Chem. Soc., Faraday Trans. II* **1981**, *77*, 17-32.

20.	Wright, P. A.; Thomas, J. M.; Cheetham, A. K.; Nowak, A. K. *Nature (London)* **1985**, *318*, 611-614.

21.	Fitch, A. N.; Jobic, H.; Renouprez, A. *J. Phys. Chem.* **1986**, *90*, 1311-1318.

22.	Bezus, A. G.; Kocirik, M.; Kiselev, A. V.; Lopatkin, A. A.; Vasilyeva, E. A. *Zeolites* **1986**, *6*, 101-106.

23.	Nowak, A. K.; Cheetham, A. K. In *New Developments in Zeolite Science Technology*; Murakami, Y.; Iijima, A.; Ward, J. W. Eds. Kodansha: Tokyo - Elsevier: Amsterdam, 1986; pp. 475-579.

24.	Newsam, J. M.; Silbernagel, B. G.; Garcia, A. R.; Hulme, R. *J. Chem. Soc., Chem. Commun.* **1987**, 664-666.

25.	Nowak, A. K.; Cheetham, A. K.; Pickett, S. D.; Ramdas, S. *Molec. Simulation*, in press.

26.	Silbernagel, B. G.; Garcia, A. R.; Newsam, J. M.; Hulme, R., in preparation.

27. Binder, K. Ed. *Monte Carlo Methods in Statistical Physics*, Second Edition; Springer-Verlag: New York, 1987.

28. Fieldler, K.; Grauert, B. *Adsorption Sci. Technol.* **1986**, *3*, 181-187.

29. Kono, H.; Takasaka, A. *J. Phys. Chem.* **1987**, *91*, 4044-4055.

30. Heerman, D. *Computer Simulation Methods in Theoretical Physics*; Springer-Verlag: New York, 1986.

31. Demontis, P.; Suffritti, G. B.; Alberti, A.; Quartieri, S.; Fois, E. S.; Gamba, A. *Gazz. Chim. Ital.* **1986**, *116*, 459-466.

32. Yashonath, S.; Thomas, J. M.; Nowak, A.; Cheetham, A. K. *Nature (London)* **1988**, *331,* 601-604.

33. Newsam, J. M. *Physica* **1986**, *136B*, 213-217.

34. Newsam, J. M. *Materials Science Forum* **1987**, *27/28*, 385-396.

35. Taylor, J. C. *Zeolites*, **1987**, *7*, 311-318.

36. Richardson, J. W.; Pluth, J. J.; Smith, J. V. *J. Phys. Chem.* **1988**, *92*, 243-247.

37. van Koningsveld, H.; van Bekkum, H.; Jansen, J. C. *Acta Cryst.* **1987**, *B43*, 127-132.

38. Baerlocher, Ch.; Meier, W. M. *Helv. Chim. Acta* **1969**, *52*, 1853-1860.

39. Jarman, R. H.; Melchior, M. T. *J. Chem. Soc., Chem. Commun.* **1984**, 414-415.

40. Melchior, M. T.; Vaughan, D. E. W.; Jarman, R. H.; Jacobson, A. J. presented at *Rocky Mtn. Conf. Applied Spectroscopy*, Denver CO, Aug. 8, 1984.

41. Hayashi, S.; Suzuki, K.; Shin, S.; Hayamizu, K.; Yamamoto, O. *Chem. Phys. Lett.* **1985**, *113*, 368-371.

42. Derouane, E. G.; Nagy, J. B. *Chem. Phys. Lett.* **1987**, *137,* 341-344.

43. Meier, W. M.; Olson, D. H. *Atlas of Zeolite Structure Types*; Butterworths: London, 1987.

44. Fraissard, J.; Ito, T.; Springuel-Huet, M.; Demarquay, J. In *New Developments in Zeolite Science Technology*; Murakami, Y.; Iijima, A.; Ward, J. W. Eds.; Kodansha: Tokyo - Elsevier: Amsterdam, 1986; p. 393.

45. Ripmeester, J. A.; Ratcliffe, C. I., in preparation.

46. Eckman, R. R.; Vega, A. J. *J. Am. Chem. Soc.* **1983**, *105*, 4841-4842.

47. Eckman, R. R.; Vega, A. J. *J. Phys. Chem.* **1986**, *90*, 4679-4683.

48. Hasha, D. L.; Miner, V. W.; Garces, J. M.; Rocke, S. C. *ACS Symp. Ser.* **1985**, *288*, 485-497.

49. Zibrowius, B.; Caro, J.; Pfeifer, H. *J. Chem. Soc., Faraday Trans. II* **1988**, *84*, 2347-2356.

50. Egelstaff, P. A.; Stretton Downes, J.; White, J. W. In *Molecular Sieves*; Barrer, R. M. Ed.; Soc. Chem. Ind.: London, 1968; pp. 306-318.

51. Howard, J.; Waddington, T. C.; Wright, C. J. *J. Chem. Soc. Faraday Trans. II* **1977**, *73*, 1768-1787.

52. Dutta, P. K.; Del Barco, B.; Shieh, D. C. *Chem. Phys. Lett.* **1986**, *127*, 200-204.

53. Ratcliffe, C. I.; Waddington, T. C. *J. Chem. Soc., Faraday Trans. II* **1976**, *72*, 1935-1956.

54. Brun, T. O.; Iton, L. E.; Kleb, R.; Newsam, J. M.; Beyerlein R. A.; Vaughan, D. E. W. *J. Am. Chem. Soc.* **1987**, *109*, 4118-4119.

55. Brun, T. O.; Trouw, F.; Curtiss, L. A.; Iton, L. E.; Newsam, J. M., in preparation.

56. Cohen de Lara, E.; Kahn, R.; Mezei, F. *J. Chem. Soc., Faraday Trans. II* **1983**, *79*, 1911-1920.

57. Jobic, H.; Bee, M.; Renouprez, A. *Surf. Sci.* **1984**, *140*, 307-320.

INCLUSION OF ORGANOMETALLICS IN ZEOLITE HOST STRUCTURES

Thomas Bein,* Karin Moller, and Aticha Borvornwattananont

Department of Chemistry
University of New Mexico
Albuquerque, New Mexico 87131 (USA)

SUMMARY

An overview is given on new synthetic strategies to stabilize organometallic fragments in the cage system of large-pore zeolites. A common theme is the utilization of intrazeolite bridged hydroxyl groups as reactive centers for surface chemistry. The intracavity chemistry of $[CpFe(CO)_2]_2$ (Fp2), $COTFe(CO)_3$ (COT), $CpFe(CO)_2CH_3$ and ferrocene in different acid forms of zeolite Y has been studied with EXAFS, in situ FTIR, and TPD-MS spectroscopies. Depending on the stoichiometry of zeolite protons vs. the amount of starting complex, different reaction routes are observed, including oxidative cleavage or protonation of Fp2 and ligand rearrangement of COT. The stability of all complexes is influenced by the intracavity concentration of the zeolite bridged hydroxyl groups. Upon treatment at higher temperatures under vacuum, carbonyl and other ligands split off from the complexes and the remaining fragments attach to the zeolite host structure via oxygen coordination. These fragments are stabilized against migration and agglomeration.

INTRODUCTION

Much recent work has been devoted to the immobilization of organometallic catalysts on solid supports. The incentive for these research efforts is based upon the notion that it should be possible to combine the advantages of homogeneous catalysts, such as high selectivity, mild reaction conditions, or the potential utilization of all metal atoms, with those of heterogeneous systems, i.e., facile product separation, facile recovery of the expensive catalyst, and inherent stability. The new 'hybrid' systems derived from this strategy could even offer potential new, desirable features such as stabilization against aggregation, or greater flexibility in the choice of reaction media. However, hybrid systems can also present complications, including the almost ubiquitous instability against leaching of the catalyst metal into solution, agglomeration resulting in (undesired) metal particles, and failure to achieve true site isolation (on organic supports).

Inclusion Phenomena and Molecular Recognition
Edited by J. Atwood
Plenum Press, New York, 1990

In contrast to their amorphous counterparts, zeolite molecular sieves are highly crystalline oxides with well-defined pore sizes of typical molecular dimensions. They offer an enormous variety in pore structures and dimensions, allow controlled modifications of the internal surface, and in many cases show substantial thermal and chemical stability. In addition to the potential advantages of conventional hybrid systems, intrazeolite catalysts offer diffusional shape selectivity for substrate and product molecules, selectivity of polar vs. nonpolar substrates, and in favorable cases so-called 'transition state selectivity' which affects the transition state of the catalytic reaction. The following intrazeolite deposition concepts for catalytically active centers can be distinguished:

1. *Physisorption.* Neutral metal carbonyls have been adsorbed in large-pore zeolites, primarily in faujasite-type structures [1,2]. We have recently shown that zeolite metal cations such as Na^+ provide weak binding sites for the carbonyl ligands [3]. However, this interaction does not prevent carbonyl complexes such as $Ni(CO)_4$ [4], $Fe(CO)_5$, or $Mo(CO)_6$ from diffusion, migration out of the pore system, and eventual agglomeration at elevated temperatures [5,6].

2. A stronger, electrostatic binding mode to the anionic framework is achieved with *cationic transition metal species*. Here, transition metal cations are introduced into the zeolite framework via aqueous ion exchange and subsequently exposed to CO to form carbonyl complexes. Intrazeolite Ru [7], Ir [8], and particularly Rh carbonyl complex cations have been studied in great detail [9-11]. Migration of intrazeolite Rh-carbonyl species and eventual formation of extrazeolite Rh(0) particles appears to occur under the experimental conditions used for catalytic hydroformylation reactions [12]. The structural instability of the Rh-CO-zeolite system presents major limitations towards utilization of the intrazeolite pore structure for shape-selective catalytic reactions. The ion exchange and dehydration steps applied in metal-ion precursor concepts do not offer much control over the siting of the metal species because the metal ions typically migrate into stable positions in smaller cages upon dehydration.

3. Recently, the immobilization of intrazeolite complexes by *diffusional blocking* ("ship in the bottle") has been explored. This approach requires a rigid ligand sphere as present in phthalocyanine (Pc) and other chelate complexes. Intrazeolite Co(II)salen prepared by Co(II) ion exchange and successive reaction with the ligand was found to bind molecular oxygen [13], and intrazeolite Co- [14] and Fe-Pc [15] complexes showed catalytic activity in selective olefin oxidations. Critical issues include potential pore-blocking by the large complex (only three-dimensional or two-dimensional networks can be used with very low concentrations of the complex), and limited substrate access to the catalytically active metal center. In some cases, the severe synthesis conditions can cause local lattice breakdown and inhomogeneous siting.

4. Anchoring concepts based upon *bridged zeolite hydroxyl groups*. In view of the preceding discussion it becomes clear that a qualitative improvement is needed to form stable

inclusions of organometallics in the zeolite pore system. We explore an anchoring concept which utilizes the bridged hydroxyl groups present in the acid forms of many zeolite structure types. The protons located at the zeolite Si-O-Al bridges are highly acidic. The acid sites represent an attractive, largely unexplored type of reactive site for attaching organometallic fragments, with the following potential advantages: Since small species can be anchored, pore clogging can be avoided and a homogeneous site distribution is likely. A variety of metals, ligands, and attachment chemistry can be explored, and the mild reaction conditions allow to direct the metal siting into the large cage systems.

In analogy to the surface chemistry of transition metal allyl complexes on amorphous oxide supports [16-18] the reaction of Rh(allyl)$_3$ with partially proton-exchanged X and Y type zeolite has been reported [19-22]. The formation of intrazeolite Rh-CO and -hydride species as inferred from infrared data indicated that the anchored Rh-allyl fragment reacted similarly to the complex in solution. Preferential hydrogenation of n-olefins vs. large cycloolefins suggested that the catalytically active sites were indeed located and accessible in the zeolite pore structure. Structural information about these systems is still lacking.

We study the different relative reactivities and stabilities of ligands at a metal center with respect to the bridged intrazeolite hydroxyls as a basis for the rational design of zeolite based hybrid catalysts. Reactions of [CpFe(CO)$_2$]$_2$ [23], COTFe(CO)$_3$ [24], CpFe(CO)$_2$CH$_3$ and ferrocene in different acid forms of zeolite Y are described in the following.

EXPERIMENTAL

Sample preparation: Four different supports based upon Y zeolite were used in this study: NaY (commercial Linde LZ-Y52, [Na$_{57}$Al$_{57}$Si$_{135}$O$_{384}$] x 235 H$_2$O), partially proton exchanged H2Y derived from LZ-Y52 via ion exchange with 2 NH$_4^+$ per supercage, and highly acidic H6Y (6 H$^+$/sc, sc = supercage) derived from Linde LZ-Y62, [(NH$_4$)$_{45}$Na$_{10}$Al$_{55}$Si$_{137}$O$_{384}$] x 235 H$_2$O. Heating under vacuum at 1K/min up to 700 K gave the desired acid form of the zeolite. EXAFS samples were derived from proton-exchanged, thermally stabilized zeolite Linde LZ-Y72 with a Si/Al ratio of 2.55 (HY). With [CpFe(CO)$_2$]$_2$ and FeCp$_2$, the EXAFS samples showed IR spectra corresponding to those obtained with H2Y. With COTFe(CO)$_3$ they were equivalent to those obtained with H6Y. Loading with the complexes was accomplished by stirring a slurry of 0.500 g of zeolite with the required amounts of organometallic compound in 50 mL of hexane for twelve hours under nitrogen. The three supports were loaded at an average level of 0.5 molecules of [CpFe(CO)$_2$]$_2$ (Fp2), 1 molecule of COTFe(CO)$_3$ (COT), and 1 molecule of CpFe(CO)$_2$CH$_3$ (MeFp) per supercage. Ferrocene (FeCp$_2$) concentrations were formally 2 molecules per supercage.

FTIR-TPD-MS: FTIR data were obtained with a Mattson Polaris spectrometer at 4 cm^{-1} resolution. Thin dispersions of the zeolite samples on Si wafers were heated at 1 K/min in an in-situ cell connected to an ultrahigh vacuum thermodesorption apparatus combined with a quadrupole mass spectrometer (Dycor M200, 1-200 amu).

EXAFS: EXAFS measurements of sealed samples were performed at NSLS Brookhaven National Laboratories at beamline X-11A with a stored electron energy of 2.5 GeV and ring currents between 60-110 mA. Fe K-edge data at 7112 eV were collected at about 100 K in transmission using a Si(400) monochromator. The EXAFS data were analyzed following standard procedures [25].

RESULTS AND DISCUSSION

Intrazeolite Reactions of [CpFe(CO)$_2$]$_2$ at Room Temperature

The adduct between Fp2 and NaY is characterized by a shift of the CO stretching frequencies with respect to the unsupported complex (Figure 1a). This indicates an interaction between the Na$^+$ ions and the CO ligands (see discussion for COT). The oxidative cleavage reactions of [CpFe(CO)$_2$]$_2$ in homogeneous medium have been reported [26-30]. The dimer reacts in HCl/CHCl$_3$ to CpFe(CO)$_2$Cl, which forms the tricarbonyl cation CpFe(CO)$_3$$^+$ under CO pressure [31]. Proton exchanged zeolites (pK$_a$ < -3) represent solid acids of strength comparable to concentrated mineral acids [32]. It is found in this study that the tricarbonyl cation forms in the pores of the acid zeolite hosts H2Y and H6Y via oxidation of the precursor dimer at room temperature (IR spectra are shown in Figure 2). The EXAFS data of sample H2Y/Fp2 agree very well with this reaction path as described in the following: the

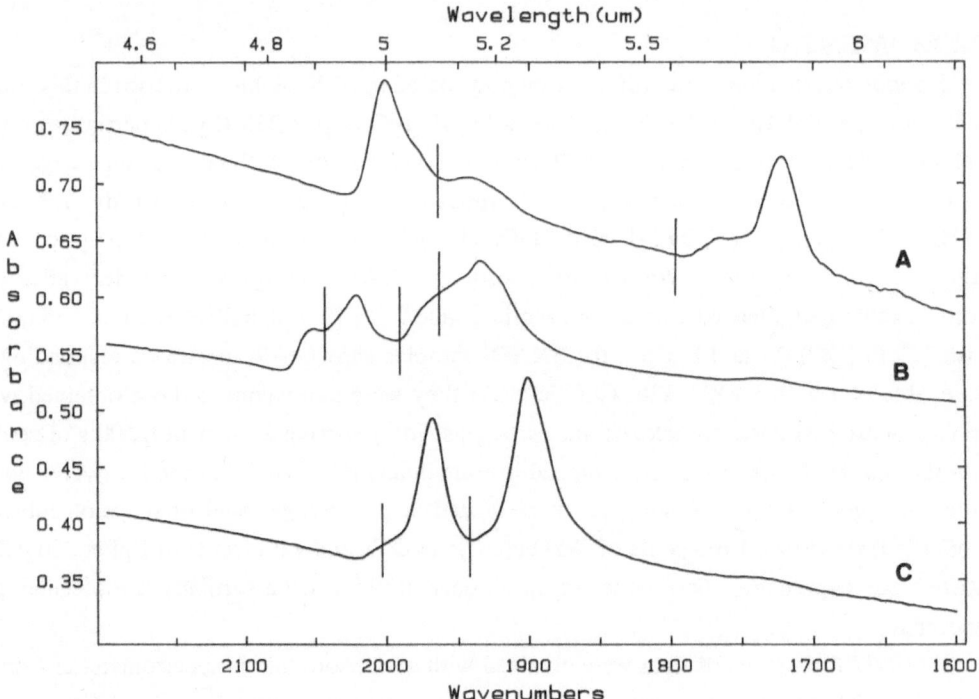

Figure 1. FTIR spectra of a) [CpFe(CO)$_2$]$_2$, b) COTFe(CO)$_3$ and c) CpFe(CO)$_2$CH$_3$ in the neutral zeolite matrix NaY at room temperature. Markers indicate the literature values of the unsupported complexes in KBr.

intensity of the iron-iron scatterer pair observed in the EXAFS spectrum of the unsupported complex is reduced to 20% after adsorption into the zeolite, indicating cleavage of the metal-metal bonds of the dimer. Coordination distances and numbers are in line with the formation of $CpFe(CO)_3^+$. The EXAFS data are consistent with the presence of about 40% $CpFe(CO)_3^+$, 40% $CpFe(OZ)_2$, and a remainder of 20% of the dimer (OZ: zeolite oxygen atoms). The formation of the cationic complex $CpFe(CO)_3^+$ is not only determined by the relative concentration and strength of intrazeolite acid groups present, but it is also kinetically controlled (adsorbed water does not affect this process).

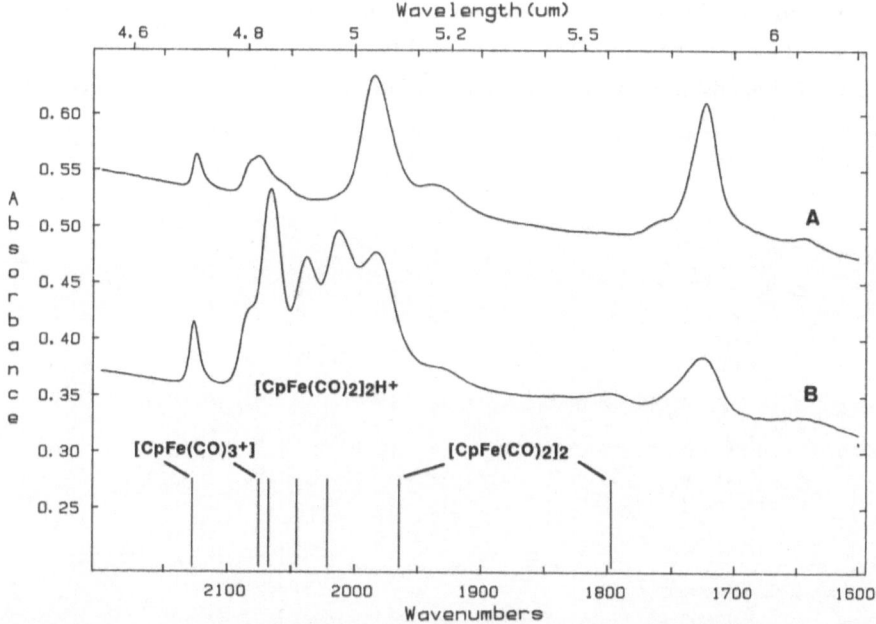

Figure 2. FTIR spectra of $[CpFe(CO)_2]_2$ in acid supports at room temperature. A) in H2Y and b) in H6Y. Markers indicate the literature values for $CpFe(CO)_3^+$ (H_2SO_4 film), $[Cp(CO)_2Fe]_2H^+$ (CH_2Cl_2), and $[CpFe(CO)_2]_2$ (KBr).

Based upon the combined EXAFS and FTIR/TPD-MS data it is demonstrated that the reactivity of the original complex $[CpFe(CO)_2]_2$ is very dependent upon the concentration of acid hydroxyls offered by the solid reaction medium. An increasing relative amount of protons favors the oxidation of the dimer to the monomeric cation. The zeolite host H6Y with even higher proton activity opens a reaction path to form the protonated dimer $[Cp(CO)_2Fe]_2H^+$ (Figure 2b). The analogous reaction in homogeneous medium has been described in a recent publication [33]. The intrazeolite reactions of $[CpFe(CO)_2]_2$ at room temperature can be understood according to the tentative reaction pathways shown in Equations (1) and (2):

medium acidity (H2Y/Fp2):

$$[CpFe(CO)_2]_2 + 2\,ZOH + OZ \quad \rightarrow \quad [CpFe(CO)_3]^+OZ^- + H_2 + CO + CpFe(OZ)_2 \tag{1}$$

(ZOH, zeolite bridged hydroxyls)

high excess of protons (H6Y/Fp2):

$$[CpFe(CO)_2]_2 + x\,ZOH \quad \rightarrow \quad \{[Cp(CO)_2Fe]_2H\}^+OZ^- + (x-1)\,ZOH \tag{2}$$

Intrazeolite Reactions of [CpFe(CO)₂]₂ at Elevated Temperatures

If sample H2Y/Fp2 is heated at 473 K for 10 hours, striking changes of the EXAFS data can be observed (compare Figure 3a,b). The small Fe-Fe backscattering contribution and that from linear CO ligands present at room temperature are eliminated. Two peaks near the

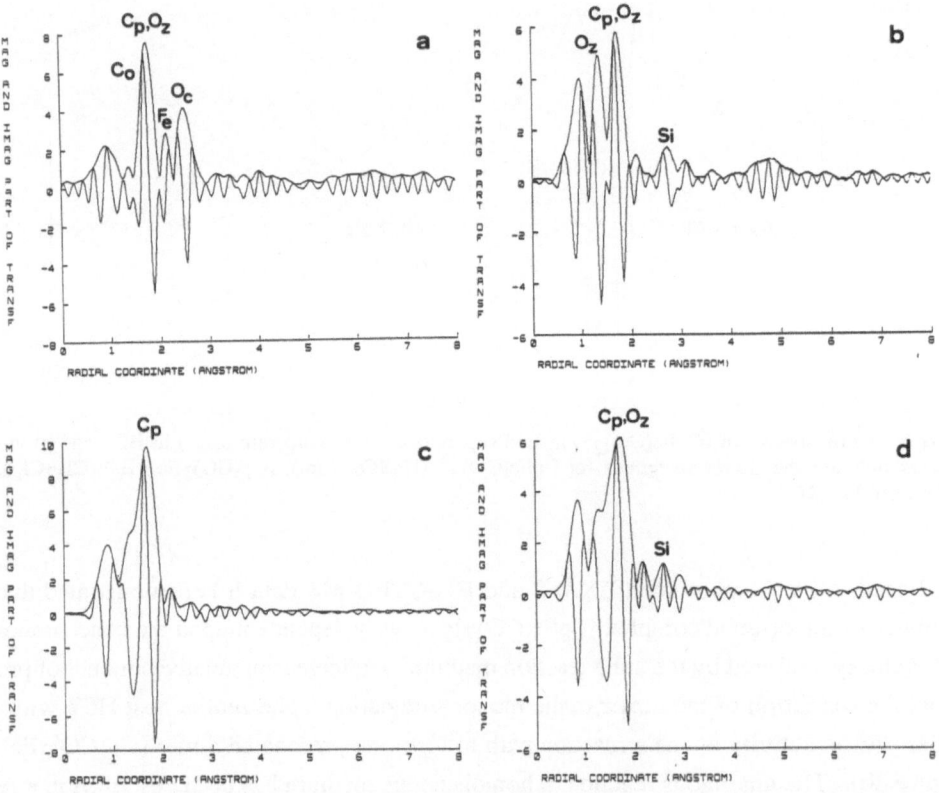

Figure 3. Iron EXAFS spectra of a,b) [CpFe(CO)₂]₂ at room temperature and 473 K, and c,d) ferrocene at room temperature and 573 K, in acid support HY. The magnitudes and imaginary parts of the k^3 weighted Fourier transformations are shown. The spectra are uncorrected for phase shifts. Labels indicate the following backscatterers: Co = Fe-CO, Cp = Fe-Cp, Fe = Fe-Fe, Oz = Fe-O(zeolite), Oc = Fe-CO, Si = Fe-Si/Al. Some peaks are convoluted and labels indicate the positions obtained from the EXAFS analysis.

position of the original Fe-Cp shell are now dominant (Figure 3b). The fitting routine deconvolutes these peaks, indicating as the major contribution an Fe-Cp fragment at 2.07 Å. An optimum fit was derived with two additional Fe-O contributions, indicating the presence of two species, i.e. bare iron ions at 1.93 Å plus $CpFe(OZ)_n$ at 2.16 Å. A smaller peak at about 2.8 Å is clearly visible at this temperature and is assigned to backscattering from the zeolite framework metals Si or Al. The combined appearance of these new peaks is a strong indication for coordination of the resulting complex fragments to zeolite cation sites [34]. As confirmed by FTIR measurements, the tricarbonyl cation thus splits off all CO ligands at higher temperatures. However, no Cp fragments are detected in the vapor phase, and the EXAFS analysis indicates that the Cp ligands are still coordinated to the iron metal to form the "half-sandwich" $CpFe^+(OZ)n^-$, anchored to the oxygen rings present in the zeolite supercage. The EXAFS data suggest that n is between two and three, depending on the fraction of bare iron ions. The limited accuracy of EXAFS coordination numbers and the potential presence of several sites does not allow precise determination of the partition of the CpFe fragments between 4-ring (SIII) and 6-ring (SII) sites. Ferrocene was found to give the same fragment at 473 K. Only two zeolite oxygens were coordinated to the Fe in this case, indicating attachment to the SIII sites (Figure 3 c,d). The combined presence of Fe^{2+} and $CpFe(OZ)_2$ fragments and the results discussed above are consistent with the following decomposition reactions of Fp2:

$$[CpFe(CO)_2]_2 + 4\ HOZ \quad \xrightarrow{470\ K} \quad 4\ CO(g) + 2\ CpH + 2\ Fe^{2+}(OZ)_2\ + H_2$$

$$[CpFe(CO)_3]^+OZ^- + OZ \quad \xrightarrow{470\ K} \quad 3\ CO(g) + CpFe(OZ)_2$$

Intrazeolite Chemistry of COTFe(CO)$_3$

As observed with $[CpFe(CO)_2]_2$, the chemistry of COTFe(CO)$_3$ in the large pore zeolite Y environment is determined by the concentration of framework protons present. In NaY, the complex is believed to reside in two forms. A majority of the complex molecules is shielded from interactions with the sodium cations by other molecules and thus exhibits only slightly perturbed CO vibrations (Figure 1b). A closer contact to the zeolite "walls", however, leads to a stronger interaction between the CO ligands and the Na^+ cations such as Na···OC-Fe which resembles contact ion-pair interactions of carbonyl complexes with alkali cations in solution [35] and results in a lower symmetry of the complex, split peaks and lower frequencies. Similar effects of lowered symmetry and band splitting were found in an earlier study of nickel carbonyl complexes in different zeolite supports [3]. As indicated by the IR and EXAFS data, in sample H6Y/COT the absence of sodium ions and the high level of protons result in the exclusive formation of bicyclo [5.1.0] octadienyl iron tricarbonyl cation $[C_8H_9Fe(CO)_3]^+$ (homotropylium iron tricarbonyl; Figure 4a, 5b).

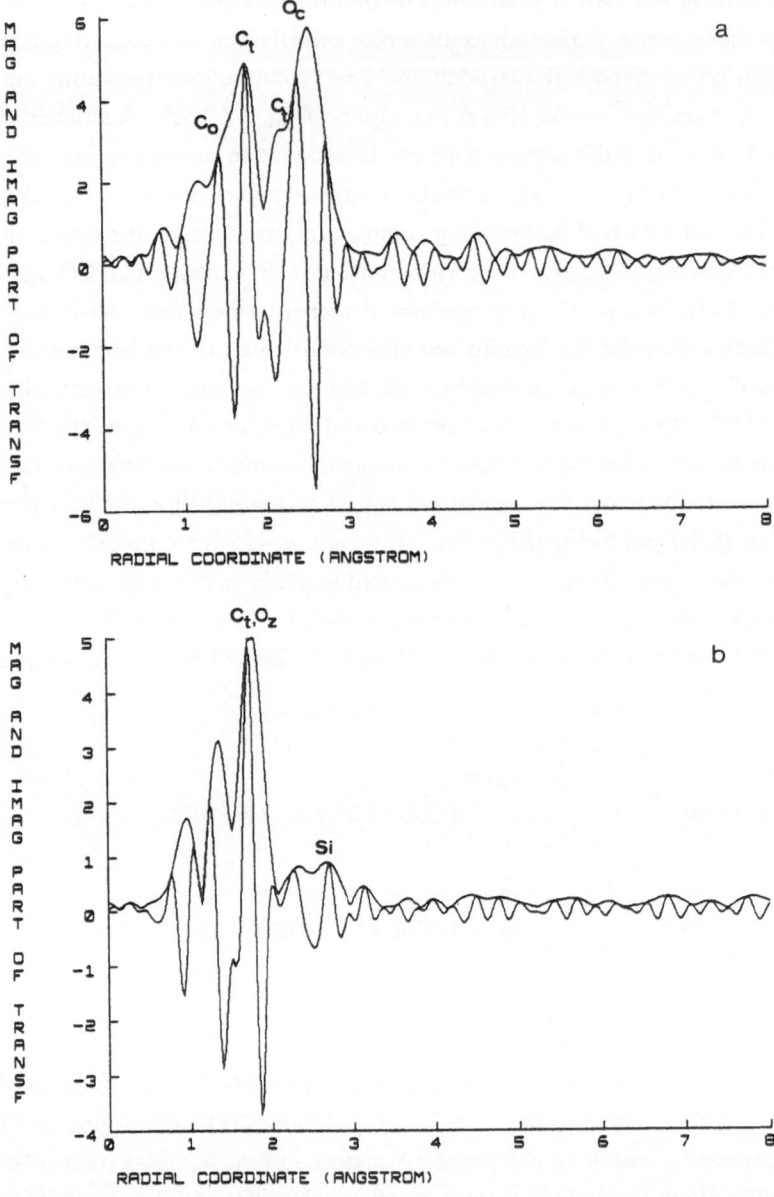

Figure 4. Iron EXAFS spectra of COTFe(CO)$_3$ in acid support HY at a) room temperature and b) 473 K. Labels indicate specific backscatterers as in Figure 3. Ct = Fe-COT.

In partially acidic H2Y/COT, coexistence of COTFe(CO)$_3$ and homotropylium ions is observed (Figure 5a).

It appears that a chemical reaction of the zeolite support with the occluded complex occurs only in the presence of acid groups, resembling the reaction path found in homogeneous medium with non-coordinating acids. Upon heating these samples, TPD-MS experiments do not show any significant fragments related to a loss of the C$_8$H$_9$ ligand or the desorption of any other iron-containing fragments. Therefore, these species have to be trapped in the zeolite cage system. The EXAFS analysis of the heated sample shows an organic fragment at R = 2.07 Å in addition to oxygen coordination at R = 2.16 Å (Figure 4b). This product can be described as a $[FeC_8H_9]^+$ fragment which is coordinated to two zeolite oxygens in the large faujasite supercage:

$$[C_8H_9Fe(CO)_3]^+ + (OZ)_2^- \xrightarrow{\Delta T} C_8H_9Fe(OZ)_2 + 3\,CO\,(g)$$

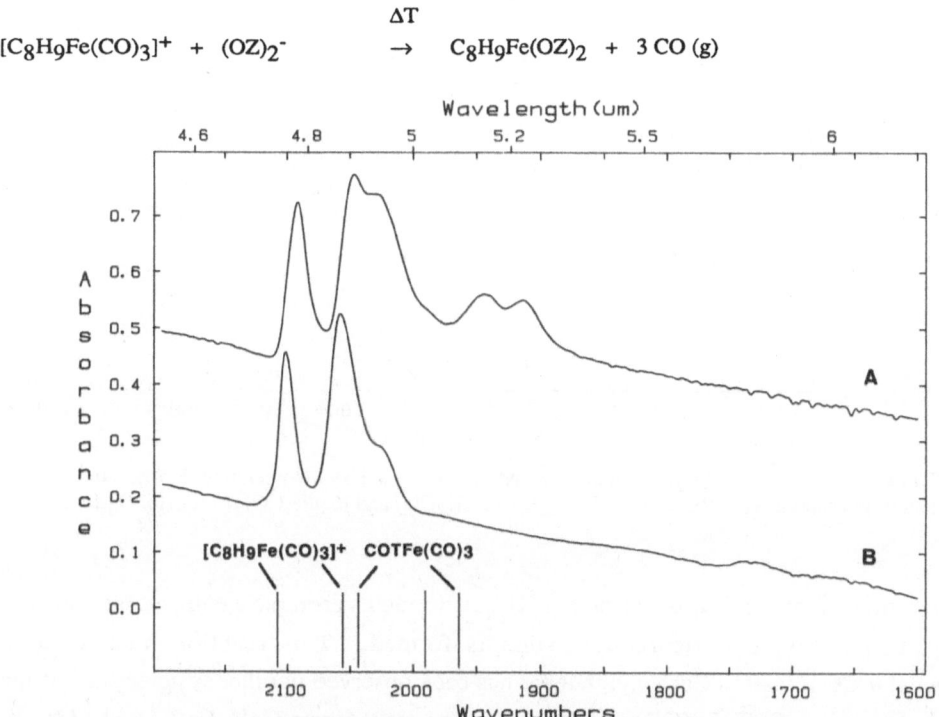

Figure 5. FTIR spectra of COTFe(CO)3 in a) H2Y and b) H6Y, at room temperature. Markers indicate the wavenumbers of unsupported complex COTFe(CO)3 (KBr) and the homotropylium iron tricarbonyl $[C_8H_9Fe(CO)_3]^+$ (in CS_2/CCl_4).

Upon cleavage of the carbonyl ligands, the zeolite oxygens of the supercage SIII four-ring positions fill the coordination sphere of the remaining organometallic iron fragment.

Intrazeolite Chemistry of CpFe(CO)₂CH₃

As with the other iron complexes, the reaction of the intrazeolite hydroxyls with $CpFe(CO)_2CH_3$ results in the formation of cationic iron-species, which under thermal treatment decompose to anchored fragments. The IR spectrum of the NaY/MeFp sample shows the presence of $CpFe(CO)_2CH_3$ which interacts with the zeolite sodium cations (Figure 1c). The complex reacts in both moderately and highly acidic zeolites to yield the dicarbonyl cation $CpFe(CO)_2^+$ as well as the tricarbonyl cation $CpFe(CO)_3^+$ as protonation products (Figure 6a,b). The latter is relatively more stable at elevated temperatures. Treatment of $CpFe(CO)_2CH_3$ in solution with acids having a $pK_a < 1$, e.g. CF_3COOH [36], resulted in the

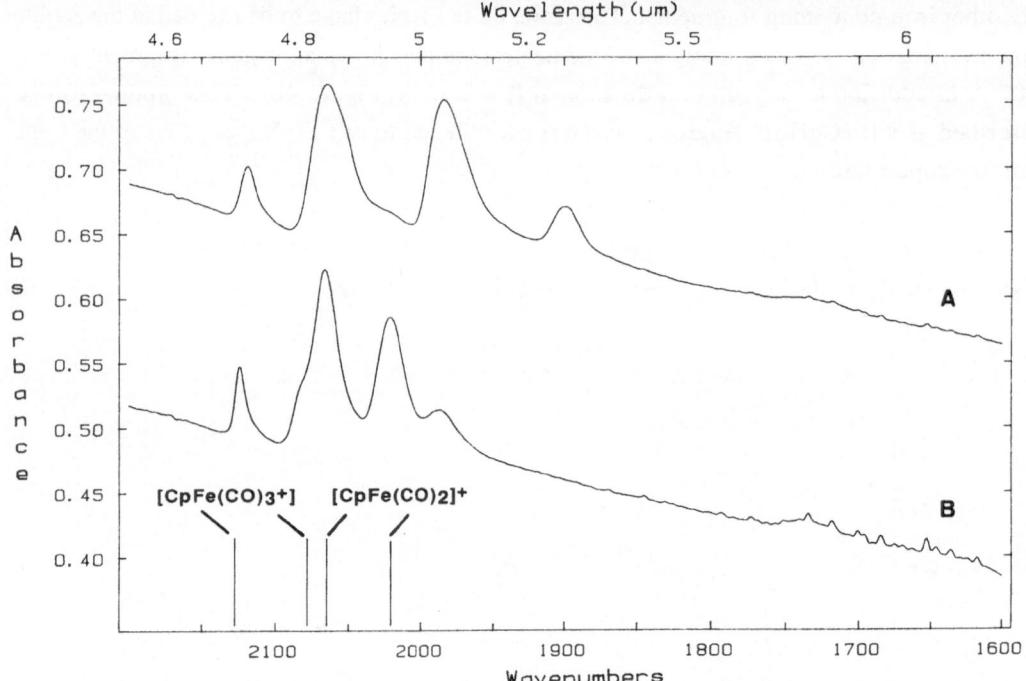

Figure 6. FTIR spectra of CpFe(CO)$_2$CH$_3$ in a) H2Y and b) in H6Y, at room temperature. Markers indicate the wavenumbers of the cationic complexes [CpFe(CO)$_2$]$^+$ (in CH$_2$Cl$_2$) and [CpFe(CO)$_3$]$^+$ (in H$_2$SO$_4$ film).

exclusive formation of the dicarbonyl cation. The acid zeolites promote a comparable reaction, however, in addition the tricarbonyl cation is formed. This reaction must involve intermolecular CO transfer. Carbonyl transfer has been observed in other systems such as the unsaturated complex C$_4$H$_7$Fe(CO)$_3^+$ which spontaneously forms C$_4$H$_7$Fe(CO)$_4^+$ [37]. In the zeolite, the formation of tricarbonyl cations must involve partial fragmentation of the dicarbonyl. It is suggested that this fragment is CpFe$^+$(OZ)$_n$, similar to intrazeolite products of Fp2 (see above). The absence of CH$_4$ in subsequent TPD-MS experiments indicates that the methyl ligands have already been split off by the zeolite hydroxyls during the loading period of CpFe(CO)$_2$CH$_3$. These results can be summarized in the following tentative reaction pathway at room temperature:

$$CpFe(CO)_2CH_3 + ZOH \rightarrow [CpFe(CO)_2]^+OZ^- + CH_4 \quad \text{(in slurry)}$$
$$3 [CpFe(CO)_2]^+ + (OZ)_n^- \rightarrow 2 [CpFe(CO)_3]^+ + CpFe^+(OZ)_n^-$$

TPD-MS data at elevated temperatures suggest that the loss of carbonyl ligands from both the di- and tricarbonyl cations results in the formation of additional intrazeolite fragment CpFe$^+$:

$$[CpFe(CO)_n]^+ + (OZ)_m^- \xrightarrow{> 473 \text{ K}} CpFe(OZ)_m + n\ CO\ (g) \qquad n,m = 2,3$$

CONCLUSION

The anchoring strategy demonstrated above is based upon the different reactivities of ligands at a transition metal center towards the bridged hydroxyl groups present in acid large-pore zeolites. It could be shown that under certain conditions the metal attaches to the zeolite oxygen rings while only the most stable ligands are retained at the metal. This technique allows us to obtain control of the *siting* of the immobilized metal fragment. Since the large neutral complexes cannot enter the smaller sodalite and double six-ring cages present in zeolite Y, they are confined to positions in the large supercage. In contrast, the traditional stepwise assembly of intrazeolite complexes from metal ions and small ligands such as CO relies upon preceding aqueous ion exchange and dehydration under drastic conditions. Thus, a large fraction of the metal is forced to migrate into the smaller cavities which are less desirable positions in the context of catalytic applications and which make characterization of the resulting mixture of intrazeolite products very difficult. The alternative procedure developed in the present work results in migration-stabilized fragments chemically anchored into accessible zeolite pores. Our studies are presently being extended to other transition metal/ligand combinations which potentially retain reactive coordination sites after being attached to the zeolite, and to bimetallic complexes with different ligand reactivities.

Acknowledgements. Acknowledgement is made to the Donors of the Petroleum Research Fund, administered by the American Chemical Society, and to the Sandia University Research Program (DOE), for partial support of this research. The operational funds for NSLS beamline X-11A are supported by DOE grant DE-AS0580ER10742.

References

1. Ballivet-Tkatchenko, D.; Coudrurier, G.; Mozzanega, H.; Tkatchenko, I.; Kinnemann, A. *J. Mol. Cat.* **1979**, *6*, 293.

2. Suib, S. L.; Kostapapas, A.; McMahon, K. C.; Baxter, J. C.; Winiecki, A. M. *Inorg. Chem.* **1985**, *24*, 858.

3. Bein, T.; McLain, S. J.; Corbin, D. R.; Farlee, R. F.; Moller, K.; Stucky, G. D.; Woolery, G.; Sayers, D. *J. Am. Chem. Soc.* **1988**, *110*, 1801.

4. Herron, N.; Stucky, G. D.; Tolman, C. A. *Inorg. Chim. Acta* **1985**, *100*, 135.

5. Bein, T.; Schmiester, G.; Jacobs, P. A. *J. Phys. Chem.* **1986**, *90*, 4851.

6. Yang, Y. S.; Howe, R. F. In *New Developments in Zeolite Science and Technology*; Murakami, Y.; Iijima, A.; Ward, J. W., Eds.; Kodansha: Tokyo, 1986; p. 883.

7. Verdonck, J. J.; Schoonheydt, R. A.; Jacobs, P. A. *J. Phys. Chem.* **1983**, *87*, 683.

8. Gelin, P.; Naccache, C.; Ben Taarit, Y.; Diab, Y. *Nouv. J. Chim.* **1984**, *8*, 675.

9. Bergeret, G.; Gallezot, P.; Gelin, P.; Ben Taarit, Y.; Lefebvre, F.; Naccache, C.; Shannon, R. D. *J. Catal.* **1987**, *104*, 279.

10. Davis, M. E.; Schnitzer, J.; Rossin, J. A.; Taylor, D.; Hanson, B. E. *J. Mol. Cat.* **1987**, *39*, 243.

11. Shannon, R. D.; Vedrine, J. C.; Naccache. C.; Lefebvre, F. *J. Catal.* **1984**, *88*, 431.

12. Rode, E. J.; Davis, M. E.; Hanson, B. E. *J. Catal.* **1985**, *96*, 574.

13. Herron, N. *Inorg. Chem.* **1986**, *25*, 4714.

14. Diegruber, H.; Plath, P. J.; Schulz-Ekloff, G. *J. Mol. Cat.* **1984**, *24*, 115.

15. Herron, N.; Stucky, G. D.; Tolman, C. A. *J. Chem. Soc., Chem. Commun.* **1986**, 1521.

16. Yermakov, Yu. I. *Catal. Rev.-Sci. Eng.* **1976**, *13*, 77.

17. Foley, H. C.; DeCanio, S. J.; Tau, K. D.; Chao, K. J.; Onuferko, J. H.; Dybowski, C.; Gates, B. C. *J. Am. Chem. Soc.* **1983**, *105*, 3074.

18. Ward, M. D.; Schwartz, J. *J. Mol. Cat.* **1981**, *11*, 397.

19. Huang, T. N.; Schwartz, J.; Kitajima, N. *J. Mol. Cat.* **1984**, *22*, 389.

20. Huang, T. N.; Schwartz, J. *J. Am. Chem. Soc.* **1982**, 104, 5244.

21. Corbin, D. E.; Seidel, W. C.; Abrams, L.; Herron, N.; Stucky, G. D.; Tolman, C. A. *Inorg. Chem.* **1985**, *24*, 1800.

22. Taylor, D. F.; Hanson, B. E.; Davis, M. E. *Inorg. Chim. Acta* **1987**, *128*, 55.

23. Moller, K.; Borvornwattananont, A.; Bein, T. *J. Phys. Chem.* **1989**, *93*, 4562.

24. Borvornwattananont, A.; Moller, K.; Bein, T. *J. Phys. Chem.* **1989**, *93*, 4205.

25. Lee, P. A.; Citrin, P. H.; Eisenberger, P.; Kincaid, B. M. *Rev. Mod. Phys.* **1981**, *53*, 769.

26. Johnson, E. C.; Meyer, T. J.; Winterton, N. *J. Chem. Soc., Chem. Commun.* **1970**, 934.

27. Johnson, E. C.; Meyer, T. J.; Winterton, N. *Inorg. Chem.* **1971**, *10*, 1673.

28. Dombeck, B. D.; Angelici, R. J. *Inorg. Chim. Acta* **1973**, *7*, 345.

29. Boyle, P. F.; Nicholas, K. M. *J. Organomet. Chem.* **1976**, *114*, 307.

30. Catheline, D.; Astruc, D. *J. Organomet. Chem.* **1984**, *266*, C11.

31. Davison, A.; Green, M. L. H.; Wilkinson, G. *J. Chem. Soc.* **1961**, 3172.

32. Beaumont, R.; Barthomeuf, D.; Trambouze, Y. *Adv. Chem. Ser.* **1971**, *102*, 327.

33. Legzdins, P.; Martin, D. T.; Nurse, C. R.; Wassink, B. *Organometallics* **1983**, *2*, 1238.

34. Woolery, G.; Kuehl, G.; Chester, A.; Bein, T.; Stucky, G. D.; Sayers, D. E. *J. Physique.* **1986**, *47*, C8-281.

35. Darensbourg, M. Y.; Barras, H. L. C. *Inorg. Chem.* **1979**, *18*, 3286.

36. De Luca, N.; Wojcicki, A. *J. Organomet. Chem.* **1980**, *193*, 359.

37. Gibson, D. H.; Vonnahme, R. L. *J. Am. Chem. Soc.* **1972**, *94*, 5090.

PHOTOCHEMISTRY IN ZEOLITE CAVITIES

V. Ramamurthy

Central Research and Development Department
E. I. du Pont de Nemours and Company, Experimental Station
P.O.Box 80328
Wilmington, Delaware 19880-0328 (USA)

INTRODUCTION

Although the use of zeolites as catalysts has been fairly well established [1], their utility as molecular reaction vessels has not attracted wide attention [2]. It is this less explored aspect of zeolites that is of interest to us and forms part of this presentation [3]. Zeolites may be regarded as open structures of silica in which aluminum has been substituted in a fraction (X/X+Y) of the tetrahedral sites [4]. The framework thus obtained contains pores, channels and cages. As the trivalent aluminum ions replace tetravalent silicon ions at lattice positions, the network bears a net negative charge which must be compensated by other counter ions. The latter are mobile and may occupy various exchange sites depending upon their radius, charge and degree of hydration. They can be replaced, to varying degrees, by exchange with other cations. If zeolite water is removed, many other organic and inorganic molecular entities can be accommodated in the intracrystalline cavities.

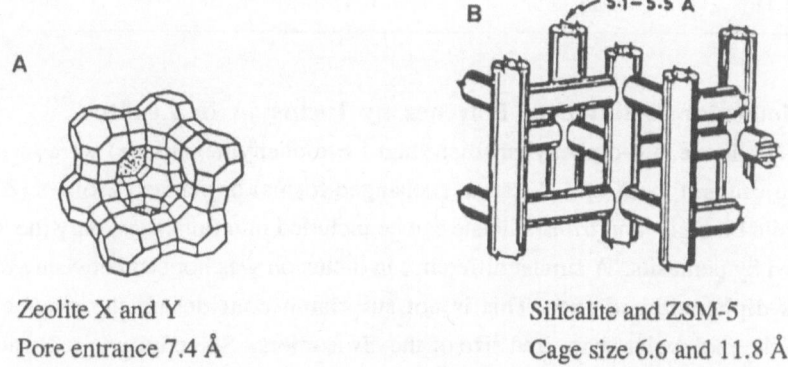

Zeolite X and Y
Pore entrance 7.4 Å

Silicalite and ZSM-5
Cage size 6.6 and 11.8 Å

Figure 1. Differences in structure between faujasites (A) and pentasils (B).

In this work we have utilized faujasite [4] and pentasil [5] zeolites as hosts to carry out phototransformations of several organic molecules. Although they possess completely interconnecting three-dimensional pore structures, these two types of zeolites have fundamentally different void space topologies (Figure 1). While the former consists of relatively large spherical cages (diameter ≈ 13 Å; entrance pore diameter ≈ 8 Å), the latter contains only interconnecting channels (diameter ≈ 5.5 Å). Our interest in zeolites extends the use of other organic host systems to control and modify the photochemical and photophysical behavior of organic guests [6]. In this context zeolites offer several advantages. For example, zeolites offer a variety of internal structures in which the guest can be accommodated; micropolarity and the internal pore/cavity dimensions can be fine tuned; zeolites are photoinert and are easily prepared in large quantities.

IMPORTANCE OF CAVITY/CHANNEL SIZE IN INFLUENCING THE BEHAVIOR OF INCLUDED GUESTS - CONSEQUENCE OF TIGHT AND LOOSE INCLUSION

With the two examples below, we illustrate the importance of the critical matching of the channel/cavity dimensions of the host and the guest to achieve maximum influence of the host on the guest's behavior. The key transformations in both the examples require rotation of a C-C bond - one involves a π-bond and the other a σ-bond.

Table 1. Photolysis of stilbene in zeolites: isomer distribution after several hours of irradiation.

Medium	Initial	Photostationary state mixtures	
		trans	cis
Benzene	trans	28	72
	cis	26	74
Li-X	trans	56	44
	cis	12	88
Cs-X	trans	73	27
	cis	34	66
ZSM-5	trans	100	–
ZSM-8	trans	100	–
ZSM-11	trans	100	–

Arrested Molecular Rotation of Polyenes by Inclusion in Zeolites

Polyenes (stilbene, 1,4-diphenylbutadiene and 1,6-diphenylhexatriene) were incorporated into either faujasites (type X and Y, cation exchanged forms) or pentasil zeolites (ZSM-5, -8 and -11). While both cis- and trans-stilbene can be included into faujasites, only the latter was accommodated by pentasils. A similar difference in inclusion was noticed between trans,trans- and trans,cis-diphenylbutadiene. This is not surprising considering the channel size of pentasils and the molecular shape and size of the cis isomers. Selectivity in inclusion is also reflected in the photobehavior of the included polyenes (Tables 1 and 2). Direct excitation of the trans (and all trans) isomers of the above three olefins incorporated in pentasils resulted in

Table 2. Photolysis of 1,4-diphenylbutadiene in zeolites: isomer distribution after several hours of irradiation.

Medium	Initial	Photostationary state mixtures		
		trans,trans	*trans,cis*	*cis,cis*
Benzene	*trans,trans*	18	75	6
	trans,cis	17	76	7
Li-X	*trans,trans*	76	20	3
	trans,cis	41	45	12
Cs-X	*trans,trans*	73	17	10
	trans,cis	35	44	20
ZSM-5	*trans,trans*	100	--	−
ZSM-8	*trans,trans*	100	--	−
ZSM-11	*trans,trans*	100	--	−

no change suggesting that their inclusion in pentasils fully arrested the rotation of π-bonds (Figure 2). However, both *cis* and *trans* isomers underwent geometric isomerization inside the super cages of faujasites. It is of interest to note that the *cis,cis*-1,4-diphenylbutadiene, reported to be unobtainable by solution photolysis [7], is obtained in significant yield (≈20%) in the cavities of faujasites, especially in Cs-X. The above restriction of molecular motion is also indicated by drastic changes in the photophysical properties of the included guests as outlined below.

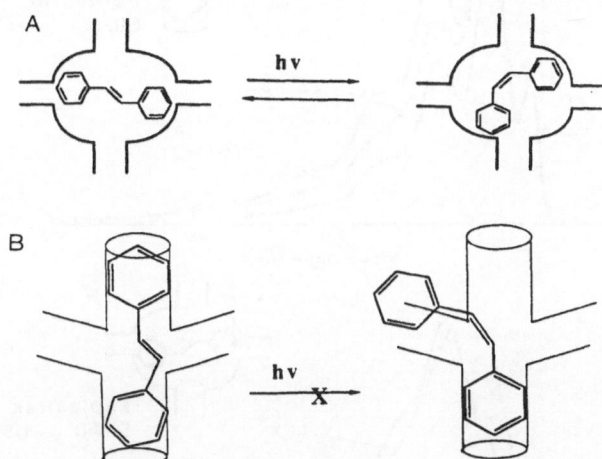

Figure 2. A model for inclusion of stilbene in pentasil and faujasite. Both isomers of the alkene were included in faujasite (A) and were isomerized upon excitation, whereas only the *trans*-isomers were included in pentasils (B) and were photoinert.

In the extremely confining space of the ZSM channels, the polyenes examined all exhibit enhanced fluorescence lifetimes (Table 3). Such lifetimes are significantly longer than in fluid solution or in the supercages of faujasites. For *trans,trans*-1,4-diphenylbutadiene the longest lifetime we observe, 16 ns in ZSM-8, is identical within errors, to the lifetime observed in a molecular beam experiment where the equivalent temperature is 4.2 K [8]. This

Table 3. Consequnce of rotational restriction on excited singlet state lifetime.

Medium	Lifetime (nanoseconds)	
	Stilbene	1,4-diphenylbutadiene
Methylcyclohexane	0.11	0.58
Na-X	0.21 (96%)	0.60
Na-Y	0.20 (98%)	0.67
ZSM-5	1.88	12.7
ZSM-8	1.87	15.1
ZSM-11	3.80	13.2
Crystal	6.0	--
Molecular beam jet (4.2 K)	--	16.0

similarity emphasizes the rigidity of the zeolite environment. Likewise, for *trans*-stilbene in ZSMs the observed lifetimes approach those of rigid analogues [9]. We believe that the enhanced lifetime is a direct reflection of the constraint provided by the host which prevents π-bond rotation.

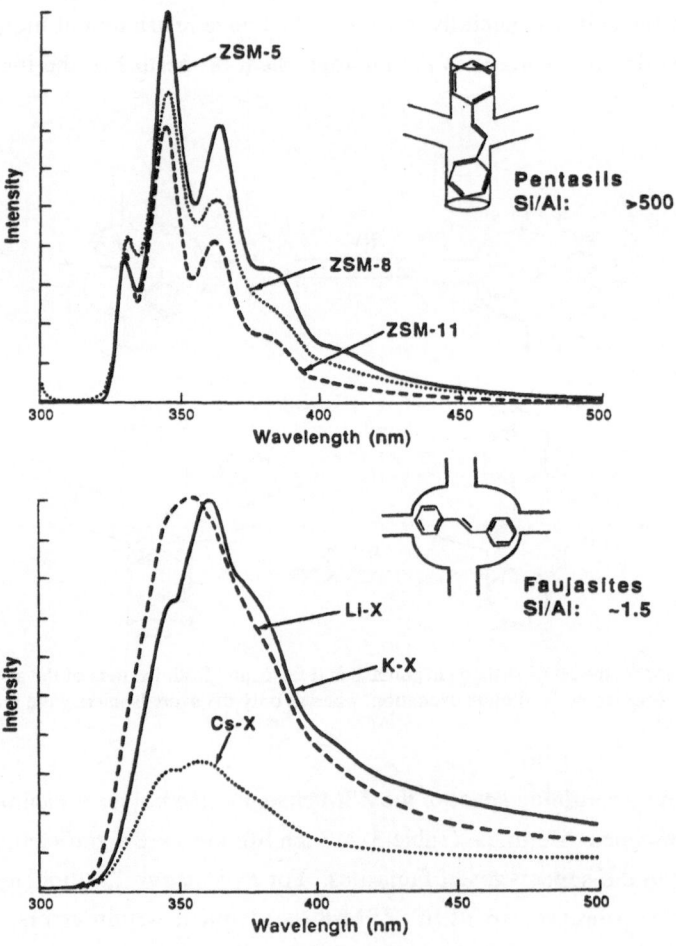

Figure 3. Fluorescence spectra of *trans*-stilbene included in various zeolites at room temperature.

Figure 3 shows fluorescence spectra of *trans*-stilbene under various conditions. Analogous behavior was found for *trans,trans*-diphenylbutadiene and *trans,trans,trans*-1,6-diphenylhexatriene. Clearly vibronic fine structure in the emission of polyenes in pentasils is better resolved than in faujasites. Even at 4 K such difference is maintained. We believe that larger cavity size (faujasite) provides for a nonhomogeneous microenvironment experienced by the polyenes. This results in a heterogeneous distribution of polyene molecules leading to a broadening and loss of fine structure of the emission spectrum.

Differences in rotational restrictions between faujasites and pentasils seen in the excited state are also observed in the ground state as revealed by the solid state ^{2}H-NMR of the perdeuterated trans-stilbene in Na-X, H-Y* and ZSM series. As shown in Figure 4, the spectrum in ZSM-5 resembles that in the crystalline state while that in Na-X and H-Y* is that of freely tumbling stilbene. Analysis of the spectrum reveals that at room temperature *trans*-stilbene included in the channels of ZSM-5 is stationary (on the NMR time scale) and does not undergo isotropic motion and possibly possesses very little freedom of motion [10]. If such is the case, absence of geometric isomerization and long excited singlet lifetime come as no surprise. NMR spectra recorded at various temperatures (Figure 5) in ZSM-5 (Si/Al ≈ 550) indicate that even at higher temperatures (350 K) stilbene molecules do not gain isotropic motion on the NMR time scale.

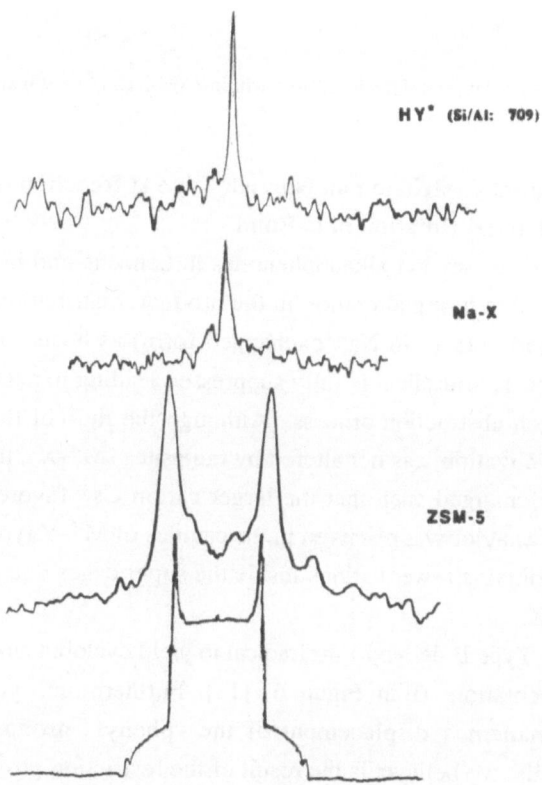

Figure 4. ^{2}H-NMR spectra of perdeuterated *trans*-stilbene included in various zeolites at room temperature.

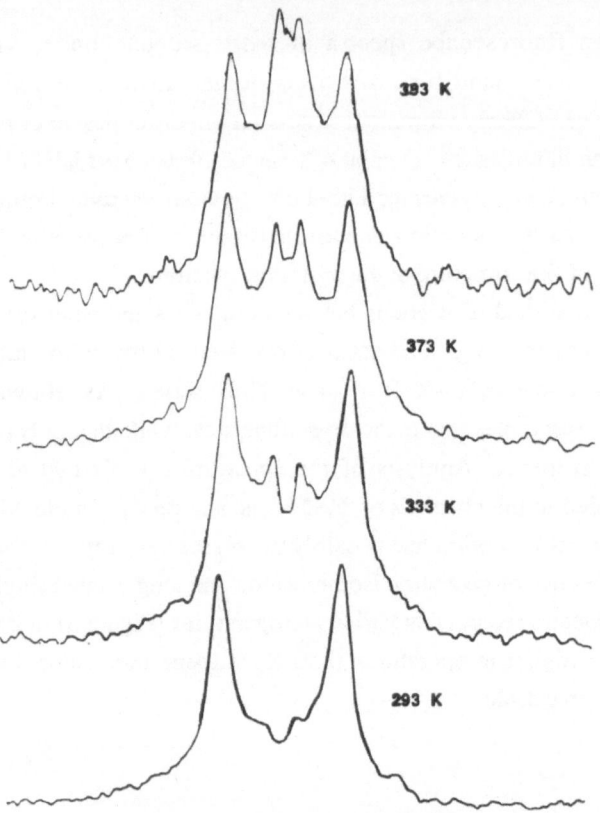

383 K

373 K

333 K

293 K

Figure 5. ^{2}H-NMR spectra of perdeuterated *trans*-stilbene within ZSM-5 (Si/Al: 490) at different temperatures.

Consequence of Rotational Restriction on Norrish Type II Reaction of Alkanophenones - Restricted Rotation of σ-Bond

Results of photolyses of several alkanophenones in benzene and in several zeolites are summarized in Table 4. A striking alteration in the product distribution was observed with pentasils (ZSM-5, -8 and -11; all in Na$^+$ exchanged form) as hosts. In these zeolites the cyclization process of the 1,4-diradical is fully suppressed leading to acetopheone as the sole product of the γ-hydrogen abstraction process. Although the ratio of the products resulting from elimination and cyclization was not altered by faujasites (M$^+$-X), the ratio of the *cis*- to *trans*-cyclobutanol was changed such that the larger cation Cs$^+$ favored the *trans*-isomer. Similar but less drastic behavior was observed in the cavities of M$^+$-Y type zeolites also. This is expected as Na$^+$-Y contains fewer cations inside the super cages due to their higher Si/Al ratio compared to Na$^+$-X.

Coupling (C) of the Type II derived 1,4-diradical to yield cyclobutanols is sterically more demanding than fragmentation (E in Figure 6) [11]. Furthermore, cyclobutanol formation requires a large permanent displacement of the phenyl group. The absence of cyclobutanols in pentasils, we believe, is the result of the restriction provided by the channel walls for the rotation of the central σ-bond (Figure 6) leaving elimination as the only mode of

Table 4. Photolysis of arylalkylketones in zeolites, product distribution: E/C.

Medium	Butyrophenone	Octanophenone	Octadecanophenone
Benzene	6.2	1.9	2.7
Li-X	3.9	1.6	2.5
Na-X	2.7	1.5	2.7
K-X	3.3	1.9	4.3
Rb-X	1.9	1.9	6.2
Cs-X	2.3	1.9	6.8
ZSM-5	73	>>100	>>100
ZSM-8	82	>>100	>>100
ZSM-11	56	>>100	>>100

decay for the above diradical. Consistent with this proposal are the results in Na-β, a pentasil zeolite with a larger channel dimension [12] (Table 4). The larger channel not only allows formation of cyclobutanols, but in fact caused them to become the major photoproduct. In the case of smaller guests, like butyrophenone, cyclization occurred to a small degree even in ZSM zeolites with a more rigid channel size than in Na-β. The alkyl chain apparently is not held rigidly in the channels of any of these zeolites as indicated by the occurrence of γ-hydrogen abstraction. The supercage of faujasite is large enough to allow the formation of cyclobutanol from the diradical. We used two approaches to induce a tighter fit. One, variation of the cation size, was moderately successful. Attempts, however, to fill the void space by lengthening the alkyl chain did not provide the necessary tight fit. We suspect that the alkyl chain extended itself into adjacent cages. When the free volume of the supercage was reduced by the introduction of a larger cation Cs$^+$ (diameter: 3.4 Å), the diradical was compelled to cyclise into the *trans* isomer.

The above results show that the photochemical and photophysical behavior of molecules can be dramatically altered by including them in a very confining environment. Tight inclusion results in restricted rotation both in the ground and in excited states.

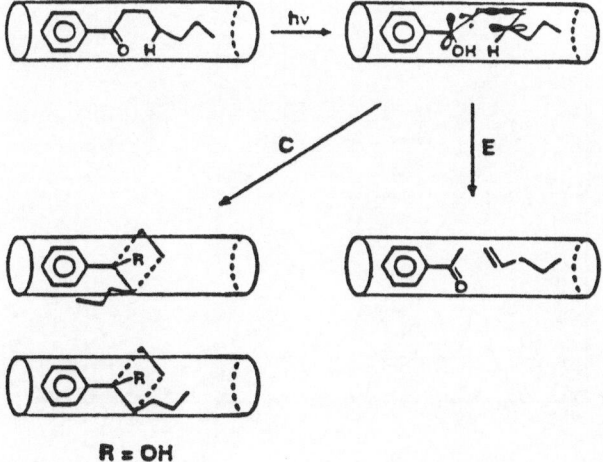

R = OH

Figure 6. A model for inclusion of arylalkylketones in ZSM-5: restricted rotation of the Type II 1,4-diradical.

IMPORTANCE OF THE Si/Al RATIO THAT CONSTITUTES THE CHANNEL/CAVITY WALLS

The zeolite framework is made up of $[SiO_4]$ and $[AlO_4]^-$ units and the ratio of these can be significantly varied during synthesis [4]. This change can alter the micropolarity of the channel/cavity, can provide varied sites for adsorption of the guests, and can possibly alter the characteristics of the internal porous structure. In order to probe such possibilities, we are pursuing photophysical and solid state NMR studies of polyenes, aromatics and arylalkylketones included within ZSM-5, -8 and -11 of widely differing Si/Al ratios. Results of one such investigation are presented below.

In Figures 7 and 8 fluorescence spectra of *trans,trans*-1,4-diphenyl-butadiene and octanophenone included in ZSM-5 of varying Si/Al ratios are shown. In both of these as well as in the spectra of several other related molecules investigated, the vibrational resolution is much dependant on the Si/Al ratio. We believe that higher aluminum content gives rise to inhomogenity within the porous structure and offers several sites for adsorption. This results in a heterogeneous distribution of the guest molecule leading to broadening and a loss of fine structure.

^{2}H NMR spectra of perdeuterated *trans*-stilbene in ZSM-5 (Si/Al: 24) at various temperatures are shown in Figure 9. These are to be compared with the spectra in ZSM-5 (Si/Al: 550) shown in Figure 5. It is obvious that at higher temperatures the behavior of *trans*-stilbene in the two zeolites is different. In samples containing high a silicon content, the stilbene molecules continue to be restricted in their motion even at higher temperatures, whereas in a high aluminum content sample, a certain amount of isotropic motion is gained by *trans*-stilbene as indicated by a single peak in the middle of the double peak. It is not clear why such is the case. At this stage we can only point out that solid state NMR and fluorescence techniques are sensitive enough to monitor the zeolite wall characteristics. Further work is underway to understand this aspect of zeolite inclusion phenomena.

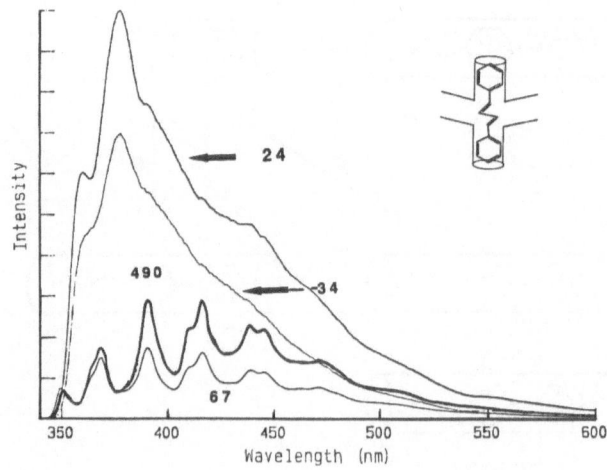

Figure 7. Fluorescence spectra of *trans,trans*-1,4-diphenylbutadiene within ZSM-5: effects due to Si/Al ratio variation.

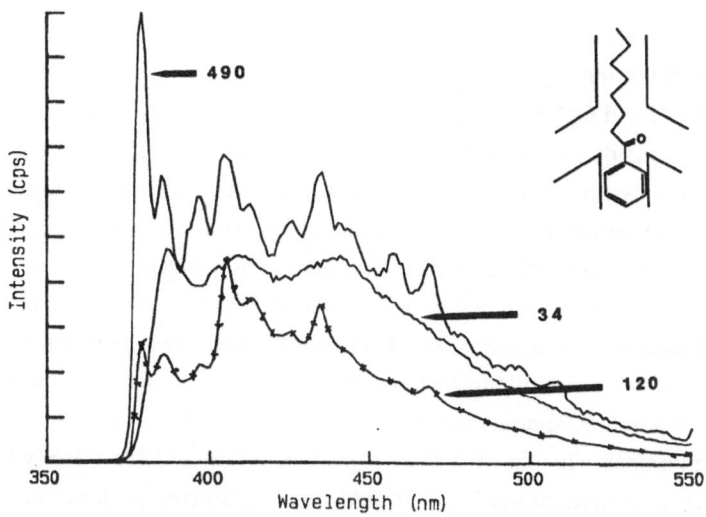

Figure 8. Fluorescence spectra of octanophenone within ZSM-5 of different Si/Al ratios.

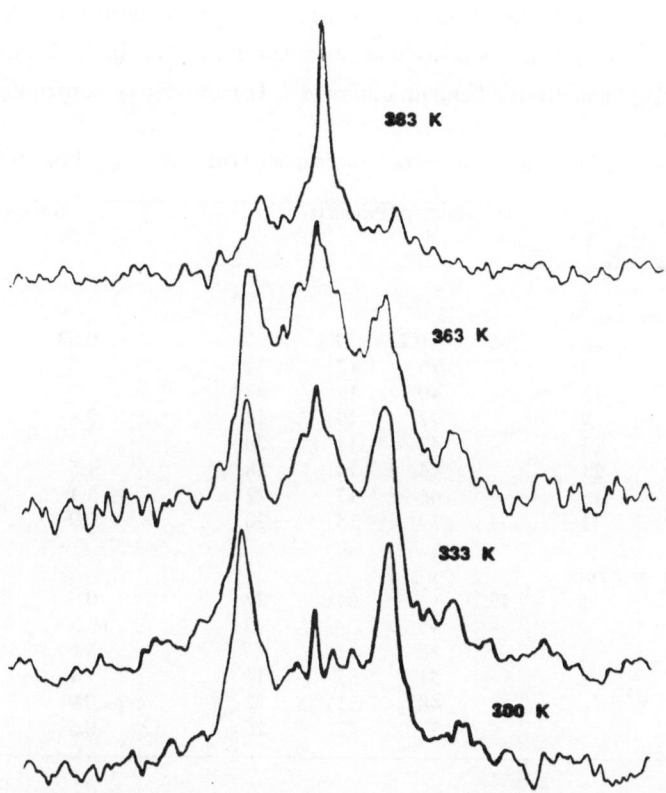

Figure 9. ^{2}H-NMR spectra of perdeuterated *trans*-stilbene within ZSM-5 (Si/Al: 24) at different temperatures.

ROLE OF EXCHANGEABLE CATIONS PRESENT INSIDE SUPERCAGES OF FAUJASITES IN CONTROLLING THE PHOTOBEHAVIOR OF THE INCLUDED GUESTS

Cations present in the supercage of faujasites can be exchanged and thus one could obtain Li^+, Na^+, K^+, Rb^+, and Cs^+ exchanged faujasites [4]. Such an exchange brings along with it reduced free space inside the supercage (due to the increase in cation size). Also, the strength of the interaction between the cation and the guest is expected to depend on the cation - larger cations such as Cs^+ would be expected to interact less strongly with the guest. A point of considerable interest to photochemists is the relative atomic weights of these cations. Heavy Rb^+ and Cs^+ can influence the reaction by the so-called "heavy atom effect" [13]. Thus, by proper choice of the cation, one might be able to alter the reactive state of the guest.

We provide here an example of how the interaction between the cation and the included guest can result in a unique alteration in the photobehavior of the guest.

The photochemical behavior of benzoin alkyl ethers (1a) and alkyl deoxy benzoins (2a) in solution is fairly well understood [14]. In benzene, the former prefers the Norrish Type II pathway while the latter gives the Norrish Type I products in high yields (as shown in the Scheme). Photolyses of alkyl benzoin ethers and alkyldeoxy benzoins in zeolites show a dramatic difference in behavior when compared to that in benzene (Table 5). Results on benzoin methyl ether and propyl deoxy benzoin in faujasite zeolite (M-X type) alone are highlighted in Table 5. Similar behavior was observed in M-Y zeolites. Identical behavior was found for other alkyl substituted benzoin ethers (1b, 1c) and deoxy benzoin (2b).

Table 5. Product distribution upon photolysis of benzoin methyl ether and propyldeoxy benzoin in zeolites.

Medium	Percentage of Products					Ratio of Products	
	3	4	5	6	7	5/6+7	3+4+5/6+7
Benzoin methyl ether							
Benzene	26	66	0.2	1.1	7	0.03	12.2
Li-X (dry)	3	-	76	13	8	3.6	3.6
(wet)	18	-	40	19	23		
Na-X (dry)	4	-	72	10	14	2.9	3.1
(wet)	22	-	47	11	20		
K-X (dry)	24	-	44	14	18	1.5	2.2
Rb-X (dry)	15	-	46	17	22	1.1	1.5
Cs-X (dry)	31	-	34	15	20	0.9	1.9
Propyl deoxy benzoin							
Benzene	5	19	-	54	22	0	0.3
Li-X (dry)	-	-	95	4	1	18.5	
Na-X (dry)	-	-	88	5	7	7.6	
K-X (dry)	-	-	51	35	14	1.1	
Rb-X (dry)	-	-	48	31	22	0.9	
Cs-X (dry)	-	-	32	42	26	0.5	

With 1a, the amount of the Type II products is considerably increased over that observed in benzene. Even more importantly, the rearrangement product, benzoylbenzyl alkyl ether (5) was predominant among the Type I products. In contrast, **2a**, which generally gives Type II products in benzene, gave the Type I derived **5** as the major product in M-X zeolite. In both cases, the yield of the rearrangement product **5** was dependent on the exchangeable cation. It is important to note that the inherent reactivity controlled product distribution has been dramatically altered by the zeolite cavity in both cases.

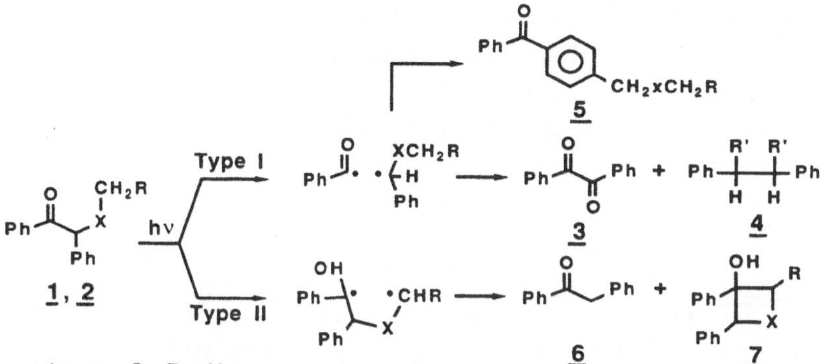

Scheme. Photobehavior of benzoin ethers and alkyldeoxy benzoins.

We believe that the preference for **5** among the Type I products is a consequence of the restriction experienced by the geminal radical pair (Scheme). Low yields of coupling products **3** and **4** in the product mixture in the case of **1a** and their absence in **2a** clearly indicate that the translational motion of the radical pair is restricted by the zeolite framework. Even more interesting is the fact that the rotational process required to produce the rearrangement product is greatly dependent upon the cavity free space. It is known that in proceeding from Li-X to Cs-X, the degree of "lebensraum" or open void space is significantly reduced [4,15]. It is of interest to note that the relative yield of **5** decreases in the same order as the cation size. Such a space dependent rotational restriction is also evident when one compares the product distribution between the dry and the wet zeolites. Water decreases the void space by coordinating to the free cations present within the zeolite cavity.

The behavior of benzoin alkyl ethers and alkyl deoxy benzoins when viewed together provides important additional information. The zeolite cavity induces **1a** to yield products derived via the Type II pathway, a minor pathway in benzene. On the other hand, zeolites do not cause **2a** to proceed via the Type II pathway, which is favored in benzene. We attribute this to the ability of the cation present in the cavity to control the conformation of the included molecules (Figure 10). The presence of an alkoxy chain in **1a** most likely directs the chelation of the cation to a conformer that is favorable for the Type II process. Similarly in **2a**, the phenyl ring directs the conformational preference in the cavity. Such a hypothesis is supported by the results on dealuminated zeolite-Y in which the Si to Al ratio is very high (>550). At very low levels of aluminum the cation concentration is also low. Therefore conformational control is expected to be minimal and indeed only the Type I products **3-5** dominated the product mixture. Furthermore, the increase in the yield of benzil from Li-X to Cs-X (Table 5) is also a reflection of the decreased interaction between the cation and the benzoyl radical. The increased yield of the Type II products in **1a** in zeolite must also be the result of the cage effect as significant ^{13}C enrichment (25%) of the oxetanol was observed.

Figure 10. Conformational control by cations present within supercages of faujasites.

CONCLUSIONS

The results presented above illustrate the important role zeolites can play in selective phototransformations. Proper choice of zeolite to achieve a tight fit is essential to obtain the desired selectivity. Further, one should pay attention to the Si/Al ratio as it might influence the internal homogeneity. We have also illustrated the utility of exchangeable cations present in the supercages of faujasites to control the behavior of the included guests. Since zeolites of varied and tailor-made frameworks can be prepared, the future for zeolites as a reaction medium appears bright.

Acknowledgements. It is a pleasure to thank Ms. A. Pittman for technical assistance. Dr. D. R. Corbin provided much needed zeolite samples, encouragement and support. Pleasant interactions with Dr. J. V. Caspar, Dr. D. F. Eaton, Professor C. Dybowski and Mr. J. Kauffman have been very valuable.

REFERENCES

1. Minachev, K. M.; Kondratev, D. A. *Russ. Chem. Rev.* **1983**, *52*, 1921.
2. Turro, N. J. *Pure Appl. Chem.* **1986**, *58*, 1219; Sciano, J. C.; Casal, H. L.; Netto-Ferreria, J. C. In *Organic Phototransformations in Non-homogeneous Media*; Fox, M. A. ed.; American Chemical Society, 1985; pp 211-222.
3. Corbin, D. R.; Eaton, D. F.; Ramamurthy, V. *J. Am. Chem. Soc.* **1988**, *110*, 4848; Corbin, D. R.; Eaton, D. F.; Ramamurthy, V. *J. Org. Chem.*, in press.; Corbin, D. R.; Eaton, D. F.; Ramamurthy, V. *J. Am. Chem. Soc.*, submitted.

4. Breck, D. R. *Zeolites and Molecular Sieves - Structure, Chemistry and Use*; Wiley-Interscience: New York, 1974.

5. Jacobs, P. A.; Beyer, H. K.; Valyon, J. *Zeolites* **1981**, *1*, 161.

6. Ramamurthy, V.; Eaton, D. F. *Acc. Chem. Res.* **1988**, *21*, 300.

7. Atom Yee, W.; Hug, S. J.; Kliger, D. *J. Am. Chem. Soc.* **1988**, *110*, 2164.

8. Heimbrook, L. A.; Kohler, B. E.; Spiglanin, T. A. *Proc. Nat. Acad. Sci. USA* **1983**, *80*, 4580.

9. DeBoer, C. D.; Schlessinger, R. H. *J. Am. Chem. Soc.* **1968**, *90*, 803; Saltiel, J.; Zafirious, O. C.; Megarity, E. D.; Lamola, A. A. *J. Am. Chem. Soc.* **1968**, *90*, 4759.

10. See for example, Kustanovich, I.; Fraenkel, D.; Luz, Z.; Vega, S. *J. Phys. Chem.* **1988**, *92*, 4134.

11. Wagner, P. J. *Top. Curr. Chem.* **1976**, *66*, 1; Scaiano, J. C. *Acc. Chem. Res.* **1982**, *15*, 252.

12. Treacy, M. M.; Newsam, J. M. *Nature (London)* **1988**, *332*, 249.

13. Turro, N. J. *Modern Molecular Photochemistry*; Benjamin/Cummins; Menlo Park CA, 1978.

14. Lewis, F. D.; Lauterbach, R. L.; Heine, H. G.; Hartmann, W.; Rudolph, H. *J. Am. Chem. Soc.* **1975**, *97*, 1519; de Mayo, P.; Nakamura, A.; Tsang, P. W. K.; Wong, S. K. *J. Am. Chem. Soc.* **1982**, *104*, 6824; Dasaratha Reddy, G.; Usha, G.; Ramanathan, K. V.; Ramamurthy, V. *J. Org. Chem.* **1986**, *51*, 3085; Heine, H. G.; Hartmann, W.; Kory, D. R.; Magya, J. G.; Hoyle, C. E.; McVey, J. K.; Lewis, F. D. *J. Org. Chem.* **1974**, *39*, 691; Lewis, F. D.; Hoyle, C. E.; Magyar, J. G.; Heine, H. G.; Hartmann, W. *J. Org. Chem.* **1975**, *40*, 488.; Dasaratha Reddy, G.; Ramamurthy, V. *J. Org. Chem.* **1987**, *52*, 5521.

15. Turro, N. J.; Zhang, Z. *Tetrahedron Lett.* **1988**, *28*, 5637.

ARTIFICIAL PHOTOSYNTHESIS IN ZEOLITE-BASED MOLECULAR ASSEMBLIES

Jonathan S. Krueger, Cuiwei Lai, Zhuyin Li, James E. Mayer, and Thomas E. Mallouk*

Department of Chemistry
University of Texas at Austin
Austin, TX   78712 (USA)

SUMMARY

Zeolites are crystalline aluminosilicates which admit molecules of appropriate size and charge into their internal void spaces. This selectivity allows zeolites to be used as templates for self-organizing molecular electron transport chains. In zeolites Y and L, the framework is negatively charged and electroactive cations of critical dimension 8 Å or less are admitted to the internal volume. Larger, multiply charged cations adhere via ion-exchange to the external surface in roughly monolayer quantities. The surface ions mediate electron transfer between molecules in the bulk and a metal electrode; with surface and bulk ions of appropriate relative redox potential, e.g. metalloporphyrins and viologens, current rectification and charge trapping effects can be observed electrochemically. Three-component systems for light induced charge separation can be prepared by exchanging donor-acceptor molecules onto the outer surface of zeolite particles which contain secondary acceptors. We have studied the photochemistry of covalently linked $Ru(bpy)_3$-diquat cations exchanged onto the surface of zeolite L particles containing benzylviologen (BV^{2+}) ions in the bulk. Following visible light excitation, two sequential electron transfers ($Ru(bpy)_3^{2+} \rightarrow DQ^{2+} \rightarrow BV^{2+}$) occur within 100 ns to yield a Ru^{3+}-$BV^{+\cdot}$ state with a 35 μs lifetime. The series of electron transfers in this self-assembling triad mimics the primary electron transfer steps in bacterial photosynthesis, in that a long-lived charge separated state, in which the electron and hole reside on spatially separated molecules, is formed photochemically.

The primary electron acceptor in these zeolite-based molecular triads may be replaced by a semiconductor particle with a suitable flat-band potential. Hydrolysis of titanium isopropoxide in zeolite L yields quantum size TiO_2 particles with optical bandgaps of 3.4-3.6 eV. Zinc tetrakis(4-carboxyphenyl)porphyrin and ruthenium tris(4,4'-dicarboxy-2,2'bipyridine) adsorb strongly onto these TiO_2/zeolite particles, and transfer electrons to the TiO_2 conduction band when excited by visible light at pH < 5. These electrons are transferred back to the adsorbed dye within 50 ns, and are not transferred to ions such as TTF^+ and BV^{2+} contained within the zeolite.

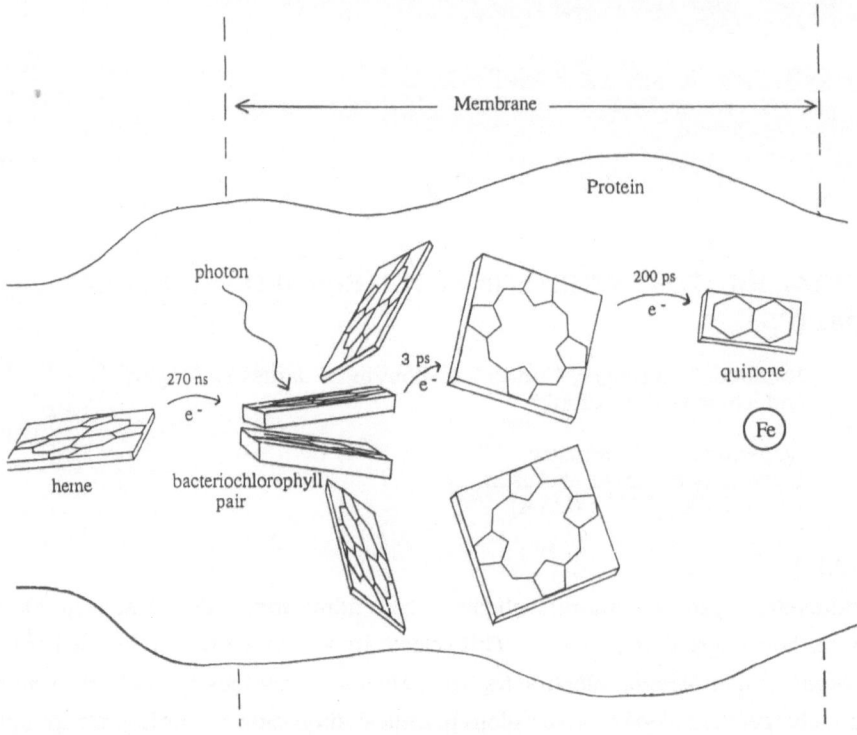

Figure 1. Photosynthetic reaction center structure of the purple bacterium Rhodopseudomonas Viridis, from reference 1; electron transfer kinetics are taken from reference 3.

INTRODUCTION

Natural photosynthesis involves light driven separation of oxidizing and reducing equivalents along a molecular chain. A fraction of the light energy absorbed in this process is stored as electrochemical potential energy, because the ultimate electron acceptor in the chain is a stronger reducing agent than the ultimate electron donor. The electron transfer events which occur in the photosynthetic reaction center of purple bacteria *rhodopseudomonas viridis* and *rhodobacter sphaeroides* are now understood in considerable detail, thanks to recent crystal structures [1,2] and to time resolved spectroscopic examinations of the kinetics of light-induced electron transfer [3]. The results of these studies are summarized in Figure 1. The first step occurs in less than 3 ps, and involves electron transfer from the photoexcited bacteriochlorophyll special pair to a pheophytin primary acceptor. Subsequent, slower electron transfers occur between pheophytin and quinone molecules, and the heme donor and special pair. The net effect is light-induced motion of an electron across the photosynthetic membrane, against an electrochemical potential gradient.

To an electrochemist this reaction center looks simply like a chain of four redox molecules held in a special energetic and spatial arrangement. In order to design a functioning analogue of this system, one needs to find a way to organize molecules in space so that the ultimate electron acceptor is well separated from the electron donor. Short circuits in the chain must be avoided. Additionally, in order to attain, as nature does, a high quantum efficiency for charge separation, each forward electron transfer step must be faster than the reverse step which returns the system to its ground state. Experimental approaches to this problem have involved the synthesis of model compounds containing two to four redox-active components [4-9] and the preparation of Langmuir-Blodgett multilayer films containing an ordered array of light absorber and electron acceptor molecules [10].

We have explored the use of zeolites as templates for self-organizing molecular electron transport chains. Zeolites are crystalline, microporous aluminosilicates which admit neutral and cationic molecules to their internal volume. Entry to the internal space is restricted to molecules of critical dimensions less than 7-8 Å in zeolites L and Y; electroactive cations larger than this adhere via ion exchange, at approximately monolayer coverage, to the external surface. This combination of ion-exchange and size-exclusion properties allows one to prepare, through judicious choice of excluded and included cations, simple molecular redox chains. Electron transfer at the zeolite/solution interface may then be induced and studied both electrochemically and photochemically.

ELECTROCHEMICAL SYSTEMS

Several strategies have been developed in recent years for studying electron transfer reactions in zeolites electrochemically [11]. Following the method of Gemborys and Shaw [11a], we prepared electrodes coated with zeolite Y particles and an inert polystyrene binder. These electrodes were ion-exchanged first with small electroactive cations (metallocenes, viologens) and then with size-excluded cations (such as tris(2,2'-bipyridyl)M^{2+}, M=Ru, Os, or M(II)tetrakis(N-methyl-4-pyridyl)porphyrin^{4+}, M=Zn, Co) [12]. Cyclic voltammetry of these electrodes in aqueous solution containing only supporting electrolyte gave results like that shown in Figure 2. In this experiment the surface-bound $Ru(bpy)_3^{2+}$ cations are harder to reduce than the $CoCp(CpCOOCH_3)^+$ ions contained in the bulk of the zeolite. Cathodic current is observed at a potential near the formal potential of the $Ru(bpy)_3^{2+/+}$ couple, but there is no anodic current on the reverse sweep; UV-visible spectra recorded concurrently with the cyclic voltammetry show that essentially all the current passed corresponds to reduction of zeolite-bound cobalticenium ions within 1 μm of the electrode surface. Since the cathodic current onsets not at the cobalticenium potential, but at the $Ru(bpy)_3^{2+/+}$ potential, we know that reduction of the cobalticenium is mediated by $Ru(bpy)_3^{2+/+}$. The reverse process cannot occur readily because it would involve electron transfer from cobaltocene to $Ru(bpy)_3^{2+}$, which is energetically unfavorable. Changing the substituents on the cobalticenium ion, in order to bring its redox potential into coincidence with $Ru(bpy)_3^{2+/+}$, results in an electrode at which both cathodic and anodic currents are observed [12a].

The current-rectifying behavior manifested in Figure 2 is a consequence of spatial ordering at the molecular level. A single monolayer of size-excluded cations blocks direct electron transfer between the electrode and the zeolite-encapsulated cobalticenium ions; however, the surface bound $Ru(bpy)_3^{2+}$ ions exchange electrons readily with either the electrode or the cobalticenium. This ordering has an important implication for artificial photosynthesis with zeolites. It suggests the possibility of making molecular chains two or more units long in which there is no short circuit between an electron donor and electron acceptor on opposite sides of a molecular monolayer at the zeolite outer surface.

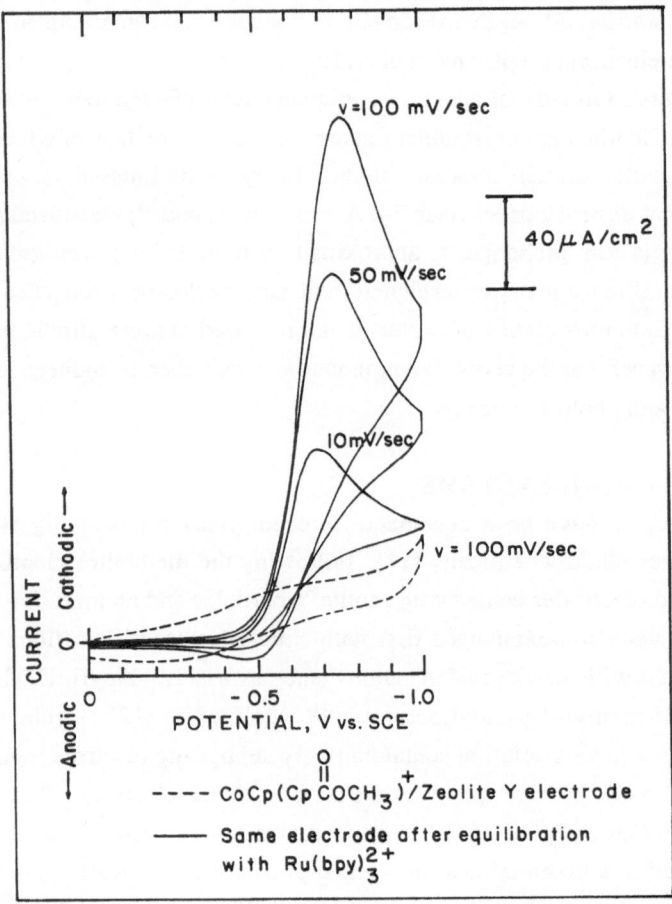

Figure 2. Cyclic voltammetry of a $CoCp(CpCOOCH_3)^+$ - zeolite Y / SnO_2 electrode in 1 mM aqueous KH_2PO_4 solution, before and after 5-min equilibration with 10 mM $Ru(bpy)_3^{2+}$. Electrode held 1 min at 0.0 V between scans; scan rate 100 mV/sec.

The scan rate dependence of the peak cathodic current in Figure 2 provides a measure of the charge-transport diffusion coefficient (D_{ct}) of ions within the zeolite. This diffusion coefficient is 2×10^{-10} cm^2/sec, and implies a timescale of 10^{-5} - 10^{-6} sec for isoenergetic electron hopping between cobalticenium ions in hydrated zeolite Y [12]. Knowledge of this timescale provides a guideline for kinetic design of an artificial photosynthetic system. If reverse electron transfer within a photogenerated charge transfer state at the zeolite outer surface is slower than 10^5 sec^{-1}, efficient separation of charge via electron transfer diffusion into the bulk can be expected to occur.

TRIMOLECULAR REDOX CHAINS

Electrode-bound three-component chains are prepared by providing an anion exchange site between the zeolite particle and the electrode surface. Sb-doped SnO_2 or Pt electrodes functionalized with **1** at 1-5 monolayer coverage can bind a layer of 1 μm diameter zeolite Y particles [13]. It is postulated that the zeolite particles bind to surface-confined **1** via siloxane linkages. Since the electrode now contains three ion-exchange sites associated with the quaternary ammonium groups of **1**, the zeolite outer surface, and the zeolite internal pore

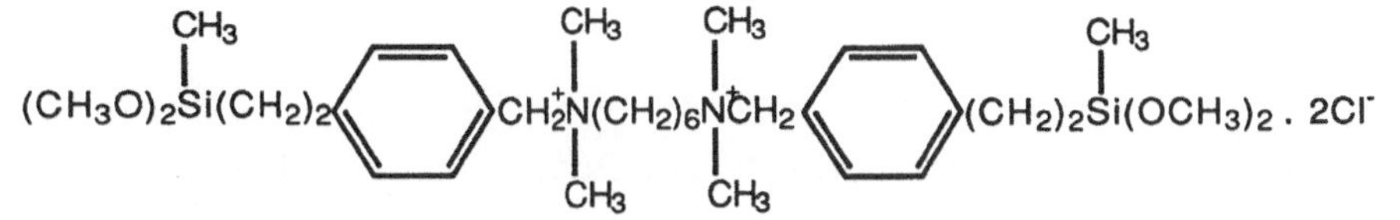

1

structure, three different electroactive ions may be bound as shown in Scheme 1. Here the ferrocyanide ions lie closest to the electrode; zeolite-surface bound $Os(bpy)_3^{2+}$ and bulk-exchanged $Fe(Cp)CpCH_2N(CH_3)_3^+$ ions occupy sites farther from the electrode surface.

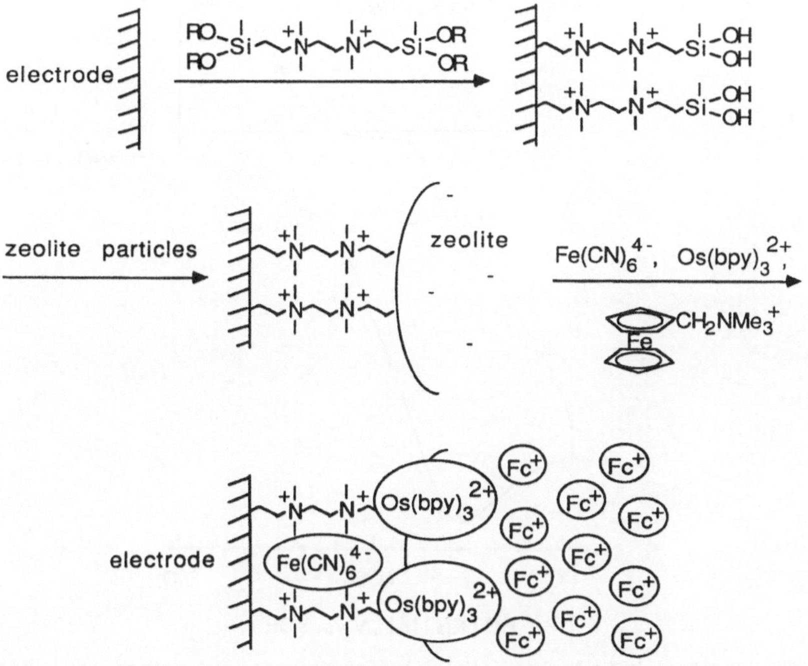

Scheme 1

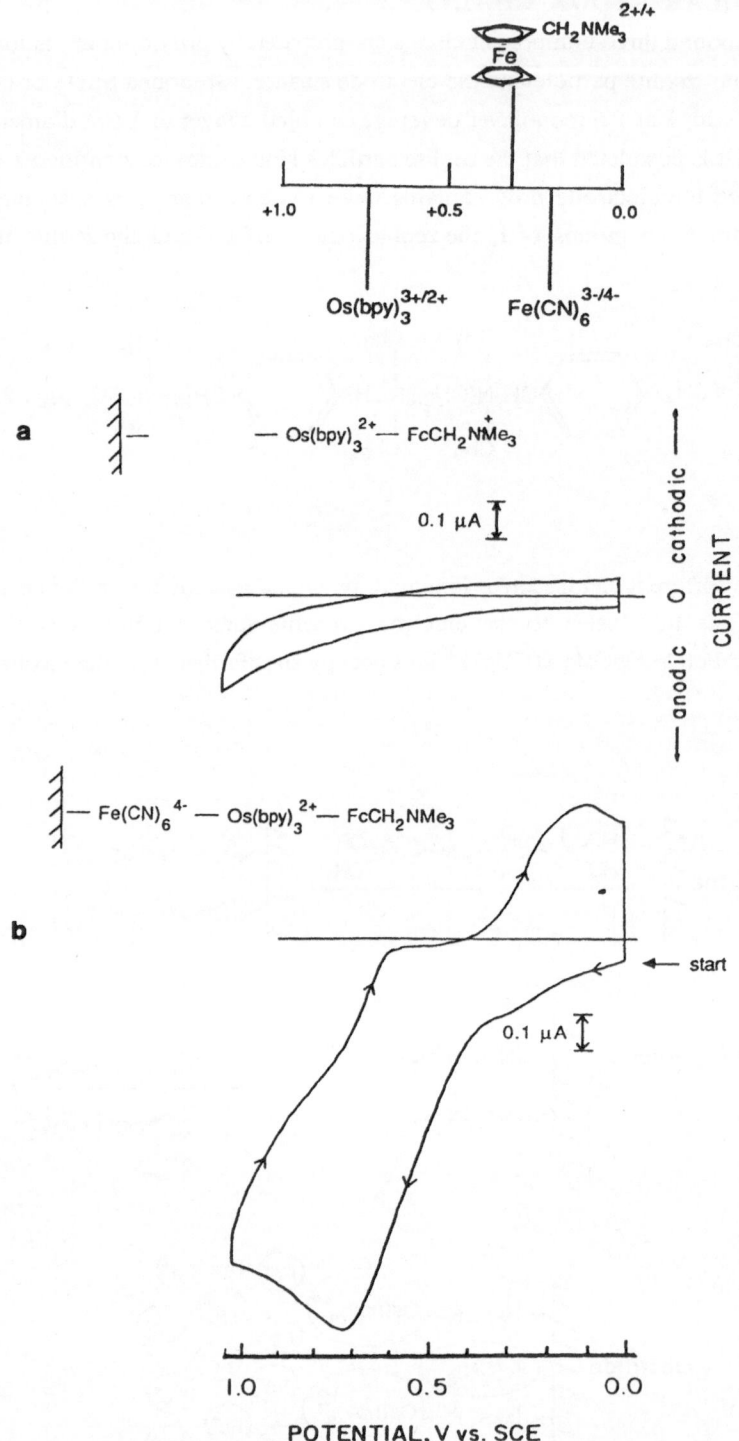

Figure 3. Cyclic voltammetry in 1 mM aqueous KH_2PO_4 of an electrode functionalized with 1 and then zeolite Y; (a) electrode ion-exchanged with 10 mM $Fe(Cp)CpCH_2N(CH_3)_3{}^+I^-$, 10 mM $Os(bpy)_3(ClO_4)_2$; (b) same electrode soaked 16 h in 1 mM $K_4Fe(CN)_6$. Scan rate 10 mV/sec. Formal potentials of the redox-active ions are indicated at the top of the figure.

By leaving out one or more of these electroactive ions (i.e., replacing them with inert ions from the supporting electrolyte), we may determine whether the three-molecule chain contains short circuits. Figure 3a shows two of the eight possible combinations involving $Fe(CN)_6^{4-}$, $Os(bpy)_3^{2+}$, $Fe(Cp)CpCH_2N(CH_3)_3^+$, and/or ions from the supporting electrolyte. With ferrocyanide as the only electroactive ion present, a reversible oxidation-reduction wave is observed. Leaving out $Fe(CN)_6^{4-}$ but including both $Fe(Cp)CpCH_2N(CH_3)_3^+$ and $Os(bpy)_3^{2+}$ (Figure 3a), no current is seen. Presumably both of these ions are too far from the electrode surface for direct electron transfer to occur on the electrochemical timescale. When all three electroactive ions are present we obtain the cyclic voltammetry shown in Figure 3b. On the positive sweep a small anodic wave corresponding to the $Fe(CN)_6^{4-} \rightarrow Fe(CN)_6^{3-}$ interconversion and a large wave attributed to $Fe(Cp)CpCH_2N(CH_3)_3^+ \rightarrow Fe(Cp)CpCH_2N(CH_3)_3^{2+}$ are observed. UV-visible spectra, Figure 4, confirm this last assignment. The large anodic wave onsets near the $Os(bpy)_3^{2+/3+}$ potential, consistent with mediation of the ferrocene oxidation by $Os(bpy)_3^{2+/3+}$ ions at the zeolite outer surface. That the ferrocene/ferricenium ions within the zeolite do *not* communicate directly with $Fe(CN)_6^{4-}$ or the electrode is shown by the negative sweep in Figure 3b and corresponding UV-visible spectra in Figure 4. Two cathodic waves are observed, but the spectra show that no $Fe(Cp)CpCH_2N(CH_3)_3^{2+}$ is reduced; hence the two waves correspond to reduction of $Os(bpy)_3^{3+}$ and $Fe(CN)_6^{3-}$. $Os(bpy)_3^{2+}$ is not a sufficiently strong reducing agent to effect the conversion of $Fe(Cp)CpCH_2N(CH_3)_3^{2+}$ to $Fe(Cp)CpCH_2N(CH_3)_3^+$, so the oxidizing equivalents are trapped within the zeolite. Repetitive scans give progressively smaller anodic currents, because less $Fe(Cp)CpCH_2N(CH_3)_3^{2+}$ is available for mediated oxidation.

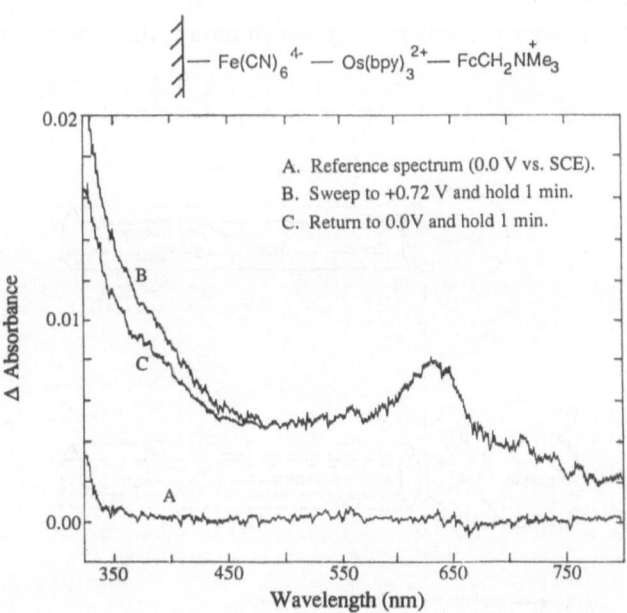

Figure 4. Transmission UV-visible spectra of the SnO_2/glass electrode used in Fig. 3b. Peak at 640 nm corresponds to oxidation of approximately 10^{-8} mol/cm^2 $Fe(Cp)CpCH_2N(CH_3)_3^+$.

It is apparent that this three-molecule chain is not perfect, since both anodic and cathodic $Os(bpy)_3^{2+/3+}$ waves are observed when $Fe(CN)_6^{3-/4-}$ ions are present. This direct oxidation and reduction implies that $Os(bpy)_3^{2+/3+}$ ions can migrate to the electrode surface, presumably via ion pairing with $Fe(CN)_6^{3-/4-}$. Other combinations of multiply charged cations and anions (e.g., $Fe(phen)_3^{2+}$ and $W(CN)_8^{4-}$) show similar effects in these zeolite Y/quaternary ammonium siloxane films [13].

PHOTOCHEMICAL SYSTEMS

The insight gained from electrochemistry into the siting of large and small cations, and the rates of electron transfer between them, can guide us in the design of zeolite-based photochemical assemblies. Elegant examples of the use of photochemistry to differentiate internal and external surface reactivity, and to probe intrazeolite molecular motion, were provided in this conference by Prof. Turro and by Dr. Ramamurthy. We have studied assemblies in which only outer-sphere electron transfer (and no bond breaking) is induced photochemically. These are again three-component redox chains; two kinds of self-organizing chains, shown in Scheme 2, have been prepared and studied by flash photolysis/transient diffuse reflectance techniques [14-16].

In the donor-acceptor-secondary acceptor molecular triad (top of Scheme 2), self-assembly is achieved via exchange of a covalently linked $Ru(bpy)_3^{2+}$-diquat^{2+} cation, 2, onto the surface of zeolite L or Y. The kinetics of light-induced electron transfer in this system are summarized in Figure 5. In fluid solution 2 is not luminescent, indicating that electron transfer from the $Ru(bpy)_3^{*2+}$ MLCT state occurs on a timescale much shorter than the luminescence lifetime of $Ru(bpy)_3^{*2+}$, about 650 ns. Picosecond transient absorbance studies have shown, in molecules like 2, that electron transfer from $Ru(bpy)_3^{*2+}$ to the diquat moiety occurs in about 300 ps and that reverse electron transfer, which returns the molecule to its ground state, is equally fast [17].

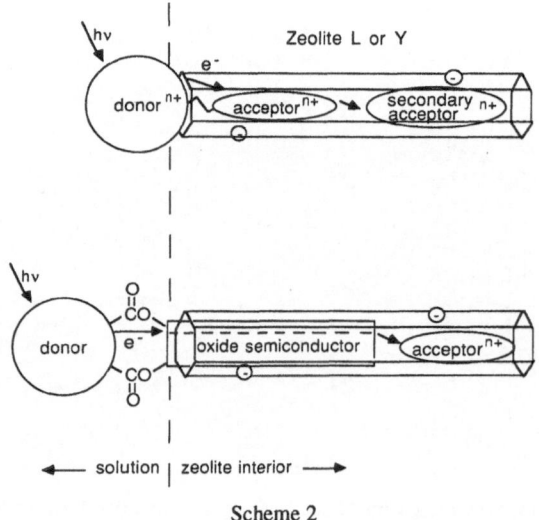

Scheme 2

R = CH₃

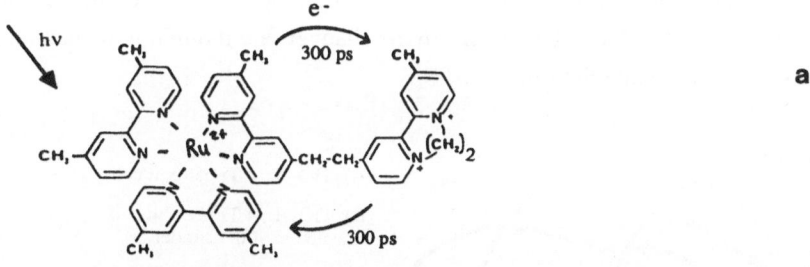

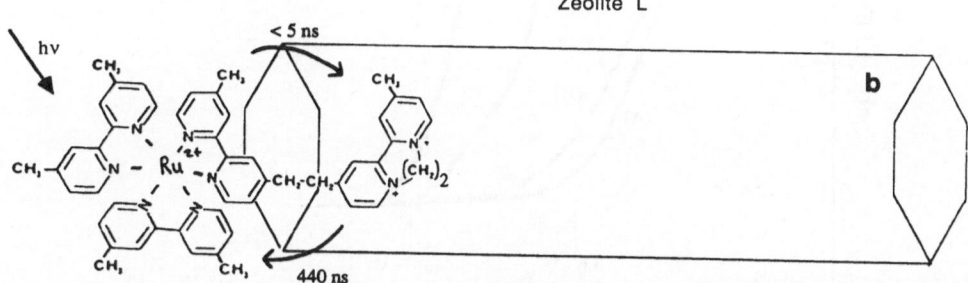

Zeolite L

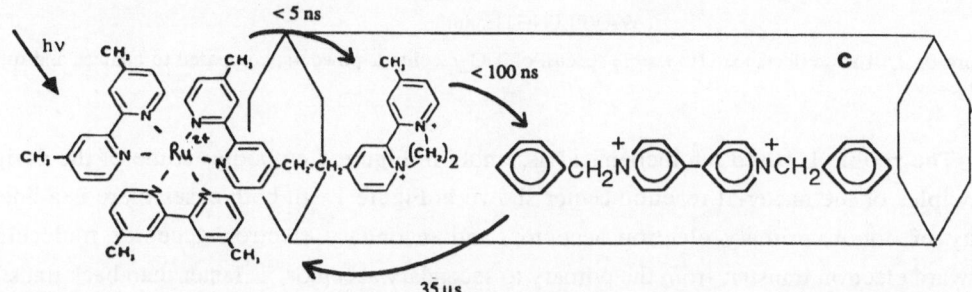

Figure 5. Kinetics of light-induced electron transfer reactions of **2** (a) in acetonitrile solution (reference 18); (b) ion-exchanged on zeolite L powder, aqueous suspension; (c) on zeolite L which was previously exchanged with 1.5×10^{-4} mol BV^{2+}/ g zeolite.

2 can be immobilized on the outer surface of zeolite L particles by ion exchange. Flash photolysis of an aqueous suspension of these particles reveals a charge-separated state $(Ru(bpy)_3^{3+}$-diquat$^{+\cdot})$ which decays with a lifetime of 440 ns. This dramatic decrease in the rate of reverse electron transfer, relative to **2** in solution, probably comes about because of restricted molecular motion on the zeolite surface: the two ends of the molecule are kept apart because the smaller diquat moiety can fit into the 7Å pore opening. The postulated orientation of **2** relative to the one-dimensional zeolite channels is shown in Figure 5. Inclusion of a secondary electron acceptor, benzylviologen (BV^{2+}), into these channels results in rapid light-induced formation of an oxidized ruthenium-reduced benzylviologen state [15]. Since it is doubtful that direct electron transfer from the $Ru(bpy)_3^{2+}$ moiety of **2** to zeolite-encapsulated BV^{2+} can compete with intramolecular electron transfer quenching in **2**, it is most likely that a two-step electron transfer process $(Ru(bpy)_3^{*2+} \rightarrow$ diquat $\rightarrow BV^{2+})$ occurs. This Ru^{3+}-$BV^{+\cdot}$ state persists for 35 μs before decaying back to the ground state. Both the physical separation of Ru^{3+} and $BV^{+\cdot}$ imposed by the zeolite and isoenergetic electron hopping between $BV^{+\cdot}$ and BV^{2+} ions in the channels are thought to contribute to this unusually long Ru^{3+}-$BV^{+\cdot}$ state lifetime.

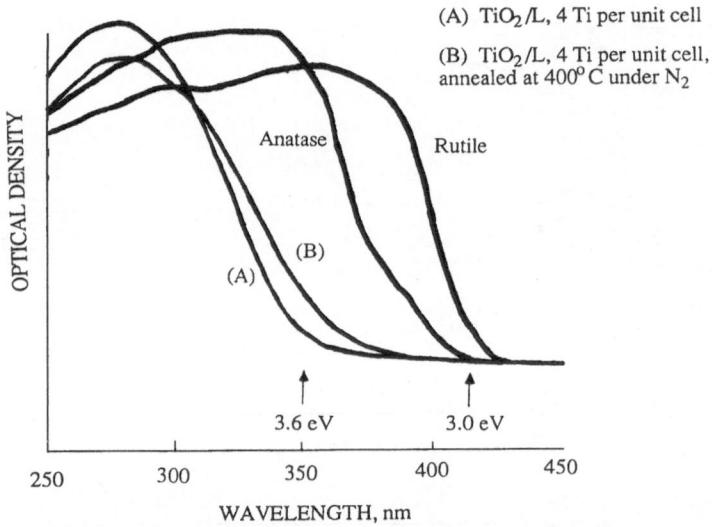

Figure 6. Diffuse reflectance UV-visible spectra of TiO_2/zeolite L powders, compared to anatase and rutile TiO_2.

The zeolite L-based photochemical assembly in Figure 5 embodies some of the design principles of the bacterial reaction center shown in Figure 1. In both cases there is a linear array of donor, primary electron acceptor, and secondary electron acceptor molecules. Forward electron transfer, from the primary to secondary acceptor, is faster than back transfer to the donor. In both cases this arrangement results in a high quantum efficiency for charge separation (15% in our system, 99+% in natural photosynthesis) and a long-lived charge-separated state.

The lower diagram in Scheme 1 shows another possibility for artificial photosynthesis using zeolites. We have prepared TiO_2 particles in zeolite L by sol-gel techniques [16]. Diffuse reflectance UV-visible spectra of these particles are compared, in Figure 6, to those of bulk anatase and rutile TiO_2 powders. The 0.4 eV shift to the blue of bandgap absorption is attributed to a quantum size effect [18]; from the magnitude of the shift and the electron effective mass [19] of TiO_2, we calculate an average particle diameter of 10-15 Å. This is consistent with encapsulation of the particles by zeolite L, since 13 x 7.5 Å cavities are linked together into linear tunnels in that structure [20]. These non-intersecting tunnels are well separated from each other (center-to-center distance 18.4 Å).

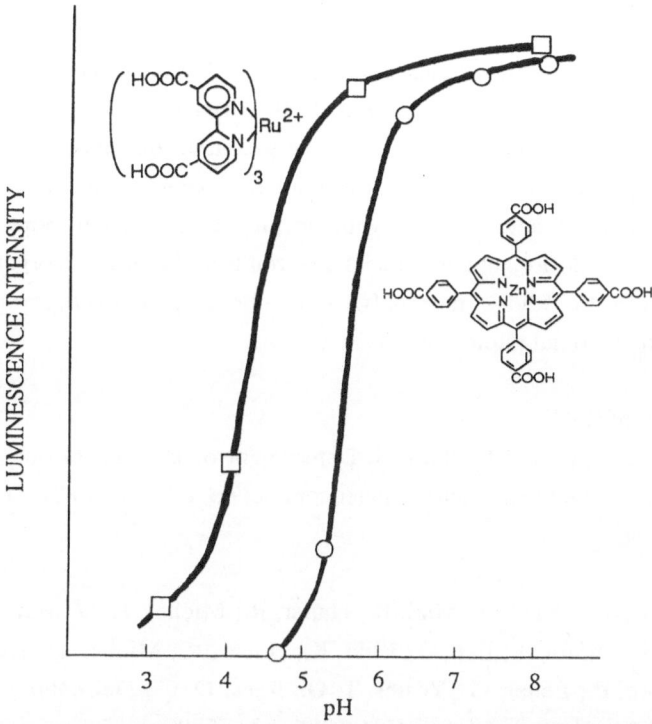

Figure 7. pH-dependent luminescence intensity of $Ru(bpy[COOH]_2)_3^{2+}$ and ZnTPPC (3 x 10^{-6} mol/ g zeolite) in aqueous suspensions of TiO_2/zeolite L.

TiO_2 may be sensitized by a number of dyes which absorb in the visible. Of particular interest to us were size-excluded complexes such as $Ru(bpy[COOH]_2)_3^{2+}$ and zinc tetrakis(4-carboxyphenyl)porphyrin (ZnTPPC). These dyes, when photoexcited, inject electrons with 20-70% quantum efficiency into the conduction band of oxide semiconductors such as TiO_2 and $SrTiO_3$ [21,22]. The oxidized dye can be reduced by any of a variety of anionic electron donors [22,23] including I^-, Br^-, $Fe(CN)_6^{4-}$, and $Mo(CN)_8^{4-}$. Figure 7 shows luminescence spectra of TiO_2/zeolite L powders in water with these dyes present at approximately monolayer coverage (3 x 10^{-6} mol/g zeolite). At high pH these anionic dyes do not adsorb onto the TiO_2, and are strongly luminescent. At low pH the protonated TiO_2

surface is positively charged, causing the dye to adsorb, and the luminescence is quenched efficiently. Since the fluorescence lifetime of ZnTPPC is 1.8 ns [24], this quenching event takes place on a subnanosecond timescale. By transient diffuse reflectance techniques we cannot detect bleaching of these adsorbed dyes on a 50 ns timescale; therefore the electron injected into the TiO_2 is transferred back to the dye very rapidly. Consistent with the rapidity of this back transfer, we do not see forward electron transfer to cationic electron acceptors such as tetrathiofulvalenium (TTF^+) or BV^{2+} which have been ion-exchanged into the zeolite. We are currently experimenting with doped TiO_2, and with zeolites which have a higher degree of channel connectivity (such as zeolite Y) to see if more efficient charge separation can be effected.

CONCLUSIONS

We have shown that zeolites are excellent templates for self-organizing molecular redox systems. Two- and three-component vectorial electron transport chains are easily prepared at electrode surfaces, and similar assemblies can be prepared and photochemically driven in aqueous zeolite suspensions. Some of these systems are able to mimic kinetically the primary electron transfer steps of natural photosynthetic processes. When more is known about molecular orientation, motion, and interfacial electron transfer in these organized systems, it should be possible to devise more efficient schemes, involving molecules and/or semiconductors, for artificial photosynthesis in zeolites.

ACKNOWLEDGMENT

This work was supported by the U.S. Department of Energy, Office of Basic Energy Sciences, Division of Chemical Sciences under contract no. DE-FG05-87ER13789.

REFERENCES

1. (a) Deisenhofer, J.; Epp, O.; Miki, K.; Huber, R.; Michel, H. *J. Mol. Biol.* **1984**, *180*, 385; (b) Deisenhofer, J.; Epp, O.; Miki, K.; Huber, R.; Michel, H. *Nature* **1985**, *318*, 618; (c) Allen, P.; Feher, G.; Yeates, T. O.; Rees, D. C.; Deisenhofer, J.; Michel, H.; Huber, R. *Proc. Natl. Acad. Sci. USA* **1986**, *83*, 8589.

2. Chang, C. H.; Tiede, D.; Tang, J.; Smith, U.; Norris, J.; Schiffer, M. *FEBS Lett.* **1986**, *205*, 82.

3. Martin, J. L; Breton, J.; Hoff, A. J.; Migus, A.; Antonetti, A. *Proc. Natl. Acad. Sci. USA* **1986**, *83*, 957; (b) Breton, J.; Martin, J. L.; Migus, A.; Antonetti, A.; Orszag, A. *Proc. Natl. Acad. Sci. USA* **1986**, *83*, 5121; (c) Kaufman, K. J.; Dutton, P. L.; Netzel, T. L.; Leigh, J. S.; Rentzepis, P. M. *Science* **1975**, *188*, 1301; (d) Gunner, M. R.; Robertson, D. E.; Dutton, P. L. *J. Phys. Chem.* **1986**, *90*, 3783.

4. Gust, D.; Moore, T. A.; Liddell, P. A.; Nemeth, G. A.; Makings, L. R.; Moore, A. L.; Barrett, D.; Pessiki, P. J.; Bensasson, R. V.; Rougee, M.; Chachaty, C.; De Schryver, F. C.; Van der Auweraer, M.; Holzwarth, A. R.; Connolly, J. S. *J. Am. Chem. Soc.* **1987**, *109*, 846, and references contained therein.

5. Wasielewski, M. R.; Niemczyk, M. P.; Svec, W. A.; Pewitt, E. B. *J. Am. Chem. Soc.* **1985**, *107*, 5562.

6. Danielson, E.; Elliott, C. M.; Merkert, J. W.; Meyer, T. J. *J. Am. Chem. Soc.* **1987**, *109*, 2519.

7. (a) Westmoreland, T. D.; Schanze, K. S.; Neveux, P. E.; Danielson, E.; Sullivan, B. P.; Chen, P.; Meyer, T. J. *Inorg. Chem.* **1985**, *24*, 2596; (b) Chen, P.; Westmoreland, T. D.; Danielson, E.; Schanze, K. S.; Anthon, D.; Neveux, P. E.; Meyer, T. J. *Inorg. Chem.* **1987**, *26*, 1116.

8. (a) Elliott, C. M.; Freitag, R. A. *J. Chem. Soc., Chem. Commun.* **1985**, 156; (b) Elliott, C. M.; Freitag, R. A.; Blaney, D. D. *J. Am. Chem. Soc.* **1985**, *107*, 4647.

9. For recent reviews of light-induced electron transfer in donor-acceptor molecules see (a) Turro, N. J.; Kavarnos, G. *J. Chem. Rev.* **1986**, *86*, 401; (b) Fendler, J. H. *J. Phys. Chem.* **1985**, *89*, 2730.

10. (a) Fromherz, P.; Arden, W. *Ber. Buns. Phys. Chem.* **1980**, *84*, 1045; (b) Arden, W.; Fromherz, P. *J. Electrochem. Soc.* **1980**, *127*, 370; Arden, W.; Fromherz, P. *Ber. Buns. Phys. Chem.* **1978**, *82*, 868; Kuhn, H. *Pure Appl. Chem.* **1979**, *51*, 341; Moebius, D. *Acc. Chem. Res.* **1981**, *3*, 63.

11. (a) Gemborys, H. A.; Shaw, B. R. *J. Electroanal. Chem.* **1986**, *208*, 95; (b) Murray, C. G.; Nowak, R. J.; Rolison, D. R. *J. Electroanal. Chem.* **1984**, *164*, 205; (c) Susic-Milenko, V. *Electrochim. Acta* **1979**, *24*, 535; (d) Periera-Ramos, J.; Messina, R.; Perichon, J. *J. Electroanal. Chem.* **1983**, *146*, 157; (e) de Vismes, B.; Bedioui, F.; Devynck, J.; Bied-Charreton, C. *J. Electroanal. Chem.* **1985**, *187*, 197.

12. (a) Li, Z.; Mallouk, T. E. *J. Phys. Chem.* **1987**, *91*, 643; (b) Li, Z.; Wang, C. M.; Persaud, L.; Mallouk, T. E. *J. Phys. Chem.* **1988**, *92*, 2592.

13. Li, Z.; Lai, C.; Mallouk, T. E., submitted for publication.

14. Willsher, C. J. *J. Photochem.* **1985**, *28*, 229.

15. Krueger, J. S.; Mayer, J. E.; Mallouk, T. E. *J. Am. Chem. Soc.* **1988**, *110*, 8232.

16. Krueger, J. S.; Mallouk, T. E., in preparation.

17. Cooley, L. F.; Headford, C. E. L.; Elliott, C. M.; Kelley, D. F. *J. Am. Chem. Soc.* **1988**, *110*, 6673.

18. (a) Brus, L. E. *J. Chem. Phys.* **1983**, *79*, 5566; (b) Brus, L. E. *J. Phys. Chem.* **1986**, *90*, 2555.

19. Brus, L. E. *Nouv. J. Chim.* **1987**, *11*, 123.

20. Barrer, R. M.; Villiger, H. *Z. Kristallogr.* **1969**, *128*, 352.

21. (a) Rotzinger, F. P.; Munavalli, S.; Comte, P.; Hurst, J. K.; Grätzel, M.; Pern, F.-J.; Frank, A. J. *J. Am. Chem. Soc.* **1987**, *109*, 6619; (b) Liska, P.; Vlachopoulos, N.; Nazeeruddin, M. K.; Comte, P.; Grätzel, M. *J. Am. Chem. Soc.* **1988**, *110*, 3686; (c) Desilvestro, J.; Grätzel, M.; Kavan, L.; Moser, J.; Augustynski, J. *J. Am. Chem. Soc.* **1985**, *107*, 2988; (d) Kalyanasundaram, K.; Vlachopoulos, N.; Krishnan, V.; Monnier, A.; Grätzel, M. *J. Phys. Chem.* **1987**, *91*, 2342.

22. Dabestani, R.; Bard, A. J.; Campion, A.; Fox, M. A.; Mallouk, T. E.; Webber, S. E.; White, J. M. *J. Phys. Chem.* **1988,** *92,* 1872.

23. Vlachopoulos, N.; Liska, P.; Augustynski, J.; Graetzel, M. *J. Am. Chem. Soc.* **1988,** *110,* 1216.

24. Kalyanasundaram, K.; Neumann-Spallart, M. *J. Phys. Chem.* **1982,** *86,* 5163.

SILVER SODALITES: NOVEL OPTICALLY RESPONSIVE NANOCOMPOSITES

Geoffrey A. Ozin,† Andreas Stein,† Galen D. Stucky,§ and John P.Godber#

†Lash Miller Chemical Laboratories
University of Toronto
Toronto M5S 1A1 (Canada)
§Department of Chemistry
University of California
Santa Barbara, California 93106 (USA)
#Albright and Wilson Americas
Islington, Ontario MSB 1R1 (Canada)

SUMMARY

A range of novel silver sodalites has been synthesized. These solid state microstructures are viewed as "packaged" silver salts comprised of nanoassemblies of silver cations tetrahedrally organized with various charge balancing anions. A collection of physicochemical characterization techniques (UV-VIS reflectance/emission/excitation; luminescence lifetimes; FT-Far/Mid-IR; XPS; TGA-MS; ^{29}Si, ^{27}Al, ^{23}Na MAS-NMR; Rietveld XRD profile analysis) have been employed to investigate the structure and properties of the parent silver sodalites, as well as the chemical and physical transformations of the encapsulated silver salts that relate to a number of interesting transducer effects. Intercavity communication between entrapped silver microaggregates and expanded-metal superlattice ideas will be considered briefly. The utilization of the silver sodalites in high resolution imaging/printing and high density write/read/erase data storage applications are also considered.

INTRODUCTION

The myriad of one, two and three dimensional, molecular size channel and cage structures found in the class of crystalline framework materials referred to as zeolites, confers upon them a unique spatial quality displayed by few other classes of solid state materials. With this in mind, recall that today's world of advanced solid state materials is unrelentingly driven by the ever increasing demands for smaller, faster, more selective and efficient products and processes. Thus one can justifiably ask whether or not the well known structural, physical and chemical properties of zeolites together with their size and shape discrimination power, traditionally applied in catalysis, selective adsorption, molecular sieving, water softening and waste water clean-up to name a few, can be effectively exploited in emerging fields of solid state chemistry.

Inclusion Phenomena and Molecular Recognition
Edited by J. Atwood
Plenum Press, New York, 1990

One of the points that will emerge from this paper, as well as others presented at this conference, is that microporous molecular electronic materials with nanometer dimension window, channel and cavity architecture represent a "New Frontier" of solid state chemistry with great opportunities for innovative research and development [1].

SILVER SODALITES: NOVEL, OPTICALLY RESPONSIVE, PACKAGED SILVER SALTS

A range of novel silver sodalites has recently been synthesized [1]. These solid state microstructures are viewed as "packaged" silver salts comprised of nanoassemblies of silver cations tetrahedrally organized with various charge balancing anions [2]. A collection of physicochemical characterization techniques have been employed to investigate the structures and properties of the parent silver sodalites, as well as the chemical and physical transformations of the encapsulated silver salts that relate to a number of transducer effects [3]. An interesting question in these materials concerns intercavity communication between aggregates of the entrapped silver salt and the connection of their properties with those of the analogous bulk phase material (such as electronic transport, photoconductivity, photoaggregation and photoluminescence). Intracavity redox processes that produce encapsulated silver clusters and expanded-metal-superlattice phenomena above the percolation threshold loading level are also intriguing avenues of enquiry within the silver sodalite family of materials.

In this summary paper we will briefly explore the synthesis and characterization of a few key silver sodalites and examine their possible utilization in high resolution imaging/printing and high density data storage applications.

PURE ALUMINATE SODALITE (Si/Al = 0) [6]	$Sr_6\{(AlO)_{12}\}.2SrX_2$
NORMAL SODALITE AT THE LOEWENSTEIN LIMIT (Si/Al = 1) [2]	$Na_6\{(AlO_2)_6(SiO_2)_6\}.2NaX$
SILICA-RICH TMA SODALITE (Si/Al = 5) [7]	$(CH_3)_4NAlSi_5O_{12}$
POROUS, CRYSTALLINE SILICA SODALITE (Si/Al = ∞) [8]	$\{(SiO_2)_{12}\}.2C_2H_4(OH)_2$

Scheme 1

A summary of some of the attractive features of silver sodalites in solid state chemistry is presented schematically in Figure 1. The regular, all space filling, cubic framework of sodalite (a Federov solid) provides a homogeneous microporous matrix of 6.6 Å sodalite cavities, suitable for stabilizing small isolated molecules, atoms and clusters as neutral species, ions or even radicals [2] (Figure 2). The unit cell dimensions, charge balance requirements and cage-filling can all be tuned by incorporating a large variety of anions during the sodalite synthesis. Various sodalite cage anion packing schemes are illustrated in Figure 3.

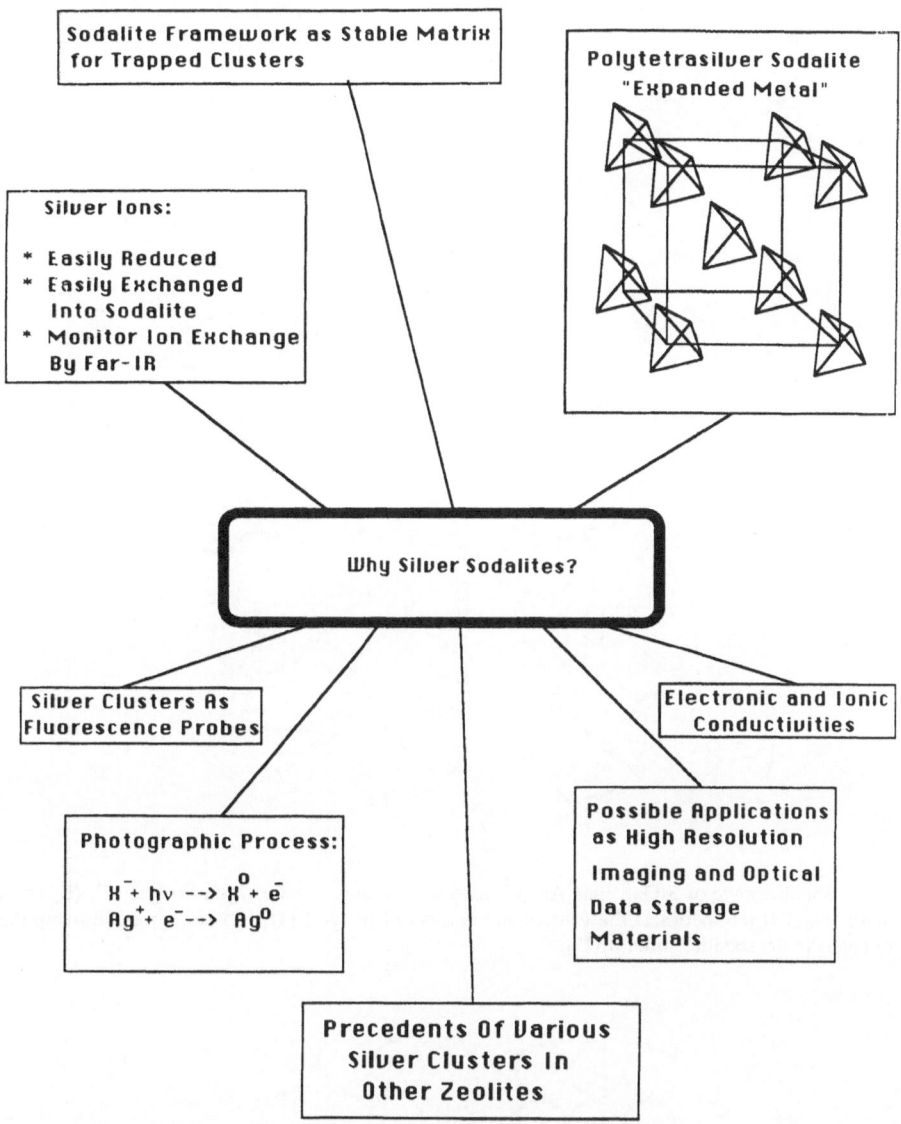

Figure 1. Summary of features of silver sodalites of interest in solid state chemistry.

Variations in the Si/Al ratio from pure aluminate to pure silica sodalite are also possible as summarized in Scheme 1.

This impressive range of sodalite cage compositions attests to the remarkable isomorphous replacement qualities of the sodalite framework topology [2]. Substitution of the framework atoms (Si^{4+}, Al^{3+}) for other elements, such as Ge^{4+}, Ga^{3+}, B^{3+}, P^{5+}, Fe^{3+} allows further control over the material properties. Sodalites with sodium cation guests are readily synthesized by hydrothermal reactions of silicate and aluminate sources (e.g. kaolin, or silica gel and sodium aluminate) with the appropriate anion salt in concentrated NaOH solution [2,3].

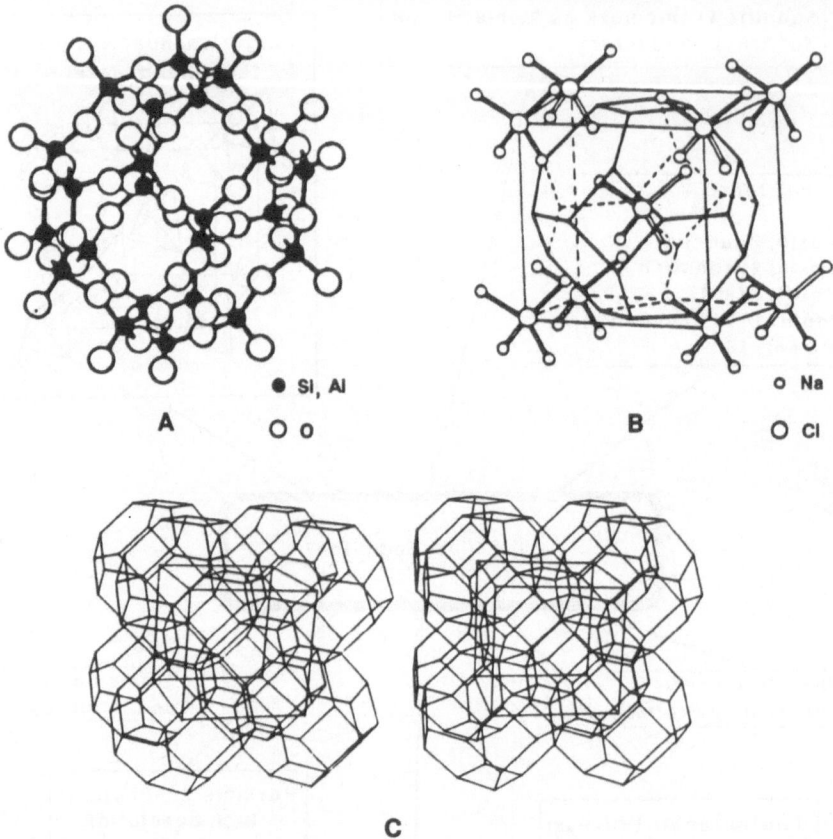

Figure 2. (A) Sodalite cage of SiO_4^{4-} and AlO_4^{5-} tetrahedra showing the oxygen bridges [4], (B) unit cell of sodalite with [Na$_4$Cl] tetrahedra at the centers and corners of the cell [4], (C) stereoplot showing the close packing of cages in the sodalite structure [5].

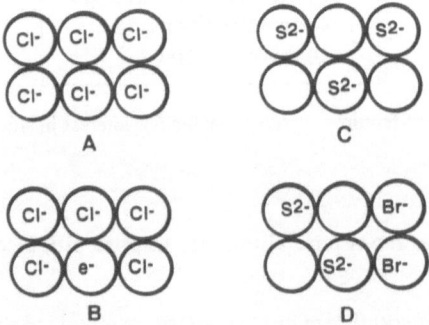

Figure 3. The effect of charge restrictions in a 1:1 aluminosilicate sodalite with monovalent cations. An idealized layer of sodalite cages is shown in each case. Extraframework cations are left out for clarity; (A) each cage is occupied by a monovalent anion, (B) a trapped electron (F-center) can replace an anion in photochromic or cathodochromic sodalites, (C) divalent anions require that half of the cages do not contain negative ions, (D) anion mixing allows some control over negative ion deficiencies in the sodalite cages.

Sodium ions are exchanged by silver ions in a silver salt melt, or in some instances, in a silver salt aqueous solution. The framework structure is generally maintained after exchange, as confirmed by powder X-ray diffraction and mid-IR of the framework vibrations [3]. The silver sodalites prepared to date in our laboratory are listed in Table 1. The silver exchange process is conveniently monitored by intensity changes of diagnostic sodium and silver ion translatory modes in the far-IR as well as shifts in the frequencies of far-IR anion translations [9] (Figures 4,5).

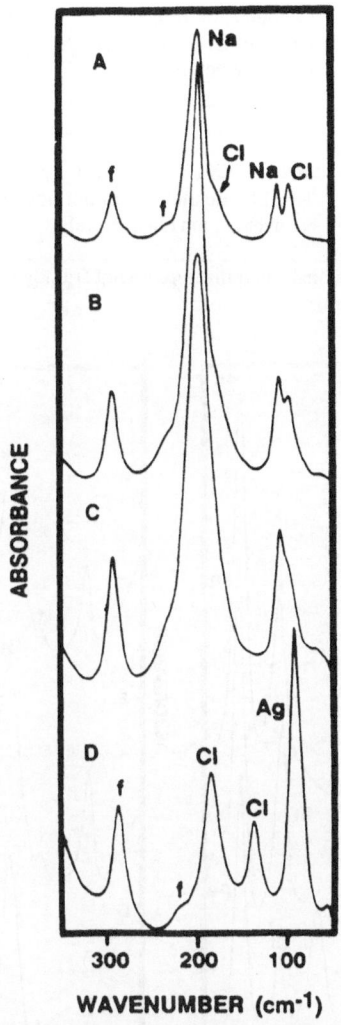

Figure 4. Far-IR spectra of chloro-sodalites with various silver loadings. (A) Na,Cl-SOD, (B) Na,Ag,Cl-SOD (1 Ag/unit cell), (C) Na,Ag,Cl-SOD (2 Ag/unit cell), (D) Ag,Cl-SOD (8 Ag/unit cell). The spectra are not all shown on the same absorbance scale. "f" denotes a framework absorption.

Table 1. Silver sodalites prepared in this study.

Sodalite type	Reason for Study
Ag,Cl-SOD	Analogues to photochromic and cathodochromic sodalites
Ag,Br-SOD	Similarities with photographic process
Ag,I-SOD	Occluded guest species are photoconductors, semiconductors, fast ion conductors
	Study variation in properties down a group
Ag,S-SOD[a]	Sensitization
Ag,Br,S-SOD	Photographic imaging
Ag,OH-SOD	Some hydroxide present in most hydrothermally prepared sodalites
Ag_6-SOD	Cages contain only three silver ions and no anion
	Occluded metal clusters
	Expanded metal
	Quantum size effects
Ag,CO_3-SOD	Intracage anion decomposition possible
Ag, HCO_2-SOD ≡ Ag,Fo-SOD	Redox switching
Ag,C_2O_4-SOD ≡ Ag,Ox-SOD	
Ag,ClO_4-SOD	
Ag,SO_x-SOD	
Na,Ag,X-SOD	Mixed cation systems to study silver concentration effects
	Tunable insulators/semiconductors (NaX: Insulators, AgX: Semiconductors, ion conductors)

[a] An XRD powder diffraction analysis indicated that significant framework decomposition occurred during the synthesis of Ag,S-SOD by melt exchange.

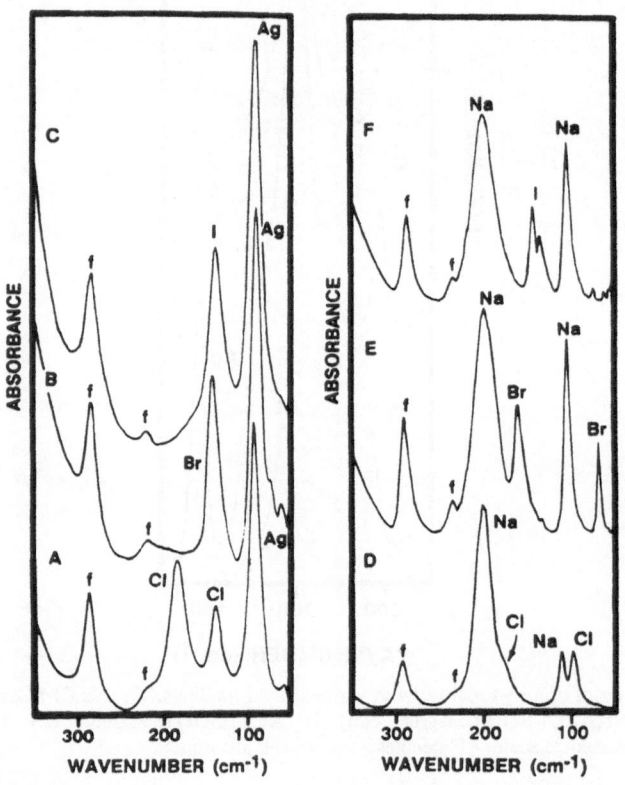

Figure 5. Far-IR spectra of halo-sodalites; (A) Ag,Cl-SOD, (B) Ag,Br-SOD, (C) Ag,I-SOD, (D) Na,Cl-SOD, (E) Na,Br-SOD, (F) Na,I-SOD. "f" denotes a framework absorption.

Many of the silver sodalites prepared in our studies are novel materials. Interesting responses in color as well as luminescence properties are observed upon exposing several compositions to a variety of physical stimuli, including heat, light, pressure, moisture, X-rays and e-beams. These effects are categorized in Table 2. Some of these changes were reversible or semi-reversible [1], the details of which will be published elsewhere.

Table 2. Qualitative responses of silver sodalites to various physical stimuli.

BAROCHROMIC:
  White Ag,OH-SOD or Ag,HCO$_2$-SOD samples darken upon application of pressure

HYDROCHROMIC:
  Ag,OH-SOD undergoes a reversible color change upon dehydration

PHOTOCHROMIC:
  Color changes and/or sample darkening can be produced by irradiation of Ag,Ox-SOD, Ag,OH-SOD, Ag,CO$_3$-SOD, Ag,HCO$_2$-SOD and sulfur-doped silver sodalites with various light sources

THERMOCHROMIC:
  Na,Ag,Ox-SOD, Ag,OH-SOD and sulfur-doped silver sodalites undergo various color changes upon heat treatment

X-RAY SENSITIVITY:
  White Ag,Ox-SOD turns yellowish-green upon exposure to X-rays

FLUORESCENCE CHANGES:
  Heat, light and X-rays can induce Na,Ag,Ox-SOD samples to fluoresce under UV-light

---

These effects are related to the ease of silver reduction and the ability of the silver to form small clusters or expanded superclusters within the sodalite framework. In silver hydroxo-, formato- and oxalato-sodalites this is facilitated by intrazeolitic redox reactions:

$$2\,Ag^+(Z) + H_2O(Z) \quad \rightarrow \quad 2\,Ag^0(Z) + 1/2\,O_2 + 2H^+(Z) \tag{1}$$

$$2\,Ag^+(Z) + OH^-(Z) \quad \rightarrow \quad 2\,Ag^0(Z) + 1/2\,O_2 + H^+(Z) \tag{2}$$

$$2\,Ag^+(Z) + HCO_2^-(Z) \quad \rightarrow \quad 2\,Ag^0(Z) + CO_2(Z) + H^+(Z) \tag{3}$$

$$2\,Ag^+(Z) + C_2O_4^{2-}(Z) \quad \rightarrow \quad 2\,Ag^0(Z) + 2\,CO_2 \tag{4}$$

Reaction 4 can be effected both thermally and photolytically. Note that these equations are intended to illustrate simple stoichiometric redox reactions and are not meant to represent balanced unit cell reactions. Depending upon the level of silver exchange, aggregates of the type $(Ag_nNa_{4-n})^{q+}$, n = 0 to 4, may be formed within the cages, where charge variations of q in the range q = (4-n) to 4 can lead to the observed optical responses.

By increasing the silver loading in the parent sodalite, it is possible to form cluster and extended supercluster structures in the quantum size regime. Such superclusters have been reported in the case of semiconductor materials in faujasite and zeolite A [10]. One important advantage of sodalites over the other zeolite hosts is that the sodalite lattice is not comprised of coexisting sodalite and supercages but is instead entirely composed of close-packed sodalite cages, thereby permitting direct or through-bond interaction between silver guests in all cages. In addition, sodalites are unique due to the presence of guest-anions, which can not only be used as internal reagents (e.g. for the intra-sodalite cage redox reactions) but also provide great flexibility in the composition of trapped species. By judicious selection of the anion and cation one can form packaged insulators, semiconductors as well as metals within the sodalite framework. Control of the guest-concentration in sodalites allows transitions from molecular type species to expanded bulk materials above a percolation threshold (PT) (Scheme 2).

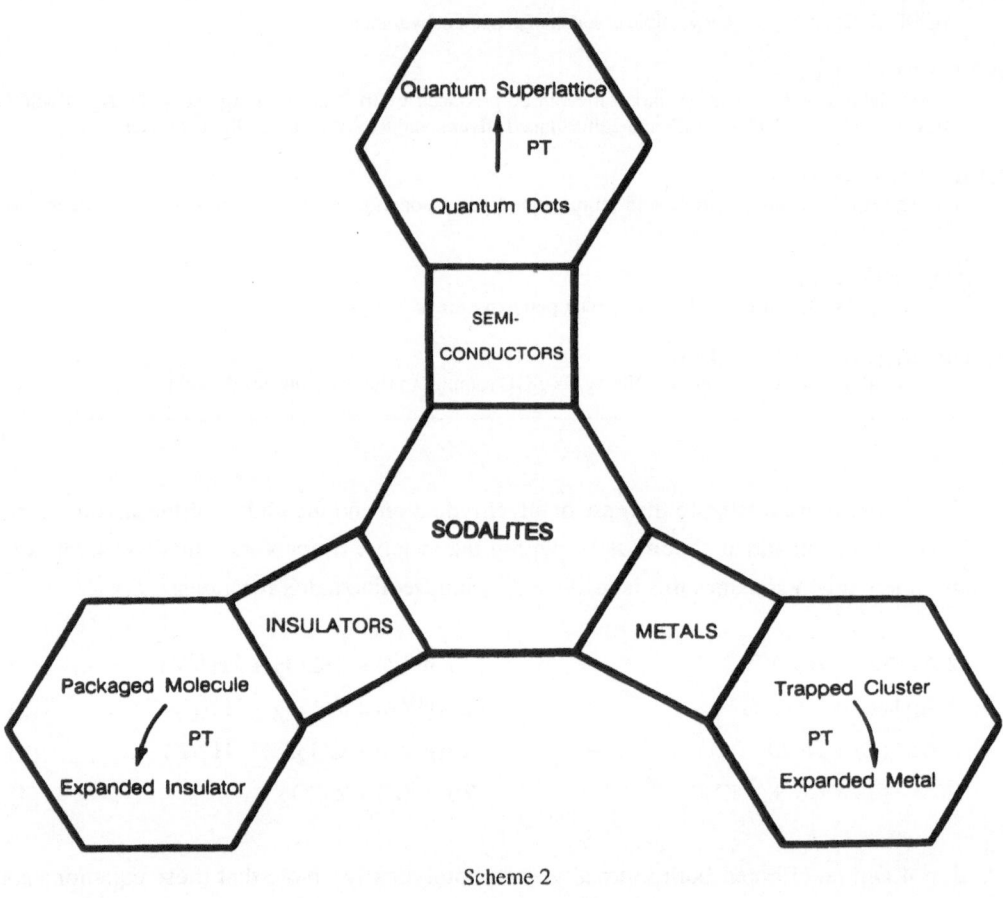

Scheme 2

The oxalato system has been studied in greatest detail with various silver loadings. The results of a full-profile Rietveld analysis [11] of X-ray powder diffraction data for completely silver ion-exchanged Ag,Ox-SOD are illustrated in Figure 6. The data were refined [12], allowing for three types of silver, each statistically distributed. A good refinement was obtained which is reflected in the small value of $\chi^2 = 2.16$ and final R factors of $wR_p = 0.140$, $R_p = 0.107$. In accordance with the -2 charge on the oxalate anion, both anion filled Ag_4Ox and empty Ag_4 cages are observed displaying a tetrahedral disposition of Ag atoms with each Ag coordinated to three oxygen atoms in adjacent six-rings. The intra-cage Ag-Ag distance is ca. 6 Å for oxalate ion-containing cages and 5.5 Å for empty Ag_4 cages, with intercage distances of 4.5 Å and 4.6 Å, respectively. For those cages exhibiting some loss of AgOH (extracted in a hot washing procedure) residual Ag_3 triangles are observed with an intracage Ag-Ag distance of 4.5 Å and an intercage Ag-Ag distance of 4.9 Å [11]. These interatomic distances are too long for direct overlap of the silver orbitals but they are well within the range of through-bond interaction involving the sodalite framework. Such long

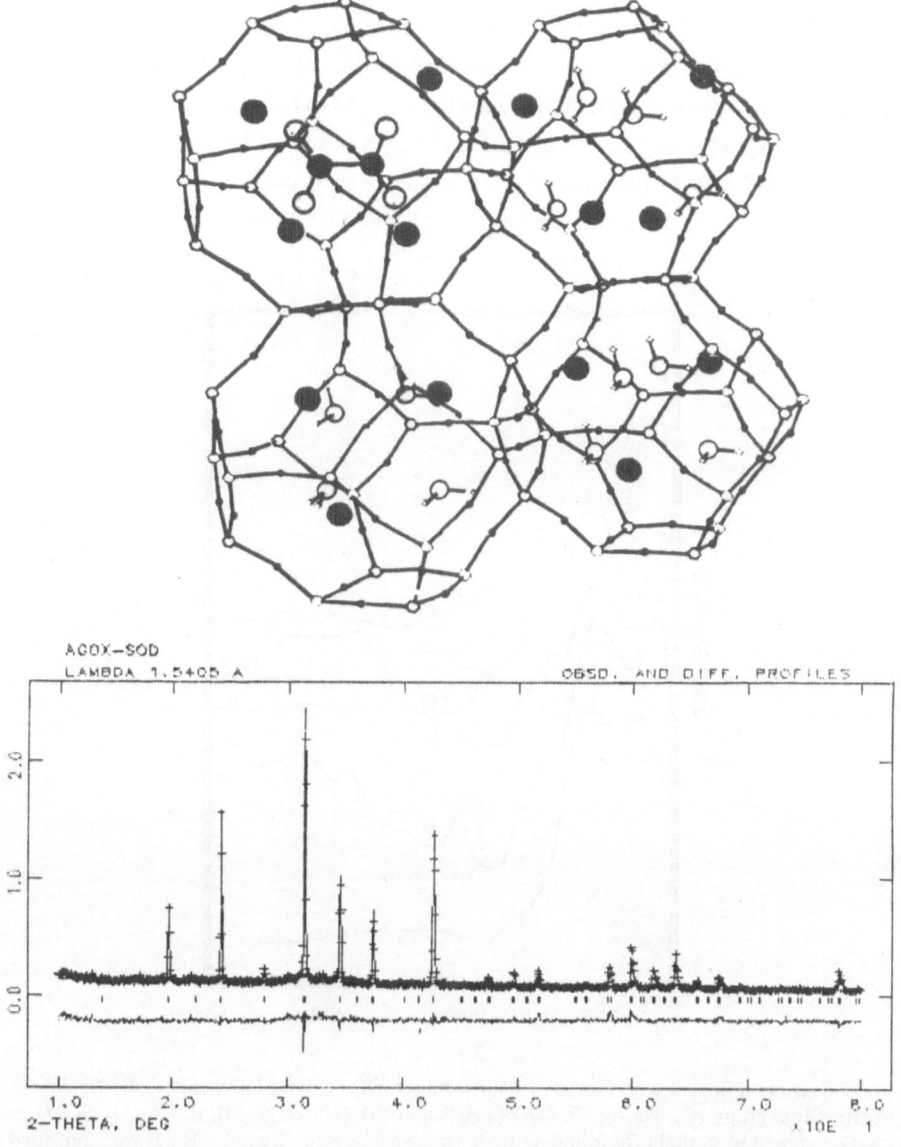

Figure 6. Four cages of Ag,Ox-SOD showing the framework of SiO_4^{4-} and AlO_4^{5-} units, silver ions, the oxalate anion, and water molecules. Except for hydrogen atoms and the oxalate ion, all atoms were located by a Rietveld profile analysis of powder X-ray diffraction data ($wR_p = 0.140$, $R_p = 0.107$) [11].

range interactions would allow the formation of an expanded silver supercluster, similar to a recently reported $(CdS)_4$ supercluster formed in the sodalite cages of zeolite Y above the percolation threshold, with Cd to Cd distance of ca. 6 Å between adjacent sodalite cages [10].

After aqueous silver exchange, sodium/silver oxalato-sodalites are white. Following thermal treatment at a temperature allowing for oxalate decomposition, the materials absorb light in the UV and visible regions, the absorption spectrum depending on the silver concentration (Figure 7). Concentration effects are also exhibited by the position of the most intense emission bands [3]. Variations in both the identities of trapped silver clusters and the unit cell sizes can be responsible for the shifts in transition energies.

It is important to note that in the virgin Na,Ag,Ox-SOD samples, one does not observe a linear increase of the unit cell dimension with Ag^+ loading (Figure 8). The largest increase in unit cell size occurs after the introduction of what appears to be less than a single Ag^+ per unit cell. One deduces therefore that the observed changes in unit cell size are not purely $R(Na^+)$ and $R(Ag^+)$ spatial effects but instead contain subtle contributions arising from electronic, bonding and coordination differences between Na^+ and Ag^+ cations.

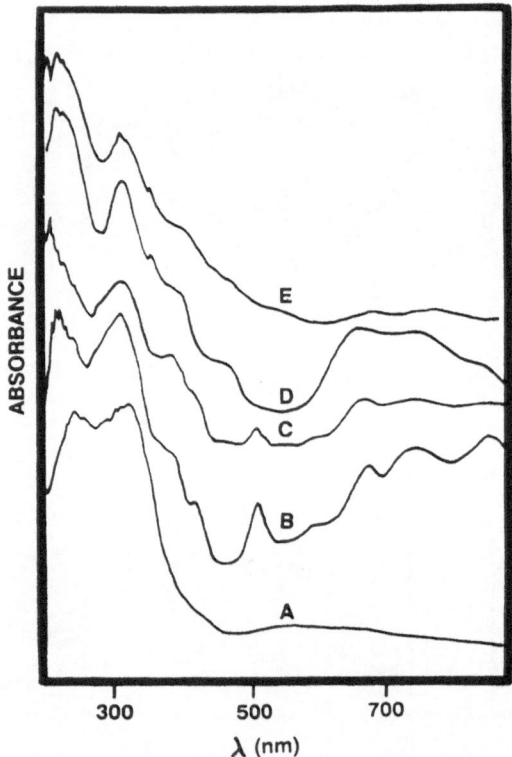

Figure 7. UV-visible reflectance spectra of oxalato sodalites; (A) Na,Ox-SOD (white), (B) Na,Ag-Ox-SOD (1 Ag/unit cell), (light grey), (C) Na,Ag,Ox-SOD (2 Ag/unit cell), (bluish-green), (D) Ag,Ox-SOD (green), (E) Ag,Ox-SOD, exposed to sunlight for four days (pale yellowish green). Samples B - E were prepared by melt exchanges.

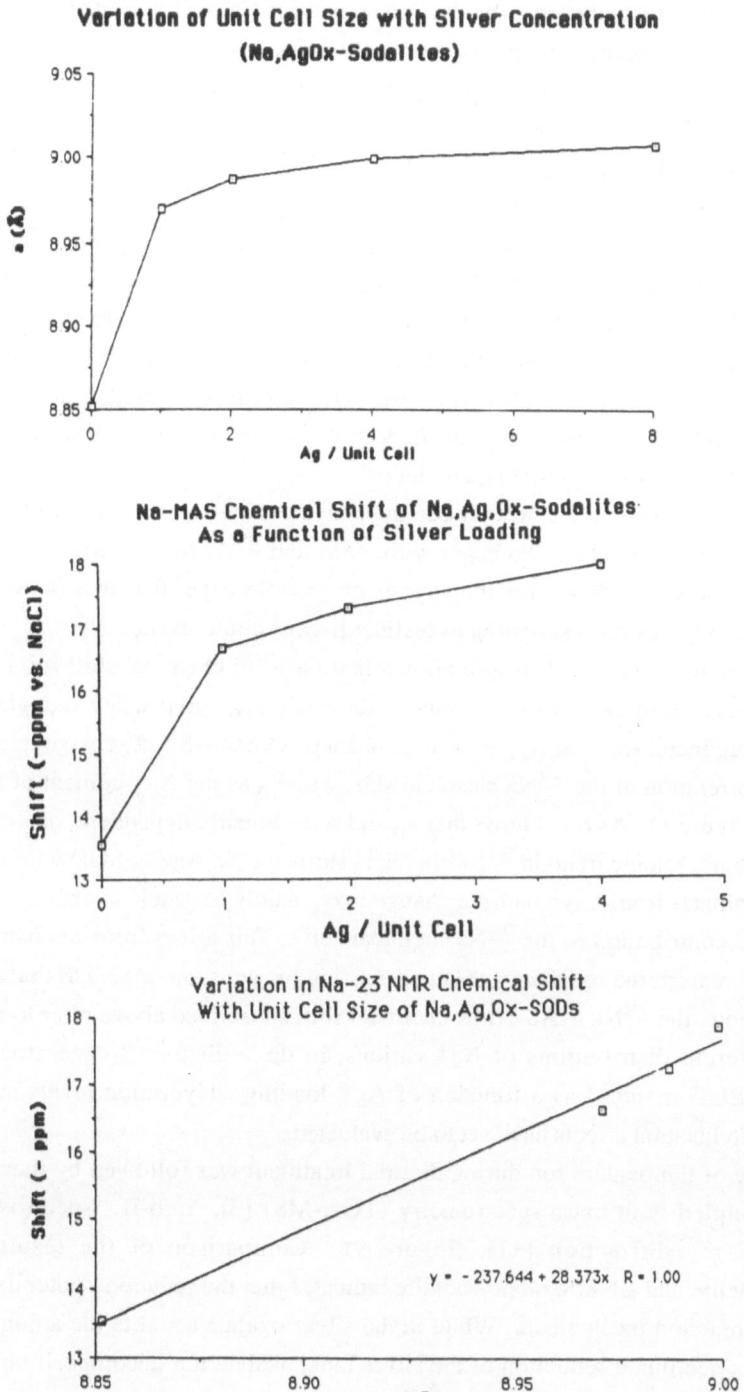

Figure 8. Variation in the unit cell sizes of Na,Ag,Ox-SOD and ^{23}Na MAS-NMR chemical shifts as a function of silver loading.

Another point that emerges from the XRD data for Na,Ag,Ox-SOD as a function of Ag^+ unit cell content, concerns the question of the distribution of silver in sodalites with mixed Na^+/Ag^+ compositions. Three models require serious consideration:

a) domains of $4Na^+$ and $4Ag^+$

b) ordered $(Na^+)_n(Ag^+)_{4-n}$ (n fixed)

c) statistical $(Na^+)_n(Ag^+)_{4-n}$ (n = 0-4).

From the XRD results, one can eliminate ordered model (b) as no superlattice reflections were observed. Also no splitting or broadening of XRD lines expected for a domain model (a) were noticed after silver addition. Furthermore, a_0 does not appear to change abruptly with Ag^+ loading, indicative of an alteration from one ordered phase to another. On these grounds one can conclude that the observed smooth, monotonic change in a_0 with Ag^+ in Na,Ag,Ox-SOD implies the statistical model (c) and the existence of a "solid solution" of β-cage encapsulated $(Na^+)_n(Ag^+)_{4-n}$ moieties.

In this context the silver concentration dependence of ^{23}Na MAS-NMR chemical shifts is particularly interesting [11]. To begin with, ^{29}Si and ^{27}Al MAS-NMR data for the parent Na,Ox-SOD shows evidence for three types of sodalite cage, that is Si(4 Al) and Al(4 Si) tetrahedral groupings corresponding to distinct β-cage contents, namely $β(C_2O_4^{2-})$, $β(OH^-)$ and β(empty), cf. Figure 6. It is also known that the ^{29}Si chemical shift for a large range of sodalites directly correlates with the unit cell dimension a_0 and the SiOAl angle α [13]. Here ^{29}Si shielding increases with α, paralleling enhanced $O(p\pi) \rightarrow Si(d\pi)$ charge transfer. We find a smooth correlation of the ^{23}Na chemical shift with a_0 as the Ag^+ content of the unit cell is increased (Figure 8). As one knows that a_0 and a are linearly dependent, one can deduce that the observed monotonic trend in ^{23}Na chemical shifts for Na,Ag,Ox-SOD with increasing Ag^+ content originates from a sympathetic change in α, mainly traceable to alterations in the Lamb diamagnetic contribution to the ^{23}Na chemical shift. This arises from a constant decrease in charge density from the sodalite cage lattice six-ring oxygens to $Na^+(3s)$ as that to Si increases. On a final note, the ^{23}Na MAS-NMR chemical shifts described above refer to average values for the different distributions of Na^+ cations, in three distinct β-cages for the hydrated Na,Ag,Ox-SOD material as a function of Ag^+ loading. Hydration levels and quadrupole broadening/relaxation effects have yet to be evaluated.

The fate of the oxalate ion during thermal treatment was followed by thermogravimetric analysis coupled with mass spectrometry (TGA-MS) [3], mid-IR spectroscopy [3] and powder X-ray diffraction [11], (Figure 9). Comparison of the results for sodium oxalato-sodalite and silver oxalato-sodalite indicates that the reduction potential of the cation directs the reaction mechanism. While in the silver oxalate sodalite the anion is oxidized to CO_2 with concomitant reduction of the silver ions, oxalate ion decomposition in the sodium oxalate sodalite produces CO and CO_3^{2-}, with no redox reaction occuring [3]. In mixed sodium/silver oxalato sodalites the degree of silver exchange can thus be used to control not only the optical and structural properties, but also the thermal behavior of the sodalite.

The question of the extent of silver reduction following thermal decomposition of Ag,Ox-SOD has been addressed from the TGA-MS analysis of CO_2 and O_2 evolution [3]. This provides an "ideal case" estimate of the maximum degree of silver reduction. With reducing electron equivalents derived from intra-β-cage $C_2O_4^{2-}$ decomposition (Equation 4) and autoreduction by intra-β-cage water and hydroxide (Equations 1 and 2) one calculates that the maximum change in oxidation state of silver amounts to an average of only $0.5e^-$ per unit cell, corresponding to 6% per silver ion. This nicely accounts for our XPS analysis [3] of the parent and thermally decomposed Ag,Ox-SOD which shows very little change in the silver ion core level ionization energies.

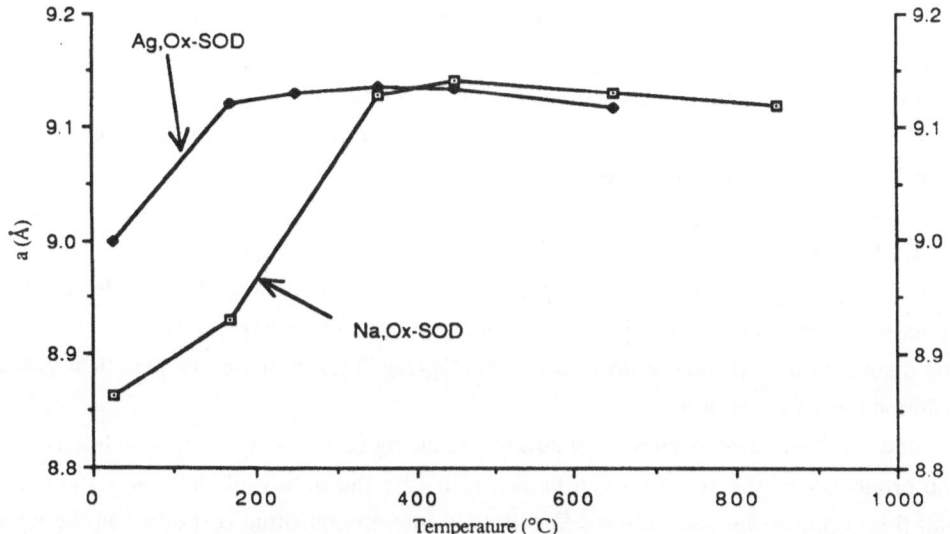

Figure 9. Variation in the unit cell size of Na,Ox-SOD and Ag,Ox-SOD with temperature. A drastic cell expansion occurs as water is lost. When the sodalites release CO or CO_2, respectively, the unit cell size decreases slightly.

In concert with the TGA-MS data it is intriguing to note the corresponding alterations in unit cell dimension a_o in Ag,Ox-SOD and its parent Na,Ox-SOD with temperature [11] (Figure 9). In both cases a drastic cell expansion occurs as water is driven out of the sodalite cages paralleling the results reported for Na_6-SOD.$4H_2O \rightarrow Na_6$-SOD [14]. This unusual phenomenon is ascribed to the loss of structural hydrogen-bonding between the encapsulated H_2O and the sodalite framework oxygens, thereby allowing the cage to "flex and expand". Of particular importance is the measurable decrease in the unit cell dimension of Ag,Ox-SOD and Na,Ox-SOD around 350°C and 450°C roughly corresponding to the thermal decomposition of $C_2O_4^{2-}$ to $2CO_2$ and CO/CO_3^{2-}, respectively [11]. This is nicely in line with our TGA-MS observations which show the evolution of CO_2 or CO in these temperature regions [3].

Although the number of electron equivalents involved in intracavity silver ion reduction is relatively small, the alterations in silver cluster optical properties that ensue are quite spectacular (Figure 7). As a result, the silver sodalites lend themselves to applications in high resolution imaging and optical data storage [1], because the regular close-packed framework structure contains small entrance windows with 2.2 - 2.6 Å diameters and single size cages with 6.6 Å diameters, which can trap and stabilize clusters formed inside. During a write cycle these clusters can be manipulated by the physical or chemical treatments described above (Table 2) to change their optical properties and produce a mark on the sodalite samples. The presence or absence of an optically absorbing or fluorescing mark defines a binary state. The size of the mark is affected mainly by the silver distribution, the sodalite particle size and the writing mechanism employed. The sodalite may be in the form of a self-supporting pressed disk, suspended in another material (e.g. glass or polymer), or supported on a substrate as a thin film [1].

Further research efforts will be directed at characterizing the silver clusters responsible for the optical properties of silver sodalites, and at fine-tuning the compositions of these compounds to optimize their responses.

CONCLUSIONS

Assembling the information from the present study of silver sodalites one can deduce that:

i) silver sodalites can be synthesized with a variety of optochromic responses;

ii) the distribution of β-cage encapsulated $(Na^+)_n(Ag^+)_{4-n}$ moieties is statistical (solid-solution like as a function of n);

iii) framework destruction occurs under strongly reducing (e.g. H_2 at 350°C) conditions;

iv) the properties of the sodalites can be controlled by the nature of the anion and cation and their relative concentrations. Specifically, the optochromic responses in the silver sodalites are determined by:

v) the ease of silver reduction;

vi) the ease of anion decomposition if a redox reaction is involved.

Acknowledgements. The generous financial support of Alcan, Canada and the Natural Sciences and Engineering Council of Canada is deeply appreciated. G.D.S. would like to thank the Office of Naval Research for their support. Acknowledgements also go to Dr. David Creber and Bob Lazier (Alcan, XPS), Dr. Hellmut Eckert (University of California, Santa Barbara, solid-state NMR), Bob Ramik (Royal Ontario Museum, TGA-MS), Dr. Peter Lea and Battista Calvieri (University of Toronto, Medical Sciences, TEM), and Bill Mercer (Union Carbide, chemical analysis) for their expertise and advice in carrying out analytical measurements.

References

1. Ozin, G. A.; Godber, J. P.; Stein, A. *U.S. Patent*, filed August, 1988; Ozin, G. A.; Stein, A.; Kuperman, A. *Adv. Materials* **1989**, *101*, 374.

2. Barrer, R. M. *Hydrothermal Chemistry of Zeolites*; Academic Press: London, 1982.

3. Stein, A., M.Sc. Thesis, University of Toronto, 1988.

4. Miller, M. F.; Bradley, E. B.; Todd, L. T. *Infrared Physics* **1985**, *25*, 531.

5. Richardson Jr., J. W.; Pluth, J. J.; Smith, J. V.; Dytrych, W. J.; Bibby, D. M. *J. Phys. Chem.* **1988**, *92*, 243.

6. Depmeier, W.; Schmid, H.; Setter, N.; Werk, M. L. *Acta Cryst.* **1987**, *C43*, 2251; Depmeier, W. *Acta Cryst.* **1984**, *C40*, 226.

7. Baerlocher, C.; Meier, W. M. *Helv. Chim. Acta* **1969**, *52*, 1853; Jarman, R. H. *J. Chem. Soc., Chem. Commun.* **1983**, 512.

8. Meinhold, R. H.; Bibby, D. M. *Zeolites* **1986**, *6*, 427.

9. Ozin, G. A.; Godber, J. P. *J. Phys. Chem.* **1988**, *92*, 2841, 4980.

10. Wang, Y.; Herron, N. *J. Phys. Chem.* **1988**, *92*, 4988; Wang, Y.; Herron, N. *J. Phys. Chem.* **1987**, *91*, 257; Herron, N.; Wang, Y.; Eddy, M. M.; Stucky, G. D.; Cox, D. E.; Moller, K.; Bein, T. *J. Am. Chem. Soc.* **1989**, *111*, 2569.

11. Ozin, G. A.; Stein, A.; McDougall, J. E.; Stucky, G. D.; Eckert, H., manuscripts in preparation.

12. The Generalized Structure Analysis System was used for Rietveld refinement. This program was kindly provided by A. C. Larson and R. B. Von Dreele, LANSCE, Los Alamos National Laboratory.

13. Weller, M. T.; Wong, G. *J. Chem. Soc., Chem. Commun.* **1988**, 1103.

14. Felsche, J.; Luger, S.; Baerlocher, Ch. *Zeolites* **1986**, *6*, 367.

SURFACE ENHANCED LUMINESCENCE WITH SILVER EXCHANGED ZEOLITES

James F. Tanguay and Steve L. Suib*

Department of Chemistry and Institute of Materials Science and
Department of Chemical Engineering
University of Connecticut
Storrs, CT 06268 (USA)

SUMMARY

Silver clusters have been produced in zeolites by a variety of different synthetic routes by several research groups. The types of clusters produced are quite dependent on the type of zeolite, concentration of silver, and type of activation. We have been studying the enhancement of luminescence of luminophores such as rhodamine in the presence of a variety of silver clusters in zeolites. The enhancement of luminescence has been studied with steady state and lifetime methods. The type of silver clusters present in the zeolites has been characterized by diffuse reflectance, luminescence, and electron paramagnetic resonance methods. The data suggest that several silver clusters are present when surface enhanced luminescence is observed. Since deposition of rhodamine onto the silver exchanged zeolites causes a change in the types of silver species it is necessary to characterize the clusters after deposition of the luminophore. Factors influencing the surface enhanced luminescence will be discussed.

INTRODUCTION

The phenomenon of surface enhanced luminescence was reported by Glass, Liao, Bergman and Olson [1] in 1980 for silver islands that were vapor deposited on glass. When luminescent dyes were deposited onto the silver islands it was found that partial overlap of the dye and the silver clusters led to an enhancement in luminescence intensity. Other researchers [2] have verified such behavior and suggest that direct contact of the silver and the dye leads to luminescence quenching; a large separation of the silver and dye leads to a superposition of the bands of the dye and silver species; and that partial overlap indeed leads to an enhancement in luminescence intensity. Placing spacer layers of 10 to 100 Angstrom thickness between the dye and silver islands leads to 10 to 100 fold enhancements of luminescence.

Inclusion Phenomena and Molecular Recognition
Edited by J. Atwood
Plenum Press, New York, 1990

Several clusters of silver having nuclearities between 1 and 6 have been isolated in molecular sieves [3-6]. Charges on such clusters are also variable ranging from no charge to charges as high as 5+. There has been a tremendous amount of research in this area and the exact composition and charge of such clusters is not always known with certainty. Nevertheless, it is clear that zeolites offer unique environments for isolation of such unique silver clusters.

The purpose of this research is to prepare silver clusters in zeolites and to study the enhancement of luminescence of dye molecules. The specific approach is to try to correlate size and charge of silver clusters with luminescence enhancement of dyes. By choosing a large luminescent dye that will not enter the pores of the zeolite, and by preparing silver clusters inside the zeolite then the zeolite itself can act as a spacer layer for separation of these two entities. The final objective is to maximize the surface enhanced luminescence and to understand the factors that control it.

EXPERIMENTAL

Silver exchanged zeolites were prepared according to literature procedures [3-6]. Zeolites used in this study were USY, NaY, NaA and NaX. The first was purchased from Davison and the latter three were purchased from Alfa Ventron. Exchanges were done in the dark and samples were filtered, washed and dried after exchange.

After exchange of silver, rhodamine dye was added to the surface of the zeolites by incipient wetness of 1×10^{-5} M dye in ethanol solution. Samples were initially evaporated to dryness in a rotary evaporator. Thermal treatments were carried out either on a vacuum line or in an oven exposed to air.

Samples were characterized with diffuse reflectance ultraviolet-visible spectroscopy [7], electron paramagnetic resonance [8], luminescence emission [9], and lifetime methods [10]. Details of such experiments can be found elsewhere [7-11].

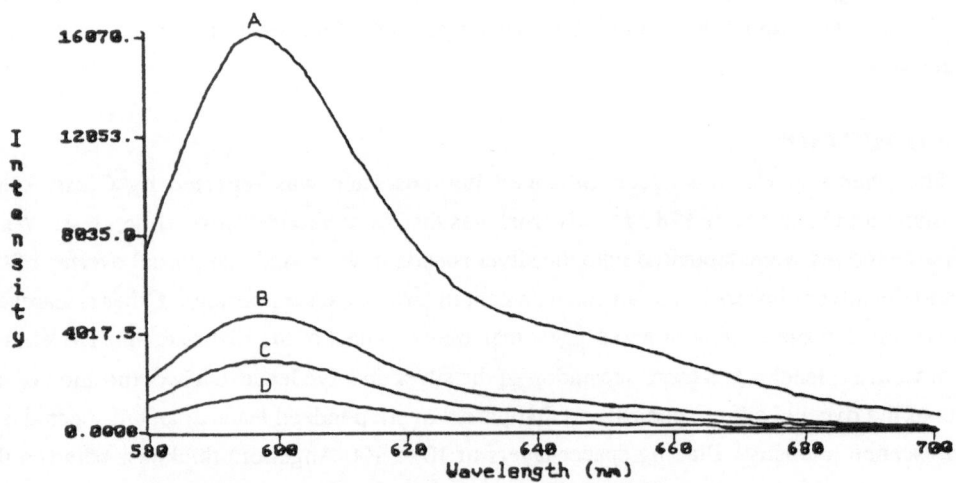

Figure 1. Luminescence emission spectra of rhodamine deposited on zeolites, vacuum treated: (a) NaX; (b) AgX, 500°C, 6 hr, vac; (c) a plus O_2, 3 hr, 450°C, vac; (d) AgA, salmon color.

Table 1. Luminescence intensity ratios of rhodamine on silver zeolites.

Sample	I_S/I_O
AgX, 500°C, vacuum[a], 6 hr	0.1
AgX, 500°C, vacuum, 1 hr	0.01
AgX, 500°C, vacuum, 6 hr; O_2, 3 hr, 450°C	0.2
AgA, Salmon Color	0.3
AgX, 500°C, vacuum, 6 hr; O_2, 3 hr, 450°C; vacuum, 18 hr, 450°C	0.2
AgX, <1 % Ag, 450°C, air, 6 hr	2.5 - 3.8
AgX, 5 % Ag, 450°C, air, 6 hr	2.5
AgX, 10 % Ag, 450°C, air, 6 hr	0.7
AgUSY, <1% Ag, 450°C, air, 6 hr	8.3 - 32.9
AgUSY, 1 % Ag, 450°C, air, 6 hr	31.2
AgUSY, 3 % Ag, 450°C, air, 6 hr	31.8
AgUSY, 9 % Ag, 450°C, air, 6 hr	75.8

[a]vacuum = vacuum treatment at 1×10^{-5} torr

RESULTS

The luminescence emission spectrum of rhodamine deposited on NaX zeolite is shown in Figure 1a. Figure 1b shows the luminescence of rhodamine deposited on a silver exchanged NaX zeolite that was heated in vacuum at 500°C for 6 hrs. The luminescence intensity is considerably less than that of the sample in Figure 1a. After treating the same sample shown in Figure 1b with O_2 for 3 hrs and then heating it to 450°C for 18 hrs the luminescence intensity decreases again as shown in Figure 1c. For comparison, the luminescence of rhodamine deposited on AgA zeolite is given in Figure 1d.

The ratio I_S/I_O which is the ratio of the luminescence of rhodamine in the presence of silver to that in the absence of silver for a variety of preparations and zeolites is given in Table I. All of the samples of Figure 1 are listed in Table 1 and have I_S/I_O ratios between 0.01 and 0.3.

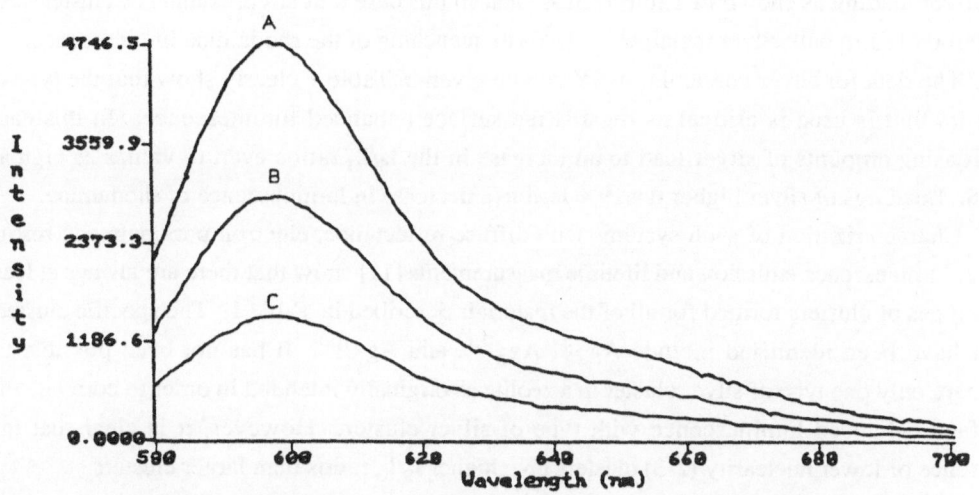

Figure 2. Luminescence emission spectra of rhodamine deposited on zeolites, thermally treated in air, λ_{exc} = 550 nm: (a) 1% Ag in NaX; 450°C, 6 hr in air (b) a not heated; (c) NaX + rhodamine, 1×10^{-5} M.

The luminescence of rhodamine deposited on 1% load Ag in zeolite NaX is shown in Figure 2a and b. For reference purposes the luminescence of rhodamine deposited on NaX zeolite is given in Figure 2c. There is an obvious increase in luminescence of the rhodamine dye when silver is present in the NaX zeolite. In fact, heating the AgNaX zeolite in air to 450°C substantially increases the luminescence intensity of the rhodamine dye. The ratios of the luminescence intensity in the presence of silver to that in the absence of silver are also listed in Table 1. The I_S/I_0 ratios for these systems range from 0.7 to 0.38.

Also listed in Table 1 are the I_S/I_0 ratios for rhodamine deposited on ultrastable Y (USY) zeolites that contain variable loadings of silver. As the silver loading increases from <1% Ag by weight to 9% by weight, the I_S/I_0 ratios increase from 8.5 to 75.8.

DISCUSSION

The data of Figure 1 clearly show that silver exchanged zeolites treated in vacuum do not show any enhancement of luminescence of deposited rhodamine dye with respect to rhodamine deposited on a sodium exchanged zeolite. The luminescence of rhodamine is in fact quenched by the presence of silver as is expected for bulk silver metal. Treatment of the silver zeolites with hydrogen at elevated temperatures [11] also leads to a quenching of luminescence of rhodamine.

On the other hand, if silver exchanged zeolites are heated in air to temperatures of 450°C and rhodamine is deposited on the surface, there is a slight enhancement of luminescence intensity of the rhodamine with respect to the sodium exchanged zeolites (Figure 2). It is also clear from Figure 2 that dehydration of the silver zeolite X sample prior to deposition of the rhodamine causes a slight increase in luminescence of the dye with respect to not heating the zeolite. These data suggest that the state of hydration of the zeolite is an important factor in the surface enhanced luminescence. In the NaX zeolite system as the weight percent silver is increased from 1 % to 10 % there is a decrease in I_S/I_0 to a quenching value of 0.7 for the 10 % silver loading as shown in Table 1. It is clear in this case that silver islands or cluster have sintered to form bulk silver metal which leads to quenching of the rhodamine luminescence.

The data for silver containing USY zeolite given in Table 1 clearly show that the type of zeolite that is used is critical as regards the surface enhanced luminescence. In this case increasing amounts of silver lead to an increase in the I_S/I_0 ratios even to values as high as 75.8. Loadings of silver higher than 9% lead to a decrease in luminescence of rhodamine.

Characterization of such systems with diffuse reflectance, electron paramagnetic resonance, luminescence emission and lifetime measurements [11] show that there are always at least two types of clusters formed for all of the materials described in Table 1. The specific clusters that have been identified include Ag_2^+, Ag_3^{2+}, and Ag_6^{n+}. It has not been possible to prepare only one type of silver cluster in a zeolite as originally intended in order to correlate the surface enhanced luminescence with type of silver cluster. However, it is clear that the presence of lower nuclearity (2-3) clusters give higher I_S/I_0 ratios than larger clusters.

The difficulty in preparing single cluster species and characterizing such systems may stem from the fact that thermal treatment leads to migration of silver ions to the surface of the zeolites

[12]. Ion migration is well documented in the literature for Li^+ ion zeolites as well [13]. In addition, it is clear that X-rays [12], ultraviolet light [3], and other radiation [4-6] that are used in characterization experiments cause the reduction of silver in such systems. For these reasons it would be very difficult to prepare a stable zeolite system for the detection of luminescence of pollutants or contaminants in a practical device.

References

1. Glass, A. M.; Liao, P. F.; Bergman, J. G.; Olson, D. H. *Optics Letters* **1980**, *5*, 368-370.
2. Chang, R. K.; Owen, J. F.; Barber, P. W.; Dorain, P. B. *Phys. Rev. Lett.* **1981**, *47*, 1075-1078.
3. Jacobs, P. A.; Uytterhoeven, J. B. *J. Chem. Soc., Chem. Commun.* **1977**, 128-129.
4. Kevan, L.; Narayana, M.; Li, A. S. W. *J. Phys. Chem.* **1985**, *89*, 132-135.
5. Ozin, G. A.; Hugues, F. *J. Phys. Chem.* **1982**, *86*, 5174-5179.
6. Ozin, G. A.; Baker, M. D.; Godber, J. *J. Phys. Chem.* **1985**, *89*, 305-311.
7. Carrado, K. A.; Suib, S. L.; Skoularikis, N. D.; Coughlin, R. W. *Inorg. Chem.* **1986**, *25*, 4217-4221.
8. Carrado, K. A.; Kostapapas, A.; Suib, S. L. *Sol. State Ionics* **1988**, *26*, 77-86.
9. Occelli, M. L.; Psaras, D.; Suib, S. L. *J. Catal.* **1985**, *96*, 363-370.
10. Tanguay, J. F.; Suib, S. L. *Catal. Rev. Sci. Eng.* **1987**, *29*, 1-40.
11. Tanguay, J. F. Ph.D. Thesis, University of Connecticut, 1988.
12. Willis, W. S.; Suib, S. L. *J. Am. Chem. Soc.* **1986**, *108*, 5657-5659.
13. Fyfe, C. A.; Kokotailo, G. T.; Graham, J. D.; Browning, C.; Gobbi, G. C.; Hyland, M.; Kennedy, G. J.; DeSchutter, C. T. *J. Am. Chem. Soc.* **1986**, *108*, 522-523.

SIZE QUANTIZED SEMICONDUCTORS IN POROUS HOSTS –
QUANTUM DOTS

Norman Herron and Ying Wang

Central Research and Development Department
E. I. du Pont de Nemours & Co.
Experimental Station, P. O. Box 80328
Wilmington, DE 19880-0328 (USA)

SUMMARY

The construction of discreet clusters of semiconductors inside the pore structure of zeolite hosts has generated a new class of materials where confinement effects on the semiconductor optical properties are pronounced. In addition to the size-quantization effects, novel intercluster phenomena become manifest as the individual semiconductor clusters reach a volume density above the percolation limit and begin to interact three-dimensionally. This interaction is modulated by the zeolite framework topology and hence leads to an ordered array of quantum dots in what we have termed superclusters. Novel absorption, emission and excitation behaviors of these materials, dominated by defect sites, result. Detailed characterization of the semiconductor species responsible (by X-ray powder diffraction and EXAFS) reveal a cubane-like $(CdS)_4$ unit as the basic building block of the structure.

INTRODUCTION

Small metal and semiconductor clusters, having hybrid molecular and bulk properties, represent a new class of materials and are under intensive investigation [1]. The basic problem facing this area of study is the control of surface reactions of the particles so as to arrest their growth at the small cluster stage. Many approaches have been explored for the preparation of these small clusters including the use of micelles [2], colloids [3], polymers [4] and glasses [5] to control the aggregation problem. In all cases, however, the cluster sizes are poorly defined and one would like to find an approach to this class of materials which produces a mono-dispersion of cluster sizes in a well defined and characterizable array. These criteria would seem well met by an inclusion type approach using a zeolite host lattice as the template within which the clusters could be constructed and confined. This paper describes the synthesis and characterization of CdS clusters in zeolites Y, X and A and the resulting effects of size confinement on the semiconductor optical properties.

Inclusion Phenomena and Molecular Recognition
Edited by J. Atwood
Plenum Press, New York, 1990

WHY SMALL SEMICONDUCTING PARTICLES ARE INTERESTING

It is important to understand why there is the current interest in very small particles of semiconductors [1]. The concept of an all optical or opto-electronic computer technology has attracted attention because of its potential for extreme speeds and parallel processing capabilities in such areas as image recognition. Such a technology requires several basic optical materials for the construction of devices which mimic their electronic counterparts. One such fundamental computing element is the optical transistor or bistable device which acts as a light switch or valve/amplifier. Basic requirements placed on materials for such a device are that they have a very rapid switching speed (ideally picosecond) and extreme photostability in order to perform trillions of switching operations/sec for years at a time. One realization of such a material could involve the use of third order non-linear optical properties, χ^3, to effect a transient refractive index change. Thus illumination of such a material with intense laser light will cause a change of its refractive index leading to a switch from an opaque to transmissive state in an interferometer type bistable device. While semiconductor materials themselves will perform this kind of switching at their band edge frequencies the speed of the effect is slow – usually as a consequence of a long free-carrier lifetime. This speed can be increased by providing more sites for efficient removal of these free carriers – in other words more defect sites. One can view surface sites on a semiconductor particle as such defect sites and one way to increase their concentration is obviously to go to very small particles. The commercial color filters of Schott and Corning based on CdS/Se nanoparticulates in a silica matrix have verified the utility of this kind of material for non-linear-optical devices [5]. We would like to explore a wide range of other semiconductors and matrices for these purposes and the zeolite host provides an almost ideal starting point.

CdS IN ZEOLITE Y [6]

The zeolite Y occurs naturally as the mineral faujasite and consists of a porous network of aluminate and silicate tetrahedra linked through bridging oxygen atoms (Figure 1). The structure consists of truncated octahedra, called sodalite units, arranged in a diamond net and linked through double six-rings [7]. This gives rise to two types of cavity within the structure – the sodalite cavity of ≈ 5Å diameter with access through ≈ 2.5Å windows and the supercage of ≈ 13Å diameter with access through ≈ 7.5Å windows. Whenever an aluminum atom occurs in the framework it introduces one negative charge onto the zeolite skeleton which is compensated by loosely attached cations, giving rise to the well known ion-exchange properties of zeolites.

Cadmium ion exchange of the zeolite is carried out by slurrying 10g of zeolite LZY-52 (sodium zeolite Y from Linde) in 1 L of distilled water and adjusting the pH to five with nitric acid. A calculated amount of cadmium nitrate designed to give a specific exchange level is stirred into the slurry and the mixture is stirred at room temperature overnight. Collection of the exchanged zeolite by filtration and washing with distilled water is followed by drying and

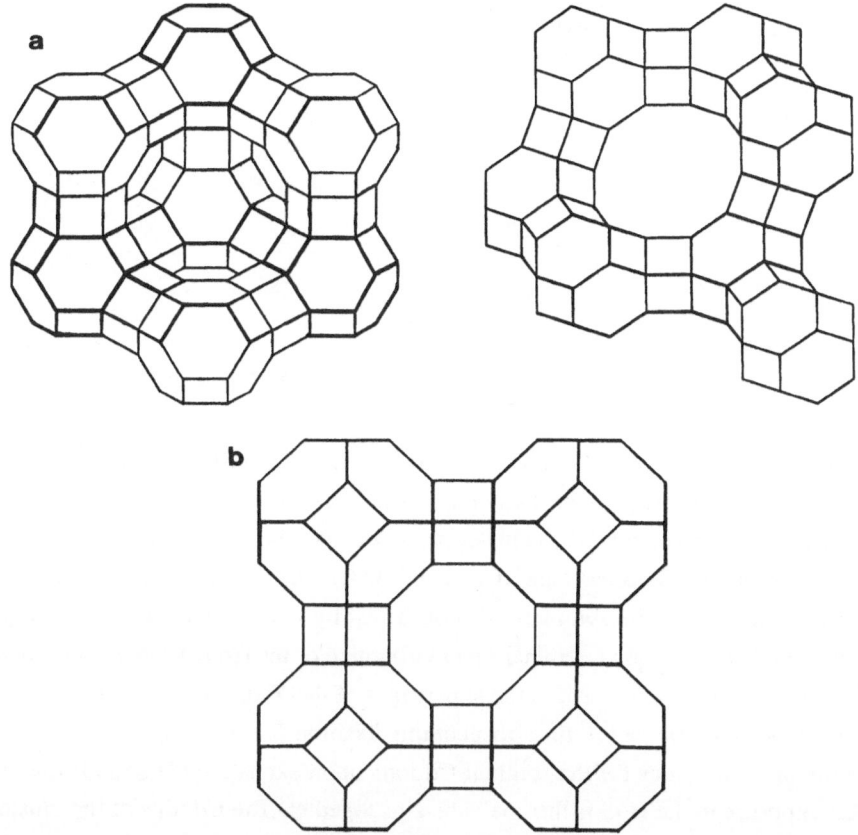

Figure 1.   Representative zeolite structures where the open framework is represented by sticks joining the Si or Al atoms. Oxygen bridge atoms lie roughly at the mid-point of these atoms and are omitted for clarity, a) zeolite Y b) zeolite A.

calcination. The powder is heated to 400°C at 3°/min in flowing dry oxygen (100 mL/min) then cooled in vacuo to 100°C. The zeolite is then exposed to flowing hydrogen sulfide (40 mL/min) at 100°C for 30 min. Finally the still white zeolite is evacuated at 100°C for 30 min, then sealed and transferred to an inert atmosphere dry box for handling and storage. The zeolite turns pale yellow/cream during the final evacuation step. All zeolites prepared in this manner are moisture sensitive becoming deep yellow (zeolite Y or X) or pale yellow (zeolite A) on prolonged exposure to the atmospheric humidity. Chemical analysis confirms Cd and S are present in from 0 to 25 wt% depending on exchange conditions. XPS shows that there is no detectable Cd on the exterior surface of the zeolite crystallites. IR spectra show no SH groups but there are the expected OH groups attached to the zeolite framework [6]. The exact nature of the CdS cluster units is revealed by a combined application of optical spectroscopies and X-ray techniques.

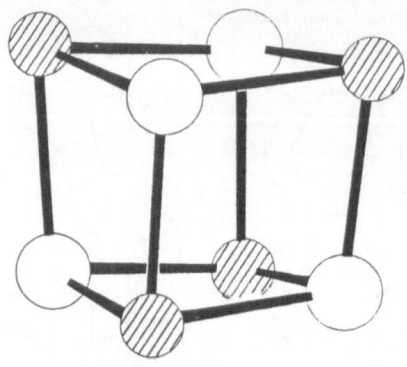

Figure 2. Structure of the $(CdS)_4$ unit located within the zeolite sodalite units (hatched circles = Cd; open circles = S).

STRUCTURE OF CdS IN Y AND ITS OPTICAL CONSEQUENCES

Detailed analysis of the powder X-ray diffraction data on a series of CdS loaded zeolite Y samples reveals the fundamental CdS cluster present consists of interlocking tetrahedra of Cd and S atoms forming a distorted cube (Cd-S = 2.47Å) (Figure 2). This structure, which is heavily dictated by the zeolite symmetry, is confirmed by EXAFS data at the Cd edge which reveals the local symmetry and coordination environment of the Cd [6]. To our initial surprise, these Cd_4S_4 cubes were not located in the supercages of the Y structure but were instead sited within the smaller sodalite cages. In retrospect this location is entirely reasonable since these cages are the preferred sites for the original Cd ions upon exchange [8] and all that needs to occur upon exposure to the H_2S is that the cube zips together. The Cd ions of the cluster are in octahedral coordination to 3 sulfur atoms of the cube and 3 oxygen atoms of the zeolite framework six-ring window (Figure 3). The sodalite cage seems to have been made for this CdS cluster!

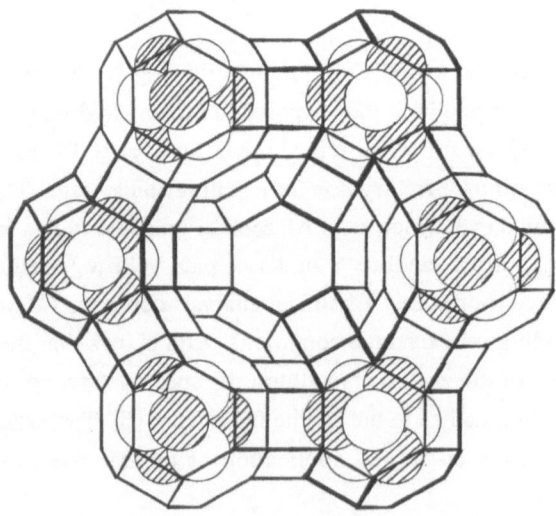

Figure 3. The hyperlattice arrangement of CdS clusters in adjacent sodalite units of the zeolite Y (hatched = Cd; open = S).

The evolution of the optical spectra as a function of CdS loading density is particularly revealing. At loading densities of <5 wt% the sample has an optical absorption spectrum with a shoulder at 280 nm. This represents a shift of the band edge by 230 nm and correlates well with the extremely small clusters Cd_4S_4 found in the X-ray analysis. However, as the loading rises above 5 wt% a new absorption feature at 350 nm begins to appear and grows as the loading increases up to the maximum yet attained of 25 wt%. While still quantum confined CdS, the structural nature of the material responsible for this new absorption feature was unknown. The band edge observed is similar to that of colloidal CdS particles with 25-30 Å diameters [9] yet there are no void spaces of this kind of dimension within zeolite Y. The real clue as to the nature of this species comes from the threshold loading density of 4 ±1% – this corresponds to the concentration at which, statistically, CdS cubes must now populate adjacent sodalite units. Simple calculations based on the bulk density of CdS and the sodalite pore volume of Y reveal that the 4 wt% loading density corresponds to filling of 14 wt% of the available sodalite volume. Percolation theory [10] predicts the percolation threshold in such systems to be at this 15% volume i.e. above 5 wt% of CdS the individual clusters must begin to interact with one another in adjacent sodalite cages in a percolative fashion. In the limit, when all sodalite units are occupied (28 wt% CdS) the structure would be such as that represented in Figure 3 and a hyperlattice of CdS clusters dictated by the zeolite Y topology results.

This unique arrangement of individual clusters into an interconnected network gives rise to the luminescence and excitation spectra shown in Figure 4. As anticipated, the luminescence is not band-gap recombination in nature but rather is dominated by Cd related defects.

Since the novel optical behavior at higher loading densities of CdS is a consequence of the interaction between individual clusters as modulated by the zeolite structure it would be interesting to study a zeolite with a different connectivity between the same kinds of CdS/sodalite clusters. In this case the spectral behavior should be very similar to zeolite Y at low loading densities but differ significantly as the interconnected CdS hyperlattice develops. Zeolite A is just such a system [7]. Now the sodalite cages are connected via double 4-rings (Figure 1) in a cubic arrangement and the distance between Cd_4S_4 cubes is projected at ≈9 Å, i.e. the cluster-cluster interaction must be much weaker in A than in Y. The optical behavior follows these predictions precisely. While at low loading the absorption spectra of A and Y look very similar (isolated clusters) as the loading increases there is a smaller shift in the absorption edge of the A zeolite material (to 320 nm) indicating a much weaker interaction between clusters than in Y and so no significant development of an electronically cogent hyperlattice.

PHOTOCATALYSIS

CdS is a well known photocatalyst for a number of transformations of organic molecules [11] and the CdS/Y system offers an opportunity to study the effect of zeolite encapsulation of the CdS upon the selectivity of these transformations. Since all of the CdS resides in the small sodalite units of the structure the larger supercage pores are still available for absorption of moderately sized organic molecules. A competitive photooxidation of styrene and 1,1-diphenylethylene reveals that the smaller styrene is more favored as a substrate in the zeolite (oxidation ratio 1:1 styrene:DPE) than in free solution with bulk CdS powder as the photocatalyst (ratio 1:2) [11]. These novel semiconductor/porous host composites offer an interesting size and shape selective medium in which to perform this kind of photochemistry and this area promises to become an important aspect of inclusion phenomena in the future (see paper by T. Mallouk in this volume).

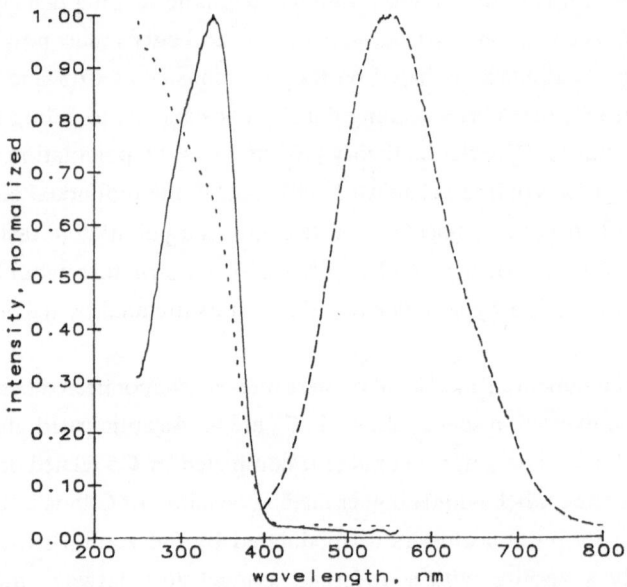

Figure 4. Absorption (dotted), excitation (solid) and emission (dashed) spectra of 6.5 wt% CdS in zeolite Y. The absorption spectrum was taken at room temperature and the others at 77K.

CONCLUSIONS

Zeolites provide a novel host for the generation of semiconductor hyperlattices within their pore volume. The control of the connectivity between the quantum dots of semiconductor is unparalleled in any other host medium and so has allowed a detailed study of the optical consequences of such connectivity. However, from the practical standpoint, such materials have some severe drawbacks − most notably the lack of single crystals of sufficient size to produce viable optical devices such as optical transistors or spatial light modulators.

Future directions of this kind of research will focus on other types of host materials with intrinsically good optical properties and especially those which allow a careful control of the surface chemistry of the quantum dots involved. Either rational synthesis of clusters whose surface is deliberately terminated with a covalently attached capping group (e.g. thiophenol groups attached to a CdS core) [12] or use of the host pores to terminate the surface physically is likely to be of extreme importance. In either case, inclusion phenomena are likely to be intimately involved in the next generation of such optical materials.

ACKNOWLEDGEMENTS

The fine technical assistance of J. B. Jensen and S. H. Harvey is gratefully acknowledged. Collaborations with J. E. Macdougall on synthesis and photocatalysis, Drs. M. M. Eddy and G. D. Stucky on X-ray powder analysis and Drs. K. Moller and T. Bein on EXAFS were key contributions to this work.

References

1. Brus, L. E. *J. Phys. Chem.* **1986**, *90*, 2555 and references therein; Brus, L. E. *J. Chem. Phys.* **1984**, *80*, 4403.

2. Weller, H; Schmidt, H. M.; Koch, U.; Fojtik, A.; Baral, S.; Henglein, A; Kunath, W.; Weiss, K.; Dieman, E. *Chem. Phys. Lett.* **1986**, *124*, 557.

3. Tricot, Y-M; Fendler, J. H. *J. Phys. Chem.* **1986**, *90*, 3369.

4. Wang, Y.; Mahler, W. *Opt. Comm.* **1987**, *61*, 233.

5. Borelli, N. F.; Hall, D. W.; Holland, H. J.; Smith, D. W. *J. Appl. Phys.* **1987**, *61*, 5399.

6. Wang, Y.; Herron, N. *J. Phys. Chem.* **1988**, *92*, 4988; Herron, N.; Wang, Y.; Eddy, M.; Stucky, G. D.; Cox, D. E.; Bein, T.; Moller, K. *J. Am. Chem. Soc.*, in press.

7. Breck, D. W. In *Zeolite Molecular Sieves*; Wiley: New York, 1974.

8. Calligaris, M.; Mardin, G.; Randaccio, L.; Zangrando, E. *Zeolites* **1986**, *6*, 439.

9. Rossetti, R.; Hull, R.; Gibson, J. M.; Brus, L. E. *J. Chem. Phys.* **1985**, *83*, 1406.

10. Kirkpatrick, S. *Rev. Mod. Phys.* **1973**, *45*, 574 and references therein.

11. Fox, M. A.; Chen, C-C.; Park, K-H.; Younathan, J. N. *A.C.S.Symp. Ser.* **1985**, *278*, 69; Fox, M. A. *Acc. Chem. Res.* **1983**, *16*, 314.

12. Herron, N.; Wang, Y.; Eckert, H. submitted for publication.

INDEX

Semiconductors, size quantized, 401-406
Shape selective oxidation, 209-213
Spherand, 218
    cryptahemi, 220, 221
    hemi, 220
Spermidine, 78
Spermine, 77
Starch, 261
Surfactants, 251-258

Template, 24
    twin, 24
Thyroxin, 195
Torand, 50
    stability constants of alkali metal complexes, 51, 52
    stability constant for complex with urea, 53, 55
TREN, 11
Triton X-100, 263
Tröger's base, 28, 30, 31

Vesicle membranes, 145-150
Vesicle-forming surfactants, 251-258
Viologen, 251
    benzyl, 374
    electrochemistry of, 252-254
Vitamin B_{12}, 107, 108, 112-115

Zeolites, 277-286
    artificial photosynthesis in, 365-376
    benzene as guest, 327, 328
    calculation of activation energies for diffusion, 281
    calculation of dynamics of sorbates, 284-285
    calculation of heats of adsorption, 279-281
    composition, 290
    Cd_4S_4, 401-406
    dibenzyl ketone adsorbate, 293
    diffusional blocking, 340
    diffusion in, 291
    electrochemistry, 367-372
    effect of adsorbed molecules on structure of, 299-321

effect of temperature on structure of, 299-321
Faujasite void space, 290
^2H NMR, 331, 332, 355, 358, 359
iron EXAFS spectra, 344, 346
inelastic neutron scattering, 332-334
intrazeolite reactions, 342-349
ion locations in Faujasite, 292, 293
location of adsorbed molecules, 278, 279
low Si/Al ratio, 302
mobility of adsorbed molecules, 291, 325-334
molecular modelling using atom-atom potentials, 326-328
molecular rotation of polyenes in, 352-355
neutron scattering, 325-329
NMR studies of, 299-321
Norrish Type reaction in, 356, 357, 360-362
organometallic molecules as guest, 339-349
paradigm, 290
photocatalysis, 405, 406
photochemistry of organic molecules, 289-296, 351-362
photochemical paradigm, 293-295
photolysis of adsorbed ketones, 295
p-xylene as guest, 315, 317-319, 331
pyridine as guest, 278, 327
radical motion in, 296
rhodamine dye, 396
$Ru(bipy)_3^{2+}$ photochemistry in, 372-376
semiconductors, 401-406
^{29}Si NMR spectra, 303-307, 309-318, 319-320
silver exchanged, 395-399
silver sodalites, 379-392
sodalite cage, 300, 379-392
surface enhanced luminescence, 395-399
TiO_2 as guest, 375, 376

Zwitterions
    binding, 6
    hydrogen bonds, 119